AF388658

Data Driven Decision-Making for Complex Production Systems

Data Driven Decision-Making for Complex Production Systems

Editors

Zaoli Yang
Yuchen Li
Ibrahim Kucukkoc

Basel • Beijing • Wuhan • Barcelona • Belgrade • Novi Sad • Cluj • Manchester

Editors

Zaoli Yang
School of Economics and
Management
Beijing University
of Technology
Beijing
China

Yuchen Li
School of Economics and
Management
Beijing University
of Technology
Beijing
China

İbrahim Kucukkoc
Industrial Engineering Department
Balikesir University
Balikesir
Turkey

Editorial Office
MDPI
St. Alban-Anlage 66
4052 Basel, Switzerland

This is a reprint of articles from the Special Issue published online in the open access journal *Systems* (ISSN 2079-8954) (available at: www.mdpi.com/journal/systems/special_issues/dddm).

For citation purposes, cite each article independently as indicated on the article page online and as indicated below:

Lastname, A.A.; Lastname, B.B. Article Title. *Journal Name* **Year**, *Volume Number*, Page Range.

ISBN 978-3-0365-9629-7 (Hbk)
ISBN 978-3-0365-9628-0 (PDF)
doi.org/10.3390/books978-3-0365-9628-0

Contents

About the Editors vii

Preface ix

Jiekun Song, Zeguo He, Lina Jiang, Zhicheng Liu and Xueli Leng
Synergy Management of a Complex Industrial Production System from the Perspective of Flow
Structure
Reprinted from: *Systems* **2023**, *11*, 453, doi:10.3390/systems11090453 1

Hewen Gao, Fei Li, Jinhua Zhang and Yu Sun
A Study of the Strategic Interaction in Environmental Regulation Based on Spatial Effects
Reprinted from: *Systems* **2023**, *11*, 62, doi:10.3390/systems11020062 46

Shouzhen Zeng, Wendi Chen, Jiaxing Gu and Erhua Zhang
An Integrated EDAS Model for Fermatean Fuzzy Multi-Attribute Group Decision Making and
Its Application in Green-Supplier Selection
Reprinted from: *Systems* **2023**, *11*, 162, doi:10.3390/systems11030162 59

Xuan Wei, Hongyu Wu, Zaoli Yang, Chunjia Han and Bing Xu
Simulation of Manufacturing Scenarios' Ambidexterity Green Technological Innovation Driven
by Inter-Firm Social Networks: Based on a Multi-Objective Model
Reprinted from: *Systems* **2023**, *11*, 39, doi:10.3390/systems11010039 82

Yun Zhang and Chaoxia Qin
A Gaussian-Shaped Fuzzy Inference System for Multi-Source Fuzzy Data
Reprinted from: *Systems* **2022**, *10*, 258, doi:10.3390/systems10060258 100

Xueqin Huang, Xianqiang Zhu, Xiang Xu, Qianzhen Zhang and Ailin Liang
Parallel Learning of Dynamics in Complex Systems
Reprinted from: *Systems* **2022**, *10*, 259, doi:10.3390/systems10060259 120

Miloš Gligorić, Zoran Gligorić, Suzana Lutovac, Milanka Negovanović and Zlatko Langović
Novel Hybrid MPSI–MARA Decision-Making Model for Support System Selection in an
Underground Mine
Reprinted from: *Systems* **2022**, *10*, 248, doi:10.3390/systems10060248 138

Yuanying Chi, Zhaoxuan Qiao, Yuchen Li, Mingyu Li and Yang Zou
Type-1 Robotic Assembly Line Balancing Problem That Considers Energy Consumption and
Cross-Station Design
Reprinted from: *Systems* **2022**, *10*, 218, doi:10.3390/systems10060218 159

Francisco Javier Álvarez García and David Rodríguez Salgado
An Approach for Predictive Maintenance Decisions for Components of an Industrial Multistage
Machine That Fail before Their MTTF: A Case Study
Reprinted from: *Systems* **2022**, *10*, 175, doi:10.3390/systems10050175 173

Cong Xu and Yu-Min Wang
The Regulatory Architecture of Digital Platforms: A Perspective of Life Cycle and Risk
Management
Reprinted from: *Systems* **2022**, *10*, 145, doi:10.3390/systems10050145 193

Shouzhen Zeng, Zitong Fang, Yuhang He and Lina Huang
An Integrated Entropy-COPRAS Framework for Ningbo-Zhoushan Port Logistics
Development from the Perspective of Dual Circulation
Reprinted from: *Systems* **2022**, *10*, 131, doi:10.3390/systems10050131 . . . . **209**

Hui Li, Bo Zhang and Xiangyu Ge
Modeling Emergency Logistics Location-Allocation Problem with Uncertain Parameters
Reprinted from: *Systems* **2022**, *10*, 51, doi:10.3390/systems10020051 . . . . **224**

Jing-Yu Ma, Quan-Lin Li and Li Xia
Optimal Asynchronous Dynamic Policies in Energy-Efficient Data Centers
Reprinted from: *Systems* **2022**, *10*, 27, doi:10.3390/systems10020027 . . . . **241**

Dawei Wang, Jun Guan, Chunxiu Liu, Chuke Jiang and Lizhi Xing
Simulation of Cooperation Scenarios of BRI-Related Countries Based on a GVC Network
Reprinted from: *Systems* **2022**, *10*, 12, doi:10.3390/systems10010012 . . . . **289**

About the Editors

Zaoli Yang

Zaoli Yang is an Associate Professor at the Beijing University of Technology, Beijing China, and visiting scholar at the University of Edinburgh and Heriot-Watt University, UK. His research interests include intelligence decision supporting system, expert systems and decision support, information management, artificial intelligence, fuzzy set theory, data mining, and their application in various fields. His research has been published in refereed international journals (over 50 SCI/SSCI indexed papers), including the *European Journal of Operational Research, IEEE Transactions on Fuzzy Systems, IEEE Transactions on Engineering Management, Technological Forecasting & Social Change, Journal of Business Research, International Journal of Production Economics, Renewable and Sustainable Energy Reviews* and *Applied Energy*. He has served as the academic editor of more than 10 SCI/SSCI indexed journals, including the Associate Editor of "Journal of Intelligent & Fuzzy Systems", "Heliyon", the Editorial Board Member of "Journal of Organizational and End User Computing", "International Journal of Information Technologies and Systems Approach", "Discrete Dynamics in Nature and Society". In addition, he has been recognized among the 'World's Top 2% Scientists in 2022 and 2023,' as identified by Elsevier BV and Stanford University.

Yuchen Li

Yuchen Li received a B.S. degree in reliability and systems engineering from Beihang University in 2010, an M.S. degree in operations research from Columbia University in 2012, and a Ph.D. degree in industrial and systems engineering from Rutgers University in 2016. He is currently an associate professor with the College of Economics and Management, Beijing University of Technology. His research has been published in some academic journals, including *European Journal of Operational Research*, the *International Journal of Production Research, Computers and Industrial Engineering*. His research interests include production optimization, game theory, and engineering economics.

Ibrahim Kucukkoc

Dr. Ibrahim Kucukkoc is an Associate Professor in the Industrial Engineering department at Balikesir University (Turkey) where he has been a faculty member since December 2009. He completed his Ph.D. at the University of Exeter (UK) in 2015 upon completing his thesis on modelling and optimisation of complex assembly lines (funded by a 4-year scholarship from Turkish Council of Higher Education). His research interests lie in the area of the modelling and optimisation of modern manufacturing systems (assembly line balancing in particular), production planning of additive manufacturing machines and application of modern algorithms on sophisticated combinatorial optimisation problems. He has collaborated actively with researchers in several other disciplines, such as computer science and mechanical engineering. His publications appeared in various esteemed journals, including the *European Journal of Operational Research, Computers & Operations Research, International Journal of Production Economics, International Journal of Production Research, Computers & Industrial Engineering* and *Production Planning & Control*. His research has been cited more than 1400 times (based on Scopus metrics) by national and international scholars.

Preface

The complexity of production processes is on the rise due to evolving technology and changing lifestyles. This complexity can stem from factors within the production system or be influenced by characteristics or events external to the system. The former is termed technological complexity, associated with the inherent intricacies of the system and its technologies, encompassing both products and overall systems. The latter is environmental complexity, depicting the coordination between the system and related industries or customers, such as raw material suppliers and retailers. This complexity presents a significant challenge to production systems.

With the advancement of big data technology, data-driven decision-making (DDM) algorithms offer novel tools and perspectives for humans to delve deeper into the complexities and uncertainties within production systems. Specifically, technologies like deep learning and artificial intelligence can be employed to analyze multifaceted complexity factors in production processes and control systems, leading to the development of multi-source DDM algorithms. This approach emerges as a crucial method for addressing the aforementioned challenges. For instance, big data mining technology can be applied to extract diverse features from each component within complex production systems. This facilitates the exploration of correlations between features, aiding in the diagnosis of issues present in intricate production systems. Utilizing multi-source features, causal relationships between features can be constructed to identify the root causes and mechanisms of problems. Additionally, the integration of information fusion theory and comprehensive decision technology is introduced to evaluate the performance of complex production systems, among other applications.

This reprint aims to conduct rigorous research focused on the application of data-driven decision-making algorithms to address diverse challenges within complex production systems.

Zaoli Yang, Yuchen Li, and Ibrahim Kucukkoc
Editors

Article

Synergy Management of a Complex Industrial Production System from the Perspective of Flow Structure

Jiekun Song *, Zeguo He, Lina Jiang, Zhicheng Liu and Xueli Leng

School of Economics and Management, China University of Petroleum, Qingdao 266580, China; z21080318@s.upc.edu.cn (Z.H.); s21080048@s.upc.edu.cn (L.J.); z22080013@s.upc.edu.cn (Z.L.); z22080018@s.upc.edu.cn (X.L.)
* Correspondence: songjiekun@upc.edu.cn

Abstract: Modern industry has become very complex and requires an equally complex engineering technology system, which includes resource utilization, energy conversion, product research and development, technological innovation, environmental protection and industrial ecology, and other aspects of the system. Continued development of large-scale, streamlined, and continuous processes is critical; however, there are also problems such as data redundancy, overcapacity, redundant construction, and waste of resources. Based on the system synergy theory, this paper introduces the system analysis method from the perspective of flow structure, with the purpose of solving the management defects of complex industrial production systems. First, we analyze the complex industrial production system as a collaborative structure of three subsystems: material flow, energy flow, and information flow. The following concepts are clarified: "material flow is the main body, energy flow is attached to and drives material flow, material flow and energy flow generate information flow, and information flow reversely drives material flow and energy flow". Secondly, the collaborative evolution process of the complex industrial production system is divided into three periods, which are the generation period, the stalemate period, and the maturity period, and a synergy degree evaluation model is established, which considers the Theil index and subsystem gray correlation method, and extends the dynamic differential equation model of three-stage collaborative evolution. Subsequently, we used MATLAB numerical simulation to demonstrate that the collaborative evolution of production systems is related to four aspects. They are the self-organizing ability of the system, the dominant role of order parameters, the competition and cooperation between order parameters, and whether mutations can become order parameters. At the same time, it was also found that it is basically independent of other factors, such as attenuation inertia. Then, the self-organizing map network (SOM) algorithm was used for the rapid identification of mutation data. Finally, we use the empirical research of SG enterprises to show that their production level and management system are advanced, but they were in a non-cooperative state from 2014 to 2021. In 2022, they had the basic conditions and trends to enter the synergistic generation period, and a synergistic management model is required. At the end of the article, we give a collaborative management method for complex industrial enterprises with a good management foundation. These include the management mechanism based on flow structure collaboration and the management path based on collaborative evolution. Of course, the management countermeasures given in this study are also applicable to other complex process-based industrial enterprises.

Keywords: industrial production system; flow structure; synergy theory; collaborative management; system dynamics

Citation: Song, J.; He, Z.; Jiang, L.; Liu, Z.; Leng, X. Synergy Management of a Complex Industrial Production System from the Perspective of Flow Structure. *Systems* **2023**, *11*, 453. https://doi.org/10.3390/systems11090453

Academic Editors: Zaoli Yang, Yuchen Li and Ibrahim Kucukkoc

Received: 29 June 2023
Revised: 18 August 2023
Accepted: 28 August 2023
Published: 1 September 2023

1. Introduction

With the continuous improvement of technology and manufacturing, modern industry is developing towards large-scale, complex, process-oriented, and continuous development. In many fields, such as metallurgy, chemical industry, etc., it is no longer a single integrated system but is composed of many coordinated and integrated subsystems. There are numerous contradictions within complex industrial production systems, such as frequent dynamic fluctuations, and the operation process must consider the global performance. In addition to external environmental interference, the system itself also has strong nonlinear characteristics; however, the supply, use, and allocation of resources and energy, as well as their management methods, directly determine the operational status of the production system, which in turn affects production efficiency, product quality, product costs, and so on. The current integrated management system has been implemented for several years. The method is to use manufacturing platforms such as ERP and MRP to achieve information sharing, thereby achieving order-driven production and on-time delivery, and to prevent unexpected problems by implementing separate lean improvement projects. Although integrated management has achieved many results, there is still a widespread problem of overcapacity. A large amount of redundant data reduces the efficiency of management work and has poor management accuracy. Due to the wide variety of products, frequent batch changes, and the lack of a unified model for equipment data management, problems of redundant construction and resource waste exist. The construction of manufacturing platforms is still based on modular or departmental management, lacking a management mechanism compatible with the coordinated development of enterprise informatization.

For any system in the objective world, matter is the carrier, providing tangible entities, such as production resources; energy is the executor, providing production power; and information is the conductor and the soul that forms the process. The static and dynamic operational structures of production systems at different levels are closely related to material flow, energy flow, and information flow. The same applies to enterprise production systems; therefore, in recent years, the development of interdisciplinary cooperation within such systems as engineering, dynamics, and management has brought new directions to industrial production management. Many scholars analyze industrial production systems from the perspective of flow structure to save management effort and simultaneously achieve cost-effectiveness and environmental benefits.

In spite of this, there are many different analytical methods for production systems, such as process flow, capital flow, value flow, green energy flow, material flow, energy flow, information flow, capital flow, and other structures. It is worth noting that there are analysis defects, such as inconsistent subsystem levels and unclear flow structure characteristics. For instance, both cash flow and information flow have the functions of scheduling, controlling, and commanding production; however, the optimization of cash flow aims to reduce costs, while the research focus of information flow is to shorten delivery times. The fundamental goal of both is to improve efficiency. Another issue is the lack of distinction between spatial flow paths (material flow and energy flow, etc.) and temporal flow paths (information flow and financial flow, etc.). Moreover, when analyzing the "potential" and "resistance" of the "flow" structure, as well as the "drive" and "dissipation" of the system, the absence of combining dynamic principles results in relatively single model construction and incomplete variable analysis.

Nowadays, complex industrial production systems involve various elements, such as humans, machines, the environment, and management. Managing these systems' internal structure, organizational form, and functional processes is a complex task. People are increasingly paying attention to the dynamic utilization of energy in industrial enterprises; however, most studies are limited to energy flow and overlook the characteristics of complex industrial production systems, such as nonlinearity, dynamism, openness, and orderliness. The changes in material flow parameters and the driving force of information flow are the root causes of changes in energy flow; therefore, whether solving subsystem problems or large-scale system problems, research should focus on the operational mechanism of the overall large-scale system and the role of the relationships between subsystems in the large-scale system. Scholars emphasize the concepts of "coordinated progress" and "collaborative innovation", aiming to provide a theoretical analysis and practical guidance for industrial enterprises using the concept of synergy to improve economic efficiency.

Starting from the research on the composition of enterprise systems, this paper innovatively explores the connections between various subsystems within the enterprise and the external environment and proposes new solutions for the organizational form of the enterprise. There is limited research on the mechanism and influencing factors of subsystem synergy, and there is no indication of the path through which this advanced concept should be implemented. The research on collaborative management in industrial enterprises coincides with the research direction of collaborative theory from the perspective of system flow structure; therefore, it is extremely important to deepen the understanding of the collaboration among various subsystems of the production system as a whole and to reflect the optimization results in management decisions in order to build a management mechanism and path that is easy to implement and promote. In the current context, industrial enterprises need to comprehensively consider factors such as human, machine, environment, and management to achieve efficient and collaborative operation of production systems, thereby achieving the goal of improving economic efficiency.

This study provides a collaborative research framework for material flow, information flow, and energy flow in complex industrial production systems. We have constructed a collaborative evaluation and collaborative evolution model, which can clarify the conditions for collaborative evolution, identify the direction of enterprise collaborative management, and take an important step towards achieving a clear path from comprehensive integrated management to collaborative management. This method is easy to generalize to other industrial enterprises that already have advanced management system practice environments and is especially suitable for complex nonlinear process-oriented industrial enterprises. The main innovation points cover the following aspects:

(1) Based on the theory of synergetics, we provide a combination of system dynamics research methods and flow structure co-evolution research.

(2) We apply system dynamics methods to analyze the collaborative evolution of flow structures and construct evolution models. Additionally, we extend the evolution model to a three-stage system of equations.

(3) When establishing a synergy evaluation model, we considered the Taylor index and the system's gray relationship. Furthermore, we utilize the self-organizing mapping network (SOM) algorithm to identify sudden disturbance data in industrial production systems.

The remaining content of this article is structured as follows: In Section 2, a literature review is presented, encompassing system synergy, system flow perspective, and complex industrial management. Section 3 outlines the construction of a collaborative model for the flow structure within complex industrial production systems. Section 4 delves into a collaborative model of the production system's flow structure, employing SG enterprise as a case study. Lastly, Section 5 provides a comprehensive summary of the entire text.

2. Literature Review

2.1. System Synergetics

Haken proposed the collaborative theory in the 1970s, which is based on system theory and control theory, using a combination of dynamic and statistical analysis methods. Its main focus is to study the cooperation, coordination, and synchronization mechanisms of various components or subsystems within complex systems during operation (Haken, 2013) [1]. As an important branch of systems science theory, the research methods of collaborative theory are applicable in various systems, especially for complex, large-scale systems. Meng (2000) [2] pointed out that the collaboration of complex systems is a process of achieving the overall effect of the system through internal self-organization and external regulation and management activities. Meng emphasized that a subsystem analysis should be conducted from a reasonable perspective. Peng (2009) [3] regards logistics and manufacturing as two subsystems of the modern economy. By utilizing the sequential parameter evolution process of synergy theory, the mechanism of symbiotic evolution of these two subsystems was explored, emphasizing that the interdependence and cooperation between modern logistics and advanced manufacturing are natural outcomes of industrial evolution. Zheng et al. (2010) [4] applied the principles and methods of collaborative theory to study the supply and demand relationship between automotive manufacturers and suppliers, providing decision support for optimizing the automotive supply chain. Anbanadam et al. (2011) [5] considered various variables, such as management commitment, information sharing, cooperative trust, relationship risk, and return sharing. They constructed a supply chain collaboration model with manufacturers and retailers as subsystems. The degree of collaboration is measured to evaluate the degree of supply chain cooperation, while also providing potential cooperation opportunities for other elements outside the supply chain subsystem.

After Haken introduced the term "order parameter" into synergetics, the concept of the order parameter was extensively expanded in fields such as management, economy, and society. The order parameter plays a dominant and pivotal role in coevolution, exerting a synergistic effect on the system (Leydesdorff et al., 2013) [6]; however, the order parameter identification method based on Haken's classical model mostly remains at the macro-level and lacks a detailed analysis of the information in the collaborative evolution process. Wanger (1994) [7] improved Haken's three algorithms, and Schanz and Pelster (2012) [8] used this method to identify the order parameters of nonlinear time-delay systems. May et al. (2015) [9] provided an identification method and indicator system for order parameters in energy systems. Hryshchuk et al. (2016) [10] updated the dynamic meaning of order parameters: due to self-organizing processes, the system reduces its dynamic parameters to a limited number of variables, which can maintain stability under disturbances. Xu et al. (2017) [11] used this method to identify the system order parameters in the coevolutionary model. Additionally, there are the main melody analysis method based on objective programming (Warm et al., 2011) [12], the optimized relaxation coefficient method (Zheng et al., 2013) [13], and the order parameter identification model constructed based on the gray system theory (Wu et al., 2017) [14]. With the expansion of the research scope, order parameter analysis has entered more fields. Wang Haiyan et al. (2017) [15] proposed a method for identifying order parameters in food quality chain collaborative systems based on the gray correlation degree and attribute reduction, and provided a method for solving order parameters considering the overlap degree. The focus of this method is to determine the collaborative elements, establish parameters that represent the system state, and define overlapping relationships between the parameters. Wen et al. (2020) [16] analyzed the formation process of multi-order parameters from the perspective of output/input. Combined with its impact on the system evolution process, an improved data envelopment analysis method was applied to establish an efficiency-oriented multi-order parameter identification model for the system evolution process.

Based on the identification of order parameters, research on system collaboration issues is mainly conducted from two perspectives: complex system theory and economics. For example, research covers aspects such as system collaborative optimization, collaborative mechanisms, collaborative operational performance, and collaborative degree measurement. Li et al. (2012) [17] proposed a composite system synergy model based on order parameters and made significant progress in evaluating the synergy levels. Additionally, Tang et al. (2010) [18] introduced the Euclidean distance method into the evaluation model of system collaborative development and analyzed the Chinese economic and technological system through empirical research. Cui (2016) [19] emphasized that the collaborative level of the system cannot be simply evaluated using the general method of assessing the development level of the system. The development level primarily measures the evolution trajectory of the system from a vertical perspective, while the level of collaboration emphasizes the consistency of relationships between various elements within the system from a horizontal perspective. Deng et al. (2016) [20] pointed out that the synergy of various subsystems in a large system can impact the synergy of a composite system. Chen Lilan (2016) [21] studied the integrated management of engineering project elements based on collaboration theory and constructed a model to measure the degree of internal element collaboration in engineering projects. This model also calculates collaboration based on order degree and proposes using a large sample survey method to correct the identification results of order parameters. Luo and Dong (2017) [22] divided the synergy of regional economic systems into three stages: primary stage (synergy degree of 0–0.4), intermediate stage (synergy degree of 0.4–0.7), and advanced stage (synergy degree of 0.7–1) and expanded the synergy model into dynamic and static parts. Li et al. (2016) [23] believe that the low synergy of the macroeconomic system is less than 0.3. Chen et al. (2016) [24] studied regional intellectual property management systems and improved the traditional collaborative evaluation model, proposing the concept of subsystem "consistency". When the consistency of each subsystem is high, the synergy will fall within the [0, 1] range; conversely, it will fall within the [−1, 0] range. Zhang et al. (2017) [25] extended the concept of synergy to the research field of regional development, analyzing the Beijing–Tianjin–Hebei greater system from the perspective of five subsystems and studying the spatial differences in the order of collaborative development. Li et al. (2017) [26] also measured the level of coordinated development between Beijing, Tianjin, and Hebei. They also discussed the development and synergy of the system, pointing out the positive impact of development strategies on the collaborative process. They believe that every game synergy mutation process pushes the collaborative development of urban agglomerations to a higher level of synergy and presents a phased pattern.

The development of systems science has expanded the range of applications for synergetics. In recent years, scholars have predominantly employed synergetics to examine self-organizing processes in macroscopic systems. Lv Tong et al. (2002) [27] proposed that the energy economy environment system also qualifies as a dissipative structure. During system evolution, the influence of fluctuation mechanisms can give rise to phenomena such as synergistic and non-synergistic spiral escalation, aligning with the fundamental tenets of synergy theory. Bao et al. (2014) [28] devised a bilevel programming mathematical model to optimize the allocation of resources for product customization by analyzing collaborative manufacturing resource allocation for such customization. Employing an optimized hybrid genetic algorithm, they attained the optimal solution for collaborative manufacturing resource allocation in product customization. Fang (2017) [29] asserts that the collaborative process of socio-economic systems follows a nonlinear spiral progression involving game, collaboration, mutation, re-game, re-collaboration, and re-mutation. This analysis, too, rests on synergy theory. Lychkina (2016) [30] delved into the collaborative development of socio-economic subsystems and noted constraints in traditional socio-economic models. To create effective models, interdisciplinary perspectives, such as system dynamics and strategic decision simulation, must be applied to study cyclic collaborative evolution phenomena from a dynamic standpoint, bridging macro- and micro-system collaborative research. Meynhardt et al. (2016) [31] examined service ecosystems at micro- and macro-levels utilizing synergy theory, emphasizing the value characteristics inherent in dynamic collaborative evolution. Using nine collaborative attributes, including critical points, stability, endogenous variables, nonlinearity, feedback, and finite prediction as starting points, pathways for enhancing value can be identified. Zheng et al. (2017) [32] expanded the concept of collaborative management to the realm of industrial economy, illustrating the dual effects of supply chain collaborative management using steel enterprises as an illustration. These effects involve achieving both financial and ecological performance, as well as internal and external performance for a single enterprise. Yang et al. (2019) [33] identified that the original classical model could solely address the limitations of static system synergy. They advanced a speed feature model based on improvements, quantifying the quality and effectiveness of system synergy evolution through co-driving the "evolution speed state" and "evolution speed trend". This model integrates weight information and accounts for the "functionality" and "coordination" of order parameters. Guo et al. (2019) [34] employed collaborative analysis methods in the e-commerce and big data industries. They established an e-commerce big data system (EBDS) collaborative evolution model founded on the classic Haken model. Through quantitative evaluation of order and synergy, they presented managerial recommendations for the secondary industry encompassing collaborative application, industrial chain, risk, external environment, and ecosystem dimensions.

2.2. System Flow Perspective

In the 1960s, Reiter (1966) [35] pioneered the study of production systems, introducing the concept of "Lot Streaming" and examining the transfer of product batches. This focus on production batch flow and transfer modes continued with subsequent researchers, such as Jacobs (1984) [36], Graves, and Kostreva (1986) [37], who identified batch transfer modes in their respective fields. The 1980s saw the emergence of the just-in-time production theory and optimal production technology. Truscott (1985) [38] utilized heuristic methods to address continuous flow challenges in general workshops with equal sub-batches. Potts (1989) [39] applied the method of equal quantity batch to flow shop problems and employed heuristics for multi-batch solutions. Vickson (1995) [40] employed a fast polynomial algorithm to solve flow shop multi-batch problems, considering the job setting time and sub-batch transfer time. Etinkaya (2006) [41] extended this by incorporating units and independent settings to optimize batch problems in dual-machine process workshops, termed batch flow integration optimization. Despite these advancements, such research mainly focuses on small-scale device environments, with limited exploration of multi-flow integration at the production line level.

In recent years, international scholars have primarily explored complex large-scale systems through the lens of "flow" structure, yielding extensive applications in industrial production systems. Ruth (1995) [42] delved into the interplay between material flow, energy flow, and information flow, utilizing the industrial balance theory. She noted that energy propels physical state changes alongside information transmission. Simons et al. (2003) [43] examined the production value stream, shifting from a time-centric approach to a sustainable value stream, thus analyzing energy use and identifying the least environmentally friendly links, particularly in relation to carbon dioxide emissions. This tool has proven highly effective in energy consumption analysis through practical application. Long et al. (2008) [44] emphasized that the analysis of "flow" primarily concerns the flow rate and velocity, providing an initial depiction of material flow, energy flow, and information flow characteristics and relationships within extensive systems. Bascur et al. (2009) [45] explored production material flow and energy flow, utilizing historical data and shared production information to foster continuous enhancement grounded in quality monitoring, process control, and variable analysis. William et al. (2014) [46] expanded the original value stream map by incorporating raw material utilization and energy consumption. This comprehensive approach vividly highlights the waste and potential pollution aspects of a company, furnishing a global perspective for enhancing the company's sustainability.

Taulo et al. (2016) [47] analyzed the material and energy flows within the tea industry supply chain from a collaborative stance. They argued that this approach can illuminate environmental issues early in production and offer a basis for prioritizing factory improvement projects. Suominen et al. (2016) [48] introduced a nonlinear optimization scheduling scheme tailored for production, grounded in material and energy flow networks within production systems. This scheme optimizes production efficiency via process simulation, furnishing effective production conversion plans and establishing predictive mathematical models for equipment parameters. Yin (2016) [49] viewed the production process as a multifaceted network encompassing various manufacturing processes. His emphasis lay in addressing energy flow concerns accompanying material flow to ensure minimal production costs. Collaborative research on flow structure, from an energy management perspective, becomes pivotal for modern steel mills, supplying solutions and system support for energy-saving decisions. Zhang et al. (2017) [50] merged an environmental values stream analysis with the Flexsim simulation tool, enhancing the value-added ratio of energy consumption during production. Zheng et al. (2017) [51] proposed a collaborative approach to material and energy flows in steel enterprises. They underscored the need to plan the coupling of material and energy flows at unit equipment and process network levels, supported by digitization and informatization, to optimize resource utilization. Yu et al. (2018) [52] identified a coupling synergy between material, value, and energy flows, calculating the synergy coefficients between them. Li et al. (2018) [53] constructed a multi-stream collaborative model encompassing "material flow, energy flow, information flow, and capital flow" in information physics energy systems. Their approach considered production costs, energy consumption, collaborative scheduling, and market aspects. Huang et al. (2019) [54] analyzed the structure and attributes of "flow", introducing a collaborative framework for the security of the "four flows" within a system, and conceptualizing a model. Collectively, these scholars' research highlights three perspectives within system collaboration from the flow standpoint: micro (human–machine systems), meso (organizational systems within enterprises, encompassing various flows), and macro (energy economy environment system synergy within social and economic systems).

2.3. Complex Industrial Management

The study of complex industrial production control management theory and industrial system dynamics are complementary. Qi et al. (2008) [55] used the system dynamics method (SD) to study the knowledge transitive model within industrial enterprises and obtained the change law of knowledge potential energy under different factors. Based on the industrial system dynamics modelling method, Jiang (2011) [56] proposed two breakthrough management deficiency optimization paths to improve the performance level of enterprises. Wang et al. (2012) [57] constructed a dynamics model of the logistics operation of an industrial enterprise and proposed a simulation optimization scheme for the crafts industry, demonstrating the feasibility and applicability of system dynamics applied to production cost control.

In the realm of industrial production's general management methods, key solutions encompass factors influencing management system optimization (Love et al., 2002) [58], enhancing efficiency and reducing costs across all production aspects (Zhang et al., 2007) [59], employing dynamic programming for management strategies (Lee et al., 2006) [60], embracing the enterprise development life cycle and self-organizational learning mode (Hu Bin et al., 2006) [61], and more. A dynamic simulation analysis of enterprise business model operations, conducted by Dai and Chen (2014) [62], explored developmental drivers through an organizational structure lens. Gary et al. (2018) [63] introduced an innovative dynamic model to investigate the cost, performance, and development strategy matters within automotive manufacturing enterprises. Hanafi et al. (2019) [64] crafted a quantitative system dynamics model for the smelting industry, tackling intricate production investment competition challenges.

The energy revolution has substantially boosted productivity while inducing shifts in industrial management and production structures (Stan et al., 2015) [65]. Larson et al. (2004) [66] introduced the "Lean and Green" concept, highlighting their intricate interrelation. Lean manufacturing, with its well-established history and widespread application in developed countries, has yielded impressive outcomes, leading to a focus on the environmental impacts of unnoticed lean practices. This initial research phase explored how lean manufacturing, through waste elimination, enhances environmental performance (Zhang et al., 2018) [67]. Kurdve et al. (2014) [68] devised a comprehensive bottom-up system incorporating value-driven and operational-driven approaches, providing an expansion path for implementing this new system alongside existing structures. Yang (2015) [69] outlined the evolution of enterprise management in the low-carbon era, emphasizing the integration of energy conservation and consumption reduction into core values and developmental goals. This involves a shift in traditional management paradigms, aiming for mutually beneficial outcomes between energy preservation and profitability, guiding institutional innovation through energy transformation responsibility, and aligning carbon emission mandates with market dynamics. Modern management techniques, such as benchmarking, technology outsourcing, standard operations, and personnel training, are believed to reduce energy expenses for enterprises. Tetiana et al. (2018) [70] stress the significance of informed green lean decision-making, particularly for industries heavily reliant on conventional energy sources. Taking power enterprises as an illustration, they present specific conditions for implementation, including structural reformation of the energy sector and technological transitions in the industrial realm. Jarrahi et al. (2019) [71] discussed the connection and difference between the two production modes of "Lean manufacturing" and "Industry 4.0". They connected two important research fields of the industrial production management system and provided a collaborative integration scheme of the two.

Various management methods, including system engineering, optimization theory, lean manufacturing, sustainable development, and enterprise production, have been deeply integrated into production management. This has led to the inclusion of input–output analysis, production energy conservation control, energy consumption analysis, and regulation (Javied et al., 2019) [72]. Dues et al. (2013) [73] pointed out that challenges exist between managing energy resources and production. Scholars continuously strive to introduce new methods and modeling ideas. Lee et al. (2014) [74] introduced Six Sigma, a lean quality management tool, into the energy plan. They pioneered the Six Sigma energy management approach, now a part of modern production management systems. Pampanelli et al. (2014) [75] created models for implementing environmental management based on lean manufacturing. This integrated lean concept with environmental sustainability improves resource use and reduces the environmental impact. Zhang et al. (2016) [76] developed an energy demand prediction model for industrial enterprises. It considers how production systems and energy management interact across time and space. This model was optimized under uncertain decision-making and applied to power demand management. Li et al. (2017) [77] built a system dynamics model with five subsystems: economic development, primary and secondary aluminum production, carbon dioxide emission intensity, and policy formulation. This model can analyze trends in carbon dioxide emissions from the aluminum industry. Hilorme et al. (2019) [78] established adaptive and multiplication models for energy technology implementation based on seasonal factors and overall energy management trends. They used a space analysis matrix. Laura et al. (2020) [79] created a system dynamics model combining technological advancements and economic evaluation. This model analyzes the cost impact of implementing carbon dioxide capture technology in cement plants. It evaluates the economic benefits of the cement industry under different carbon capture scenarios.

In terms of modern innovation in management systems, Chang (2005) [80] and Sarin (2008) [81] both studied production management from the perspective of flow by using batch flow to study the comprehensive integration of just-in-time production management methods. Zeng (2018) [82] believes that a good management system is the foundation for the healthy development of enterprises. It helps to break down technical, institutional, and market barriers between energy subsystems within enterprises. Hillman et al. (2018) [83] studied the value of enterprises as drivers of low-carbon transformation, believing that the driving force of enterprises for sustainable development in the world is currently underestimated. In order to address contemporary ecological challenges, industrial enterprises should adopt a continuous improvement organizational model. Domestic and foreign scholars have gradually constructed an effective management system based on continuous improvement and focused on energy management, referencing the architecture of ISO. It has been widely applied in the implementation monitoring and optimization management of industrial production processes (Zhang et al., 2019) [84]. Until the end of the twentieth century, innovative industrial production management ideas continued to emerge, including total quality management, Lean Six Sigma, etc. Subsequently, enterprises all over the world favored this low-risk organizational change mode and built an enterprise operation management system with this as the core. In the 21st century, the production management system centered on lean concepts has been widely applied and has evolved into a world-class manufacturing (WCM) system. Over the past thirty years, the management system has undergone continuous development, leading to improved technological efficiency and a focus on achieving efficient resource utilization through production management.

Industrial enterprises have advanced their production management thinking, progressing from individual equipment resource management to the strategic planning and control of production process resources.

2.4. Commentary

Based on an analysis of the existing literature, comprehensive discussions have taken place on collaborative research within large-scale systems, both domestically and internationally. These studies encompass separate examinations of elements such as material flow, energy flow, and information flow within large systems. They also delve into collaborative interactions between multiple flows of elements, spanning micro-, meso-, and macro-levels. The existing research primarily focuses on empirical studies concerning subsystem analyses, collaborative evaluation methods, and complex interrelationships within large-scale systems. The synergy theory-based evaluation method emphasizes guiding and constraining factors in system development, highlighting the significance of order parameters and collaborative trends. This stands in contrast to traditional system evaluation methods, which prioritize enhancing indicator systems and collecting numerical values; however, due to the extensive data support and specialized nature required for order parameter identification, the current models are complex and demand significant effort. With the increasing complexity of industrial production systems and the infusion of diverse research methods and perspectives from various disciplines, scholars have expanded the application of synergy theory and flow structure. This expansion ranges from industrial production systems to macroeconomic systems. Despite this progress, certain challenges persist. These include inconsistent subsystem levels, unclear characteristics of flow structures, and analytical limitations. Additionally, the current model construction is relatively singular due to the insufficient integration of dynamic principles, resulting in an incomplete variable analysis. It is noteworthy that several methods used to address energy flow issues in production systems are not directly adaptable to complex industrial production systems. The latter exhibit characteristics such as nonlinearity, dynamism, and openness, posing unique challenges for research.

Amidst the ongoing trends of intelligent manufacturing and green manufacturing, scholars have put forth a perspective that seeks to advance and break new ground through comprehensive integrated management. They emphasize concepts such as "coordinated progress" and "collaborative innovation", with the goal of offering both a theoretical analysis and practical guidance to industrial enterprises using the principles of synergetics. This approach begins by examining the composition of enterprise systems and innovatively exploring the relationships between various subsystems within the enterprise and their external environment; however, there is currently a lack of sufficient research on the mechanisms and influencing factors of subsystem synergy. Furthermore, there is no clear indication of the precise path to implement this advanced concept. Correspondingly, the research focus on collaborative management within industrial enterprises aligns with the direction of collaborative theory from the standpoint of system flow structure. Deepening the understanding of collaboration among diverse subsystems of the production system is of the utmost importance. This understanding should be reflected in management decisions, enabling the creation of management mechanisms and pathways that are both practical and promotable.

3. Construction of Flow Structure Collaborative Model for Complex Industrial Production Systems

3.1. Preparation of Flow Structure Modeling for Complex Industrial Production Systems

3.1.1. Analysis and Modeling Assumptions of Complex Industrial Production Systems

The complex industrial production system is an open large-scale system that has both energy exchange and material exchange with the outside world. P represents the set of all processes and equipment in the system, R signifies the comprehensive network of interrelations within the same system, and the complex industrial production system can be simply expressed as Formula (1) [85].

$$\sum = \langle P, R \rangle \tag{1}$$

In explaining Formula (1), we can understand "R" as encompassing the storage and transportation of raw materials and energy within the production layer of complex industrial systems. This layer involves various stages, such as raw material processing, forming, and packaging, resembling both continuous and intermittent production processes. It can also be viewed as a system made up of different subsystems, each with distinct functions. Encompassing dimensions of material flow, energy flow, and information flow, this process system operates based on changes in the state and properties of matter. A complex industrial production system fundamentally constitutes a large-scale engineering construct. It is marked by high nonlinearity, multivariability, and limited information. Its intricate nature comprises various components, multiple tiers, openness, nonlinearity, and dynamic orderliness, ultimately giving rise to a significant attribute: structural complexity. Analyzing its dynamic structure involves addressing three key facets. The first is the dynamic essence of the system's flow structure, which encompasses characteristics of propulsion and dissipation. Second are the factors governing subsystem behavior (comprising material flow, energy flow, and information flow). Last are the interrelations among subsystems and their collective impact on the overall system.

Based on the situation described above, in order to construct a more scientific mathematical model for evaluating the collaborative degree and dynamic collaborative evolution of complex industrial production systems, we make the following modeling assumptions:

① A complex industrial production system functions as an open system, facilitating continuous exchanges of materials, energy, and information between internal and external components. This dynamic interaction occurs not only among various subsystems within the system itself, but also among the diverse elements within each subsystem.

② Within a specific time and space range, the system displays a relatively stable state, known as dynamic stability. During this state, the diverse resources of the production system (including labor, capital, raw materials, technology, fuel, etc.) remain at a consistent scale. A value range denoted as N characterizes the level of collaborative evolution in production systems, and this value is positively linked to the system's resource count. In simpler terms, collaborative evolution does not have a final point, but it does lead to recognizable phased results. Following a certain disturbance (such as the emergence of a problem), the system gradually transitions to a stable and organized state, referred to as the collaborative evolution process of the system.

③ The timing of system state reduction and feedback is allowed, as the system state is not only related to the resource situation at a specific time, but is also influenced by factors such as the supply chain environment, policy requirements, and management methods.

3.1.2. Determination of Order Parameters

Based on the theory of collaboration given herein, the collaborative order parameters of complex industrial systems represent the behavior between subsystems of material flow, energy flow, and information flow. They are also the dominant factors that cause subsystems to compete and make large systems tend to collaboratively evolve.

In the context of identifying order parameters, Haken introduced three methodologies in his works, "Introduction to Synergy" and "Advanced Synergy". These methodologies comprise the relaxation coefficient method, the maximum information entropy method, and the adiabatic elimination method. In recent years, both domestic and international scholars have proposed novel approaches. These include a theme analysis, identification based on target programming, and identification rooted in the input–output analysis. Prior research has effectively synthesized a spectrum of feasible order parameters. Furthermore, the Ministry of Industry and Information Technology of the People's Republic of China issued the "Evaluation Specification for the Integration of Informatization and Industrialization of

Industrial Enterprises" (National Standard GB/T23020) in September 2013. This standard establishes the evaluation criteria for industrial enterprises across aspects such as materials, information, and management, accentuating the interplay and synergy between resource efficiency and subjective production factors. Drawing on the available data, this study has curated an illustrative list of order parameters, as outlined in Appendix A Table A1. By referencing the research findings of May et al. (2015) [9], Wu et al. (2017) [14], Zheng et al. (2017) [51], and Wen et al. (2020) [16] in the domain of order parameter identification, and in conjunction with Appendix A Table A1, a comprehensive compilation of order parameters suited for intricate industrial production systems is presented in Table 1.

Table 1. Order parameter of industrial production system (comprehensive identification).

Number	Parameter	Unit
1	Yield of finished products	%
2	Production reliability	ND
3	Comprehensive heat production rate	%
4	Equipment production efficiency	%
5	Wastewater recycling capacity	m^3
6	Information management index	ND
7	Production plan completion rate	%
8	Unrecognized energy loss	kgce/t
9	Water consumption per unit product	L/m^3
10	Product one-time qualification rate	%
11	Production defect loss ratio	%
12	Energy cost loss ratio	%
13	Comprehensive energy consumption per unit product	MWh/m^2
14	Air pollution per unit product	kg/m^3
15	Cost proportion of information technology construction	%
16	Manage controllable OEE ratios	%
17	Product fragment recycling rate	%
18	Flexibility (inventory turnover days)	d
19	Equipment overall efficiency (OEE)	%

3.2. Collaborative Evaluation Model Based on Order Parameters

3.2.1. Collaborative Evaluation Considering the Taylor Index

Synergy is a dynamic indicator that is difficult to measure at all times. Based on the Haken model and referring to the synergy measurement model proposed by Meng et al. (2000) [2] and Li et al. (2016) [23], it refers to the use of the order degree at a certain moment to measure the collaborative development results of the system from the initial moment to that moment. Assuming that the production system is S, the material flow subsystem, energy flow subsystem, and information flow subsystem are S_1, S_2, and S_3, respectively; then, $S = (S_1, S_2, S_3)$, where S_{ij} represents the j-th element in the i-th system, such as S_{21} representing the first element in the energy flow subsystem. The order parameters of the material flow subsystem, the energy flow subsystem, and the information flow subsystem are represented by μ_M, μ_E, and μ_I, respectively:

$\mu_M = (\mu_1, \mu_2, \ldots, \mu_m)$; the corresponding value is $q_M = (q_1, q_2, \ldots, q_m)$;

$\mu_E = (\mu_{m+1}, \mu_{m+2}, \ldots, \mu_{m+n})$; the corresponding value is $q_E = (q_{m+1}, q_{m+2}, \ldots, q_{m+n})$;

$\mu_I = (\mu_{m+n+1}, \mu_{m+n+2}, \ldots, \mu_{m+n+l})$; the corresponding value is $q_I = (q_{m+n+1}, q_{m+n+2}, \ldots, q_{m+n+l})$.

According to the index characteristics, the order parameters can be divided into three types. The larger the value, the more favorable the system synergy is. For example, the overall efficiency of equipment, yield, etc., are called positive effect order parameters, and their effect value is expressed as OD_{ij}^+. Another thought is that the smaller the value is, the more favorable the system synergy is. For example, the comprehensive energy consumption per unit product is called the negative effect order parameter, and its effect value is expressed as OD_{ij}^-. Another is that the closer its value is to a certain target value, the more beneficial it is to the system synergy. In other words, it is not the greater the better or the smaller the better, for example, the proportion of wastewater recycling and information construction costs, whose effect value is expressed as OD_{ij}^*. This type of order parameter is called a moderate order parameter. We define α_i as the minimum value of order parameters for each subsystem, and β_i as the maximum value of order parameters for each subsystem. Based on the information above, calculate the fffect size of the three order parameters according to Formula (2):

$$\begin{cases} OD_{ij}^+ = \frac{q_{ij}' - \alpha_i}{\beta_i - \alpha_i} \\ OD_{ij}^- = \frac{\beta_i - q_{ij}'}{\beta_i - \alpha_i} \\ OD_{ij}^* = \frac{|q_{ij} - q_i^*| \, |q_{ij} - q_i^*|_{max}}{|q_{ij} - q_i^*| \, |q_{ij} - q_i^*|_{minmax}}, \end{cases} \quad \begin{aligned} \alpha_i &= min\{q_{i1}', q_{i2}', \ldots, q_{ik}'\} \\ \beta_i &= max\{q_{i1}', q_{i2}', \ldots, q_{ik}'\} \end{aligned} \tag{2}$$

The number of subsystems is expressed in m relative to the long coevolution process, and there can be n order parameters in each subsystem. Indeed, during infinitesimal time intervals, each subsystem possesses only one order parameter. The effect value of the order parameter is expressed as x_{ij}. If the number of order parameters in a subsystem is less than n, the lesser part of the effect value is expressed as 0, then the order parameter eigenvector matrix of the system can be expressed as Formula (3):

$$X = \begin{bmatrix} X_1 \\ X_2 \\ \vdots \\ X_m \end{bmatrix} = \begin{bmatrix} x_{11} & x_{12} & \cdots & x_{1n} \\ x_{21} & x_{22} & \cdots & x_{2n} \\ \vdots & \vdots & \ldots & \vdots \\ x_{m1} & x_{m2} & \cdots & x_{mn} \end{bmatrix} \begin{array}{l} i = 1, 2, \ldots, m \\ ' j = 1, 2, \ldots, n \end{array} \tag{3}$$

The standardization of the data in the matrix is performed according to Formula (4):

$$x_{ij}' = \frac{x_{ij}}{\sum_{j=1}^n x_{ij}} \tag{4}$$

After obtaining the new transformation matrix $[X_i']$, considering that the order parameters of each subsystem play different roles in the collaborative development process of the dominant system, it is necessary to calculate the weights of the order parameters of each subsystem. Since the numerical value of the order parameter represents a type of information about the state of the subsystem, the Thiel index is introduced here, which is an indicator based on the information entropy calculation method to measure the gap between objects. The larger the Thiel index, the greater the amount of information provided by the order parameter (symbol). We calculate the Theil index according to Formula (5) to convert the negative effect size to a positive value.

$$T_i' = \ln(n) - \sum_{j=1}^n x_{ij}' \ln \frac{1}{x_{ij}'} \quad i = 1, 2, \ldots, m; \, x_{ij}' = \lim_{\sigma \to 0} x_{ij}' + \sigma; \, x_{ij}' = 0 \tag{5}$$

The weight of the order parameter is defined as Formula (6):

$$\omega_i = \frac{T_i'}{\sum_{i=1}^m T_i'}, \omega_i > 0 \tag{6}$$

The greater the effect size of the order parameter in each subsystem is, the greater is the contribution value of the order parameter to the coordinated order state of the subsystem. Here, we define the contribution of the three subsystems to the coordination degree of the large system as the order degree. At any time t, the order degree of the material flow subsystem, energy flow subsystem, and information flow subsystem is calculated as Formula (7):

$$OD_j^t = \begin{cases} OD_1^t(S_1) = \sum\limits_i^m \omega_i OD_i^t, \sum\limits_i^m \omega_i = 1 \\ OD_2^t(S_2) = \sum\limits_m^{m+n} \omega_i OD_i^t, \sum\limits_m^{m+n} \omega_i = 1 \\ OD_3^t(S_3) = \sum\limits_{m+n}^{m+n+l} \omega_i OD_i^t, \sum\limits_{m+n}^{m+n+l} \omega_i = 1 \end{cases} , j = 1,2,3 \tag{7}$$

The order degree of the large system is:

$$OD^t = \prod_{j=1}^{3} OD_j^t \tag{8}$$

According to the order degree of the three subsystems and the order degree of the large system, the synergy degree of the large system is calculated as:

$$SD_\mu = \pm \sqrt[4]{\left| \left[\prod_{j=1}^{3} \left(OD_j^{t+1} - OD_j^t \right) \right] * (OD^{t+1} - OD^t) \right|}, \quad t, t+1 \in \{1,2,\ldots,k\} \tag{9}$$

Among them, when $OD_1^{t+1} > OD_1^t$, $OD_2^{t+1} > OD_2^t$, $OD_3^{t+1} > OD_3^t$ is simultaneously valid, SD_μ takes a positive value, otherwise it takes a negative value. The larger the value of $SD_\mu \in [-1,1]$ is, the higher is the level of coevolution, and vice versa. The extremely non-coevolutive state when SD_μ is -1 and the extremely cooperative state when SD_μ is 1 are relatively rare. The value of SD_μ reflects the collaborative state of the system at a certain time, displaying the static state of the system's collaborative measure and the evolution results during a certain period. Formula (14) takes into account the comprehensive situation of all subsystems. Even if the degree of order of a certain subsystem increases significantly, it cannot eliminate the impact of the decrease in the degree of order of other subsystems. This is reflected in $SD_\mu \in [-1,0]$, which means that the collaboration of a large system is based on the cooperation of two subsystems, and the large system is ultimately in a stable and orderly state.

3.2.2. Collaborative Analysis Considering Gray Correlation

The collaborative evaluation method for large systems is based on the calculation of the following two factors: subsystem order degree and Thiel index weighting. For situations where there is both clear and unclear information in production, the evolution process and trend should be clarified, and the collaborative mechanism between subsystems should also be studied. Wu Yuying et al. (2017) [14] and Wang Haiyan et al. (2017) [15] both used a gray correlation analysis when studying collaborative problems, but it is limited to identifying the order parameters of subsystems, that is, clearly identifying the order parameters that dominate the trend of subsystems based on the gray relationship between the subsystems. In this study, the gray synergy analysis is carried out for the subsystem. First, the original data are dimensionally removed and standardized, such as in Formula (10), and then the data are normalized according to Formula (11).

$$x_i' = 1 - \frac{|x_i - x_{best}|}{\max\{|x_i - x_{best}|\}} \tag{10}$$

$$a_{ijt} = \frac{x_{ijt} - \min\limits_{t} x_{ijt}}{\max\limits_{t} x_{ijt} - \min\limits_{t} x_{ijt}} \tag{11}$$

The Taylor index method is still used to calculate the weight ω_{ij}, and the subsystem information balance is calculated according to Formula (12).

$$z_{it} = \sum_{j=1}^{m} \omega_{ij} a_{ijt} \tag{12}$$

Use a_{im}^{+} to represent positive ideal points and a_{im}^{-} to represent negative ideal points; the positive ideal points are ideal programming values for each order parameter, negative ideal points are the worst values for each order parameter, A^{+} is the set of positive ideal points, and A^{-} is the set of negative ideal points. According to the ideal collaborative state, the equilibrium value of the information intensity of each subsystem is used as the evaluation variable, and the ideal value is equal to the actual value of the information equilibrium degree of another subsystem (in fact, the information equilibrium degree of one subsystem is used as the ideal value of another subsystem). For example, take Formula (13):

$$\left(a_{1t}^{+}, a_{2t}^{+}\right)^{T} = (z_{2t}, z_{1t})^{T}, A^{+} = \left(a_{i1}^{+}, a_{i2}^{+}, \ldots, a_{im}^{+}\right), A^{-} = \left(a_{i1}^{-}, a_{i2}^{-}, \ldots, a_{im}^{-}\right) \tag{13}$$

The distance between the subsystem and the positive and negative ideal points is calculated by Formula (14):

$$X_{it}^{+} = \sqrt{\sum_{j=1}^{m} \left(a_{ij}^{+} - a_{ijt}\right)^{2}}; X_{it}^{-} = \sqrt{\sum_{j=1}^{m} \left(a_{ij}^{-} - a_{ijt}\right)^{2}} \tag{14}$$

X_{it}^{+} and X_{it}^{-} represent the distance between subsystem i and the positive and negative ideal points during period t, which is the j-th value of the A^{+} and A^{-} vectors. Taking the effect of each order parameter of the energy flow subsystem on the material flow subsystem as an example, we can calculate the correlation coefficient so that:

$$a = \min_{i}\min_{t}\{X_{Mt} - X_{Eit}\}, i = 1,2,3; \ b = \max_{i}\max_{t}\{X_{Mt} - X_{Eit}\}, i = 1,2,3 \tag{15}$$

Define the correlation coefficient:

$$\phi(M, E_i) = \frac{a + \rho b}{|x_M - x_{Ei}| + \rho b} \tag{16}$$

ρ is the resolution coefficient, usually taken as 0.5, defining the gray correlation degree:

$$\varphi(M, E_i) = \frac{1}{t}\sum_{1}^{t}\phi(M, E_i) \tag{17}$$

We calculate the absolute correlation between the subsystems. The larger the value is, the higher is the correlation level. Formula (18) is as follows:

$$\mu_{ij} = \frac{1 + |\varphi_i| + |\varphi_j|}{1 + |\varphi_i| + |\varphi_j| + |\varphi_i - \varphi_j|}, 0 \leq \mu_{ij} \leq 1 \tag{18}$$

Based on the evaluation of system synergy, we can combine the gray correlation coefficient to clarify such a relationship, the interaction relationship between the subsystems. We can also learn how the subsystems achieve pairwise collaboration through competition and collaboration, and then drive the overall system's collaboration.

3.3. Dynamically Based Collaborative Evolution Model

We can use Formula (19) to concisely describe the state of the entire process of collaborative co-evolution of the system [44]:

$$\frac{dx(t)}{dt} = rx(t) \tag{19}$$

Formula (19) sets the level of system coevolution as a function of time $x(t)$. When no resistance is considered, the growth rate of the system coevolution level is r. The level of system coevolution will continue to increase over time, indicating that $dx(t) > 0$. It can be seen that the overall trend of system coevolution shows an exponential curve growth trend, and further analysis of its evolution state curve is needed.

3.3.1. Explanation of Coevolutionary Variables

Based on the dynamic analysis of the flow structure of the three subsystems of material flow, energy flow, and information flow, as well as the analysis of collaborative evolution, we considered the factors that affect the collaborative evolution process and set them as variables when modeling. The dynamic and practical significance of these variables are shown in Table 2.

Table 2. The dynamic significance and practical significance of each variable.

Number	Variable	Dynamic Significance	Realistic Meaning
1	S_i	Subsystem (status)	Substance flow, energy flow, information flow subsystem states
2	x_i	Subsystem order parameter	Key production indicators for leading system collaboration trends
3	y	Large system status	Collaborative evolution status of production systems
4	a	Status parameters	Collaborative evolution speed of production systems
5	b	Action coefficient 1	The effect of the relationship between two subsystems on another subsystem
6	μ	Action coefficient 2	Feedback or reverse effect of b
7	α	Trend index	The driving effect of key indicators on production collaboration
8	η	Damping coefficient	Non-collaborative factors, such as production and operation obstacles
9	β	Correlation coefficient	Production correlation or resource allocation contradiction between subsystems
10	ε	Attenuation coefficient	The degree of attenuation of key indicators on production driving effects
11	γ_i	Impact coefficient 1	The comprehensive driving effect of key indicators on production collaboration
12	γ_i	Impact coefficient 2	The role of subsystem competitive behavior in collaborative evolution process
13	φ	Random variable	Conflicts between subsystems or non-directional interference from external environments
14	γ_i	Impact coefficient 3	The impact of cooperative behavior between subsystems on evolutionary results
15	γ_i	Impact coefficient 4	The impact of competitive behavior between subsystems on evolutionary results

Table 2. *Cont.*

Number	Variable	Dynamic Significance	Realistic Meaning
16	γ_i	Impact coefficient 5	The impact of collaboration between subsystems on large-scale system collaboration
17	γ_i	Self-feedback coefficient	Collaborative self-organization capability of production systems
18	m_i	Ideal evolution result	Ordered state of collaborative evolution of production systems
19	ε_i	Coevolution bias	Deviation between actual evolution stage results and ideal results

In the process of co-evolution, the system exhibits distinct characteristics that allow for the division of evolutionary stages. For instance, it can be segmented into the initial phase, competitive phase, cooperative phase, and coordinated phase based on the transformation of entities from micro- to macro-levels. Alternatively, it can be categorized into an independent stage, integrated stage, and intelligent collaborative stage based on the progression of informationization in production. Similar classification methods have also been proposed by other scholars. For example, Zhou (2013) [86] divided the evolution of manufacturing integration and industrialization based on informationization into the starting stage, single coverage stage, integrated enhancement stage, and innovative breakthrough stage. Drawing inspiration from the research achievements of Long (2008) [44], Miao et al. (2013) [87], and Li et al. (2018) [53], this study presents an evolutionary process description model and an evolutionary self-organizing control model. The co-evolution process is categorized into the collaborative generation phase, collaborative equilibrium phase, and collaborative maturity phase based on the different sources of dynamics in each stage. As depicted in Figure 1, during the initial system state, material and energy flows jointly generate information flow, and interactions among three subsystems lead to competition and cooperation, driving the system away from an ordered state and establishing the initial conditions for evolution. In this stage, exchange occurs between matter, energy, and information with the external environment, which introduces disruptive factors influencing the ordered state. With the enhancement of energy utilization efficiency, material and energy flows enter the equilibrium phase, where the critical point order parameter determines whether the system progresses toward an ordered structure. Subsequently, material, energy, and information flows harmonize to form a coordinated operational mechanism, marking the transition into the collaborative maturity phase. In the subsequent steps, we will establish a model based on a set of dynamic equations to describe these three stages of the co-evolution model.

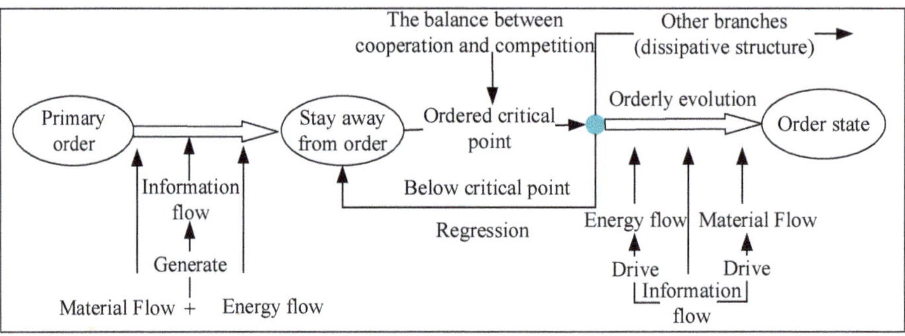

Figure 1. Synergetic evolution path of industrial production system.

3.3.2. Equations for the Generation Period of Collaborative Evolution

Let S_1, S_2, and S_3 represent the three subsystems of material flow, energy flow, and information flow, respectively. The relationship between the subsystems during the evolutionary generation period is shown in Formula (20).

$$\begin{cases} \frac{dS_1}{dt} = -a_1 t + b_1(S_1, S_2, S_3) \\ \frac{dS_2}{dt} = -a_2 t + b_2(S_1, S_2, S_3) + \varphi(t) \\ \frac{dS_3}{dt} = -a_3 t + b_3(S_1, S_2, S_3) \end{cases} \tag{20}$$

a_i is the state parameter of S_i that changes over time; b_1 represents the impact of the interaction between the material flow subsystem, energy flow subsystem, and information flow subsystem on the material flow subsystem itself, b_2, b_3, and so on; and $\varphi(t)$ represents the impact of random sudden disturbances, during which the system continuously undergoes energy conversion with the outside world, thus the random disturbances act more directly on the energy flow. Since the order parameter is the leader of the behavior of the subsystem, which determines the competition and cooperation state among the subsystems, and then determines the trend and result of the system's co evolution, Equation (20) can also be regarded as the equation of state of the order parameter of the three subsystems. The action state of order parameters of each subsystem is analyzed in detail below, and the system dynamics equation of state is constructed as Equation (21).

$$\begin{cases} \frac{dS_1}{dt} = -\eta_1 x_1 + \alpha_1 x_1 - \varepsilon_1 x_1^2 + \mu_1 x_2 x_3 \\ \frac{dS_2}{dt} = -\eta_2 x_2 + \alpha_2 x_2 + \beta_1 x_1 - \varepsilon_2 x_2^2 + \mu_2 x_1 x_3 + \varphi(t) \\ \frac{dS_3}{dt} = -\eta_3 x_3 + \alpha_3 x_3 + \beta_2 x_1 x_2 - \varepsilon_3 x_3^2 + \mu_3 x_1 x_2 \\ \frac{dy}{dt} = \gamma_1 x_1 + \gamma_2 x_2 + \gamma_3 x_3 + \gamma_4 x_1 x_2 x_3 + \varphi(t) \end{cases} \tag{21}$$

x_1 is the order parameter of the material flow subsystem, x_2, x_3, and so on; η_1, η_2 is the damping coefficient; α_1 is the synergistic trend index of the material flow subsystem, which is the contribution of x_1 to the ordered trend of the system (α_2, α_3 analogies); this index includes the dual effects of cooperation and competition. Note that the difference between competition and damping is that competition is the behavior between the subsystems (the fundamental reason is the behavior of order parameters between subsystems), which plays a positive role in the orderly trend of the system, while damping plays a negative role. β_1 is the driving effect of the material flow subsystem on the energy flow subsystem, β_2 is the generation effect of material flow and energy flow on information flow, ε_1 represents the degree of attenuation of x_1's effect on the ordered tendency of the system (ε_2, ε_3 analogies), μ_1 represents the impact of the interaction intensity between the energy flow order parameter and the information flow order parameter on the material flow (μ_2, μ_3 analogies), γ_1 represents the degree of influence of S_1 subsystem on the collaborative evolution process (γ_2, γ_3 analogies), γ_4 represents the impact of the competing behavior of each subsystem on the overall system during the collaborative evolution process (i.e., α_1, α_2, α_3 comprehensive index of action), and y represents the collaborative state of the large system. This differential equation system describes the contributions of the three subsystems S_1, S_2, and S_3 to the collaborative process and ordered results at a certain moment in the evolutionary generation period (reflected by the dominant role of the order parameter), and the first-order derivative equation of y with respect to t illustrates how the roles of each subsystem are carried out. In fact, the collaborative relationship between the subsystems is very complex and may not be as ideal at every moment, but this typical simplified state helps to identify the main influencing factors of the collaborative process.

3.3.3. Equations for the Stagnant Period of Collaborative Evolution

The contradictions, external disturbances, and other factors that arise between subsystems during the generation period may become factors contributing to the disorder of the system. During the period of coevolution and stalemate of the system, the key factor determining whether the system can form an ordered trend is the order parameter. Assuming that the order parameter at the end of the evolution stalemate can cause the system to produce an ordered trend, there will be two types of order parameters interacting with each other during this period: one is the competitive order parameter, and the other is the cooperative order parameter. These come from the competition and cooperation of the subsystems, promoting and coordinating each other. Using x_1 to represent the cooperative order parameter, x_2 to represent the competitive order parameter, and y to represent the collaborative state of the large-scale system, the dynamic differential equation system of Equation (22) is formulated as follows:

$$\begin{cases} \frac{dx_1}{dt} = (\alpha - \eta_1)x_1 + \beta x_1 x_2 + \varphi(t) \\ \frac{dx_2}{dt} = (\eta_2 - \alpha)x_2 - \beta x_1 x_2 - \varepsilon x_2^2 \\ \frac{dy}{dt} = \gamma_1 x_1 + \gamma_2 x_2 + \gamma_3 x_1 x_2 + \gamma_4 y \end{cases} \tag{22}$$

In the formula above, α represents the collaborative trend index. η_1, η_2 represents the damping coefficient and β represents the mutual influence coefficient between x_1 and x_2. Due to the opposing forces of x_1 and x_2, this influence is a resistance for both parties, but the higher the value is, the faster is the process of system evolution. $\varphi(t)$ represents random factors, and we assume that random factors at the same time may affect cooperation or competition. ε represents the attenuation coefficient of competitive power, γ_1 represents the coefficient of influence of cooperative forces on the collaborative evolution of the system, γ_2 represents the coefficient of influence of competitive forces on the collaborative evolution of the system, γ_3 represents the coefficient of influence of the combined effect of the two forces on the evolution of the subsystem, and γ_4 represents the self-feedback coefficient of the system, which represents a self-organizing ability independent of two forces and is a characteristic of the system itself. The first two equations of Formula (22) describe the guiding effect of two different order parameters on the evolution trend of the system, while the third equation describes the results of the system under these two forces.

3.3.4. Equations in the Mature Stage of Collaborative Evolution

In the mature stage of collaborative evolution, the collaborative relationships between subsystems are clearer and more stable, with mutual coordination, constraints, and cooperation. At this time, the factors that affect collaborative evolution mainly come from within the system, and according to the evolution process, external forces, such as the environment, are difficult to produce substantial disturbances on the system. Each subsystem generates a self-control mechanism to achieve the optimal overall goal of the entire system based on relevant goals. Here, the system state is still described using order parameters and various parameters, and the dynamic equation system is shown in Formula (23).

$$\begin{cases} \frac{dS_1}{dt} = A_{11}x_1 + B_1\mu_1 + \sum_{j=2}^{3} A_{1j}x_j \\ \frac{dS_2}{dt} = A_{22}x_2 + B_2\mu_2 + \sum_{j=1,3}^{3} A_{2j}x_j \\ \frac{dS_3}{dt} = A_{33}x_3 + B_3\mu_3 + \sum_{j=1}^{2} A_{3j}x_j \end{cases} \tag{23}$$

A_{ij} is the object parameter matrix, and vector μ_1, μ_2, μ_3 represents the parameter matrix of each subsystem, where the flow velocity of each flow can be used as a parameter. B_i represents the feedback matrix of the synergistic effect on the subsystem ($i = 1, 2, 3$ corresponding to the material flow, energy flow, and information flow subsystems, respectively). $\sum_{j=2}^{3} A_{1j}x_j$ represents the actual synergistic relationship between the energy flow and information flow subsystems and material flow subsystems, and other analogies. Let the variable m_i represent the ideal coevolution result, that is, the stable state of the subsystem in the later stage of evolution. At the end of collaborative maturity, the evolution results are analyzed as follows: we define synergy bias ε_i as a difference between the ideal collaboration relationship and the actual collaboration relationship. See Formula (24) for details:

$$\varepsilon_i = m_i - \sum_{j\neq i}^{3} A_{ij}x_j \tag{24}$$

The smaller the collaborative deviation, the more mature the collaborative evolution stage of the system, so the minimum value of the objective function of the system can be expressed as $\varepsilon_i \to min$. It can be seen that when $m_i \approx \sum_{j\neq i}^{3} A_{ij}x_j$, $\varepsilon_i \approx 0$, the objective function value can reach the minimum, and set the total objective J, then:

$$minJ = \left(m_1 - \sum_{j=2}^{3} A_{1j}x_j\right) + \left(m_2 - \sum_{j=1,3}^{3} A_{2j}x_j\right) + \left(m_3 - \sum_{j=1}^{2} A_{3j}x_j\right) = 0 \tag{25}$$

3.4. Collaborative Disturbance Recognition Based on SOM Algorithm

According to the analysis of the collaborative evolution process, the contradictions between the subsystems and the contradictions between the system and the outside are the reasons for the formation of ordered new trends. Identifying the internal and external disturbances of the flow structure allows for source management of the production system, without the need to analyze the impact of disturbances; however, due to the nonlinear and liquidity characteristics of complex industrial production systems, when non-surface disturbances occur, the essence of the problem can be sorted out from the massive data, relying on material flow. Tracing the direction of energy flow and information flow to the source of a problem is a complex and time-consuming task, especially in the multi-stream fusion process. We refer to the literature of Kohonen T (1982) [88], Chen (2020) [89], Li et al. (2013) [90]. The self-organizing map (SOM) neural network algorithm can be used to effectively identify the disturbance links and the links affected by the disturbance and reflected in the data. The distinction depends on the characteristics and internal links of the data themselves, without establishing an index system and preset categories. The process is as follows:

① When there is an unrecognizable disturbance in the production system, all links may be affected and reflected in the data; therefore, first assign random decimals to the weights of all the connection weights W_{ij} from input neurons to output neurons. Assuming the number of recognized objects is m, as the input layer dimension, in order to explore the nonlinear relationship between the interacting links from actual production data, the discriminant function adopts the Euclidean distance method. For each input object data x_i, the Euclidean distance between it and all output divine elements is calculated using Formula (26).

$$d_j(X) = \sum_{i=1}^{m}(x_i - w_{ij})^2, i \in \{1,2,\ldots,m\}, j \in \{1,2,\ldots,n\} \tag{26}$$

② Compare all the distance values and assume that the neuron with the minimum Euclidean distance is N_{j*}. As the winning neuron, adjust the weight according to Formula (27) and output it as "1", while other neurons output it as "0", as shown in Formula (28). Define the winning neighborhood as $NE_{j*}(t)$, and directly adjust the weights of the winning neurons within the geometric neighborhood according to Formula (29). This ensures that data with similar relationships can receive enhanced responses every time, and the position of the mapping points can reflect the clustering and distance relationships among the data. Among them, α indicates the speed of learning, and the value range is $0 < \alpha \leq 1$, whereas $S_{jN_{j*}}$ represents the distance between the winning node N_{j*} and the nodes in its geometric neighborhood:

$$\begin{cases} w_{j*}(t+1) = \hat{w}_{j*}(t) + \Delta w_{j*} = \hat{w}_{j*}(t) + \alpha(\hat{X} - w_{j*}) \\ \qquad\quad w_j(t+1) = \hat{w}_j(t), j \neq j^* \end{cases} \tag{27}$$

$$y_j(t+1) = \begin{cases} 1, j = j^* \\ 0, j \neq j^* \end{cases} \tag{28}$$

$$w_{jN_{j^*}} = exp\left(-S_{jN_{j^*}}^2 / 2\sigma^2\right) \tag{29}$$

③ When iterating the process above, the number of iteration steps can be sequentially set to [10 50 100]. Then, determine the optimal number of iterations based on the results. When α is attenuated over time, the magnitude of weight adjustment decreases, and the geometric neighborhood $NE_{j^*}(t)$ continues to contract. Finally, when α attenuates to 0, shrink the neighborhood to 1, train only the neuron N_{j^*} itself to achieve self-organizing feature mapping, and the process ends.

In the application scenario of an industrial production system, although the algorithm has slow rate of convergence, it has low complexity and is easy to establish. The autonomous learning process without a mentor also corresponds well to the self-organizing ability of the mature stage of collaborative evolution of production system flow structure. One can quickly find the root cause of the problem and solve it without even identifying what type of problem it is. It can also provide excellent management decision support. For example, mutation data have the potential to become an order parameter for system collaborative evolution, and controlling this potential can achieve the "leading" and "guiding" of collaborative management. This algorithm can be used not only for rapid detection and source finding of abnormal data, but also for rapid resolution of production random problems, fault classification management of energy utilization equipment, and evaluation of production data balance management.

4. Empirical Study on Collaborative Model of Production System Flow Structure

4.1. Overview of Empirical Case SG Enterprises

SG's main business purpose is to provide environmentally friendly, innovative, and safe automotive glass for major automobile manufacturers. It is a continuous process chemical company, with mixed input and output processes of raw materials and fuels in certain production stages, with a complex relationship between the material and energy flows that are numerous and stable.

4.2. Evaluation of Collaboration Degree of Production System

4.2.1. Calculation of Collaboration Degree of SG Enterprise Production System

Considering the data availability of various parameters in the production system of SG Enterprise, and combined with Appendix A Table A2, each subsystem selects three parameters from Table 1 as order parameters, as shown in Table 3.

Table 3. Order parameters in SG enterprise production system.

Subsystem Name	Code	Order Parameter Name	Unit
Material flow subsystem	M1	Equipment overall efficiency (OEE)	%
	M2	Yield of finished products	%
	M3	Production reliability	ND
Energy flow subsystem	E1	Comprehensive energy consumption per unit product	MWh/m^2
	E2	Air pollution per unit product	kg/m^3
	E3	Wastewater recycling capacity	m^3
Information flow subsystem	I1	Flexibility (inventory turnover days)	D
	I2	Cost proportion of information technology construction	%
	I3	Information management index	ND

We used the evaluation model in Section 3 to conduct a collaborative evaluation of the production system of SG enterprise, with q_i^* as its appropriate data reference value. The final value used was the specified value in the SG enterprise standard operating instructions. The effect size of the order parameter of each subsystem is calculated (see Appendix A Table A3). The weight process of the order parameter and the standardized effect size of the order parameter are calculated (see Appendix A Tables A4 and A5). The final obtained Taylor exponents of each order parameter, the weight of the order parameter in the system, the weight of the order parameter in the subsystem, and the weight of each subsystem in the large system are shown in Table 4:

Table 4. Weight of order parameters in SG production system.

Order Parameter	Code	Theil Index	Proportion to System Weight	Proportion to Subsystem Weight	Comprehensive Weight of Subsystems
Order parameter of	M1	0.2997	0.1276	0.4599	
material flow	M2	0.2016	0.0859	0.3094	0.2776
subsystem	M3	0.1504	0.064	0.2308	
Energy flow	E1	0.3116	0.1327	0.4015	
subsystem order	E2	0.1596	0.068	0.2056	0.3305
parameter	E3	0.305	0.1299	0.3929	
Order parameter of	I1	0.2823	0.1202	0.3067	
information flow	I2	0.4969	0.2116	0.54	0.3919
subsystem	I3	0.1411	0.0601	0.1533	

We calculate the order degree of the subsystem for each year according to Formula (7); we calculate the order degree of the production system according to Formula (8); and we calculate the collaboration degree of the production system according to Formula (9). The results are shown in Table 5.

Table 5. Order degree and synergy degree of SG company's production system.

Year \ System	Ordering Degree of Material Flow Subsystem	Order Degree of Energy Flow Subsystem	Ordering Degree of Information Flow Subsystem	Order Degree of Production System	Collaboration Degree of Production System
2013	0.481	0.411	0.210	0.041	—
2014	0.377	0.882	0.114	0.038	−0.064
2015	0.870	0.015	0.393	0.005	−0.250
2016	0.728	0.225	0.479	0.078	−0.117
2017	0.161	0.472	0.402	0.031	−0.150
2018	0.676	0.313	0.636	0.135	−0.212
2019	0.563	0.485	0.451	0.123	−0.080
2020	0.540	0.709	0.324	0.124	−0.029
2021	0.610	0.565	0.401	0.138	−0.057
2022	0.850	0.718	0.448	0.274	0.124

From Table 5, it can be observed that the material flow subsystem, energy flow subsystem, and information flow subsystem of the SG enterprise production system experienced significant fluctuations from 2013 to 2019; however, since 2020, the order of these subsystems has shown an upward trend, and the order of the entire production system has gradually increased. In terms of synergy, there is a period of continuous increase and continuous decrease for several years, and from 2021 to 2022, the trend of order and synergy became more consistent. Except for 2015, which was significantly affected by energy flow, the orderliness of the material flow subsystem showed a similar trend to that of the overall production system in other years, although the fluctuations were more significant. Based on relevant research results (referring to Li and Zhang (2016) [23], Chen et al. (2016) [24], and Luo et al. (2017) [22]), we have classified the degree of order and synergy, as shown in Table 6. Next, we will further analyze these levels.

Table 6. Ranking of orderliness and levels of collaboration.

Orderliness	Ordered Level	Synergy	Collaboration Level
0~0.1	Unordered state	<0	Uncooperative
0.1~0.5	Low-level order	0~0.3	Low-level collaboration
0.5~0.7	Intermediate order	0.3~0.6	Moderate synergy
0.7~1.0	Advanced order	0.6~0.8	Highly collaborative
		0.8~1.0	Extreme synergy

Since 2015, except for 2017, the material flow subsystem has been in a medium-to high-order state, while the energy flow subsystem has fluctuated between low- and medium-order states with significant fluctuations, and the information flow subsystem has always been in a low-order state. From 2013 to 2017, the production system remained in a state of disorder. Starting in 2018, the production system gradually transformed into an orderly state and improved to some extent. Until 2021, the production system had not achieved collaboration, while in 2022, a low degree of collaboration occurred. It is worth noting that there is a close correlation between the orderliness of the material flow subsystem, energy flow subsystem, and information flow subsystem, as well as the orderliness and synergy of the production system. These three factors collectively affect the degree of collaboration in enterprise systems. More specifically:

① The significant difference in order between the material flow and energy flow subsystems indicates that the system is in a state of non-synergy. This phenomenon further confirms the analysis conclusion in the previous section: the material flow subsystem plays a dominant role in the production system, and the dynamic mechanisms of energy flow and information flow revolve around the material flow.

② Over the past 5 years, the orderliness of the production system of SG Enterprise has improved. This can be attributed to the effective management measures adopted by the enterprise; however, this trend is not stable because when any of the three subsystems changes, it will worsen the ordered state of the entire system. For each subsystem, the state at time "t" is usually influenced by the other two subsystems. For example, in 2015, the orderliness of both the material flow subsystem and the information flow subsystem increased, resulting in an increase in the orderliness of the energy flow subsystem in 2016; however, due to their interrelationships, it is difficult to achieve an increase in the order of the three subsystems simultaneously, and conflicts often arise. This may be one of the reasons for the instability of the increase in order.

③ The ordered state of the energy flow subsystem has a significant impact on the production system and manifests as a significant "negative impact". For example, in 2015, the orderliness of the energy flow subsystem was very low. Even though the orderliness of both the material flow subsystem and the information flow subsystem increased, the orderliness and synergy of the production system remained at a relatively low level. Similarly, during the period from 2018 to 2020, although the orderliness of the material flow and information flow subsystems decreased, the orderliness of the energy flow subsystem increased, but this did not lead to an increase in the orderliness of the production system.

④ The orderly situation of the information flow subsystem also has a significant impact on the production system and presents a clear "positive impact". The order degree of the information flow subsystem is relatively stable and has a similar fluctuation trend as the material flow subsystem. This reflects the dependence and driving effect of information flow on material flow. Except for 2015, the orderliness trend of the information flow subsystem has always been consistent with that of the production system. It is worth noting that the trend of collaboration in the production system lags behind the trend of order in the information flow subsystem by one year, which means that during the period from 2013 to 2022, the synergy effect of the information flow subsystem on the production system has always maintained a positive effect; however, during the period from 2018 to 2020, the orderliness of the information flow subsystem decreased, resulting in a slight decrease in the orderliness and synergy of the production system in 2020. This may be because SG Enterprise had introduced a new management information system, which is still in the adaptation stage. However, in the long run, the benefits of this state outweigh the costs (by observing the overall increase in order and synergy trends from 2018 to 2022).

Overall, the level of collaboration in the production system of SG Enterprises shows a relatively chaotic trend from 2013 to 2014. Subsequently, in 2015, due to the implementation of the new management plan, the system had the basic conditions to enter the collaborative generation period. In the following 2015–2021 period, the conditions for co-evolution gradually deepened. By 2022, the system showed a basic trend towards entering the collaborative generation phase; however, it is currently unknown whether the ideal collaborative maturity period can be further reached, which is to form a stable collaborative state. Currently, there is still a considerable gap between the goal of collaborative management and the highly collaborative state of the production system.

4.2.2. Analysis of Gray Collaborative Relationship between Subsystems

Assume that each order parameter in the same subsystem is an independent order parameter, and there is no correlation. The original data in Appendix A Table A6 were still used to calculate the gray correlation between the two subsystems. The interval is [0, 1]. The larger the data are, the greater the correlation is. An interval of "0" means uncorrelation, and "1" means autocorrelation of this element. The correlation coefficient is shown in Appendix A Tables A7–A12. The variable X_{ij} in Tables 7–9 represents the impact of i on j, for example, in the correlation matrix of the material flow subsystem and the energy flow subsystem. We can see that the number in the first row and fourth column of Table 7 is 0.68. It represents the gray correlation effect size of the first order parameter M1 of the material flow subsystem to the first order parameter E1 of the energy flow subsystem, representing the degree of synergistic influence between the two, and other analogies.

Table 7. Gray effect value between material flow and energy flow subsystems.

M-E	M1	M2	M3	E1	E2	E3
M1	1	0	0	0.68	0.79	0.64
M2	0	1	0	0.67	0.78	0.67
M3	0	0	1	0.64	0.73	0.63
E1	0.81	0.66	0.75	1	0	0
E2	0.84	0.73	0.76	0	1	0
E3	0.77	0.66	0.74	0	0	1

Table 8. Gray effect value between material flow and information flow subsystems.

I-M	I1	I2	I3	M1	M2	M3
I1	1	0	0	0.64	0.53	0.80
I2	0	1	0	0.66	0.59	0.82
I3	0	0	1	0.63	0.65	0.79
M1	0.65	0.67	0.65	1	0	0
M2	0.59	0.65	0.72	0	1	0
M3	0.6	0.70	0.67	0	0	1

The gray correlation coefficients of different subsystems obtained by us are all greater than 0.5, with most of them above 0.6, indicating strong gray connections between the order parameters among the different subsystems. When observing data greater than 0.8, it can be seen that the order parameters of the energy flow subsystem have a more significant driving effect on the material flow subsystem, while the order parameters of the information flow subsystem have a more significant driving effect on the material flow subsystem and the energy flow subsystem. We can also draw the following conclusion: M_1 "equipment comprehensive efficiency" is a comprehensive management parameter that is significantly affected by energy flow. M_3 "production reliability" is an evaluative parameter that is significantly influenced by information flow. The impact of the I_3 "Information Management Index" on the E_3 "Wastewater Recycling Capacity" and the impact of E_3 on I_3 are both significant, indirectly indicating the information dependence of flow and the control power of information on flow. From the data less than 0.6, it can be seen that the order parameters M_2 "finished product yield" and E_2 "air pollution per unit product" have strong independence. In other words, they are less affected by the order parameters of the information flow subsystem and require special attention during management.

Table 9. Gray effect value between energy flow and information flow subsystems.

I-E	I1	I2	I3	E1	E2	E3
I1	1	0	0	0.70	0.56	0.79
I2	0	1	0	0.75	0.65	0.77
I3	0	0	1	0.63	0.55	0.84
E1	0.68	0.65	0.62	1	0	0
E2	0.65	0.65	0.66	0	1	0
E3	0.79	0.71	0.85	0	0	1

In Section 3, the generation relationship between the flow structure subsystems has been clarified. Specifically, the material flow subsystem serves as the main body, and the energy flow is dependent on the material flow. The material flow and energy flow generate the information flow. At the same time, there is also a reverse relationship, where the energy flow has a driving effect on the material flow, and the flow of energy usually occurs in time before the material flow. In addition, the information flow also plays a driving role in material flow and energy flow, and the development process of the information flow is closely related to the demand for production and manufacturing, thus leading to animal mass flow and energy flow. The variables in the Section 3.3.1 model can further explain that the coefficient of action μ represents the degree to which the mutual driving effect between two subsystems affects another subsystem, and the coefficient of action b represents the degree to which the driving effect between a certain subsystem and the other two subsystems affects the subsystem itself.

4.3. Evaluation of Collaboration Degree of SG Enterprise Production System

4.3.1. Collaborative Evolution Numerical Simulation

Based on the previous analysis and evaluation of the collaboration level of SG Enterprises, we can draw the following conclusion: during the generation period of the subsystem, due to the dominant influence of internal order parameters, the subsystem begins to generate a competitive effect. When there is no abnormal activity inside the system and the order parameters remain unchanged, the system is in a dynamically stable state. In a model analysis, it is necessary to find an equilibrium point (0, 0, 0) that satisfies Equation (22), where $\frac{dx_1}{dt} = 0$, $\frac{dx_2}{dt} = 0$, $\frac{dy}{dt} = 0$. This equilibrium point is of great significance for the evolution process of the system. By analyzing this equilibrium point, we can gain a deeper understanding of the dynamic behavior and evolution process of the system.

To simplify the expression, we can first define Formula (30) as follows: $U = \frac{dx_1}{dt}$, $V = \frac{dx_2}{dt}$, $W = \frac{dy}{dt}$. Its characteristic matrix is shown in Formula (30):

$$\begin{bmatrix} \frac{dW}{dy} & \frac{dW}{dx_1} & \frac{dW}{dx_2} \\ \frac{dU}{dy} & \frac{dU}{dx_1} & \frac{dU}{dx_2} \\ \frac{dV}{dy} & \frac{dV}{dx_1} & \frac{dV}{dx_2} \end{bmatrix} = \begin{bmatrix} \gamma_4 & \gamma_1 + \gamma_3 x_2 & \gamma_2 + \gamma_3 x_1 \\ 0 & \alpha - \eta_1 + \beta x_2 & \beta x_1 \\ 0 & -\beta x_2 & \eta_2 - \alpha - \beta x_1 - 2\varepsilon x_2 \end{bmatrix} \tag{30}$$

The characteristic equation at the equilibrium point $(0, 0, 0)$ is:

$$\begin{vmatrix} \lambda - \gamma_4 & \gamma_1 & \gamma_2 \\ 0 & \lambda - \alpha + \eta_1 & 0 \\ 0 & 0 & \lambda + \alpha - \eta_2 \end{vmatrix} = 0 \tag{31}$$

The eigenvalues are $\lambda_1 = \gamma_4$, $\lambda_2 = \alpha - \eta_1$, $\lambda_3 = \eta_2 - \alpha$, and the positive and negative values of the eigenvalues determine the stability of the equilibrium point of the equation. If the real parts of the eigenvalues are all negative, the system equilibrium point is stable. As long as one real part of the eigenvalues is positive, the system equilibrium point is unstable. η_1, η_2 represents the damping coefficient and $-\eta_1$ is always negative; therefore:

① $\gamma_4 < 0$ and $\eta_2 < \alpha < \eta_1$; the equilibrium point is stable;

② $\gamma_4 \geq 0$ or $\alpha \geq \eta_1$, or $\eta_2 \geq \alpha$; the equilibrium point is unstable.

It can be seen that stability is determined by γ_4, α, η_1, η_2. Assuming other unrelated parameters are fixed values, $\beta = 1.2$, $\varepsilon = 0.6$, $\gamma_1 = 0.6$, $\gamma_2 = 0.8$, $\gamma_3 = 1.6$ (obtained by multiple adjustments for clear images for easy observation), $\varphi(t)$ is set as a variable parameter, indicating that the sudden disturbance is random and non-directional. The MATLAB simulation results of this model under different numerical conditions are as follows:

① When the equilibrium point of $\gamma_4 < 0$ and $\eta_2 < \alpha < \eta_1$ (assuming $\gamma_4 = -1$, $\eta_1 = 0.6$, $\alpha = 0.48$, $\eta_2 = 0.36$) are stable, the solution curve for $\varphi(t) = 0$ and $\varphi(t) \neq 0$ is shown in Figure 2.

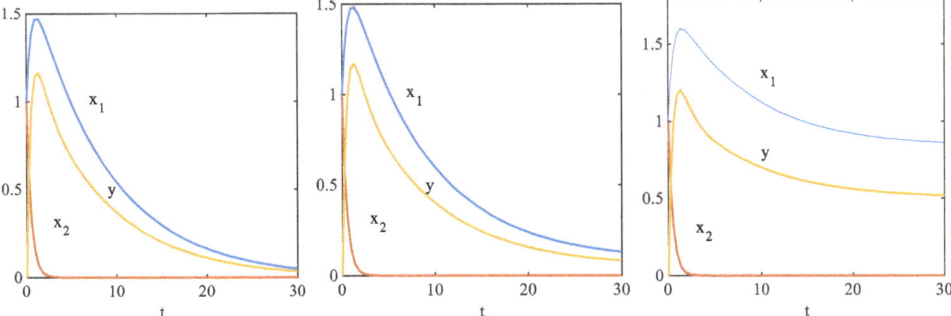

Figure 2. The solution curve of x_1, x_2, and y when $\gamma_4 < 0$ and $\varphi(t) = 0$, 0.01, and 0.1, respectively.

It can be seen that when the coevolution enters the phase of stalemate from the generation phase, the solution curve will converge, regardless of whether the mutation disturbance $\varphi(t)$ is zero. When $\varphi(t)$ is not zero, the solution curve will converge to a constant, indicating that the disturbance has caused the system to deviate and affect the internal state of the system; however, this does not affect the final collaborative outcome. If $\varphi(t)$ increases to 10 times, the results above are still the same, indicating that sudden perturbations will affect the asymptotic process of the solution curve, but will not affect the asymptotic trend, nor will they cause the system to move from "collaborative" to "non-collaborative".

② When the systems of $\gamma_4 \geq 0$ and $\eta_2 < \alpha < \eta_1$ (set as $\gamma_4 = 0.1$, $\eta_1 = 0.6$, $\alpha = 0.48$, $\eta_2 = 0.36$) are unstable, the solution curves for the cases of $\varphi(t) = 0$ and $\varphi(t) \neq 0$ are shown in Figures 3 and 4.

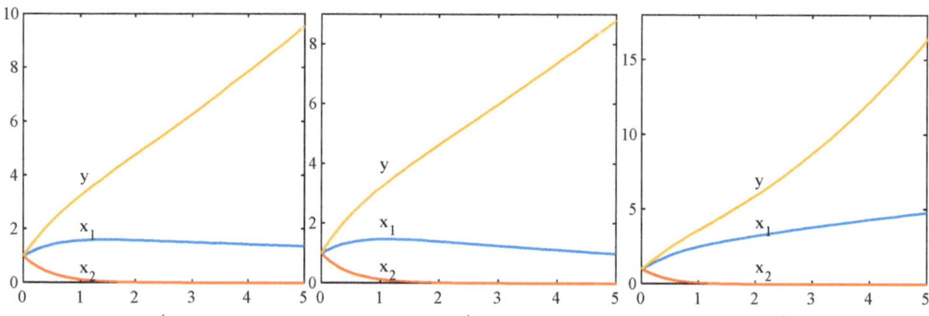

Figure 3. The solution curve of x_1, x_2, and y when $\gamma_4 = 0.1$, $\varphi(t) = 0, 0.1$, and 1, respectively.

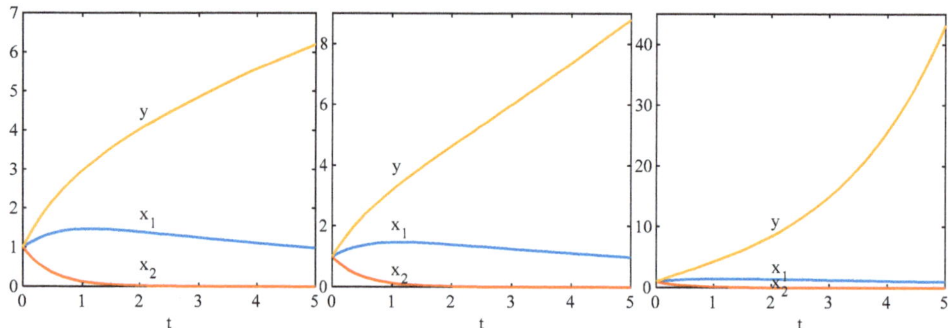

Figure 4. The solution curve of x_1, x_2, and y when $\varphi(t) = 0$, $\gamma_4 = 0, 0.1$, and 0.5, respectively.

From the previous two graphs, it can be observed that when there are no or only small abrupt perturbations, the solution curves of the two order parameters still tend to converge, but the solution curves of the large system show a divergent trend. This indicates that even if there are order parameters, if their guiding effect is lost, causing the system to develop towards dissipation, then only by generating new order parameters can the system be redirected towards the ordered direction, which may come from the internal or external environment of the system. In the third figure, the order parameter x_1 shows a divergent trend, indicating that the system is moving towards disorder more rapidly. Even in situations with significant sudden disturbances, it is difficult to change this trend. At this point, new order parameters can only be generated within the system.

When $\gamma_4 \geq 0$ and there is no random disturbance, the solution curve of the order parameter is convergent, while the solution curve of the large system is divergent. It can be seen that as long as γ_4 is not less than 0, the system will not form an ordered stable structure, and the larger γ_4, the weaker the force of the two order parameters, and the easier they are to be replaced. The situation at $\alpha \geq \eta_1$ and $\alpha > \eta_2$ is similar to this situation.

③ When the system is unstable at $\gamma_4 \geq 0$, $\eta_2 \geq \alpha$ and $\alpha < \eta_1$ (set as $\gamma_4 = 0.1$, $\eta_1 = 0.8$, $\alpha = 0.5$, $\eta_2 = 0.6$). At this time, the solution curves when $\varphi(t)$ and t take different values are shown in Figure 5.

It can be seen that because $\alpha < \eta_1$, the solution curve of the ordinal parameter x_1 in the left panel is divergent, and the system has a convergence trend under the combined effect of x_1 and x_2; however, at $t = 50$ in the middle panel, we observe that the solution curve of a large system tends to change from convergence to divergence. This means that the order parameter is guiding the orderly trend of the system structure, but due to the instability of the order parameter itself, this orderly trend cannot be sustained in the long run, leading to the system eventually developing into a dissipative structure. In the righthand figure, when $\varphi(t) = 0.1$, the solution curve of the large system clearly shows a divergent trend. This indicates that in the case of unstable order parameters, even with small sudden perturbations, the system will rapidly develop towards a dissipative structure.

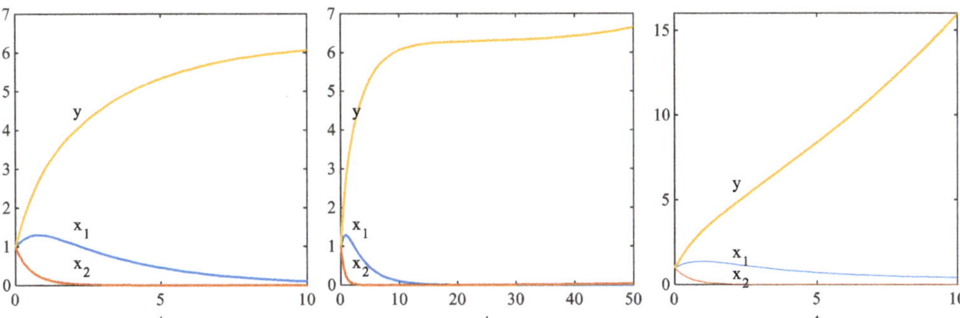

Figure 5. The solution curves of x_1, x_2, and y when γ_4 is 0, 0, and 0.1, respectively and $\varphi(t)$ and t take different values.

4.3.2. Analysis of Evolutionary Simulation Results

The numerical simulation solution curve above illustrates that during the collaborative generation period, the system undergoes structural changes due to internal contradictions or external environmental disturbances. This change may lead to the formation of a new stable structure in the system, and the trend of this new structure may not necessarily align with the ideal trend. In this scenario, the occurrence of sudden perturbations may affect whether new order parameters can be established and may also influence the direction of action of these new order parameters. In the synergistic phase holding period, when $\gamma_4 = 0$ or $\alpha = \eta_1 = \eta_2$, a boundary point will appear in the system, indicating that it is in a critical state of stability change. As time goes on, the trend of collaboration will gradually become clearer, and eventually, there may be a situation where $\eta_2 > \alpha > \eta_1$. Under new disturbances, the system may undergo a transition towards a new, higher-level ordered state.

The mechanism and path of collaborative evolution of complex industrial production systems have been clarified in Section 3. This process can be further explained here: during the generation period of collaborative evolution, the contradictions between various subsystems and the internal and external contradictions of the system become factors that disturb the system state and gradually develop into order parameters. These order parameters govern the behavior of each subsystem and guide the development of the entire large system. Under the influence of order parameters, there is strong competition and cooperation between various subsystems, which affect each other and play a positive role in the collaborative trend of the system. The self-organizing ability of the system itself enables it to gradually evolve from a disordered or low-level ordered state to a high-level ordered state during a dynamically stable evolution process, which can be referred to as the synergistic evolution phase. At this stage, the order parameter will determine the final outcome of system evolution. Even if a mutation occurs, it may affect the speed of evolution, but it will not change the direction and outcome of evolution; however, if the order parameter is not sufficient to guide the system towards an ordered state, sudden perturbations from both internal and external environments may become new order parameters. This will lead to a cyclic process in which new order parameters constantly emerge to guide the evolution of the system. The least ideal scenario is that the disordered parameters and sudden changes work together, leading the system to a dissipative disordered state. In this case, only the appearance of new order parameters within the system can guide the further evolution of the system. When the contradictions between the subsystems of the system are coordinated, the competitive effect is stable, and the system moves towards a high-level ordered state, it enters a mature period of collaborative evolution. During this period, if new contradictions or sudden disturbances occur, they may develop into the order parameters of the next coevolution. This process forms a cycle that continuously guides the system towards higher-level ordered states.

During the generation period of collaborative evolution, the system state is influenced by the relationships between subsystems and the internal and external dynamics of the system. During the phase of coevolution, the system state is determined by the dominant and competing roles of order parameters, as well as the self-organizing ability of the system. In the mature stage of collaborative evolution, the system state is primarily influenced by the self-organizing ability of the system. Overall, once collaborative evolution begins, the orderliness and various trends of the results are guided by order parameters and rely on self-organization to complete autonomous evolution. Whether a random mutation affects the system state and its participation in evolution depends on whether it becomes an order parameter.

4.4. Demonstration of SOM Algorithm in SG Enterprise Production System

SG Enterprise has a solid management foundation and extensive data accumulation, making it suitable for conducting simulation research on SOM neural network algorithms. Due to the "uniqueness" of the order parameter, it cannot occur simultaneously. To ensure the scientific validity of this empirical study, we have established the settings for the input layer data of the SOM algorithm based on comprehensive data utilization and in combination with Table 1. These settings are presented in Table 10.

Table 10. Data setting of SOM algorithm input layer.

Serial Number	Factors Contributing to the Variability of Production System Flow Structure	Use Actual Data	Unit
1	Material flow equilibrium	Material processing balance time	s
2	Accumulated dissipation of energy flow	Loss of comprehensive energy efficiency of equipment	kpcs/s
3	Information flow, generation, and operation time	Equipment, materials, and fuel input time	s
4	The synergistic efficiency of material flow and energy flow	Comprehensive efficiency of equipment technology	%
5	Collaborative efficiency of material flow, energy flow, and information flow	Comprehensive efficiency of equipment management	%
6	Production system random impact index	Unidentified equipment comprehensive loss	kpcs/s
7	Special product production flow structure impact index	Customized product quantity per process	kpcs
8	Comprehensive efficiency of production system and flow structure	Processing efficiency of each process	kpcs/d

The SOM algorithm in Section 3.4 was used to identify the perturbation factors for collaborative management, and ten major processes on the same production line were collected for calculation. When the number of training steps is 100, each object will form its own class, which represents overtraining; therefore, the number of training steps in the empirical case for SG enterprises is better than 100, and the number of training steps in the table is 50.

In Table 11, we employed fundamental data for clustering analysis, yielding the following outcomes: processes 1, 2, 6, 7, and 9 were grouped into the same category; processes 4, 8, and 10 were grouped together; and processes 3 and 5 were individually categorized. It was noted when utilizing potentially problematic data for clustering analysis, distinct outcomes emerged: process 3, 4, 5, 8, and 10 were grouped into the same category; processes 1 and 2 were classified together; processes 7 and 9 were classified together; and process 6 formed an independent category. This suggests that the problematic processes might be 5 and 3, which could potentially impact the majority of the processes.

Table 11. SOM algorithm output layer mapping point location.

Data Category Process Number	1	2	3	4	5	6	7	8	9	10
Basic data	24	24	2	1	5	24	24	1	24	1
Problem data	5	5	24	24	24	2	1	24	1	24

Similarly, the SOM neural network algorithm can be used for the flow structure equilibrium analysis of the production system to visualise and manage the production lines according to the production data mapping, which facilitates collaborative management at the plant level. Figure 6 shows the clustering of production data from production line 1 to production line 10 for SG Enterprise from January to August 2022, with a training step count of 50; the numbers within the hexagon indicate the number of mapping points that overlapped, with input objects where the mapping points overlap grouped into one category. The left panel shows the production of the 10 production lines measured by material flow, and the right panel shows the energy inputs of the 10 lines over 8 months. The 10 production lines are docked to the same assembly line and ideally should have the same production rhythm, i.e., they should achieve production balance, but the actual data (see Appendix A Tables A13 and A14) are unbalanced. The graph on the left shows that the material flow balance of the 10 production lines can be divided into 5 categories, where production lines 5, 6, 7, and 10 have similar production profiles and can be used as a unit for synergistic solutions to production problems. From the chart on the right, it is evident that during the period of January to August 2022, the energy input of 10 production lines exhibited temporal imbalances; however, this imbalance does not align with the seasonal production variations, presenting a challenge to the management of energy flow.

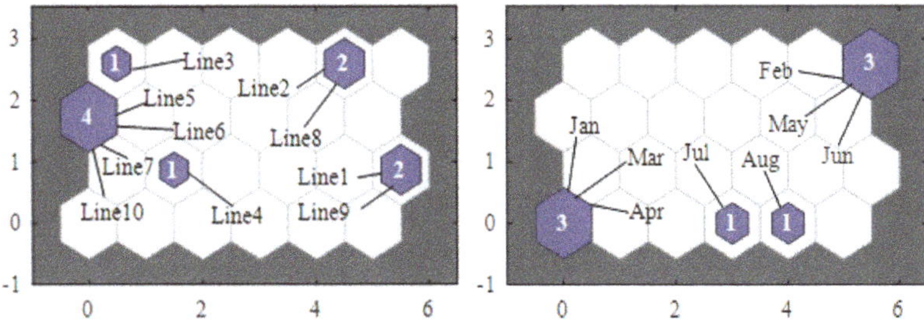

Figure 6. SOM algorithm results of 10 production lines data of SG Enterprise 2022.

4.5. SG Enterprise Production System Collaboration Management Response

SG Enterprise follows the basic framework of traditional management and uses the "competition–cooperation" relationship between material–energy–information flow subsystems as the mechanism of synergy.

The results of the synergy evaluation and coevolution simulation provide insight into the design of the co-management mechanism: firstly, the state and characteristics of the system should be clarified, and the coevolution should be guided by management tools. Each flow structure subsystem must have the initial conditions for generating the sequence parameters, and this condition needs to be generated under the guidance of the management mechanism. Burst problems are controlled, and self-organizing mapping network algorithms are used for the identification of burst data to avoid them becoming a cause of system dissipation. There are two directions of control: firstly, to make it a sequential parameter that leads the production system to a higher level of order; secondly, to solve the problem so that it does not have a large impact or participate in the evolution of the system. The second approach should be used more often.

Under specific foundational conditions, in order to attain sustainable production objectives, it is advisable to employ indicators, such as greening, informatization, and ecology, to formulate order parameters. By optimizing the collaborative levels of material flow, energy flow, and information flow within the production process at process network nodes and circulation paths, a self-organizing evolutionary process towards a contemporary green and ecological industrial system can be progressively realized. This study underscores that within the production systems of industrial enterprises, collaborative evolution should establish particular mechanisms and pathways at the management level to ensure the system's advancement towards a high-level and orderly structure.

4.5.1. Management Mechanism Based on System Collaboration

Drawing from the contents of Sections 3.2 and 4.2, the collaborative management mechanism should be propelled by information flow while also emphasizing energy management and control. Within the framework of information systems architecture, enterprises can devise production plans and allocate production tasks based on orders. Material flow, energy flow, and information flow are conveyed in their respective formats through resource planning systems, manufacturing execution systems, and energy management systems. Through the collaboration of material flow and energy flow, information flow is integrated across various systems. By considering process limitations and product requisites, a blueprint is developed for utilizing existing resources and energy, aligning the coordination of production resources and energy with the objectives of production planning and execution scheduling. Each process is executed in accordance with the planned instructions. The operational mechanism is visually depicted in Figure 7. In this mechanism, the concept of collaborative management is primarily manifested in the following dimensions: digital support for material and energy flows, harmonization between energy plans and production plans (shaping resource and energy plans via production plans), and synchronization in dynamic scheduling procedures.

Complex industrial production systems are order-oriented and customized production. Section 3.4 proposes the ultimate goal of collaborative management: resource optimization based on material flow, energy optimization based on energy flow, and information optimization based on production system synergy and on material flow–energy flow–information flow synergy for the effective allocation of production resources. Based on the management mechanism in Figure 7, the specific implementation plan is to introduce a new information co-optimization sub-system based on the current MES and ERP information system architecture, combined with the process control system of the main process equipment unit, relying on the model library and database, adding "resource planning system, energy management system", and other modules to establish optimized flows. The subsystem model of material flow, energy flow, and information flow co-optimization is driven by the information flow for the interaction and utilization of energy flow and material flow, completing the synergy between the flow structure optimization module and the production management system as a whole. In the collaborative management mechanism, the energy control method has many advantages over the traditional project-based energy control, as shown in Table 12. The results of improving comprehensive energy efficiency from the energy perspective alone are limited, and according to the driving and dissipation structure of the energy flow, with collaborative management as the control center, the supply and demand should be consistent, and the demand and consumption should be similar to the maximum.

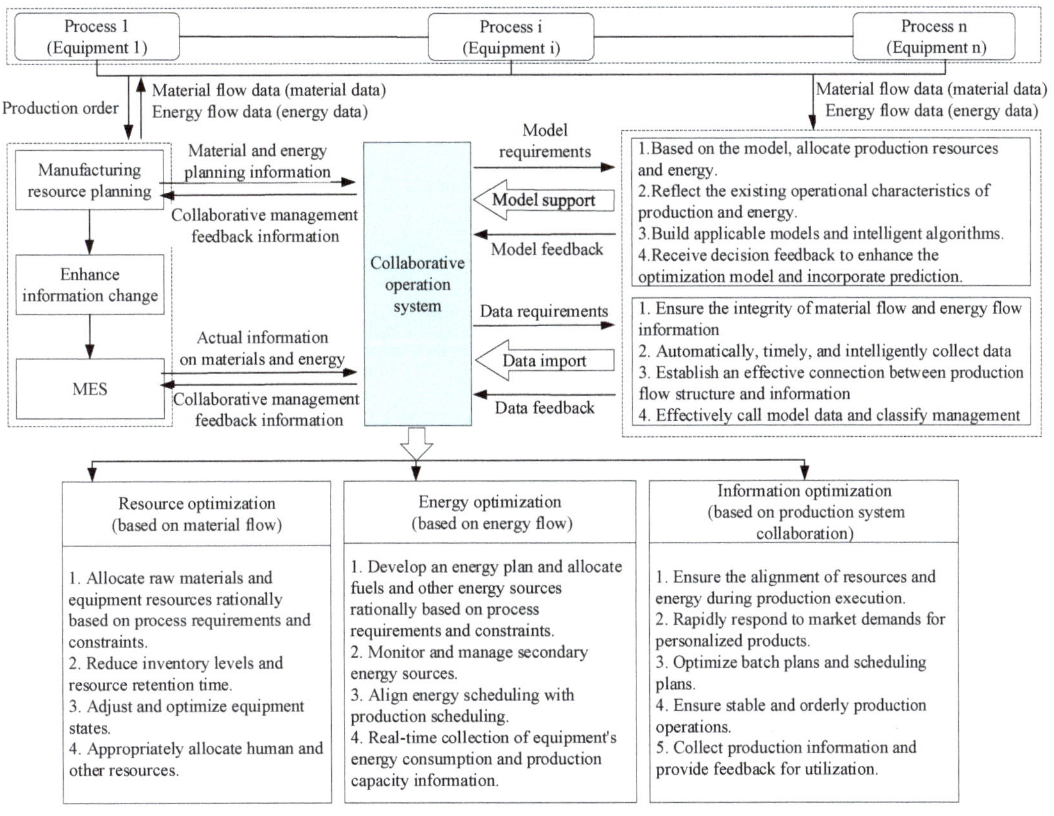

Figure 7. Synergy management mechanism in flow structure of industrial production system.

Table 12. Comparison between synergy management and project management in energy management and control.

Energy Control Projects	Project Management	Collaborative Management
Pollution control methods	Input and emission control	Full process control
Energy management methods	Planning, measurement, and post control	Self-organizing management
Pollutant generation	No changes before and after management	Reduce after management
Pollutant discharge level	Next cycle reduction	Reduction within the cycle
Energy consumption	Increased consumption of governance processes	Reduce
Energy usage costs	Increase	Reduce
Administrative expenses	Increase	Reduce
Proactive implementation	Passive	Active
Product quality	Unchanged	Increase

4.5.2. Management Path Based on Collaborative Evolution

The collaborative management mechanism emphasises the integration of flow structure modules and information systems, while the collaborative management path highlights the idea of synergy and cooperation between the elements, focusing on how to regulate the behavior of management to achieve collaborative production management. As shown in Figure 8, first of all, each department should complete business divestiture and reorganization, and refocus so as to achieve resource sharing, which is more effective in allocating resources than the traditional management method that emphasizes the division of labour, and on the basis of rational allocation and utilization of internal resources such as human, financial, material, technology, and information. External resources are also included in the scope of synergy, and the combination of internal and external meets the basic conditions of system self-organization.

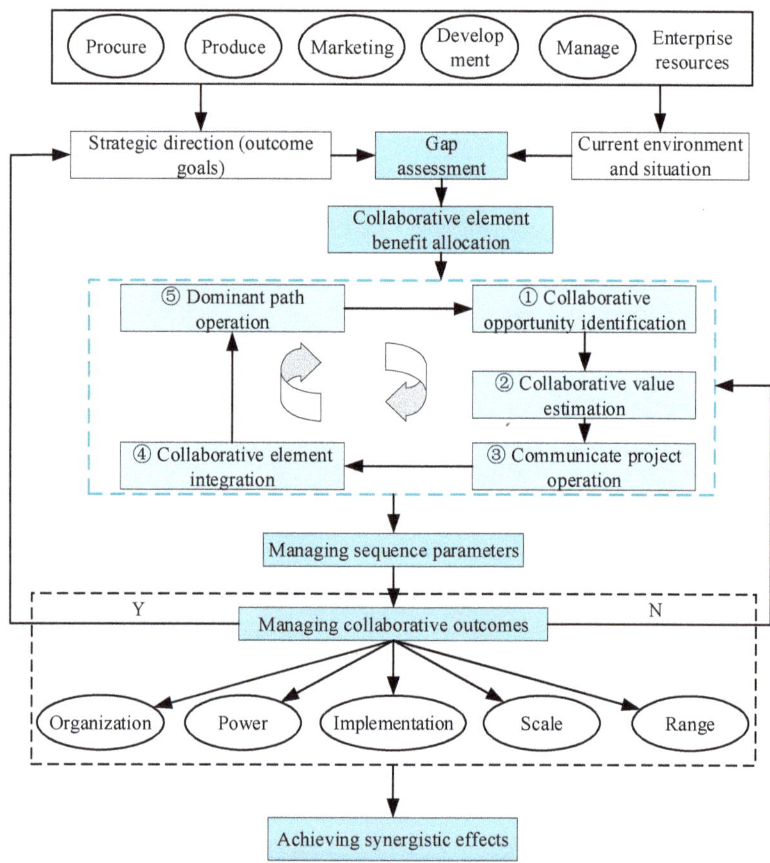

Figure 8. Synergy management path in flow structure of industrial production system.

Secondly, a comprehensive assessment of the environment in which the production system is located and the operating conditions, etc., should be performed. The items to be assessed include resources such as capital, raw materials, manpower, and technology, as well as production, marketing and service capabilities, and the industry's internal and external environments. Market demand reflects the value of customer needs. By constantly identifying gaps between the existing strategic orientations and market demand, companies are able to determine the direction of development, identify gaps between the current synergy trends and synergy goals based on the synergy evaluation, and collect relevant information, a step that determines whether management wants to make synergistic changes. If the assessment results in the need for synergistic management, the distribution of benefits between the elements should first be harmonized, as each stream structure subsystem can gain more benefits under improvement, but not necessarily automatically to the mutual benefit of the other subsystems.

Once again, assessments should start from the breakthrough point of collaborative management—identifying collaborative opportunities and opportunities that may generate order parameters in the system, that is, identifying constraints or bottlenecks. Maximizing the value of synergy is the goal of synergy management. Assessing the value of synergy in advance allows one to anticipate the effects of synergy management and uncover the value of using synergy elements. As shown in Figure 9, value opportunities mainly occur in the conversion of raw materials to products, collaborative production, and the recycling of waste. Studying the value structure at the process level, controlling all value-added and non-value-added activities, and identifying all opportunities for value appreciation is critical.

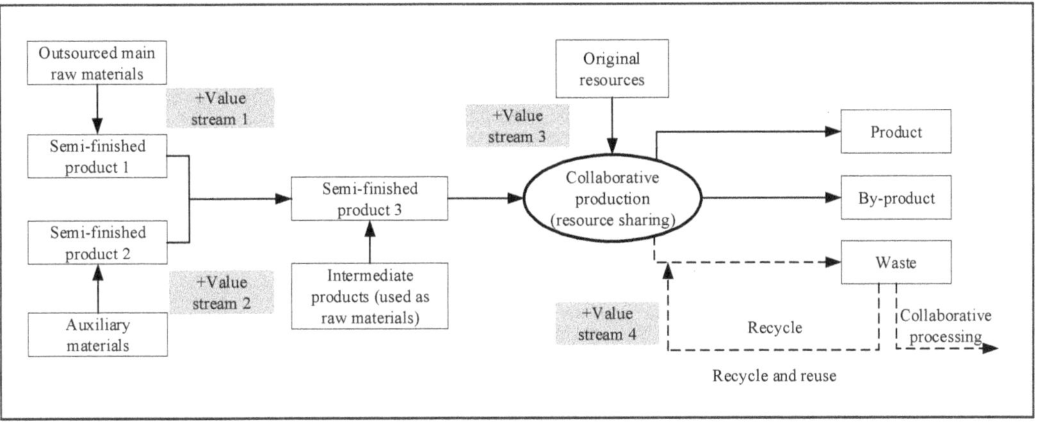

Figure 9. Value identification of synergy management in industrial production system.

SG Enterprise production system SOM calculations compare the value of synergy with the cost of synergy to obtain the actual value of synergy. While material and energy have value in themselves, a traditional value stream analysis is used to describe the activities of the material and information flows in an enterprise, visualizing both, focusing on improving productivity and shortening production cycles, emphasizing time-based improvements, and failing to consider the efficiency of energy and the impact of the production process on the environment. Collaborative value identification is therefore also an improvement on traditional value stream tools, and there is empirical evidence of the method.

Next, we need to adopt communication projects, such as two-way communication and mutual trust. Effective communication projects can enable employees and management entities in the enterprise to clearly understand, recognize, and accept collaborative values, and transform them into conscious behaviors in daily work. Integration refers to the balancing, selection, and coordination of collaborative elements. At the production level, element coordination refers to the allocation of funds, products, technology, human resources, etc. At the enterprise level, element coordination refers to the coordination of subsystems, such as research and development, procurement, production, marketing, and services. External element coordination refers to mergers and acquisitions, dynamic alliances, and industry–university research cooperation between enterprises.

The order parametric will dominate the subsystem that caused it to be generated, while reinforcing the order parametric itself. Whether its guidance of the system is compounded by the trend of synergistic management requirements must also be compared to the goal through feedback to obtain the answer. As can be seen from Figure 8, there are two types of feedback: circular feedback and judgmental feedback. The five links from opportunity identification to the domination path constitute circular feedback, with each step being self-checked; the opportunity re-identified if it cannot be passed. After the synergy result is obtained, the synergy goal is compared, and if it is inconsistent, the aforementioned link needs to be reconsidered. This is the typical path of enterprise level collaborative management.

In addition, it is important to be wary of cost increases where strategic advantages are sought from synergies. For example, SG companies have had problems with rising overheads, and pursuing procurement synergies excessively can increase costs; strategic differences between different businesses (e.g., tempered glass business pursues cost leadership, whereas smart glass systems business pursues differentiation) may make production and R&D synergies less effective. Insufficient understanding of the market activities of specific business enterprises (e.g., some companies may have procurement staff from different departments or with different responsibilities, and the same type of sales channels may not be shared because of differences in business practices) can also significantly reduce the synergy effect, and when the interests of the various business units may conflict with those arising from the company's synergy, it will prevent the synergy from being effective. SG enterprises should address the issues above in the process of identifying opportunities for organizational collaboration, analyzing costs, and communicating and learning.

5. Conclusions

This study applies a collaborative theory to complex industrial production systems. From the perspective of process dynamics, we have analyzed the dynamic structures of material flow, energy flow, and information flow in detail through graphical and mathematical forms. At the same time, we delved into the collaborative operation mechanism between these three subsystems and the collaborative evolution process of the entire production system. To this end, we have established a collaborative evaluation model based on order parameters and gray relations and extended the three-stage dynamic equation system collaborative evolution model. In our research, we also proposed a method for identifying sudden disturbances in production systems using self-organizing mapping network algorithms. We selected SG Enterprise, an international enterprise with high management levels, as the empirical object, and conducted model calculations and result analysis to obtain insight into the mechanism and path of collaborative management. Based on this work, the following conclusions have been drawn from this study:

① From the perspective of flow structure, complex industrial production systems can be analyzed as a large system formed by the synergy of three subsystems: material flow, energy flow, and information flow. The material flow subsystem is the main body, and the energy flow is dependent on and drives the material flow. The information flow is generated by and drives the material flow and energy flow. The collaborative mechanism is the overall collaboration achieved by the three subsystems through local collaboration in pairs.

② "Order" is an important representation of the collaborative process of a system in the time dimension, and the competition and cooperation between order parameters enable the system to complete the evolution process from disorder to order. The evaluation of the level of system collaboration is based on the ordered measurement at a certain moment, and the introduction of the Thiel index can optimize the weight calculation method. The self-organizing map (SOM) algorithm can be used to effectively identify production abrupt disturbance data, which can be used as a tool for collaborative management.

③ The collaborative evolution of complex industrial production systems can be categorized into three stages: the collaborative generation period, collaborative stalemate period, and collaborative maturity period, each exhibiting distinct dynamic states. During the collaborative generation period, various parameters engage in robust competition, gradually giving rise to dominant order parameters within the system. In the collaborative stalemate period, the interplay between the subsystems evolves to reveal a blend of competition and cooperation, ultimately reaching a harmonized state. The collaborative maturity period showcases the system's notable self-organization capacity, underscored by evident feedback and self-control phenomena. The essence of constructing mathematical models for these three stages lies in accurately grasping their evolutionary traits, judiciously defining variables, and utilizing equilibrium points to expound upon computational outcomes. From the MATLAB simulation curve of the model, the ensuing conclusion can be drawn: during the coevolution generation period, the system's state is influenced by the interrelationships among subsystems and the internal and external dynamics of the system. In the stalemate phase of collaborative evolution, the system's state is collectively impacted by the predominant and competitive effects of the order parameters, alongside the system's inherent self-organizing prowess. In the mature stage of collaborative evolution, the system's state is solely intertwined with its self-organization ability. The potential influence of random mutations on the system's state and their role in evolution hinges on their potential to become order parameters.

④ The collaboration degree of the production system of SG enterprise from 2014 to 2021 was less than 0, and it did not enter collaborative evolution. In 2022, a basic trend of entering the collaborative generation period was noted. SG Enterprise should optimize production collaboration management; otherwise, it cannot rely on a self-organization ability to move towards a high-level, orderly, and stable state. Collaborative management should follow clear management mechanisms and paths.

⑤ The key management strategy based on the synergy of material flow, information flow, and energy flow is utilized to strengthen energy control driven by information flow, integrate the production and operation mode of flow structure into the overall architecture, and form a stable management mechanism, achieving resource optimization based on material flow, energy optimization based on energy flow, and information optimization based on system collaboration. At the enterprise level, it should follow the path of resource restructuring, evaluating collaborative gaps, identifying collaborative opportunities, and confirming collaborative value. Compared to traditional production management, collaborative management mode fully utilizes the self-organizing characteristics of the system, grasps the evolution time point, transforms from process control to driving force control and mutation control, simplifies repetitive work and redundant data, and reduces the consumption of work links, resources, energy, and manpower.

This article also has the following shortcomings. First of all, in terms of the methods, the self-organizing map network algorithm has application value for enterprises with a good management foundation and complete data collection, but it is not the most suitable method for small- and medium-sized enterprises for the selection of enterprise parameters, identification of problems, allocation of resources, and other parts involving various systems of enterprises. In addition, the next step of research should focus on modeling related to the collaborative characteristics of industrial production process systems. On the basis of achieving the synergy of material flow, energy flow, and information flow, the collaborative management of material flow network, energy flow network, and information flow network should be studied.

Author Contributions: Conceptualization, J.S.; data curation, J.S. and Z.H.; funding acquisition, J.S.; methodology, J.S.; supervision, L.J.; visualization, Z.L.; writing—original draft, J.S. and Z.H.; writing—review and editing, X.L. All authors have read and agreed to the published version of the manuscript.

Funding: The funding for this research comes from the Shandong Provincial Natural Science Foundation project (ZR2023MG046).

Data Availability Statement: Not applicable.

Acknowledgments: We greatly appreciate the associate editor and the anonymous reviewers for their insightful comments and constructive suggestions, which have greatly helped us to improve the manuscript and guide us forward in future research.

Conflicts of Interest: The authors declare no conflict of interest.

Appendix A

Table A1. Parameter Inventory (Source Identification of Ordered Parameters).

Serial Number	Parameter	Serial Number	Parameter
1	Return scrap rate	75	Failure analysis ineffective cost
2	Final scrap rate	76	Consumption rate of compressed gas
3	Comprehensive heat production rate	77	Leakage rate of compressed gas
4	Yield of finished products	78	Emission rate of compressed gas
5	Production reliability	79	Production layout safety index
6	Labor productivity	80	Production defect loss ratio
7	Error component rate	81	Energy cost loss ratio
8	Waste recycling rate	82	Comprehensive energy consumption per unit product
9	Qualified rate of finished products	83	Air pollution per unit product
10	Thermal pollution emission rate	84	Cost proportion of information technology construction
11	Production line downtime	85	Material scrap and defect data
12	Loss reduction index	86	Quantification of complaints and material waste
13	Device setup time	87	Deviation difference in material detection
14	Material balance index	88	Energy consumption per unit production cost
15	Special machine losses	89	Control index of residual energy
16	Defect decomposition index	90	Regulatory notice or authorization index
17	Complaint decomposition index	91	SOP review interval
18	Raw material consumption rate	92	Reporting index for major accidents
19	Rejection rate of defective products	93	Labor time used for rework
20	Air monitoring index	94	Consumption and disposal of packaging materials
21	Equipment failure loss	95	Manage controllable OEE ratios
22	Short stop loss	96	Product fragment recycling rate
23	Safety production index	97	Flexibility (inventory turnover days)
24	Equipment production efficiency	98	Production dynamic risk assessment index
25	Input–output efficiency	99	Environmental factory boundary noise statistical index
26	Product added scrap rate	100	Production process exhaust gas statistical index
27	Recovery and utilization rate of surplus energy	101	Energy safety isolation practice index
28	Waste heat recovery and utilization rate	102	Management risk progress assessment index
29	Progressiveness production equipment	103	Verify available material loss index
30	Wastewater recycling capacity	104	Water demand per unit production cost
31	Information management index	105	Risk assessment program coverage index
32	Reduction in production water consumption	106	Impact index of key control points
33	Quality index decline rate	107	Generation of residual heat and energy per unit product
34	Operational index (OPI)	108	Degree of production continuity (index)
35	OEE overall efficiency	109	Equipment overall efficiency (OEE)
36	Improved production line productivity	110	Scrap quantity/high-quality product quantity
37	Machine failure reduction rate	111	Reduced product value caused by obsolescence
38	Main equipment maintenance time	112	Number of customer complaints that generate costs
39	Key product key production	113	Number of customer complaints expressing dissatisfaction
40	Material specification accuracy	114	Basic fuel consumption of process auxiliary materials
41	Reduced use of natural gas	115	Volume ratio of water consumption to water intake
42	New material safety data	116	Utilization rate of SOP and other guidance books
43	Hazardous waste exposure index	117	Consumption statistics of engine oil and lubricating oil

Table A1. *Cont.*

Serial Number	Parameter	Serial Number	Parameter
44	Logistics stability index	118	Proportion of information technology cost to output value
45	Production visualization index	119	OEE decomposition—loss failure rate
46	Isolation index of energy	120	OEE decomposition—production conversion rate
47	Control index of energy	121	Probability of converting production indicators into actions
48	Energy shutdown index	122	Probability of discovering defects during regular inspections
49	Energy release index	123	Downtime caused by operator absence
50	Production plan completion rate	124	New radiation source or laser protection index
51	Unrecognized energy loss	125	Exhaust emissions per unit production cost
52	Water consumption per unit product	126	Percentage of jobs directly exposed to hazardous waste
53	Input energy per unit product	127	Statistics of protective devices, machinery and equipment
54	Comprehensive energy consumption per unit product	128	Detection rate of energy utilization tools and equipment
55	Comprehensive electricity consumption per unit product	129	Emission rate of solid liquid waste pollutants (sulfur)
56	Organizational structure construction level	130	Emission rate of gas waste pollutants (nitrogen)
57	Information system construction level	131	Inspection frequency of water treatment equipment and network
58	Planned downtime gap	132	Downtime caused by insufficient material supply
59	Hazardous waste hazard assessment index	133	Effective index of energy isolation device for equipment
60	Production accident impact index	134	Machine protection and LOTO inspection index
61	OEE breakdown maintenance rate	135	Proportion of full-time personnel engaged in information technology work
62	Line production gap index	136	Popularity of information technology related production equipment
63	Supervision improvement index	137	Production time statistics for no less than one rotation
64	Total loss due to quality issues	138	Index for incorporating new products and procedures into the process
65	Material value range stability	139	Speed loss caused by machine operation not reaching speed
66	Material characteristic calibration coefficient	140	Amount of waste generated per unit production cost
67	Measurement accuracy deviation index	141	Risk assessment index for three or more energy sources
68	Probability of chemical leakage	142	Proportion of non-recyclable waste to total waste
69	Construction of independent websites	143	Number of times discharge water quality is measured by external agencies/year
70	Internal network application situation	144	Proportion of information security investment in informatization investment
71	Product one-time qualification rate	145	Enterprise resource planning (ERP) application index
72	Archive data exposure level	146	Critical area cleaning and lubrication inspection (CIL) rate
73	Technical level evaluation index	147	Fire/explosion index caused by combustible or flammable materials
74	Cost of undiscovered defective products	148	Complexity index of the isolation system when there are more than three energy sources

Table A2. Qualitative Identification of Ordered Parameters.

Serial Number	Parameter	Unit	Serial Number	Parameter	Unit
1	Yield of finished products	%	11	Production defect loss ratio	%
2	Production reliability	ND	12	Energy cost loss ratio	%
3	Comprehensive heat production rate	%	13	Comprehensive energy consumption per unit product	MWh/m^2
4	Equipment production efficiency	%	14	Air pollution per unit product	kg/m^3
5	Wastewater recycling capacity	m^3	15	Cost proportion of information technology construction	%
6	Information management index	ND	16	Manage controllable OEE ratios	%
7	Production plan completion rate	%	17	Product fragment recycling rate	%
8	Unrecognized energy loss	kgce/t	18	Flexibility (inventory turnover days)	d
9	Water consumption per unit product	L/m^3	19	Equipment overall efficiency (OEE)	%
10	Product one-time qualification rate	%			

Table A3. Effect Values of Flow Structural Subsystems in SG Enterprises.

Number \ Year	2013	2014	2015	2016	2017	2018	2019	2020	2021	2022
M_1	0.5285	0.1338	1.0000	0.5165	0.0000	0.9741	0.3026	0.1649	0.2572	0.8300
M_2	0.7693	0.8202	0.8642	1.0000	0.0000	0.0482	0.7048	0.8057	0.9552	0.7667
M_3	0.0000	0.2683	0.6188	0.7828	0.6979	0.9253	0.8939	0.9328	0.8496	1.0000
E_1	0.7273	1.0000	0.0000	0.1688	0.5407	0.0888	0.1802	0.5861	0.3129	0.3596
E_2	0.5770	0.4556	0.0000	0.5164	0.3633	0.4659	0.3871	0.3936	0.4518	1.0000
E_3	0.0000	0.9848	0.0389	0.1309	0.4597	0.4620	0.8468	1.0000	0.8810	0.9369
I_1	0.6831	0.0000	0.2796	0.1093	0.4423	0.3555	0.1415	0.5067	0.7315	1.0000
I_2	0.0000	0.0000	0.3855	0.5964	0.2410	0.6928	0.6024	0.2108	0.1205	0.0000
I_3	0.0000	0.7422	0.6483	0.8021	0.8884	1.0000	0.5366	0.3592	0.7285	0.9238

Table A4. Standardized Effect Values of Flow Structural Subsystems in SG Enterprises.

Number \ Year	2013	2014	2015	2016	2017	2018	2019	2020	2021	2022
M_1	0.1123	0.0284	0.2124	0.1097	0.0000	0.2069	0.0643	0.0350	0.0546	0.1763
M_2	0.1142	0.1218	0.1283	0.1485	0.0000	0.0072	0.1047	0.1196	0.1418	0.1138
M_3	0.0000	0.0385	0.0888	0.1123	0.1001	0.1328	0.1283	0.1338	0.1219	0.1435
E_1	0.1835	0.2522	0.0000	0.0426	0.1364	0.0224	0.0455	0.1478	0.0789	0.0907
E_2	0.1251	0.0988	0.0000	0.1120	0.0788	0.1010	0.0840	0.0854	0.0980	0.2169
E_3	0.0000	0.1715	0.0068	0.0228	0.0801	0.0805	0.1475	0.1742	0.1535	0.1632
I_1	0.1607	0.0000	0.0658	0.0257	0.1041	0.0837	0.0333	0.1192	0.1721	0.2353
I_2	0.0000	0.0000	0.1353	0.2093	0.0846	0.2431	0.2114	0.0740	0.0423	0.0000
I_3	0.0000	0.1120	0.0978	0.1210	0.1340	0.1508	0.0809	0.0542	0.1099	0.1394

Table A5. Theil Index Values of Flow Structural Subsystems in SG Enterprises.

Number \ Year	2013	2014	2015	2016	2017	2018	2019	2020	2021	2022
M_1	0.2455	0.1012	0.3291	0.2420	0.0000	0.3260	0.1764	0.1174	0.1588	0.3060
M_2	0.2478	0.2564	0.2635	0.2832	0.0000	0.0354	0.2362	0.2540	0.2770	0.2474
M_3	0.0000	0.1254	0.2150	0.2456	0.2304	0.2681	0.2634	0.2692	0.2566	0.2786
E_1	0.3111	0.3474	0.0000	0.1344	0.2717	0.0851	0.1405	0.2826	0.2004	0.2177

Table A5. *Cont.*

Number / Year	2013	2014	2015	2016	2017	2018	2019	2020	2021	2022
E_2	0.2601	0.2287	0.0000	0.2452	0.2002	0.2316	0.2080	0.2101	0.2276	0.3315
E_3	0.0000	0.3024	0.0339	0.0862	0.2022	0.2028	0.2823	0.3044	0.2876	0.2958
I_1	0.2938	0.0000	0.1790	0.0942	0.2355	0.2076	0.1133	0.2536	0.3029	0.3405
I_2	0.0000	0.0000	0.2706	0.3273	0.2089	0.3438	0.3285	0.1927	0.1338	0.0000
I_3	0.0000	0.2451	0.2274	0.2555	0.2693	0.2853	0.2035	0.1580	0.2427	0.2746

Table A6. Production System Ordered Parameter Raw Data.

Number / Year	2013	2014	2015	2016	2017	2018	2019	2020	2021	2022
M_1	62.0%	49.7%	76.7%	61.6%	45.5%	75.8%	54.9%	50.6%	53.5%	71.%
M_2	96.10%	96.53%	96.90%	98.04%	89.63%	90.04%	95.56%	96.41%	97.67%	96.08%
M_3	6.73	10.49	15.41	17.70	16.51	19.70	19.26	19.81	18.64	20.75
E_1	12.00	10.87	15.01	14.31	12.77	14.64	14.27	12.58	13.72	13.52
E_2	3.79%	4.51%	7.22%	4.15%	5.06%	4.45%	4.92%	4.88%	4.53%	1.27%
E_3	200.00	237.96	201.50	295.15	282.47	282.39	232.64	238.54	266.24	264.08
I_1	20.55	45.77	35.45	41.73	29.44	32.64	40.54	27.06	18.77	8.85
I_2	0.20%	0.50%	2.50%	3.20%	5.10%	3.60%	3.90%	5.20%	5.50%	5.90%
I_3	35.64%	62.40%	59.02%	64.56%	67.68%	71.70%	54.99%	48.59%	61.91%	68.95%

Table A7. Grey Correlation Coefficients of Flow Structural Subsystems in SG Enterprises (Energy Flow to Material Flow).

Year / Relationship	$E_1\text{-}M_1$	$E_1\text{-}M_2$	$E_1\text{-}M_3$	$E_2\text{-}M_1$	$E_2\text{-}M_2$	$E_2\text{-}M_3$	$E_3\text{-}M_1$	$E_3\text{-}M_2$	$E_3\text{-}M_3$
2013	0.652	0.691	0.335	0.734	0.758	0.532	0.542	0.568	0.402
2014	0.956	0.553	0.583	0.736	1.000	0.574	0.698	0.851	0.458
2015	0.620	0.701	0.566	0.597	0.455	0.422	0.358	0.565	0.694
2016	0.843	0.858	0.994	0.843	0.836	0.778	0.644	0.652	0.737
2017	0.554	0.947	0.847	0.572	0.728	0.800	0.414	0.595	0.693
2018	0.599	0.622	0.716	0.653	0.919	0.717	0.687	0.601	0.841
2019	0.616	0.795	0.712	0.733	0.846	0.886	1.000	0.820	0.534
2020	0.713	0.780	0.488	0.670	0.872	0.824	0.726	0.861	0.523
2021	0.645	1.000	0.705	0.808	0.986	0.815	0.614	0.918	0.840
2022	0.586	0.991	0.501	0.355	0.407	0.337	0.690	0.892	0.577

Table A8. Grey Correlation Coefficients of Flow Structural Subsystems in SG Enterprises (Material Flow to Energy Flow).

Year / Relationship	$M_1\text{-}E_1$	$M_1\text{-}E_2$	$M_1\text{-}E_3$	$M_2\text{-}E_1$	$M_2\text{-}E_2$	$M_2\text{-}E_3$	$M_3\text{-}E_1$	$M_3\text{-}E_2$	$M_3\text{-}E_3$
2013	0.794	0.730	0.680	0.768	0.692	0.636	0.501	0.529	0.557
2014	1.000	0.732	0.803	0.646	0.987	0.857	0.737	0.570	0.610
2015	0.769	0.587	0.504	0.776	0.379	0.633	0.724	0.420	0.796
2016	0.931	0.845	0.763	0.899	0.782	0.707	1.000	0.772	0.824
2017	0.713	0.562	0.562	0.964	0.659	0.659	0.919	0.794	0.795
2018	0.752	0.645	0.795	0.708	0.884	0.665	0.836	0.712	0.887
2019	0.766	0.728	0.990	0.852	0.794	0.835	0.833	0.880	0.676
2020	0.842	0.663	0.823	0.840	0.825	0.865	0.656	0.819	0.667
2021	0.789	0.807	0.740	1.000	0.969	0.904	0.828	0.810	0.886
2022	0.741	0.343	0.797	0.994	0.334	0.886	0.668	0.335	0.711

Table A9. Grey Correlation Coefficients of Flow Structural Subsystems in SG Enterprises (Information Flow to Material Flow).

Year	Relationship I_1-M_1	I_1-M_2	I_1-M_3	I_2-M_1	I_2-M_2	I_2-M_3	I_3-M_1	I_3-M_2	I_3-M_3
2013	0.608	0.624	0.666	0.379	0.384	0.660	0.349	0.361	0.556
2014	0.428	0.509	0.368	0.472	0.407	0.563	0.515	0.901	0.360
2015	0.872	0.783	0.693	0.523	0.690	0.770	0.453	0.936	0.827
2016	0.596	0.600	0.633	0.905	0.897	0.837	0.818	0.833	0.995
2017	0.715	0.969	1.000	0.476	0.565	0.602	0.379	0.549	0.643
2018	0.766	0.810	0.853	0.751	1.000	0.826	0.830	0.476	1.000
2019	0.550	0.609	0.761	0.826	0.955	0.990	0.998	0.766	0.492
2020	0.933	0.849	0.644	0.500	0.591	0.740	0.937	0.549	0.377
2021	0.674	0.571	0.510	0.484	0.549	0.613	0.617	0.980	0.739
2022	0.367	0.421	0.349	0.577	0.487	0.624	0.927	0.618	0.717

Table A10. Grey Correlation Coefficients of Flow Structural Subsystems in SG Enterprises (Information Flow to Energy Flow).

Year	Relationship I_1-E_1	I_1-E_2	I_1-E_3	I_2-E_1	I_2-E_2	I_2-E_3	I_3-E_1	I_3-E_2	I_3-E_3
2013	0.657	0.719	0.786	0.451	0.467	0.482	0.603	0.648	0.695
2014	0.359	0.437	0.411	0.507	0.444	0.460	0.660	0.929	0.832
2015	0.896	0.480	0.519	0.623	0.432	0.873	0.780	0.419	0.713
2016	0.560	0.464	0.665	0.803	0.965	0.712	0.984	0.746	0.833
2017	0.967	0.735	0.735	0.592	0.697	0.697	0.718	1.000	0.999
2018	1.000	0.827	0.920	0.894	0.979	0.857	0.813	0.685	0.866
2019	0.591	0.620	0.490	0.964	1.000	0.809	0.764	0.725	1.000
2020	0.925	0.685	0.899	0.572	0.652	0.578	0.789	0.624	0.772
2021	0.499	0.508	0.476	0.572	0.565	0.591	0.984	0.956	0.961
2022	0.356	1.000	0.342	0.517	0.335	0.535	0.760	0.338	0.823

Table A11. Grey Correlation Coefficients of Flow Structural Subsystems in SG Enterprises (Material Flow to Information Flow).

Year	Relationship M_1-I_1	M_1-I_2	M_1-I_3	M_2-I_1	M_2-I_2	M_2-I_3	M_3-I_1	M_3-I_2	M_3-I_3
2013	0.597	0.341	0.542	0.613	0.344	0.554	0.650	0.590	0.732
2014	0.421	0.425	0.701	0.499	0.364	0.960	0.360	0.503	0.550
2015	0.854	0.470	0.646	0.770	0.621	0.978	0.677	0.690	0.912
2016	0.585	0.813	0.909	0.588	0.810	0.924	0.618	0.750	0.998
2017	0.701	0.428	0.574	0.955	0.507	0.730	0.976	0.538	0.797
2018	0.751	0.676	0.915	0.796	0.905	0.668	0.833	0.740	1.000
2019	0.540	0.743	1.000	0.598	0.863	0.884	0.743	0.889	0.678
2020	0.913	0.450	0.971	0.835	0.531	0.731	0.629	0.663	0.568
2021	0.661	0.435	0.781	0.560	0.493	1.000	0.498	0.548	0.860
2022	0.361	0.519	0.966	0.413	0.436	0.784	0.340	0.558	0.846

Table A12. Grey Correlation Coefficients of Flow Structural Subsystems in SG Enterprises (Energy Flow to Information Flow).

Year	Relationship E_1-I_1	E_1-I_2	E_1-I_3	E_2-I_1	E_2-I_2	E_2-I_3	E_3-I_1	E_3-I_2	E_3-I_3
2013	0.678	0.341	0.599	0.812	0.467	0.738	0.790	0.359	0.679
2014	0.381	0.394	0.657	0.574	0.444	0.948	0.423	0.339	0.821
2015	0.904	0.512	0.780	0.615	0.432	0.527	0.530	0.810	0.697
2016	0.584	0.727	0.991	0.601	0.965	0.816	0.673	0.600	0.822

Table A12. *Cont.*

Year	E_1-I_1	E_1-I_2	E_1-I_3	E_2-I_1	E_2-I_2	E_2-I_3	E_3-I_1	E_3-I_2	E_3-I_3
2017	0.970	0.480	0.716	0.823	0.697	0.994	0.741	0.582	0.999
2018	1.000	0.853	0.813	0.886	0.979	0.768	0.917	0.788	0.858
2019	0.614	0.959	0.764	0.736	1.000	0.800	0.501	0.722	1.000
2020	0.932	0.458	0.789	0.786	0.652	0.718	0.897	0.452	0.759
2021	0.524	0.458	0.991	0.641	0.565	0.966	0.487	0.465	0.958
2022	0.379	0.404	0.759	0.989	0.335	0.442	0.353	0.409	0.812

Table A13. SG Enterprise 2019 Jan-Aug 11 Production Lines Energy Input Values (Unit: €).

Production Line	January	February	March	April	May	June	July	August
Line1	652,799	458,694	640,533	557,784	589,908	526,010	553,914	589,313
Line2	498,668	356,508	480,682	419,418	446,176	374,483	411,995	459,002
Line3	177,416	128,552	181,638	150,592	179,114	159,835	185,137	171,850
Line4	321,252	227,956	299,044	268,826	267,062	214,648	226,858	287,152
Line5	154,131	102,186	159,851	138,366	143,732	151,527	141,919	130,311
Line6	156,471	149,308	161,640	135,600	157,018	134,347	169,295	134,769
Line7	109,647	85,602	134,950	114,119	117,924	107,143	126,322	117,437
Line8	404,678	412,077	340,153	384,845	475,019	309,949	360,162	375,895
Line9	717,132	736,438	890,730	832,550	775,571	649,772	791,020	728,500
Line10	100,315	72,581	100,113	88,496	56,778	53,448	60,520	73,017

Table A14. SG Enterprise 2019 Jan-Aug Production Flow Structural Parameter Raw Values.

Parameter	January	February	March	April	May	June	July	August
OEE	87.1%	83.8%	86.9%	86.0%	86.2%	84.5%	87.2%	88.0%
OEE SR	69.9%	67.8%	64.8%	74.2%	78.0%	77.4%	77.9%	75.6%
OEE SL	72.0%	66.5%	72.3%	72.7%	72.3%	72.2%	73.9%	72.3%
OEE Tes	59.5%	67.1%	61.1%	71.1%	81.9%	86.4%	86.0%	81.4%
Yield WS	95.1%	95.7%	95.1%	95.6%	95.2%	95.4%	94.8%	95.3%
Yield KTL	92.9%	92.7%	92.8%	92.4%	93.1%	93.1%	93.3%	93.4%
Yield BT3	96.8%	97.8%	97.7%	97.7%	97.4%	97.1%	97.7%	97.1%
PVT KTL	16.3	16.8	15.6	16.7	15.7	17.1	15.6	18.0
PVT BT3	43.5	39.1	41.0	40.5	40.9	45.0	42.1	41.0
WS Prod	35.0	26.1	52.5	35.4	43.8	26.6	35.9	59.0
KTL Prod	108.1	67.3	73.3	59.8	111.6	84.0	77.1	102.9
BT3 Prod	353.9	165.4	347.3	331.7	207.2	226.2	283.1	326.3

Table A15. Qualitative Identification of Raw Data for Ordered Parameters.

Number	2013	2014	2015	2016	2017	2018	2019	2020	2021	2022
1	96.10%	96.53%	96.90%	98.04%	89.63%	90.04%	95.56%	96.41%	97.67%	96.08%
2	6.73	10.49	15.41	17.70	16.51	19.70	19.26	19.81	18.64	20.75
3	98.5%	98.1%	97.2%	97.3%	99.4%	99.1%	98.2%	98.6%	94.3%	98.3%
4	89.6%	95.5%	91.9%	90.4%	89.6%	89.7%	97.1%	94.9%	86.4%	93.1%
5	200.00	237.96	201.50	295.15	282.47	282.39	232.64	238.54	266.24	264.08
6	35.64%	62.40%	59.02%	64.56%	67.68%	71.70%	54.99%	48.59%	61.91%	68.95%
7	99%	92%	86%	101%	92%	102%	107%	85%	101%	100%
8	887	493	311	396	305	293	234	271	189	211
9	12.9	9.3	10.4	9.7	10.2	10.8	13.4	10.4	9.2	10.8

Table A15. *Cont.*

Number \ Year	2013	2014	2015	2016	2017	2018	2019	2020	2021	2022
10	89.6%	95.5%	91.9%	90.4%	89.6%	89.7%	97.1%	94.9%	86.4%	93.1%
11	5.6%	5.1%	2.0%	3.0%	9.3%	0.9%	5.5%	0.0%	1.1%	0.0%
12	1.9%	3.7%	3.1%	3.5%	0.6%	1.8%	0.5%	0.3%	0.3%	0.2%
13	12.00	10.87	15.01	14.31	12.77	14.64	14.27	12.58	13.72	13.52
14	3.79%	4.51%	7.22%	4.15%	5.06%	4.45%	4.92%	4.88%	4.53%	1.27%
15	0.20%	0.50%	2.50%	3.20%	5.10%	3.60%	3.90%	5.20%	5.50%	5.90%
16	36%	62%	59%	65%	68%	72%	55%	49%	71%	62%
17	63.4%	76.7%	83.0%	93.0%	90.0%	93.0%	80.0%	87.0%	89.0%	91.0%
18	20.55	45.77	35.45	41.73	29.44	32.64	40.54	27.06	18.77	8.85
19	62.0%	49.7%	76.7%	61.6%	45.5%	75.8%	54.9%	50.6%	53.5%	71.4%

Table A16. Original Data for SOM Identification of 10 Process in SG Enterprises.

Operation Sequence \ Index Number	1	2	3	4	5	6	7	8
Process 1	2236.5	407.2	53.1	63.3%	62.4%	887.3	501	1932.5
Process 2	3172.8	971.1	920.8	71.2%	59.0%	311.4	59	2280.8
Process 3	3470.9	398.2	1317.5	85.5%	64.6%	189.3	883	2853.9
Process 4	3638.2	656.9	846.7	80.3%	67.7%	234.2	501	3365.2
Process 5	3854.6	435.7	793.0	84.1%	71.7%	292.8	166	2602.6
Process 6	2956.2	1185.6	837.8	65.1%	55.0%	396.4	993	2250.2
Process 7	2612.4	511.7	1981.2	77.0%	48.6%	270.7	669	1004.4
Process 8	3811.7	664.3	688.7	81.3%	70.9%	211.4	50	3628.7
Process 9	3328.2	428.1	1315.0	82.0%	61.9%	304.7	181	1673.2
Process 10	3706.8	315.5	1140.8	87.5%	69.0%	212.9	385	3706.8

References

1. Haken, H. *Synergetics: Introduction and Advanced Topics*; Springer Science & Business Media: Berlin/Heidelberg, Germany, 2013.
2. Meng, Q.; Han, W. Research on Coordination Models of Complex Systems. *J. Tianjin Univ.* **2000**, *4*, 444–446.
3. Peng, B. Research on the Coevolution of Modern Logistics Industry and Advanced Manufacturing Industry. *China Soft Sci.* **2009**, *S1*, 149–153.
4. Zheng, D.; Li, J.; Zhang, X. Analysis of Collaborative Learning between Automobile Manufacturers and Supplier's Supply and Demand System. *China Soft Sci.* **2010**, *3*, 152–160.
5. Anbanandam, R.; Banwet, D.K.; Shankar, R. Evaluation of Supply Chain Collaboration: A Case Study of the Apparel Retail Industry in India. *Int. J. Prod. Perform. Manag.* **2011**, *60*, 82–98. [CrossRef]
6. Leydesdorff, L.; Strand, O. The Swedish System of Innovation: Regional Synergies in a Knowledge-Based Economy. *J. Am. Soc. Inf. Sci. Technol.* **2012**, *64*, 1890–1902. [CrossRef]
7. Wagner, T.; Boebel, F.G. Testing Synergetic Algorithms with Industrial Classification Problems. *Neural Netw.* **1994**, *7*, 1313–1321. [CrossRef]
8. Schanz, M.; Pelster, A. Synergetic System Analysis for the Delay-Induced Hopf Bifurcation in the Wright Equation. *SIAM J. Appl. Dyn. Syst.* **2012**, *2*, 1056–1061. [CrossRef]
9. May, G.; Barletta, I.; Stahl, B.; Taisch, M. Energy Management in Production: A Novel Method to Develop Key Performance Indicators for Improving Energy Efficiency. *Appl. Energy* **2015**, *149*, 46–61. [CrossRef]
10. Hryshchuk, R.; Molodetska, K. Synergetic Control of Social Networking Services Actors' Interactions. In *Advances in Intelligent Systems and Computing*; Springer: Cham, Switzerland, 2016; Volume 12, pp. 34–42.
11. Xu, Y.L.; Yu, L.; Wang, Y.D. Analysis of Order Parameters for Co-evolution of Regional Sci-tech Innovation and Sci-tech Finance System. *Sci. Technol. Manag. Res.* **2017**, *37*, 15–20.
12. Wen, X.; Zhao, X.; Jia, J. Research on System Order Parameter Identification Method Based on GPEM Main Melody Analysis. *Oper. Res. Manag.* **2011**, *20*, 168–175.
13. Zheng, J.; Wu, T. Three-Dimensional Regulation Mechanism and Model Research of Circular Economy Construction System Order Parameter. *Sci. Technol. Prog. Policy* **2013**, *30*, 147–150.
14. Wu, Y.; Wei, G.; He, X. Measurement and Empirical Study on Synergy Degree between High-tech Manufacturing and Factors Based on Coupling Coefficient Model. *Syst. Eng.* **2017**, *35*, 93–100.
15. Wang, H.; Meng, X.; Yu, R. Research on Order Parameter Identification in Food Quality Chain Collaborative System. *Syst. Eng. Theor. Pract.* **2017**, *37*, 1741–1751.

16. Wen, X.; Zhou, J. An Efficiency-Oriented System Multi-Order Parameter Identification Method. *Oper. Res. Manag.* **2020**, *29*, 183–189.

17. Li, T.; Chen, W. Research on Coordinated Development of Regional Intellectual Property Management System Evolution from a Longitudinal Perspective—Measurement Based on Composite System Coordinated Development Model. *Inf. Stud.* **2012**, *31*, 99–105.

18. Tang, L.; Li, J.P.; Yu, L.A.; Tan, D.H. Quantitative Evaluation Method for System Coordinated Development Based on Distance Coordinated Development Model. *Syst. Eng. Theor. Pract.* **2010**, *30*, 594–602.

19. Cui, L.; Jiang, H. PLSPM-GIA Measurement Model and Its Application in the Coordinated Development of 3E Systems. *Math. Stat. Manag.* **2016**, *35*, 984–996.

20. Deng, X.; Chen, M. Study on the Synergy Degree between Science and Technology Achievement Transformation System and Enterprises. *Sci. Res. Manag.* **2016**, *37*, 116–125.

21. Chen, L. Research on Integrated Management Model of Engineering Project Five Elements Based on Collaborative Learning. *J. Eng. Manag.* **2016**, *30*, 101–105.

22. Luo, W.; Dong, B. Research on Evaluation Model of Regional Economic and Transportation Development Synergy. *J. Highw. Transp. Res. Dev.* **2017**, *34*, 151–158.

23. Li, H.; Zhang, X. Study on Regional Ecological Innovation Synergy and Its Influencing Factors. *China Popul. Resour. Environ.* **2016**, *26*, 43–51.

24. Chen, W.; Yang, Z.; Li, J. Empirical Study on Synergy and Evolution of Regional Intellectual Property Management System. *Stud. Sci. Sci. Manag.* **2016**, *37*, 30–41.

25. Zhang, Y.; Wang, D. Quantitative Measurement of Beijing-Tianjin-Hebei Collaborative Development Based on Composite System Synergy. *Econ. Manag. Res.* **2017**, *38*, 33–39.

26. Li, J.; Fan, C.; Yuan, Q. Measurement of Collaborative Development Level of Beijing-Tianjin-Hebei Based on Distance Collaboration Model. *Sci. Technol. Manag. Res.* **2017**, *37*, 45–50.

27. Lü, T.; Han, W. Chaos Control of Regional "Economy-Resource-Environment" System Based on Coordination. *Syst. Eng. Theor. Pract.* **2002**, *3*, 8–12.

28. Bao, B.F.; Yu, Y.; Tao, Y. Resource Optimal Allocation in Product Customization Collaborative Manufacturing. *Comput. Integr. Manuf. Syst.* **2014**, *20*, 1807–1818.

29. Fang, C. Analysis of Theoretical Basis and Regularity of Collaborative Development of Beijing-Tianjin-Hebei Urban Agglomeration. *Progr. Geogr.* **2017**, *36*, 15–24.

30. Lychkina, N.N. Synergetics and Development Processes in Socio-Economic Systems: Search for Effective Modeling Constructs. *Bus. Inform.* **2016**, *1*, 66–79. [CrossRef]

31. Meynhardt, T.; Chandler, J.D.; Strathoff, P. Systemic Principles of Value Co-creation: Synergetics of Value and Service Ecosystems. *J. Bus. Res.* **2016**, *69*, 2981–2989. [CrossRef]

32. Zheng, J.; Zhou, X. Evaluation of Synergy Effects in Green Supply Chain Management of Iron and Steel Enterprises. *Sci. Res. Manag.* **2017**, *38*, 563–568.

33. Yang, Z.; Chen, W.; Li, J. Study on Speed Characteristics of Synergy and Evolution of China's Intellectual Property Management System. *J. Ind. Eng. Eng. Manag.* **2018**, *32*, 171–177.

34. Guo, H.; Fan, Z.; Li, J.; Wang, L.; Wu, H.; Yang, Y. Research on Co-evolution of E-commerce and Big Data Industries Considering Internal and External Factors. *Oper. Res. Manag.* **2019**, *28*, 191–199.

35. Reiter, S. A system for managing job-shop production. *J. Bus.* **1966**, *39*, 371–393. [CrossRef]

36. Jacobs, F.R. OPT uncovered: Many production planning and scheduling concepts and be applied with or without the software. *Ind. Eng.* **1984**, *16*, 32–41.

37. Graves, S.C.; Kostreva, M.M. Overlapping operations in material requirements planning. *J. Oper. Manag.* **1986**, *6*, 283–294. [CrossRef]

38. Truscott, W.G. Scheduling production activities in multi-stage batch manufacturing systems. *Int. J. Prod. Res.* **1985**, *23*, 315–328. [CrossRef]

39. Potts, C.N.; Baker, K.R. Flow shop scheduling with lot streaming. *Oper. Res. Lett.* **1989**, *8*, 297–303. [CrossRef]

40. Vickson, R.G. Optimal lot streaming for multiple products in a two-machine flow shop. *Eur. J. Oper. Res.* **1995**, *85*, 556–575. [CrossRef]

41. Etinkaya, F.C. Unit sized transfer batch scheduling in an automated two-machine flow-line cell with one transport agent. *Int. J. Adv. Manuf. Technol.* **2006**, *29*, 178–183. [CrossRef]

42. Ruth, M. Information, order and knowledge in economic and ecological systems: Implications for material and energy use. *Ecol. Econ.* **1995**, *13*, 99–114. [CrossRef]

43. Sa, M. lean and green: Doing more with less. *ECR J.* **2003**, *3*, 84–91.

44. Long, Y.; Huang, S.Y.; Liu, K. Basic characteristics of material flow, energy flow, and information flow in large systems. *J. Huazhong Univ. Sci. Technol.* **2008**, *36*, 87–90. [CrossRef]

45. Bascur, O.A.; Hertler, C. Collaboration at the enterprise using real time data analysis: From data to action. *IFAC Proc. Vol.* **2009**, *42*, 314–319. [CrossRef]

46. William, F.; Fazleena, B. Sustainable Value Stream Mapping (Sus-VSM): Methodology to visualize and assess manufacturing sustainability performance. *J. Clean. Prod.* **2014**, *85*, 8–18.

47. Taulo, J.L.; Sebitosi, A.B. Material and energy flow analysis of the Malawian tea industry. *Renew. Sustain. Energy Rev.* **2016**, *56*, 1337–1350. [CrossRef]

48. Suominen, O.; Mörsky, V.; Ritala, R.; Vilkko, M. Framework for optimization and scheduling of a copper production plant. In Proceedings of the 26th European Symposium on Computer Aided Process Engineering, Portorož, Slovenia, 12–15 June 2016.

49. Yin, R.Y. *Theory and Methods of Metallurgical Process Integration*; Elsevier: Amsterdam, The Netherlands, 2016.

50. Zhang, H.; Luo, W. Research on lean production improvement based on EVSM and simulation technology. *Ind. Eng.* **2017**, *20*, 64–70.

51. Zheng, Z.; Huang, S.; Long, J.; Gao, X. Synergistic method of material flow and energy flow in the context of intelligent manufacturing of steel. *J. Eng. Sci.* **2017**, *39*, 115–124.

52. Yu, Y.X.; Yao, S.; Mao, J.S. Quantitative analysis of the coupling coefficients between energy flow, value flow, and material flow in a Chinese lead-acid battery system. *Environ. Sci. Pollut. Res.* **2018**, *25*, 34448–34459. [CrossRef]

53. Li, G.; Meng, Y.; Dong, W.; Wang, T. Multi-flow modeling and analysis of information-physical-energy systems based on collaborative theory. *Electr. Power Constr.* **2018**, *39*, 1–9.

54. Huang, L.; Wu, C.; Wang, B. Construction of a system safety collaborative theory model from the perspective of "flow". *China Saf. Sci. J.* **2019**, *29*, 50–55.

55. Qi, L.; Wang, K.; Zhang, F.; Zhao, X. Research on the system dynamics model of internal knowledge dissemination in enterprises. *J. Manag. Sci.* **2008**, *21*, 9–20.

56. Jiang, C. Analysis of breakthrough paths for independent innovation traps of emerging Chinese enterprises. *J. Manag. Sci. China* **2011**, *14*, 36–51.

57. Wang, C.; Mu, D. Simulation and optimization of manufacturing enterprise logistics operation cost based on system dynamics. *Syst. Eng. Theory Pract.* **2012**, *32*, 1241–1250.

58. Love, P.; Holt, G.; Shen, L.; Li, H.; Irani, Z. Using systems dynamics to better understand change and rework in construction project management systems. *Int. J. Proj. Manag.* **2002**, *20*, 425–436. [CrossRef]

59. Zhang, L.; Han, Y.; Chen, J. Bullwhip effect and cost analysis of quantity-based integrated replenishment in supplier-managed inventory. *Comput. Integr. Manuf. Syst.* **2007**, *2*, 410–416. [CrossRef]

60. Lee, S.H.; Pena-Mora, F.; Park, M. Dynamic planning and control methodology for strategic and operational construction project management. *Autom. Constr.* **2006**, *15*, 84–97. [CrossRef]

61. Hu, B.; Zhang, D.; Zhang, J. System dynamics modeling and simulation of enterprise lifecycle. *Chin. J. Manag. Sci.* **2006**, *3*, 142–148.

62. Dai, M.; Chen, L. CSO business model and its system dynamics simulation research. *Soft Sci.* **2014**, *28*, 119–123.

63. Gary, L.; Amos, N.H.C.; Tehseen, A. Towards strategic development of maintenance and its effects on production performance by using system dynamics in the automotive industry. *Int. J. Prod. Econ.* **2018**, *200*, 151–169. [CrossRef]

64. Hanafi, M.; Wibisono, D.; Mangkusubroto, K.; Siallagan, M.; Badriyah, M.J.K. Designing smelter industry investment competitiveness policy in Indonesia through system dynamics model. *J. Sci. Technol. Policy Manag.* **2019**, *10*, 617–641. [CrossRef]

65. Shi, D.; Wang, L. Energy revolution and its impact on economic development. *Ind. Econ. Res.* **2015**, *2015*, 1–8.

66. Larson, T.; Greenwood, R. Perfect complements: Synergies between lean production and eco-sustainability initiatives. *Environ. Qual. Manag.* **2004**, *13*, 27–36. [CrossRef]

67. Zhan, Y.; Tan, K.H.; Ji, G.; Chung, L.; Chiu, A.S. Green and lean sustainable development path in China: Guanxi, practices and performance. *Resour. Conserv. Recycl.* **2018**, *128*, 240–249. [CrossRef]

68. Kurdve, M.; Zackrisson, M.; Wiktorsson, M.; Harlin, U. Lean and green integration into production system models—Experiences from Swedish industry. *J. Clean. Prod.* **2014**, *85*, 180–190. [CrossRef]

69. Yang, R.; Long, R. Reflection on enterprise management transformation under the background of low-carbon economy. *Sci. Technol. Manag. Res.* **2015**, *35*, 235–239.

70. Tetiana, H.; Karpenko, L.M.; Olesia, F.V.; Yu, S.I.; Svetlana, D. Innovative Methods of Performance Evaluation of Energy Efficiency Projects. *Acad. Strateg. Manag. J.* **2018**, *17*, 465–476.

71. Jarrahi, F.; Manenti, A.; Tortorella, G.L.; Gaiardelli, P. Facing the challenges of the future through the synergetic adoption of Industry 4.0 and Lean manufacturing. Augmented knowledge: A new era of industrial systems engineering. In Proceedings of the XXIV Summer School Francesco Turco, Brescia, Italy, 11–13 September 2019; pp. 129–135.

72. Javied, T.; Deutsch, M.; Frank, J. A model for integrating energy management in lean production. *Procedia CIRP* **2019**, *84*, 357–361. [CrossRef]

73. Dues, C.M.; Tan, K.H.; Lim, M. Green as the new Lean: How to use Lean practices as a catalyst to greening your supply chain. *J. Clean. Prod.* **2013**, *40*, 93–100. [CrossRef]

74. Lee, J.; Yuvamitra, K.; Guiberteau, K.; Kozman, T.A. Six-Sigma Approach to Energy Management Planning. *Strateg. Plan. Energy Environ.* **2014**, *33*, 23–40. [CrossRef]

75. Pampanelli, A.B.; Found, P.; Bernardes, A.M. A Lean & Green Model for a production cell. *J. Clean. Prod.* **2014**, *85*, 19–30.

76. Zhang, Q.; Grossmann, I.E. Enterprise-wide optimization for industrial demand side management: Fundamentals, advances, and perspectives. *Chem. Eng. Res. Des.* **2016**, *116*, 114–131. [CrossRef]

77. Li, Q.; Zhang, W.; Li, H.; He, P. CO_2 emission trends of China's primary aluminum industry: A scenario analysis using system dynamics model. *Energy Policy* **2017**, *105*, 225–235. [CrossRef]

78. Hilorme, T.; Honchar, O. Model of energy saving forecasting in entrepreneurship. *J. Entrep. Educ.* **2020**, *22*, 1–8.

79. Proaño, L.; Sarmiento, A.T.; Figueredo, M.; Cobo, M. Techno-economic evaluation of indirect carbonation for CO_2 emissions capture in cement industry: A system dynamics approach. *J. Clean. Prod.* **2020**, *263*, 121–457. [CrossRef]

80. Chang, J.H.; Chiu, H.N. A comprehensive review of lot streaming. *Int. J. Prod. Res.* **2005**, *43*, 1515–1536. [CrossRef]

81. Sarin, S.C.; Kalir, A.A.; Chen, M. A single-lot, unified cost-based flow shop lot-streaming problem. *Int. J. Prod. Econ.* **2008**, *113*, 413–424. [CrossRef]

82. Zeng, M. Building an integrated energy system. *China Electr. Power Enterp. Manag.* **2018**, *10*, 57–59.

83. Hillman, J.; Axon, S.; Morrissey, J. Social enterprise as a potential niche innovation breakout for low carbon transition. *Energy Policy* **2018**, *117*, 445–456. [CrossRef]

84. Zhang, S.; Wei, X. Has information and communication technology reduced energy consumption in enterprises? Evidence from survey data of Chinese manufacturing firms. *China Ind. Econ.* **2019**, *2*, 155–173.

85. Zhao, Y.Q. System Dynamics Modeling and Dynamic Simulation of the Iron and Steel Production Process. Computer Applications. Master's Thesis, Kunming University of Science and Technology, Kunming, China, 2012.

86. Zhou, J.; Chen, J. Construction of evaluation index system for the integration of the two industrializations in manufacturing enterprises. *Comput. Integr. Manuf. Syst.* **2013**, *19*, 2251–2263.

87. Miao, C.; Feng, J.; Sun, L.; Ma, L. Evolution model of enterprise capability system based on collaborative theory and self-organization theory. *J. Nanjing Univ. Sci. Technol.* **2013**, *37*, 192–198.

88. Kohonen, T. Self-organized formation of topologically correct feature maps. *Biol. Cybern.* **1982**, *43*, 59–69. [CrossRef]

89. Chen, H.; Liu, P.; Yu, H.; Xiao, X. Diagnostics of Oil Filter Clogging Fault for Electrostatic Servo Actuator Based on Improved PCA-SOM. *China Mech. Eng.* **2020**, *32*, 799–805.

90. Li, W.; Zhang, S.; He, G. Semisupervised Distance-Preserving Self-Organizing Map for Machine-Defect Detection and Classification. *IEEE Trans. Instrum. Meas.* **2013**, *62*, 869–879. [CrossRef]

Article

A Study of the Strategic Interaction in Environmental Regulation Based on Spatial Effects

Hewen Gao, Fei Li, Jinhua Zhang and Yu Sun *

School of Foreign Languages and Literature, Changchun Humanities and Sciences Collage, Changchun 130117, China
* Correspondence: sunyu@ccrw.edu.cn

Abstract: The incomplete enforcement of environmental regulations in China is a serious issue in environmental protection affairs, and this paper attempts to provide a new explanation for its prevalence from the perspective of strategic interaction. Under Chinese decentralization, environmental regulations are seen by local governments as a tool to compete for scarce resources, which leads to strategic interactions between regions. Therefore, under the theoretical framework of regional policy spillovers, this paper examines the strategic interaction behavior of local governments in environmental regulation with a spatial econometric approach research methodology based on panel data of 29 Chinese provinces (autonomous regions and municipalities directly under the central government) from 2015 to 2019, taking spatial interdependence and the strategic interaction relationship of local governments as the entry point. The study finds that the intensity of environmental regulation in a region is not only related to the characteristics of the region, but also related to the intensity of environmental regulation in competing provinces, and there is a significant strategic interaction of environmental regulation behavior between regions, which is manifested as complementary spatial strategies. If the neighboring provinces invest more in environmental regulation, the region will also strengthen its level of environmental regulation accordingly, showing the contagiousness of non-complete enforcement of environmental regulation. At the same time, the complementary strategic interaction behavior of environmental regulation between regions has weakened since 2017, which highlights the role of green environmental performance assessment. Based on this, this paper proposes to provide a policy reference to avoid the environmental regulation enforcement dilemma.

Keywords: environmental regulation; data driven decision-making; strategic interaction; spatial spillover

Citation: Gao, H.; Li, F.; Zhang, J.; Sun, Y. A Study of the Strategic Interaction in Environmental Regulation Based on Spatial Effects. *Systems* **2023**, *11*, 62. https://doi.org/10.3390/systems11020062

Academic Editors: Zaoli Yang, Yuchen Li and Ibrahim Kucukkoc

Received: 7 December 2022
Revised: 13 January 2023
Accepted: 20 January 2023
Published: 23 January 2023

1. Introduction

Over the 30 years after China's reform and opening up, the world's economy has continuously developed, and environmental problems have drawn greater attention from all over the world. In such a context, governments have taken a raft of measures to tackle the intensifying environmental problems, trying to minimize the external diseconomy caused by environmental pollution. However, the increasing environmental pollution issues have not been solved. Central governments give great importance to environmental protection, but the reality is that environmental problems are difficult to solve. Why is environmental management not always as good as it should be? The incomplete implementation of the central governments' environmental regulations is the key to solving this problem.

Theoretically, analyzing the prevalence of environmental regulatory strategy interactions requires a return to the origins of environmental regulatory competition. The first study of environmental regulation competition was conducted by Fredriksson and Milllimet (2002a), who examined the existence of interstate competition among environmental regulation strategies in the United States and confirmed the existence of a significant positive correlation between regional regulatory behavior [1]. In addition, Woods (2006) successfully verified the existence of bottom-up competition in environmental regulation using interstate data in the United States [2]. Domestic studies mainly focus on the competitive behavior of local governments' environmental regulation strategies under fiscal decentralization. These studies also confirm that there is a clear strategic game and imitation behavior among local governments in China [3–7]. The current research is focused on the strategic interactions of environmental regulation. Konisky (2007) found that it is difficult to identify a single top-by-top or bottom-by-bottom competition among local governments in environmental regulation, and the competition among local governments has a dynamic adjustment process. [8]. Similarly, a study by Ye and Zhang (2011) showed that local governments not only have obvious strategic behaviors in the current period but also tend to compete on behaviors across periods [9,10]. Existing studies have noted the explanations of local governments' behavior on the phenomenon of the incomplete implementation of environmental regulations and described some conceptual terms, such as selective implementation, symbolic implementation, negative implementation [11], and policy implementation bias [12].

In general, the behavior of competing local governments in environmental regulation has been studied in some depth, but the research perspective is limited to the dynamics of the local governments' behavior and treats each local government as an independent individual, thus ignoring the influence and constraints of local governments' behavior as competitors [13–15], which cannot satisfactorily explain the universality of the non-complete enforcement of environmental regulations. The contribution of this paper is that through the study of inter-regional environmental regulation strategy interactions, we attempt to provide a new explanation for the prevalence of the incomplete enforcement of environmental regulations in China. We highlight that under the Chinese yardstick competition, the profit-seeking local governments implement strategic interaction behaviors, which lead to the mutual imitation of environmental regulations between regions and the incomplete enforcement of environmental regulations in one region. The contagiousness explains the prevalence of the incomplete enforcement of environmental regulations. From this perspective, this paper answers questions surrounding environmental regulation strategy interactions and explores feasible paths to alleviate the incomplete implementation of environmental regulations, providing possible corresponding policies.

2. Research Design

2.1. Methodology

2.1.1. Exploratory Spatial Data Analysis Approach (ESDA)

According to the first law of geography, things that are spatially adjacent are more closely related. In this paper, we use exploratory data analysis (ESDA) to test the spatial characteristics of environmental regulations. Moran's I is used to analyze whether there is clustering or outliers between things, while the local Moran's I is used to explore spatial agglomeration and dispersion [16].

The global Moran's I is calculated as

$$I = \frac{\sum_{i=1}^{n}\sum_{j=1}^{n} \omega_{ij}(A_i - \overline{A})}{S^2 \sum_{i=1}^{n}\sum_{j=1}^{n} W_{ij}} \tag{1}$$

where I is the measure of the overall correlation of the strength of environmental regulation between regions.

$$S^2 = \frac{1}{n}\sum_{i=1}^{n}\left(A_i - \overline{A}\right)^2 \tag{2}$$

$$\overline{A} = \frac{1}{n}\sum_{i=1}^{n} A_i \tag{3}$$

Here, A_i is the weighted value of environmental regulatory enforcement in the jurisdictions, n is the number of regions, and W is the spatial weight matrix. I takes values from -1 to 1. When it approaches 1, it indicates that the intensity of environmental regulatory between the regions presents a spatially positive correlation. When it approaches -1, there is a negative spatial correlation. When it approaches 0, there is no spatial correlation.

When the spatial effects are introduced into the study of economic management processes and spatial econometric models are established for spatial statistical analyses, spatial weight matrices are needed to express the spatial interaction [17]. Accurately measuring the spatial relationships between the provinces and establishing reasonable spatial weight matrices are two key points of the empirical study in this paper. Based on the practices in the existing literature, this study established the spatial weight matrices according to geospatial standards. The details are as follows. According to geographic information, the simplest case is a binary geographic adjacency matrix, which takes the value of 1 if province i and province j share a boundary, and 0 otherwise. If two provinces share a boundary or intersect with each other, they are defined as adjacent. They can be either located along the same latitude, are the same along the same latitude, or are in a diagonal manner.

The global Moran's I represents the overall spatial autocorrelation, and this overall evaluation may neglect the atypical characteristics of local areas [18]. In addition, local correlation indexes need to be used to examine the correlation and agglomerations in the local areas. The local Moran's I is always used, and its equation is

$$I_i = \frac{(A_i - \overline{A})}{S^2}\sum\nolimits_{j=1}^{n}\omega_{ij}(A_i - \overline{A}) \tag{4}$$

where I_i is the measure of the degree of correlation between the strength of the environmental regulation in region i and that in its surrounding regions, A_i, $\overline{A}$, n, W, and S^2, are the same in the equation of the global Moran's I. An $I_i > 0$ suggests a positive correlation between region i and its surrounding regions. In other words, regions with a similar strength of regulations gather together, showing high and high (H-H) or low and low (L-L) agglomerations. An $I_i < 0$ denotes a negative correlation, which means that regions with different environmental regulations in place gather together, showing high and low (H-L) or low and high (L-H) agglomerations [19].

2.1.2. Setting of the Econometric Model

The key to the empirical analysis in this study is how to identify the spatial correlation and dependence of the governments' environmental regulation policies among regions, and spatial econometrics is a good tool for this. In this study, provincial balanced panel data in China were used to construct a spatial econometric model.

Due to spatial correlations and spatial differences, spatial constant coefficient regression models (including spatial lag models and spatial error models) are the most commonly used in spatial econometrics. Spatial lag models (SARs) are mainly used to examine whether the model variables show spillover effects in a region. The model is expressed as $y = \rho\omega y + \beta x + \varepsilon$, where y is the explained variable, x is an $n \times k$ exogenous explanatory variable matrix, ρ is the spatial autoregressive coefficient, ω is an $n \times n$ spatial weight matrix, ωy is the explained variable of the spatial lag, and ε is the random error item. Spatial error models (SERs) are mainly used to examine the explained variables of the adjacent regions on the explained variable of the region in question in terms of the direction and degree. The model is expressed as $y = \beta x + \varepsilon$, $\varepsilon = \lambda\omega\varepsilon + \mu$, where ε is the random error item, λ is an $n \times 1$ spatial error coefficient matrix, and μ is the random error vector of the normal distribution. β reflects the influence of the independent variable, x, on the dependent variable, y, and ε reflects the direction and degree of the adjacent regions' explained variables on the explained variable of the region in question.

The classical econometric model was expanded based on the above theoretical basis to obtain the following spatial econometric models:

1. The spatial lag panel data model:

$$Y_{it} = \delta \sum_{j \neq i}^{n} W_{ijt} Y_{jt} + \beta X_{it} + \mu_i + d_t + \varepsilon_{it} \tag{5}$$

where Y_{it} denotes the environmental regulation strategy formulated by province i (i = 1, 2, … , n) in the year (t = 1, 2…, T). W is the given $n \times n$ spatial weight matrix, the elements of which represent the spatial relationship between province i and province j. WY is the spatial lag term. Since the main diagonal element of W is zero, and it is a row-normalized matrix, the spatial lag variable, WY, can be explained as the weighted average level of the environmental policies of other neighboring provinces except for province i in that year. X_{it} represents a set of covariates that affect the environmental policy of a province. μ_i represents the province's fixed effects that do not change over time, and d_t represents time's fixed effects. ε_{it} denotes independent and identically distributed disturbance terms. δ and β are unknown parameters of the model. β represents the marginal influence of the economic characteristics of a province on the local environmental policy, and δ represents the strategic response coefficient of the local government's environmental regulation behavior, which is the core parameter that should be paid attention to in this study. The empirical study was conducted to estimate δ and test whether it equals 0. If $\delta \neq 0$, there is a strategic interaction between the corrupt acts of the local governments. If $\delta > 0$, the environmental regulation behaviors between the local governments show strategic complementary interactions, and when $\delta < 0$, there is a strategic substitutive interaction.

2. The spatial error panel data model:

$$Y_{it} = \beta X_{it} + \mu_i + d_t + \varepsilon_{it} \quad \varepsilon_{it} = \lambda \sum_{j \neq i}^{n} W_{ijt} \varepsilon_{it} + \vartheta_{it} \quad \vartheta_{it} \sim n\left(0, \vartheta_{it}^2\right) \tag{6}$$

where λ is the strength of the spatial correlation between the regression residuals, and the spatial weight matrix, W, is the same as in the previous equation. The parameter estimates of OLS are no longer consistent due to spatial autocorrelation. Therefore, the maximum likelihood (ML) is used for estimating the above two types of spatial models.

2.2. Data Description and Variable Selection

2.2.1. Variables

(1) Environmental regulation

It is well known that there is no direct measure of environmental regulation. In general, the established literature uses alternative proxy indicators to measure the intensity of environmental regulation, including pollution reduction rate cost, pollution control investment, the amount of supervision and inspection, and government environmental protection expenditure [20,21]. In this paper, the environmental regulation intensity is measured by the ratio of industrial pollution control investment per unit of the industrial value added to the industrial value added per unit of GDP.

(2) Decentralized indicators (FD)

In this paper, we use fiscal decentralization to measure the unique macro environment for environmental regulation decisions. According to the approach from Guo [22], the indicator is measured as follows: FD = per capita provincial fiscal expenditure/(per capita provincial fiscal expenditure + per capita federal fiscal expenditure).

(3) Public demand for environmental protection (LETTER)

Public demand for environmental protection is the subjective awareness of people and does not have direct quantifiable indicators, so the existing literature uses alternative indicators to measure it [23,24]. This paper constructed indicators of public demand for environmental protection in terms of letters and visits according to the approach of Yu [23].

(4) Other variables

Other control variables are introduced with reference to the environmental STIRPAT model [25]. The GDP per capita and its square measure the environmental Kuznets curve. Population density is the ratio of the total population to the area of the city at the end of the year. The FDI is the ratio of foreign investment to the GDP, which is used as a variable to measure local government competition.

2.2.2. Data

China issued a new environmental law in 2013 and implemented the separation of environmental law enforcement and regulation, elevating the main body exercising regulatory functions to provincial environmental protection authorities, which strengthens the effectiveness of regulation while compacting the implementation of policies on the ground. Considering the time lag of laws and policies and the availability of data, this paper uses the data of local governments at the provincial level (excluding Hong Kong, Macao, and Taiwan) in China from 2015 to 2020 as the sample. In view of the lack of data on Tibet, Tibet is excluded from the study. All data were obtained from the China Statistical Yearbook (2015–2020), the China Environmental Yearbook (2015–2020), and the China Statistical Yearbook of Industrial Economy (2015–2020). Spatial correlation tests were performed using OpenGeoDa, and the processing of basic panel data and spatial panel data estimation were performed using MATLAB 8.0.

3. Empirical Results and Analysis of the Spatial Effects
3.1. Spatial Correlation
3.1.1. Global Spatial Correlation

Table 1 shows Moran's I for the environmental regulatory stringency in 29 Chinese provinces from 2015 to 2019.

Table 1. The global Moran's I of the strength of environmental regulation in 29 provinces in China from 2015 to 2019.

Year	2015	2016	2017	2018	2019
Moran's I	0.28	0.33	0.28	0.20	0.085

All the values of the global Moran's I shown in Table 1 are positive, indicating a significant positive spatial correlation of China's environmental policies. In other words, regions with strong environmental regulations tend to be adjacent to one or more regions with similar regulations (a high–high positive correlation). Meanwhile, the values in all these years (except for the low value in 2019) are around 0.3, indicating that this spatial correlation is relatively stable.

3.1.2. Local Spatial Correlation

Anselin (1995) [26] proposed that global autocorrelation reflects the particular situation of spatial correlation among regions. To analyze the correlation among provinces more clearly, this paper used local autocorrelation analysis, LISA, and scatterplots to visually analyze the local correlation.

Figures 1–5 show the scatterplots of the local Moran's I of the strength of environmental regulation in various regions in China from 2015 to 2019. The horizontal axis represents the standardized value of the strength of environmental regulation, and the vertical axis represents the value of the spatial lag of the standardized strength of environmental regulation. The mean value is the center of the axis, which divides the plot into four quadrants. The first quadrant represents a high–high positive correlation, and the third quadrant represents a low–low positive correlation. Since the local Moran's I shows a positive correlation, the second and fourth quadrants that represent the negative correlation are atypical observation areas. The figures show that from 2015 to 2019, the spatial evolution of relative environmental regulation fluctuated. The majority is clustered in the first and third quadrants in 2015 and 2016. The clustering of relative regulation is dispersed from 2017 to 2019, which indicates a certain diffusion effect. The strategic interaction behavior of the mutual imitation of the environmental regulations between the regions significantly weakened after 2017. The weakening and disappearance of imitation, and even the change to a differentiated form of strategic interaction, implies that the central government's emphasis on environmental protection, and the inclusion of environmental indicators in the performance assessment system, have gradually corrected the GDP-only view of the performance of local government officials.

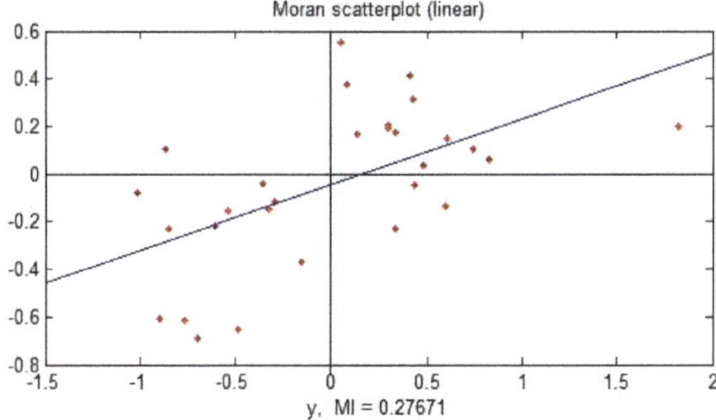

Figure 1. Scatterplot of the local Moran's *I* in 2015.

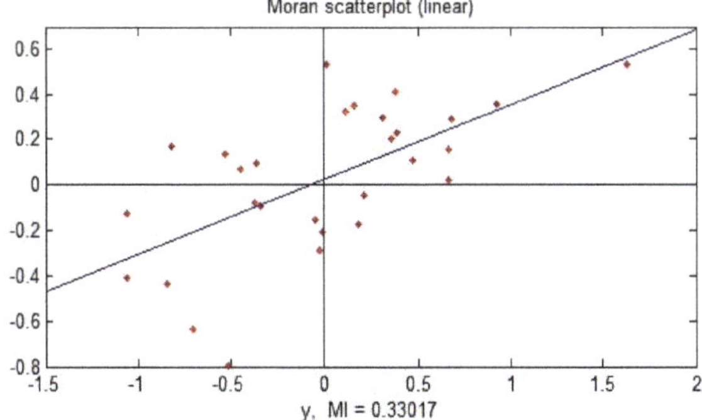

Figure 2. Scatterplot of the local Moran's *I* in 2016.

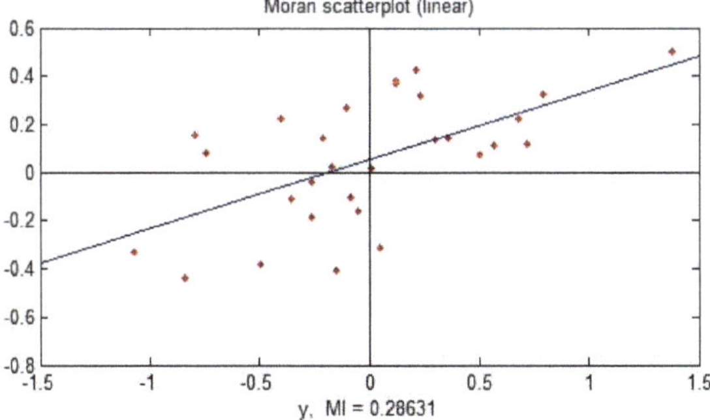

Figure 3. Scatterplot of the local Moran's *I* in 2017.

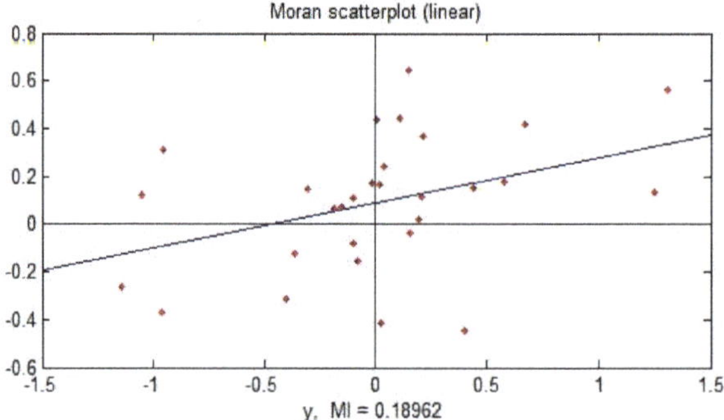

Figure 4. Scatterplot of the local Moran's *I* in 2018.

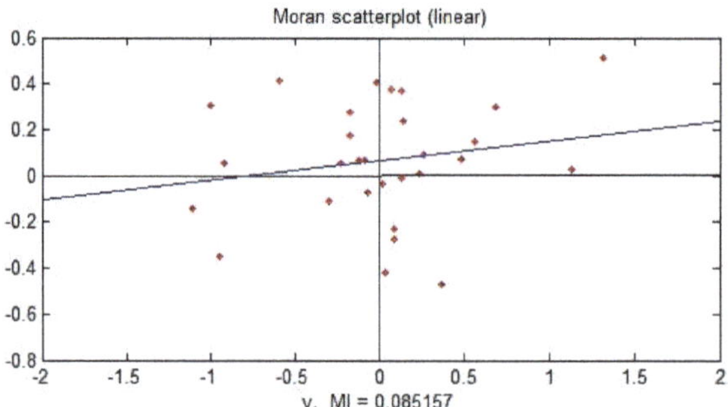

Figure 5. Scatterplot of the local Moran's *I* in 2019.

To reflect the spatial divergence characteristics of environmental regulation and local protection, we used the values of local Moran's *I* to plot the cluster maps of the strength of environmental regulation in China's various regions and compared the changes in spatial patterns from 2015 to 2019 (Figures 6–10). The agglomeration areas all passed the test at a significance level of 5%.

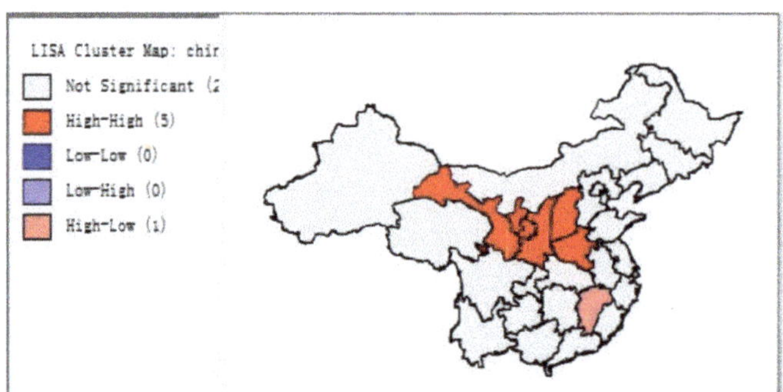

Figure 6. Cluster map of the strength of environmental regulation in 2015.

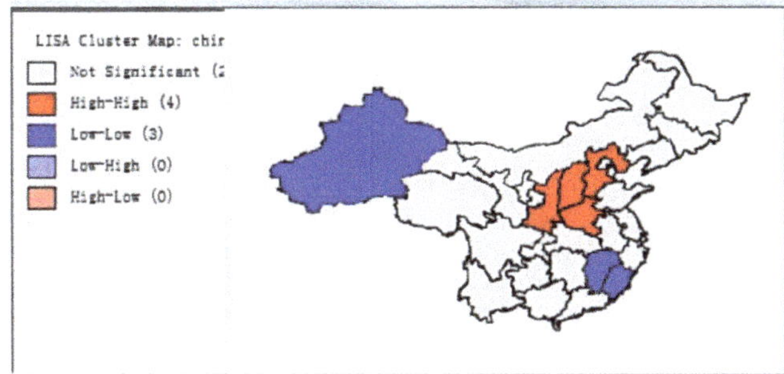

Figure 7. Cluster map of the strength of environmental regulation in 2016.

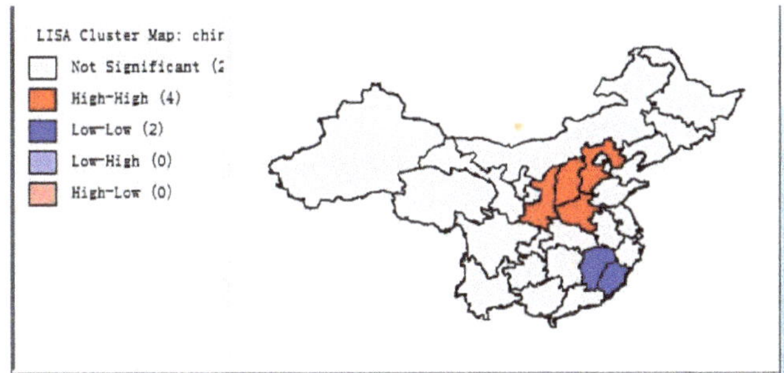

Figure 8. Cluster map of the strength of environmental regulation in 2017.

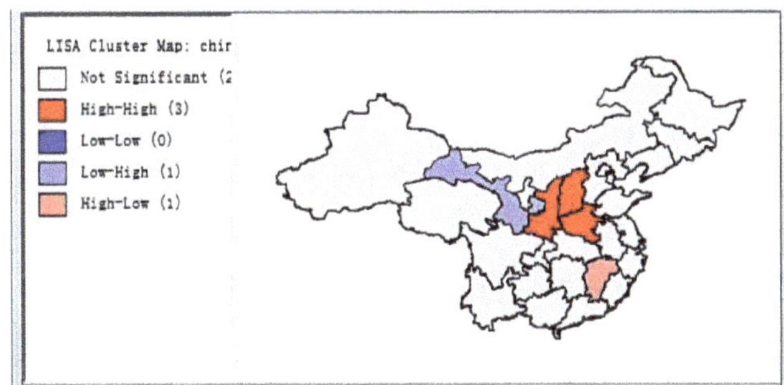

Figure 9. Cluster map of the strength of environmental regulation in 2018.

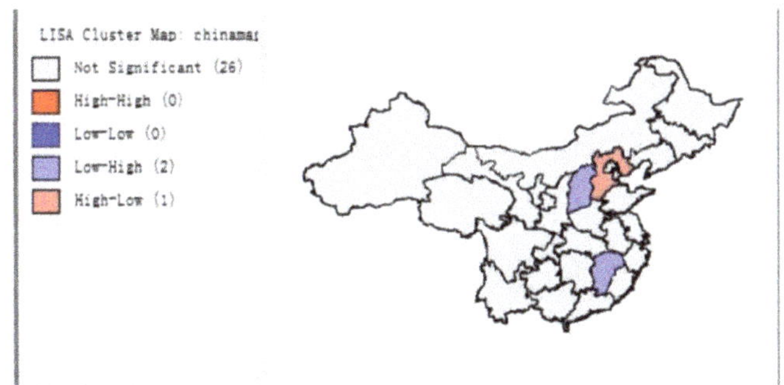

Figure 10. Cluster map of the strength of environmental regulation in 2019.

As the LISA cluster maps in Figures 6–10 show, the high–high correlations are mainly distributed in six provinces, namely Beijing, Tianjin, Hebei, Shanxi, Shaanxi, and Henan. This is mainly because in 2015, the National Development and Reform Commission issued the Beijing–Tianjin–Hebei Synergistic Development Ecological Environmental Protection Plan, and the promulgation of this policy has strongly promoted the synergistic process of ecological and environmental management in Beijing, Tianjin, and Hebei. Low–low correlations are mainly distributed in the western regions. This is mainly due to the lower level of economic development and relatively weaker environmental regulations there. In 2019, more areas showed a discrete status, which is mainly due to a document issued by the Ministry of Environmental Protection in 2018, which called for setting and strictly adhering to the red line for ecological protection. The document states the upper limit for total emissions and the bottom line for environmental access and a series of initiatives announcing the advent of the strictest environmental protection system. In this case, the environmental performance appraisal imposed a hard constraint on local government officials in terms of environmental protection tasks, which prompted local governments to differentiate their environmental regulatory behavior and optimize their strategic interactions. This highlights the important role of the continuously strengthened environmental performance appraisal and reflects that the diversification and greening of performance appraisals help shape the scientific behavioral choices of local government officials and weaken the strategic interaction behaviors of environmental regulation, thus alleviating the prevalence of the incomplete enforcement of environmental regulation.

In conclusion, there is a significant positive spatial correlation between environmental regulations in China's various regions. It is a long-term and stable correlation. The local spatial correlation shows that H-H agglomerations are concentrated in a stable state in regions with rapid local economic development. To properly protect the environment, it is necessary to thoroughly consider how significant the impact of these spatial effects on environmental regulation is. To answer this question, it is necessary to introduce a spatial econometric model to further analyze the spatial spillover effect.

3.2. Spatial Spillover Effects

Table 2 shows the coefficient results of the spatially lagged static panel regressions, which analyze the strategies of local governments' environmental regulation decisions in the absence of intertemporality. The Hausman test in MATLAB used in this study selects random effects when $p > 0.05$ (Table 3). According to the judgment of the Hausman test, the SAR was selected.

Table 2. Comparison of the three estimation results of the regression models.

	Basic Panel Data Model		Spatial Lag Panel Data Model		Spatial Error Panel Data Model	
	Model 1	Model 2	Model 3	Model 4	Model 5	Model 6
	Pooled OLS	Fixed effects ML estimation	Fixed effects ML estimation	Random effects ML estimation	Fixed effects ML estimation	Random effects ML estimation
C				0.260		3.749
				0.000		0.000
FD	−0.345	−0.511 ***	−0.415 ***	0.081 *	−0.529 ***	0.108
	(0.823)	(0.000)	(0.002)	(0.216)	(0.000)	(0.160)
LNGDP	0.556 ***	0.246	0.156	0.442 ***	0.158	0.826 ***
	(0.000)	(0.119)	(0.359)	(0.000)	(0.353)	(0.000)

Table 2. *Cont.*

	Basic Panel Data Model		Spatial Lag Panel Data Model		Spatial Error Panel Data Model	
	Model 1	Model 2	Model 3	Model 4	Model 5	Model 6
	Pooled OLS	Fixed effects ML estimation	Fixed effects ML estimation	Random effects ML estimation	Fixed effects ML estimation	Random effects ML estimation
LETTER	0.019	0.037 **	0.029 *	0.709 ***	0.039 **	0.029
	(0.000)	(0.021)	(0.100)	(0.269)	(0.052)	(0.156)
LN PEOPLE	0.087 ***	−2.300 **	−2.363 **	−0.096	−2.455 **	−0.157 **
	(0.006)	(0.015)	(0.019)	(0.212)	(0.025)	(0.044)
LNFDI	0.0275	−0.2271 ***	−0.0545	−0.257	0.00527	−0.0880
	(0.267)	(0.082)	(0.6632)	(0.264)	(0.277)	(0.400)
δ			0.227 **	0.264 ***		
			(0.023)	(0.000)		
λ					0.213 **	0.171
					(0.045)	(0.161)
Rsquare	0.209	0.471	0.853	0.798	0.846	0.772

***, **, and * denote 1%, 5%, and 10% significance levels, respectively.

Table 3. Diagnostic test for spatial correlation.

LM-Lag	Robust LM-Lag	LM-Error	Robust LM-Error
4.442	2.56	2.565	0.683
0.035	0.01	0.109	0.409

In the model, δ is greater than 0, indicating a significant spatial spillover effect of regulatory enforcement. The results of the regressions show that local governments have a significant imitative behavior in environmental regulation and will respond accordingly to the different decision-making environments. The regression results lead to the following conclusions.

(1) Environmental regulation

As can be seen from Table 2, the estimated coefficients are significantly different from zero, at least at the 5% level, which indicates that there is a strategic interaction behavior of environmental regulation expenditure between the regions, and the coefficients are significantly larger than zero. In addition, the strength of environmental regulation of a province increased by 0.264% when that of a neighboring province rose by 1%, indicating significant spatial spillover effects. Further, this indicates that the strategic interactions are complementary. This implies that, for government officials in a region, if the rival in the same position in the yardstick competition chooses to reduce the investment in pollution control, then the optimal strategy for the government officials in that region is also to reduce the investment in pollution control to a low-level equilibrium. At the same time, the estimated coefficients of the spatial lags of environmental regulations are all significantly larger than zero, indicating that there is a strategic interaction between the regions to imitate each other's environmental regulations, and this imitation makes the behavior of the incomplete enforcement of environmental regulations in one region contagious to neighboring regions, which breeds the incomplete enforcement of environmental regulations.

(2) Fiscal decentralization (FD)

The estimated coefficients of fiscal decentralization and the spatial lag factor of environmental regulation are significantly positive at the 1% level, implying that fiscal decentralization strengthens the strategic interaction behavior of environmental regulation between regions and creates incentives for the incomplete enforcement of environmental regulation.

(3)   Public environmental demands (LETTER)

The estimated coefficients of the spatial lag between public demand and environmental regulation are positive, and most of them are significant, at least at the 10% level, suggesting that public demand strengthens regulation intensity and weakens the strategic interaction behavior of inter-regional environmental regulation and forms a constraint to the incomplete implementation of environmental regulation, which supports the pivotal role of the public in environmental protection matters. At the same time, this estimation implies the need to promote the transition from unitary governmental governance to a modern environmental governance system with multi-level governance, which is also in line with the spirit of the newly revised Environmental Protection Law to establish a separate chapter on information disclosure and public participation, and ultimately to form a bottom-up external accountability and monitoring mechanism.

(4)   Other variables

There is a positive effect of the GDP per capita on the regulation and benefits of environmental regulations. This may be explained by the fact that the higher the level of economic development, the better the factor endowment, and the higher the probability that local government officials will win in the competition for a promotion, which leads to a stronger incentive for growth and a more serious spending bias for government officials in the region, and thus a greater lack of incentive for environmental spending. Population density reduces the intensity of environmental regulation. This may be due to the fact that the higher the population density, the higher the employment pressure on the public, and the tendency of local governments to loosen environmental regulations in order to alleviate this pressure. FDI has weakened environmental regulations as a whole, probably because local governments compete viciously to attract FDI and use lax environmental regulations as bait, leading to the effect of bottom-up competition for environmental regulations.

4. Conclusions

This study empirically analyzed the inter-regional strategic interaction in environmental regulatory enforcement in China. Based on the political tournament theory and the condition of capital flow, a model of the strategic interaction between the environmental regulation behaviors in two regions was built.

This paper finds that there are significant complementary strategic interactions between the regions in environmental regulation, implying that rival regions imitate each other's environmental regulations. The spillover of the incomplete enforcement of environmental regulations is explained by the fact that rival regions mimic each other's environmental regulations, which better explains the prevalence of incomplete enforcement of environmental regulations. This explains the prevalence of the non-complete enforcement of environmental regulations. After 2017, the interregional environmental regulation imitation behavior diminished, which means that the greening of performance assessment helps to weaken the strategic interaction behavior and mitigate the prevalence of non-complete implementation of environmental regulation. Based on this, in order to effectively break the prevalence of incomplete enforcement of environmental regulations and to improve the policy confidence of the central government, the following policy recommendations are made: Firstly, the greening of the performance appraisal system is conducive to alleviating the strategic interaction of environmental regulations between regions, so it should continue to reduce the weight of growth in the performance appraisal, increase the weight of environmental protection, build an effective balance mechanism between economic development and environmental regulations, and implement the same responsibility of the party and government for environmental protection and the one-vote veto system.

Secondly, China needs to further centralize its environmental management, improve the vertical management system for environmental policy enforcement below the provincial level, reduce the scope for local governments' discretion in enforcement, and expand the scope of central governments' spending on environmental protection matters. This will lead to a pattern of better matching of financial and administrative powers for environmental management. Finally, the role of the public in environmental protection matters is particularly important. The public should be given the right to make decisions on environmental management by establishing and facilitating channels for the public to express their environmental demands, thus improving the situation of insufficient government supervision. At the same time, we should incorporate social organizations into environmental protection activities to form a modern environmental governance pattern with multiple avenues of governance and social participation.

Author Contributions: H.G. Conceptualization, Methodology, Software, Validation, Visualization, and Writing—original draft. F.L. Project administration, Resources, Writing—review and editing, and Conceptualization. J.Z. Conceptualization, Validation, and Writing—review and editing. Y.S. Conceptualization, Validation, and Writing—review and editing. All authors have read and agreed to the published version of the manuscript.

Funding: This work was supported by the Jilin Province Higher Education Scientific Research Subject (JGJX2020D583).

Institutional Review Board Statement: Not applicable.

Informed Consent Statement: Not applicable.

Data Availability Statement: Not applicable.

Conflicts of Interest: The authors declare no conflict of interest.

References

1. Fredriksson, P.G.; Millimet, D.L. Strategic Interaction and the Determination of Environmental Policy Across US States. *J. Urban Econ.* **2002**, *51*, 101–122. [CrossRef]
2. Woods, N.D. Interstate Competition and Environmental Regulation: A Test of the Race-to-the-bottom Thesis. *Soc. Sci. Q.* **2006**, *87*, 174–189. [CrossRef]
3. Fan, Y.; Wang, W. Analysis of local fiscal expenditure preferences under Chinese style fiscal decentralization. *Econ. Manag. Res.* **2010**, *7*, 40–47.
4. Yang, H.S.; Chen, S.L. Local government competition and environmental policy: Evidence from Chinese provincial data. *Financ. Econ. Res.* **2019**, *4*, 46–55.
5. Yi, Z. The game behavior of local government competition and river basin water environmental protection. *Econ. Issues* **2014**, *1*, 35–43.
6. Deng, M. A study on the strategic interaction of market segmentation among regions in China. *China Ind. Econ.* **2014**, *2*, 18–29.
7. Zhang, Z.; Zhu, P. An empirical study of local environmental expenditures. *Econ. Res.* **2010**, *5*, 82–94.
8. Konisky, D.M. Regulatory Competition and Environmental Enforcement: Is There a Race to the Bottom. *Am. J. Political Sci.* **2007**, *51*, 853–872. [CrossRef]
9. Ye, Q. Research on environmental governance under the fiscal decentralization system. *Financ. Econ. Politics Law* **2011**, *3*, 37–41.
10. Zhang, K.C.; Wang, J.; Cui, S.Y. Fiscal decentralization and environmental pollution: A carbon emission perspective. *China Ind. Econ.* **2011**, *10*, 65–75.
11. Xue, L.; Wei, X.; Liu, J. Environmental regulation and its assessment in China. *China Popul. Resour. Environ.* **2010**, *9*, 70–77.
12. Brueckner, J. Strategic interaction among governments: An overview of empirical studies. *Int. Reg. Sci. Rev.* **2003**, *26*, 175–188. [CrossRef]
13. Cole, M.; Elliott, R. Determining the trade environment composition effect: The role of capital, labor and environmental regulations. *J. Environ. Econ. Manag.* **2003**, *46*, 363–383. [CrossRef]
14. Dijkstra, B. Direct regulation of a mobile polluting firm. *J. Environ. Econ. Manag.* **2003**, *45*, 265–277. [CrossRef]
15. Elhorst, J. Spatial panel data models. In *Handbook of Applied Spatial Analysis*; Fischer, M.M., Getis, A., Eds.; Springer: Berlin/Heidelberg, Germany, 2010; pp. 377–407.
16. Horrace, W.C.; Schnier, K.E. *Estimation a Production Function for Highly-Mobile Technologies*; Department of Economics, Syracuse University: Syracuse, NY, USA, 2008.
17. McFadden, D.; Train, K.E. Mixed MNL models of discrete response. *J. Appl. Econom.* **2000**, *15*, 447–470. [CrossRef]

18. Parsons, G.R.; Hauber, A.B. Spatial boundaries and choice set definition on a random utility model of recreation demand. *Land Econ.* **1998**, *74*, 32–48. [CrossRef]

19. Colmer, J.; Hardman, I.; Shimshack, J.; Voorheis, J. Disparities in PM2.5 air pollution in the United States. *Science* **2020**, *369*, 575–578. [CrossRef] [PubMed]

20. Crowder, K.; Downey, L. Inter-Neighborhood Migration, Race, and Environmental Hazards: Modeling Micro-Level Processes of Environmental Inequality. *Am. J. Sociol.* **2010**, *115*, 1110–1149. [CrossRef]

21. Glatter-Götz, H.; Mohai, P.; Haas, W.; Plutzar, C. Environmental Inequality in Austria: Do Inhabitants' Socioeconomic Characteristics Differ Depending on Their Proximity to Industrial Polluters? *Environ. Res. Lett.* **2019**, *14*, 074007. [CrossRef]

22. Guo, Q.; Jia, J. Strategic Interaction Behavior, Fiscal Spending Competition and Regional Economic Growth among Local Governments. *Manag. World* **2009**, *10*, 17–27.

23. Yu, W.; Gong, Q. Public demands, officials' incentives and regional environmental governance. *Zhejiang Soc. Sci.* **2014**, *5*, 23–35.

24. Aldeco, L.; Barrage, L.; Turner, M.A. *Equilibrium Particulate Exposure*; Technical Report; Brown University: Providence, RI, USA, 2019.

25. Allen, T.; Arkolakis, C. Trade and the Topography of the Spatial Economy. *Q. J. Econ.* **2014**, *129*, 1085–1140. [CrossRef]

26. Anselin, L. Local Indicators of Spatial Association-LISA. *Geogr. Anal.* **1995**, *27*, 93–115. [CrossRef]

Article

An Integrated EDAS Model for Fermatean Fuzzy Multi-Attribute Group Decision Making and Its Application in Green-Supplier Selection

Shouzhen Zeng [1,2], Wendi Chen [1], Jiaxing Gu [1] and Erhua Zhang [1,*]

[1] School of Business, Ningbo University, Ningbo 315211, China
[2] School of Statistics and Mathematics, Zhejiang Gongshang University, Hangzhou 310018, China
* Correspondence: zhangerhua@nbu.edu.cn

Abstract: The environment and economy benefit from the sustained growth of a high-quality green supplier. During a supplier evaluation and selection process, DMs tend to use fuzzy tools to express evaluation information due to complex practical problems. Therefore, this study explores the green-supplier evaluation method in a complex Fermatean fuzzy (FF) environment. First, a group of indicators was created to evaluate the green capabilities and the social impact of suppliers. Second, by combining the merits of the Heronian mean and power average approaches, a FF power Heronian mean and its weighted framework were developed, and their related properties and special families were then presented. Third, to acquire the relative importance of indicators, a marvelous unification of the best–worst method (BWM) and FF entropy is then introduced. The challenge of choosing a green supplier was finally solved using an integrated evaluation based on distance from the average solution (EDAS) evaluation framework in the FF environment. Finally, the presented tool's viability and robustness were confirmed by actual case analysis.

Keywords: Fermatean fuzzy set; BWM; green supplier selection; power Heronian mean; EDAS

Citation: Zeng, S.; Chen, W.; Gu, J.; Zhang, E. An Integrated EDAS Model for Fermatean Fuzzy Multi-Attribute Group Decision Making and Its Application in Green-Supplier Selection. *Systems* **2023**, *11*, 162. https://doi.org/10.3390/systems11030162

Academic Editors: Zaoli Yang, Yuchen Li and Ibrahim Kucukkoc

Received: 21 February 2023
Revised: 18 March 2023
Accepted: 20 March 2023
Published: 21 March 2023

1. Introduction

Owing to the rapid advancement in science and technology since the industrial age, domestic output has been high, resulting in an unprecedented level of material riches for a nation and its citizens. Quality lifestyle requirements are becoming extremely important with the rise in residents' standards of living. Simultaneously, people's purchasing power has increased, propelling the domestic economy to a new level. However, various significant social issues have emerged as well [1]. Rapid resource consumption has led to an increase in environmental issues such as resource wastage and air pollution, which are extremely challenging for the environment in which people live. Social unrest, economic upheaval, and other unforeseen issues can eventually result from an ecological imbalance. If the economy continues to develop without taking environmental protection measures, the environment will inevitably become polluted, posing a threat to ecological security. Environmentalism and economic growth are never mutually exclusive; rather, they are interdependent. Economic growth will eventually be affected by ecological issues. Therefore, prioritizing environmental conservation should go hand in hand with economic development [2]; this is currently the global perspective. For example, the idea of sustainable development was explained in a global report as early as 1987. China developed and implemented a sustainable development strategy in March 1994 based on its unique national circumstances [3]. Since then, a global green wave has begun, which is necessary to advance sustainable development, raise awareness among citizens on the need to safeguard the environment, and encourage synchronized economic and environmental growth [4]. In this context, the concept of green supply chain management (GSCM) has emerged [5].

All supply chain workstations and activities are intended to have a lower adverse ecological impact through the use of GSCM [6]. A crucial aspect of GSCM is selecting the right suppliers [7]. Green suppliers are in the upstream of the green supply chain, influencing the downstream of procurement and production, etc. [8]. Offering green products from green suppliers of the highest caliber can cut expenses while simultaneously preserving the environment and has the potential to attract customers for enterprises. This is highly beneficial to the enterprises and thereby enhances the abilities of both suppliers and businesses to compete [9,10]. Responding to social needs and selecting appropriate green suppliers for themselves is a strategic step toward scientific GSCM for business.

Green supplier selection (GSS) is a complex decision issue in a multi-attribute domain since it involves multiple options, attributes, and decision-makers (DMs) [11]. However, there is a significant amount of uncertainty in the green-supplier selection (GSS). DMs may not be able to provide exact evaluation values based on each assessment criterion in the context of GSS and evaluation. Fuzzy sets (FSs) [12], intuitionistic fuzzy sets (IFSs) [13], and Pythagorean fuzzy sets (PFSs) [14] were all used as tools to evaluate the green suppliers. However, the membership u and non-membership degree v of IFSs and PFSs need to satisfy the constraints $u + v \leq 1$ [15] and $u^2 + v^2 \leq 1$ [16], respectively, that prevent them from fully expressing uncertain information. Senapati and Yager [17] further reduced the limitations to $u^3 + v^3 \leq 1$, and proposed Fermatean fuzzy sets (FFSs) to describe additional information.

A set of succinct and precise evaluation index systems is crucial to the GSS. Green suppliers have numerous intricate evaluation standards. Economical details in traditional supplier selection and environmental standards had been considered [18]. Scholars are particularly concerned about the impact of green issues on GSS. For example, Tavana et al. built an environmental capability evaluation index system for suppliers that included pollution controls, pollution products, green protection, and environmental protection [19].

Research on decision-making methods and models of GSS has also received increasing attention. Ref. [20] identified the importance of each green supplier evaluation index using the best-worst method (BWM). Ref. [21] constructed an analytic hierarchy process (AHP)-based model for determining the weight of green supplier indicators for steel companies. Evaluation information from a single DM is not convincing. Several multi-attribute group decision-making (MAGDM) models are proposed to obtain reliable GSS results. After integrating the fuzzy green evaluation information using the weighted average (WA) operator, VIKOR (ViseKriterijumska Optimizacija I Kompromisno Resenje) and TOPSIS obtained the ideal green suppliers for edible oil [14] and agricultural tool [22] enterprises, respectively.

With the above review, several research gaps continue to exist regarding GSS, as follows:

(1) Uncertainty exists in the evaluation standards of green suppliers. The information conveyed by IFSs and PFSs was limited. Few GSS studies considered the usefulness of Fermatean fuzzy (FF) evaluation information.

(2) More factors need to be taken into account rather than just stating the characteristics of traditional green suppliers. Refs. [18,19] focused only on the economic and environmental benefits of green suppliers but neglected their social responsibility.

(3) In the group decision model related to GSS, the WA operator utilized to assemble assessment fuzzy data in refs. [14,22] did not take into account the link between attributes and the incorrect judgment brought on by extreme data.

(4) BWM [20] and AHP [21] were subjective weighting methods dominated by experts' subjective judgments. It is impossible to make a fully reasonable judgment on the importance of indicators without the joint participation of objective weighting methods. Furthermore, it is crucial to apply precise and consistent evaluation methods when ranking alternative solutions. Decision methods such as VIKOR [14] and TOPSIS [22] may increase the negative impact of extreme value decision results.

Driven by the above research gaps, this work's central thrust is to support the GSS using a comprehensive FF MAGDM approach, which combines the power Heronian mean (PHM) operator, BWM, the entropy weight method (EWM), and evaluation based on distance from the average solution (EDAS). In this context, information regarding expert evaluations is aggregated using the FFPHM operator. A novel BWM based on entropy measures is combined with the EWM to provide an integrated assignment approach. The best green supplier is found by using EDAS to rank the considered alternatives. The following are the study's significant contributions:

(1) The GSS problem in the FF environment will be examined, where FFSs have a broader range of information representation.
(2) Create a comprehensive set of index systems for evaluating green suppliers. This study developed a set of index systems combining traditional qualities, green attributes, and social attributes based on references and analysis of the existing index system.
(3) We propose the FFPHM and FFWPHM operators by applying the PHM operator to the FF environment. The proposed operators consider the consistency and correlation of data when aggregating evaluation information.
(4) For the GSS problem that it is unknown how important the various indicators are, an integrated weight calculation method is offered in the foundations of EWM and BWM. This integrated technique successfully lowers the disparity between subjective and objective information.
(5) A FF MAGDM framework based on the integrated weight determination model and EDAS is developed. EDAS simplifies the calculation process while reducing the impact of extreme values on decision results. The method improves and deepens fuzzy decision theory and gives specialists technical direction for resolving GSS issues.

The remaining portions of this work are divided into several parts. The literature review component is given in Section 2. Section 3 summarizes and examines the current green supplier evaluation criteria before proposing a new set of criteria that consider social issues. Section 4 provides the definitions and properties of FFPHM and FFWPHM operators, while the fundamentals of FFSs, power average (PA), and Heronian mean (HM) operators are discussed in the Appendix A. In Section 5, an extended BWM based on the entropy measure is introduced along with the EWM. In an FF environment, the unique procedure of MAGDM based on an integrated EDAS is described. Through a case study of the GSS, Section 6 offers a sensitivity and comparison analysis to show the viability of the suggested strategy. The complete text is finally summarized in Section 7.

2. Literature Review

2.1. FFSs

Since Senapati first proposed the FFSs, scholars have recommended several aggregation operators to aggregate the FF information. Zeng et al. [23] introduced a FF Dombi-weighted partitioned Muirhead mean operator and used it to aggregate Fermatean fuzzy numbers (FFNs) for evaluating the quality of online instructions while considering the complex correlation of attributes and calculation flexibility. Wei et al. [24] described Schweizer–Sklar algorithms for FFNs and created an FF Schweizer–Sklar weighted average operator; the operator has some flexibility because of its parameters. Similarly, Tan et al. [25] presented FF frank aggregation operators and operational principles of FFNs to increase the flexibility of fusing information. Mishra and Rani [26] created a FF-weighted aggregated sum product assessment (WASPAS) with the use of a unique score and entropy function to strategically support the Indian government's choice of appropriate medical waste disposal sites. The integration of FFSs with the additive ratio assessment (ARAS) and VIKOR methods by Gül [27] in response to the global health crisis allowed for the proper selection of laboratories for performing health testing because of the stark differences between the two approaches.

The above analysis clearly shows that the FFSs have been cleverly used in combination with decision methods to solve decision problems in several domains. However, the use of EDAS to solve GSS problems in the FF environment has not been studied so far.

2.2. Power Heronian Mean Aggregation Operators

Owing to cognitive bias or personal preferences, DMs in a MAGDM process may give certain irrational assessment values (maximum or minimum). Yager [28] suggested the PA operator consider the degree of support between the data to lessen the effects of erroneous data by considering the integrity between the data. Many academics have conducted studies and consider academic promotion on this topic. Researchers have extended the PA operator to various fuzzy environments to aggregate information in various fuzzy contexts.

In addition, there are often cases in which attributes are correlated but cannot be solved by the PA operator in MAGDM problems. To effectively address interrelated issues, Beliakov et al. [29] introduced the HM operator to handle similar challenges efficiently. The HM operator was subsequently expanded upon by academics. Many academics merged the PA and HM operators to propose the PHM operator, which successfully reduced the detrimental effects of linked relationships while considering the interconnected relationships and integrity of the data. Shi et al. [30] developed a power geometry Heronian average operator in an intuitionistic fuzzy environment and provided related theorems and properties. Liu et al. [31] proposed a linguistic neutrosophic PHM operator.

It is observed that many scholars have recognized that PHM operators can compensate for the shortcomings of PA and HM operators, and they are widely extended to linguistic and fuzzy sets. However, no studies using PHM operators to integrate FF information have been found.

2.3. BWM and EWM for Attribute Weights

Attribute-confirmation methods are also significantly important in MAGDM problems with unknown attribute weights. Common weight determination methods include subjective and objective weight determination methods. However, owing to the limitations of both methods, integrated weights are often used to determine attribute weights. Since the BWM introduced by Rezaei [32] offers particular benefits when deciding on subjective weights, it is often combined with the EWM to calculate the integrated weights more comprehensively. In different fuzzy environments, research using the integrated weighting method to determine evaluation attribute weights has emerged. To solve MAGDM problems with IFSs, an integrated VIKOR model to select the best biowaste recycling channel was proposed by Liu et al. [33]. To examine the various relevance of probable factors that affect GSS, Wei et al. [24] presented a fuzzy entropy suitable for the FF environment and fused it with the traditional BWM to obtain an extended BWM. Ma et al. [34] combined the subjective and objective weights using a multiplicative integration method. An integrated model in a probabilistic linguistic fuzzy environment was proposed to evaluate online recycling platforms. The integrated weights that incorporate customer value and economic goals were determined using a combination assignment approach by Feng et al. [35] in a rough set that evaluates fuzzy information without membership functions.

Although the integrated assignment method based on BWM and EWM has received little attention, the entropy measures involved in [24] make the computation more complicated.

2.4. Evaluation Methods for GSS

There is abundant research on the evaluation methods of GSS. Wang et al. [21] suggested a complex AHP to enable DMs to measure and choose the best cooperative green suppliers in light of the vagueness of appropriate analysis information. A double-hierarchy hesitant fuzzy linguistic set was given the TODIM (an acronym in Portuguese for interactive and multi-criteria decision making) treatment by Krishankumar et al. [36] to tackle the GSS problem. After using AHP to determine the index weight, Nguyen et al. [18] used VIKOR to rank the green suppliers to be evaluated and finally gave the optimal solution.

Xiong et al. [37] set elasticity as an indicator and upgraded the WASPAS to determine a solution to the multi-attribute selection problem of green suppliers based on IFSs. Liu et al. [38] utilized the BWM to assign weights to criteria based on experts' knowledge levels and familiarity with the problem. The study of a supplier selection method on the basis of a q-rung orthopair fuzzy set (QROFS) encompasses the ambiguity of the selection process. Considering the uncertainty of the selection process, a supplier selection method based on QROFS has been studied. Considering environmental criteria in traditional supply chains, Çalık [22] integrated fuzzy AHP and TOPSIS to select the most appropriate green cooperative supplier for the firm. Baki [39] employed ARAS to help with green supplier choice after investigating and determining the elements impacting GSS. Zhang et al. [40] designed an EDAS-based decision framework for optimal GSS in a picture-fuzzy environment. Zhang et al. [41] combined EDAS and cumulative prospect theory to evaluate community group purchase platforms in a probabilistic linguistic environment. Mishra et al. [42] used EDAS to select sustainable third-party reverse logistics providers. He et al. [43] extended the EDAS method to probabilistic uncertain linguistic sets to examine its applicability to GSS.

Scholars have been searching for decision models that can identify the best green suppliers. Compared with TOPSIS and VIKOR, EDAS has higher efficiency and less workload. The FFWA operator involved in the group decision of GSS [42] is prone to the loss of integrated information, which may prevent EDAS from obtaining a reasonable ranking. In summary, there is a lack of a hybrid GSS evaluation framework integrating the FFPHM operator, BWM, EWM, and EDAS.

3. Evaluation Index System for GSS

An upper-tier green supplier may support an organization's long-term expansion. The best GSS is crucial to operating a green supply chain. Contrary to typical supplier selection, the GSS should be based on both the economic and environmental benefits they may provide. It is essential to understand how to build a collection of powerful index systems. Therefore, this section presents a collection of index systems that combine environmental and economic criteria to lay the foundation for a reasonable GSS.

Although numerous studies have been conducted on choosing green suppliers, each study used a different index system. Nguyen et al. [18] set 12 sub-criteria to evaluate and select green suppliers based on five aspects: quality, cost, transportation, technology, and environment. Tavana et al. [19] proposed evaluation criteria for the best green tire recycling supplier based on the environmental dimensions, including green products, pollution output and control, and environmental management. Fazlollahtabar et al. [20] constructed a set of index systems containing eight primary elements and thirty-one sub-criteria for the GSS. To save resources and reduce pollution, Wang et al. [21] selected an optimal green supplier for Vietnamese steel manufacturing companies by considering five aspects: price, quality, transportation, service, and environment. In a study by Çalık [22], transportation, pollution control, product quality, and environmental responsibility were evaluated by different departments of the company to select optimal green suppliers. For the green limestone supplier selection, Krishankumar et al. [36] set up six benefit-type criteria containing green products and impressions and three cost-type criteria containing pollution and costs. Xiong et al. [37] determined the optimal elastic green supplier by evaluating green, elasticity, and coincidence attributes. Liu et al. [38] employed the BWM to determine the importance of the five attributes of product quality, green design, price, organizational and transportation capacity, and the supplier's environmentally friendly cooperative culture, in which product quality, transportation, and organizational capabilities were selected as the best and worst attributes, respectively. Baki [39] explored the factors that influence GSS and established eight influencing factors to be tested in three dimensions—classical, social, and environmental. The results showed that quality, social responsibility, service, cost, and green products were key variables impacting the GSS. For the textile enterprises, five primary characteristics were included in the index systems proposed by Xu et al. [44]:

cost, quality, delivery, partnership, and environmental management. Wu et al. [45] selected a MACDM method for GSS of electric vehicle charging equipment and constructed 12 sub-criteria from five levels of cost, quality, delivery, technology, and environment. Kang et al. [46] proposed seven main criteria, which include quality, cost, service cooperation, stability ability, green environment, green development, and green competition, for analyzing and choosing environmentally friendly suppliers in papermaking enterprises covering the three dimensions. For optimal GSS, Gegovska et al. [47] identified four classical criteria—quality, cost, transport, and service—and three environmental criteria, namely pollution control, green products, and environmental management. To reduce air pollution and select appropriate green suppliers for the construction of raw materials, Krishankumar et al. [48] established three benefit-type criteria, including product delivery, quality, and green design, and two cost-type criteria, including total cost, energy, and resource utilization. Table 1 summarizes the criteria in the GSS literature mentioned above and counts their frequency of occurrence.

Table 1. Evaluation criteria for the GSS.

Evaluation Criteria	[18]	[19]	[20]	[21]	[22]	[36]	[37]	[38]	[39]	[44]	[45]	[46]	[47]	[48]	Occurrence Percentage
Green design	√				√	√	√	√	√		√		√	√	64.29%
Service	√			√	√				√			√	√	√	50.00%
Green image	√				√	√	√					√	√		42.86%
Quality	√		√	√	√			√	√	√	√	√	√	√	78.57%
Environmental management	√	√	√	√		√			√	√	√		√		64.29%
Green product	√	√							√			√			28.57%
Delivery	√		√	√	√			√	√	√	√	√		√	71.43%
Cost	√		√	√		√		√	√	√	√	√	√	√	78.57%
Technology			√	√					√		√	√	√		42.86%
Pollution control		√	√		√	√	√		√		√				50.00%
Energy resource utilization				√	√						√		√	√	35.71%
Social responsibility						√					√				14.29%
Cooperation							√	√	√				√		28.57%

From the literature review and the summary in Table 1, we find that cost (78.57%), quality (78.57%), delivery (71.43%), green design (64.29%), and environment management (64.29%) are the most frequently occurring indicators. Many studies place a high priority on them. Although the attribute "social responsibility" occurs only twice, the research by Baki [39] exactly showed that among the eight factors to be tested, social responsibility had a significant impact on GSS and could not be overlooked.

Hence, following the principles of wholeness, scientificity, and representativeness in selecting indicators, as well as focusing on the frequency of citations, this study combines classic attributes with green standards and social factors to establish a relatively complete green supplier evaluation index system. Environmental indicators are divided into environmental management and green design. Implementing environmental management and green product design at all levels of the supply chain helps meet consumers' environmental protection demands and achieve environmental benefit targets. Economic indicators include quality, cost, and delivery. Economic efficiency is the primary goal of enterprises and is a necessary prerequisite for achieving environmental and social benefits. Enterprises can improve economic efficiency by improving product quality, reducing costs, and shortening delivery times. Social responsibility is a social indicator. Companies should assume social responsibility and contribute to society. The achievement of social benefit objectives drives the sustainability of a company.

Environmental management (Λ_1): To minimize potential sources of pollution at all stages of production, green suppliers should formulate corresponding policies and plans to form a complete environmental management system [36]. The implementation of the policy and its continuous monitoring can be examined.

Green design (Λ_2): This is mainly measured by product design, including the total number of environment-friendly products, the use of energy-saving and consumption-reducing technologies, and the recovery and recycling of waste equipment [22,38].

Quality (Λ_3): Green suppliers should have a complete quality assurance system. Only by providing qualified products can the economic benefits of an enterprise be realized. Enterprises can assess the quality of products based on quality inspection pass rates and durability [45]. A higher quality inspection pass rate indicates better product quality, and excellent products have a lower failure rate during the warranty period [46].

Cost (Λ_4): Whether an enterprise can minimize cost and maximize profit margin depends on the price of the product provided by the supplier [21]. The price that depends on the cost can be measured by the transportation, environmental governance, and product research and development costs of green suppliers [18].

Delivery (Λ_5): Green suppliers should respond to order demands on time. Knowing the historical supply performance of green suppliers helps determine whether they have sufficient resources and production capacity to ensure the on-time delivery of products. It can also examine whether green suppliers have the flexibility to adjust their quantity and delivery time [18,47]. If the ordered products required by enterprises are provided within the shortest production cycle, it would be helpful to establish a long-term good partnership with them and help both parties to pre-empt fierce market competition.

Social responsibility (Λ_6): From a social perspective, green suppliers' social responsibility can be measured in aspects such as employee welfare [18], vocational training, and workplace mental health concerns. Understanding the compliance with laws and regulations, credit behavior, and administrative penalty records of green suppliers is also helpful.

4. Fermatean Fuzzy Power Heronian Mean Aggregation Operators

We will review the relevant knowledge of FFSs, PAs, and HM operators in preparation for suggesting new operators in this section. The definitions of PA and HM operators and the basics of FFSs are listed in Appendix A.

The PA aggregation method focuses on the integrity of the data [28], whereas the HM operator can help solve the problem of data correlation [29]. Using the benefits of PA and HM operators as a starting point, the concept of FFPHM is given below. The FFPHM operator is introduced for aggregating FF information in this study, which focuses on both integrity and correlation.

Definition 1. *Let $\mathfrak{A}_i = (u_i, v_i)(i = 1, 2, \ldots, \lambda)$ be a set of FFNs, where $\xi, \zeta \geq 0$ and $\Psi_i = (1 + \sigma(\mathfrak{A}_i)) / \sum_{t=1}^{\lambda} (1 + \sigma(\mathfrak{A}_t))$. Then:*

$$FFPHM(\mathfrak{A}_1, \mathfrak{A}_2, \ldots, \mathfrak{A}_\lambda) = \left(\frac{2}{\lambda(\lambda+1)} \bigoplus_{i=1,i=j}^{\lambda} \left((\lambda \Psi_i \mathfrak{A}_i)^\xi \otimes (\lambda \Psi_j \mathfrak{A}_j)^\zeta \right) \right)^{\frac{1}{\xi+\zeta}} \tag{1}$$

is called FFPHM operator, where

$$\Omega(\mathfrak{A}_i, \mathfrak{A}_j) = 1 - d(\mathfrak{A}_i, \mathfrak{A}_j), \tag{2}$$

$$\sigma(\mathfrak{A}_i) = \sum_{j=1, j \neq i}^{\lambda} \Omega(\mathfrak{A}_i, \mathfrak{A}_j). \tag{3}$$

Theorem 1. *Let $\mathfrak{A}_i = (u_i, v_i)(i = 1, 2, \ldots, \lambda)$ be a set of FFNs where $\xi, \zeta \geq 0$. Hence, the aggregation result obtained by the FFPHM operator is also a FFN and can be described as:*

$$FFPHM(\mathfrak{A}_1, \mathfrak{A}_2, \ldots, \mathfrak{A}_\lambda) = (\mu, v). \tag{4}$$

$$\text{where } \mu = \left(1 - \left(\prod_{i=1,j=i}^{\chi}\left(1 - \left(1 - \left(1 - u_i^3\right)^{\chi\Psi_i}\right)^{\zeta}\left(1 - \left(1 - u_j^3\right)^{\chi\Psi_j}\right)^{\zeta}\right)\right)^{\frac{2}{\chi(\chi+1)}}\right)^{\frac{1}{3(\xi|\zeta)}} \text{ and}$$

$$v = \sqrt[3]{1 - \left(1 - \left(\prod_{i=1,j=i}^{\chi}\left(1 - \left(1 - v_i^{3\chi\Psi_i}\right)^{\zeta}\left(1 - v_j^{3\chi\Psi_j}\right)^{\zeta}\right)\right)^{\frac{2}{\chi(\chi+1)}}\right)^{\frac{2}{\xi+\zeta}}}.$$

Following Theorem 1, we will propose and demonstrate the properties of the FFPHM operator, laying the groundwork for its use.

Property 1 (Idempotency). *Let* $\mathfrak{A}_i = (u_i, v_i)(i = 1, 2, \ldots, \chi)$ *be a set of FFNs and* $\mathfrak{A}_1 = \mathfrak{A}_2 = \ldots = \mathfrak{A}_\chi = \mathfrak{A}$, *then:*

$$FFPHM(\mathfrak{A}_1, \mathfrak{A}_2, \ldots, \mathfrak{A}_\chi) = \mathfrak{A}.$$

Proof.

$$FFPHM(\mathfrak{A}_1, \mathfrak{A}_2, \ldots, \mathfrak{A}_\chi) = \left(\frac{2}{\chi(\chi+1)}\bigoplus_{i=1,i=j}^{\chi}\left(\left(\chi\Psi_i\mathfrak{A}_i\right)^{\xi}\otimes\left(\chi\Psi_j\mathfrak{A}_j\right)^{\zeta}\right)\right)^{\frac{1}{\xi+\zeta}}$$

$$= \left(\frac{2}{\chi(\chi+1)}\bigoplus_{i=1,i=j}^{\chi}\left(\left(\chi\frac{(1+\sigma(\mathfrak{A}))}{\sum_{t=1}^{\chi}(1+\sigma(\mathfrak{A}))}\mathfrak{A}\right)^{\xi}\otimes\left(\chi\frac{(1+\sigma(\mathfrak{A}))}{\sum_{t=1}^{\chi}(1+\sigma(\mathfrak{A}))}\mathfrak{A}\right)^{\zeta}\right)\right)^{\frac{1}{\xi+\zeta}}.$$

$$= \left(\frac{2}{\chi(\chi+1)}\bigoplus_{i=1,i=j}^{\chi}\left(\mathfrak{A}^{\xi+\zeta}\right)\right)^{\frac{1}{\xi+\zeta}} = \mathfrak{A}$$

$\square$

Property 2 (Commutativity). *Let* $\mathfrak{A}_i = (u_i, v_i)(i = 1, 2, \ldots, \chi)$ *be a set of FFNs and* $\mathfrak{M}_1, \mathfrak{M}_2, \ldots, \mathfrak{M}_\chi$ *be a random permutation of* $\mathfrak{A}_1, \mathfrak{A}_2, \ldots, \mathfrak{A}_\chi$, *then,*

$$FFPHM(\mathfrak{A}_1, \mathfrak{A}_2, \ldots, \mathfrak{A}_\chi) = FFPHM(\mathfrak{M}_1, \mathfrak{M}_2, \ldots, \mathfrak{M}_\chi).$$

Proof. Let $\psi_i = (1 + \sigma(\mathfrak{M}_i)) / \sum_{t=1}^{\chi}(1 + \sigma(\mathfrak{M}_t))$, then

$$FFPHM(\mathfrak{A}_1, \mathfrak{A}_2, \ldots, \mathfrak{A}_\chi)$$

$$= \left(\frac{2}{\chi(\chi+1)}\bigoplus_{i=1,i=j}^{\chi}\left(\left(\chi\Psi_i\mathfrak{A}_i\right)^{\xi}\otimes\left(\chi\Psi_j\mathfrak{A}_j\right)^{\zeta}\right)\right)^{\frac{1}{\xi+\zeta}}$$

$$= \left(\frac{2}{\chi(\chi+1)}\bigoplus_{i=1,i=j}^{\chi}\left(\left(\chi\Psi_i\mathfrak{M}_i\right)^{\xi}\otimes\left(\chi\Psi_j\mathfrak{M}_j\right)^{\zeta}\right)\right)^{\frac{1}{\xi+\zeta}}.$$

$$= FFPHM(\mathfrak{M}_1, \mathfrak{M}_2, \ldots, \mathfrak{M}_\chi)$$

$\square$

Property 3 (Boundedness). *Let* $\mathfrak{A}_i = (u_i, v_i)(i = 1, 2, \ldots, \lambda)$ *be a set of FFNs and* $f_i = \lambda \dfrac{(1+\sigma(\mathfrak{A}_i))}{\sum\limits_{t=1}^{\lambda}(1+\sigma(\mathfrak{A}_t))}\mathfrak{A}_i$, *where* $f^+ = \max\limits_{i}(f_i)$ *and* $f^- = \min\limits_{i}(f_i)$. *Then*

$$f^- \leq FFPHM(\mathfrak{A}_1, \mathfrak{A}_2, \ldots, \mathfrak{A}_\lambda) \leq f^+. \tag{5}$$

When aggregating FF decision information, it is often necessary to consider the importance of attributes. Therefore, we provide the concept of an FFWPHM operator as below.

Definition 2. *Let* $\mathfrak{A}_i = (u_i, v_i)(i = 1, 2, \ldots, \lambda)$ *be a set of FFNs, where* $\xi, \zeta \geq 0$. *The weighted vector is* $\varphi = (\varphi_1, \varphi_2, \ldots, \varphi_\lambda)^T$, *where* $\varphi_i \in [0, 1]$ *and* $\sum\limits_{i=1}^{\lambda} \varphi_i = 1$. *Then,*

$$FFWPHM(\mathfrak{A}_1, \mathfrak{A}_2, \ldots, \mathfrak{A}_\lambda)$$

$$= \left(\frac{2}{\lambda(\lambda+1)} \bigoplus_{i=1, i=j}^{\lambda} \left(\left(\frac{\lambda\varphi_i(1+\sigma(\mathfrak{A}_i))}{\sum\limits_{t=1}^{\lambda} \varphi_t(1+\sigma(\mathfrak{A}_t))}\mathfrak{A}_i \right)^{\xi} \otimes \left(\frac{\lambda\varphi_j(1+\sigma(\mathfrak{A}_j))}{\sum\limits_{t=1}^{\lambda} \varphi_t(1+\sigma(\mathfrak{A}_t))}\mathfrak{A}_j \right)^{\zeta} \right) \right)^{\frac{1}{\xi+\zeta}}$$

is called the FFWPHM operator, where $\sigma(\mathfrak{A}_i) = \sum\limits_{j=1, j\neq i}^{\lambda} \varphi_j \Omega(\mathfrak{A}_i, \mathfrak{A}_j)$ *and* $\Omega(\mathfrak{A}_i, \mathfrak{A}_j)$ *have the same definition.*

Theorem 2. *Let* $\mathfrak{A}_i = (u_i, v_i)(i = 1, 2, \ldots, \lambda)$ *be a set of FFNs, where* $\xi, \zeta \geq 0$. *The weighted vector is* $\varphi = (\varphi_1, \varphi_2, \ldots, \varphi_\lambda)^T$, *where* $\varphi_i \in [0, 1]$ *and* $\sum\limits_{i=1}^{\lambda} \varphi_i = 1$. *Then, the aggregation value obtained by the FFWPHM operator is also an FFN.*

The FFWPHM shares the equivalent characteristics of the FFPHM operator. Since their properties are identical to those for Theorem 1 and Properties 1–3, correspondingly, the proofs are omitted.

5. Fermatean Fuzzy MAGDM Model Based on the Integrated EDAS Method

In this section, we outlined the precise phases of the suggested comprehensive MAGDM model.

5.1. Integrated Weight Based on BWM and Fermatean Fuzzy Entropy

(1) Objective weight determination based on the EWM.

Let $R = (h_{ij})_{\hbar \times \lambda}$ be the comprehensive evaluation matrix of the scheme sets $\mathbb{F} = \{\mathbb{F}_1, \mathbb{F}_2, \ldots, \mathbb{F}_\hbar\}$ normalized under the criterion $\Lambda = \{\Lambda_1, \Lambda_2, \ldots, \Lambda_\lambda\}$. Then, the objective weight ϕ_j for attribute $\Lambda_j(j = 1, 2, \ldots, \lambda)$ can be calculated as:

$$\phi_j = \frac{1 - \widetilde{E}_j}{\lambda - \sum\limits_{j=1}^{\lambda} \widetilde{E}_j}, \tag{6}$$

where $\widetilde{E}_j = \frac{1}{\hbar} \sum\limits_{i=1}^{\hbar} E(h_{ij})$. $E(h_{ij})$ is the entropy value of h_{ij} and is calculated as follows [49]:

$$E(h_{ij}) = 1 - \left[\left(u_{ij}^3 - v_{ij}^3 \right) \left(u_{ij}^3 + v_{ij}^3 \right) \right]^2. \tag{7}$$

Therefore, according to the FF entropy measure calculation formula, we can obtain the objective weights $\phi_j = (\phi_1, \ldots, \phi_\chi)^T$, where $\phi_j \in [0, 1]$ and $\sum_{j=1}^{\chi} \phi_j = 1$.

(2)   Subjective weight determination method based on the BWM.

BWM significantly reduces errors that may be caused by the objective weights. Compared with AHP, BWM has fewer comparisons, less bias, and higher agreement. The BWM is widely used owing to its excellent characteristics. The specific steps of the BWM for determining the subjective weight are as follows.

Step 1. Choose the best Λ_B and worst attributes Λ_W in the attribute collection $\{\Lambda_1, \Lambda_2, \Lambda_3, \ldots, \Lambda_\chi\}$.

Step 2. Build comparison vectors $BO = (B_{B1}, B_{B2}, \ldots, B_{B\chi})$ and $OW = (W_{1W}, W_{2W}, \ldots, W_{\chi W})$, where B_{Bj} and $W_{j\chi}\,(j = 1, \ldots, \chi)$ represent the preference of the best attribute Λ_B over other attributes, and them over the worst attribute Λ_W. B_{Bj} and W_{jW} are expressed in FFNs.

Step 3. Compute the entropy values $E(B_{Bj})$ and $E(B_{jW})$ of B_{Bj} and W_{jW} based on Equation (7) and obtain the preference matrices EBO and EOW:

$$EBO = (E(B_{B1}), E(B_{B2}), \ldots, E(B_{B\chi})), \tag{8}$$

$$EOW = (E(W_{1W}), E(W_{2W}), \ldots, E(W_{\chi W})). \tag{9}$$

Step 4. The FF entropy measure is used to build the following BWM solution weight model:

$$\min x$$
$$s.t \begin{cases} \left| \frac{\varphi_B}{\varphi_B + \varphi_j} - E(B_{Bj}) \right| \leq x \\ \left| \frac{\varphi_j}{\varphi_j + \varphi_W} - E(W_{jW}) \right| \leq x \\ \sum_{j=1}^{\chi} \varphi_j = 1 \\ \varphi_j \geq 0 \end{cases}, \tag{10}$$

Equation (10) can be changed subsequently to the formula below:

$$\min x_1$$
$$s.t \begin{cases} \left| \varphi_B - (\varphi_B + \varphi_j) \times E(B_{Bj}) \right| \leq x_1 \\ \left| \varphi_j - (\varphi_j + \varphi_W) \times E(W_{jW}) \right| \leq x_1 \\ \sum_{j=1}^{\chi} \varphi_j = 1 \\ \varphi_j \geq 0 \end{cases}. \tag{11}$$

By using the LINGO 18.0 software, we can easily obtain the subjective weights $\varphi_j = (\varphi_1, \varphi_2, \ldots, \varphi_\chi)^T$.

(3)   Integrated weight determination method based on the BWM and EWM.

According to the BWM and EWM, the subjective $\varphi_j = (\varphi_1, \varphi_2, \ldots, \varphi_\chi)^T$ and objective weights $\phi_j = (\phi_1, \phi_2, \ldots, \phi_\chi)^T$ are calculated. Therefore, the integrated weights can be calculated using the Equation (12):

$$\omega_j = \phi_j \varphi_j / \sum_{j=1}^{\chi} \varphi_j \phi_j, \tag{12}$$

evidently, $\omega_j \in [0, 1]$ and $\sum_{j=1}^{n} \omega_j = 1$.

5.2. Procedure of Fermatean Fuzzy Integrated EDAS Model

Consider a MAGDM issue that requires the cooperation of FFNs. Let $\mathbb{F} = \{\mathbb{F}_1, \mathbb{F}_2, \ldots, \mathbb{F}_\hbar\}$ denote the discrete set of alternatives, while $\Lambda = \{\Lambda_1, \Lambda_2, \ldots, \Lambda_\chi\}$ denotes the finite set of evaluation attributes. The expert evaluation set is $E = \{e_1, e_2, \ldots, e_h\}$ and corresponding weight is $\lambda = \{\lambda_1, \lambda_2, \ldots, \lambda_h\}$. Assume that the evaluation information of expert $e_k(k = 1, \ldots, h)$ about alternatives $\mathbb{F}_i \in \mathbb{F}$ under the considered attribute $\Lambda_j \in \Lambda$ is denoted by FFN $h_{ij}^k = \left(u_{ij}^k, v_{ij}^k\right)$, where $u_{ij}^k, v_{ij}^k \in [0, 1]$ and $0 \leq \left(u_{ij}^k\right)^3 + \left(v_{ij}^k\right)^3 \leq 1$. Therefore, the FFN evaluation matrix provided by expert $e_k(k = 1, \ldots, h)$ can be expressed as follows:

$$R^k = \left(h_{ij}^k\right)_{\hbar \times \chi} = \begin{pmatrix} h_{11}^k & \cdots & h_{1\chi}^k \\ \vdots & \ddots & \vdots \\ h_{\hbar 1}^k & \cdots & h_{\hbar \chi}^k \end{pmatrix}. \tag{13}$$

Considering the material above, Figure 1 depicts the workflow of the FF integrated EDAS model and the precise procedures for the FF integrated EDAS model are presented below.

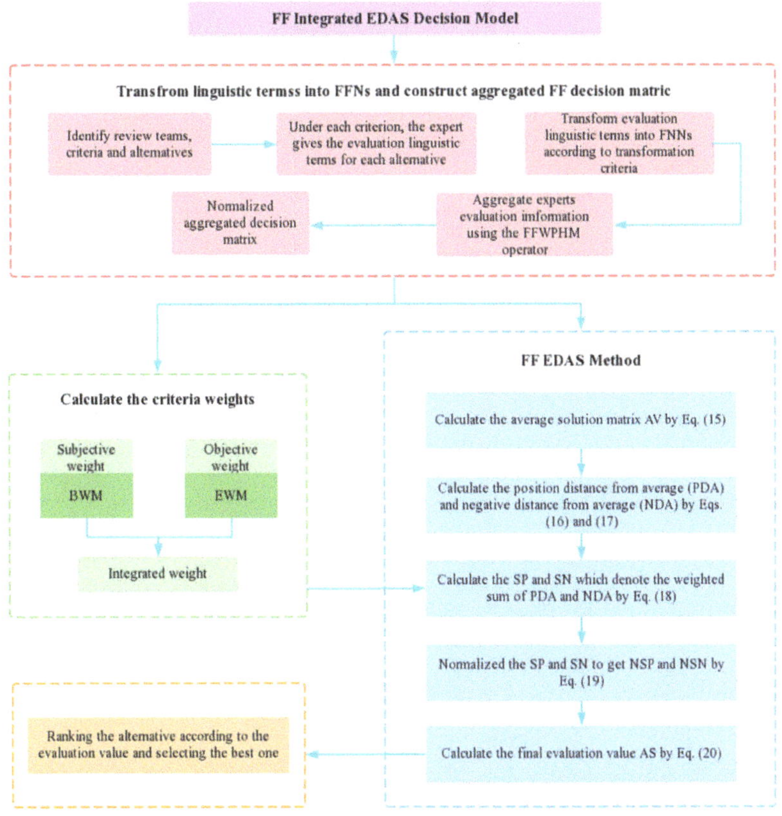

Figure 1. Process of FF's integrated EDAS model.

Step 1. To help experts better express evaluation information, each expert needs to provide personal evaluation linguistic terms for each alternative under each criterion. Using the conversion criteria in Table 2, the evaluated linguistic phrases are reduced to

FFNs. Therefore, the evaluation matrix $R^k = \left(h_{ij}^k \right)_{\hbar \times \chi}$ of the expert $e_k (k = 1, 2, \ldots, h)$ can be obtained, where $h_{ij}^k = \left(u_{ij}^k, v_{ij}^k \right)$.

Table 2. Evaluating linguistic terms and conversion criteria for FFN.

Linguistic Term	FFN
Very Eligible (VE)	$(0.9, 0.2)$
Eligible (E)	$(0.8, 0.3)$
Medium Eligible (ME)	$(0.7, 0.5)$
Medium	$(0.6, 0.6)$
Medium Unqualified (MU)	$(0.5, 0.7)$
Unqualified (U)	$(0.3, 0.8)$
Very Unqualified (VU)	$(0.2, 0.9)$

Step 2. Using the FFWPHM operator proposed in Section 4, the individual evaluations of experts are integrated to obtain a comprehensive decision matrix $\widetilde{R} = \left(\widetilde{h}_{ij} \right)_{\hbar \times \chi}$, where $\widetilde{h}_{ij} = \left(\widetilde{u}_{ij}, \widetilde{v}_{ij} \right)$.

Step 3. Make the decision matrix normalized. Unify attribute types with the conversion criterion (14) to further obtain the standardized decision $R = \left(h_{ij} \right)_{\hbar \times \chi}$, where J_b and J_c are the benefit and cost-type attribute collections.

$$h_{ij} = \left(u_{ij}, v_{ij} \right) = \begin{cases} \left(\widetilde{u}_{ij}, \widetilde{v}_{ij} \right), C_j \in J_b \\ \left(\widetilde{v}_{ij}, \widetilde{u}_{ij} \right), C_j \in J_c \end{cases} \tag{14}$$

Step 4. Determine the average solution matrix $AV = \left[AV_j \right]_{1 \times \chi}$:

$$AV_j = \frac{1}{\hbar} \overset{\hbar}{\underset{i=1}{\oplus}} h_{ij} = \left(\sqrt[3]{1 - \prod_{i=1}^{\hbar} \left(1 - \left(u_{ij} \right)^3 \right)^{\frac{1}{\hbar}}}, \prod_{i=1}^{\hbar} \left(v_{ij} \right)^{\frac{1}{\hbar}} \right) \tag{15}$$

Step 5. Compute the positive distance from the average (PDA) and negative distance from the average (NDA):

$$\left(PDA_{ij} \right)_{\hbar \times \chi} = \frac{\max \left(0, S\left(h_{ij} \right) - S\left(AV_j \right) \right)}{S\left(AV_j \right)}, \tag{16}$$

$$\left(NDA_{ij} \right)_{\hbar \times \chi} = \frac{\max \left(0, S\left(AV_j \right) - S\left(h_{ij} \right) \right)}{S\left(AV_j \right)}, \tag{17}$$

where $S\left(AV_j \right)$ and $S\left(h_{ij} \right)$ are the score values of AV_j and h_{ij} calculated by Equation (A2) in Appendix A.

Step 6. Calculate the subjective weight $\varphi_j = \left(\varphi_j, \ldots, \varphi_\chi \right)^T$ and objective weight $\phi_j = \left(\phi_1, \ldots, \phi_\chi \right)^T$ of the attributes based on the BWM and EWM proposed in Section 5, respectively. Thus, the integrated weight $\omega = \left(\omega_1, \ldots, \omega_\chi \right)^T$ of the attributes according to Equation (12) is obtained.

Step 7. Aggregate the PDA and NDA to get SP_i and SN_i.

$$\begin{cases} SP_i = \sum_{j=1}^{\chi} \omega_j PDA_{ij} \\ SN_i = \sum_{j=1}^{\chi} \omega_j NDA_{ij} \end{cases} \tag{18}$$

Step 8. Normalize SP_i and SN_i.

$$\begin{cases} NSP_i = \dfrac{SP_i}{\max(SP_i)} \\ NSN_i = 1 - \dfrac{SN_i}{\max(SN_i)} \end{cases} \tag{19}$$

Step 9. Calculate the final evaluation value AS_i of the alternative $\mathbb{F}_i$.

$$AS_i = \frac{NSP_i + NSN_i}{2} \tag{20}$$

Step 10. Sort the alternatives on the basis of AS_i. The larger the value of AS_i, the better the alternative $\mathbb{F}_i$ is.

6. Case Study

A business is currently suggesting that the best partner be selected from four environmental suppliers ($\mathbb{F}_1$, $\mathbb{F}_2$, $\mathbb{F}_3$, and $\mathbb{F}_4$). An expert panel consisting of the top management in this company as well as university academics and representatives from research institutes in the fields of supply chain and environmental management is formed. Four experts $e_k(k = 1, 2, 3, 4)$ in total with the weight vector $(0.27, 0.2, 0.3, 0.23)^T$ are invited to assess the four green suppliers according to attributes $\Lambda_j(j = 1, 2, 3, 4, 5, 6)$. Among them, Λ_1 is environmental management, Λ_2 is green design, is quality, Λ_4 is cost Λ_3, Λ_5 is delivery, and Λ_6 is social responsibility. It is easily identifiable that Λ_4 is a cost-type attribute and the rest are benefit-type attributes. The specific process steps are as follows:

Step 1. Create an individual FF assessment matrix. Based on the six constructed attributes above, experts $e_k(k = 1, 2, 3, 4)$ provided evaluations in linguistic terms of the four green suppliers that are observed in Table 3. The linguistic words are then converted into FFNs using the conversion criteria in Table 2. Finally, Table 4 summarizes the evaluation data of the four experts.

Table 3. Expert evaluation of linguistic terms.

Experts	Alternatives	Λ_1	Λ_2	Λ_3	Λ_4	Λ_5	Λ_6
e_1	$\mathbb{F}_1$	E	E	U	U	VE	E
	$\mathbb{F}_2$	M	VU	E	M	U	VE
	$\mathbb{F}_3$	E	E	U	VU	M	U
	$\mathbb{F}_4$	ME	E	E	M	M	U
e_2	$\mathbb{F}_1$	VE	E	U	M	E	ME
	$\mathbb{F}_2$	U	MU	E	ME	M	E
	$\mathbb{F}_3$	E	VE	MU	U	ME	MU
	$\mathbb{F}_4$	VE	VE	E	U	M	E
e_3	$\mathbb{F}_1$	E	E	M	U	E	E
	$\mathbb{F}_2$	U	MU	E	E	M	VE
	$\mathbb{F}_3$	VE	VE	U	U	M	U
	$\mathbb{F}_4$	E	ME	VE	M	E	U
e_4	$\mathbb{F}_1$	VE	ME	U	ME	VE	E
	$\mathbb{F}_2$	U	U	ME	VE	E	E
	$\mathbb{F}_3$	VE	E	U	MU	M	U
	$\mathbb{F}_4$	E	M	E	M	VE	M

Table 4. Expert evaluation information.

Experts	Alternatives	Λ_1	Λ_2	Λ_3	Λ_4	Λ_5	Λ_6
e_1	$\mathbb{F}_1$	(0.8, 0.3)	(0.8, 0.3)	(0.3, 0.8)	(0.3, 0.8)	(0.9, 0.2)	(0.8, 0.3)
	$\mathbb{F}_2$	(0.6, 0.6)	(0.2, 0.9)	(0.8, 0.3)	(0.6, 0.6)	(0.3, 0.8)	(0.9, 0.2)
	$\mathbb{F}_3$	(0.8, 0.3)	(0.8, 0.3)	(0.3, 0.8)	(0.2, 0.9)	(0.6, 0.6)	(0.3, 0.8)
	$\mathbb{F}_4$	(0.7, 0.5)	(0.8, 0.3)	(0.8, 0.3)	(0.6, 0.6)	(0.6, 0.6)	(0.3, 0.8)
e_2	$\mathbb{F}_1$	(0.9, 0.2)	(0.8, 0.3)	(0.3, 0.8)	(0.6, 0.6)	(0.8, 0.3)	(0.7, 0.5)
	$\mathbb{F}_2$	(0.3, 0.8)	(0.5, 0.7)	(0.8, 0.3)	(0.7, 0.5)	(0.6, 0.6)	(0.8, 0.3)
	$\mathbb{F}_3$	(0.8, 0.3)	(0.9, 0.2)	(0.5, 0.7)	(0.3, 0.8)	(0.7, 0.5)	(0.5, 0.7)
	$\mathbb{F}_4$	(0.9, 0.2)	(0.9, 0.2)	(0.8, 0.3)	(0.3, 0.8)	(0.6, 0.6)	(0.3, 0.8)
e_3	$\mathbb{F}_1$	(0.8, 0.3)	(0.8, 0.3)	(0.6, 0.6)	(0.3, 0.8)	(0.8, 0.3)	(0.8, 0.3)
	$\mathbb{F}_2$	(0.3, 0.8)	(0.5, 0.7)	(0.8, 0.3)	(0.8, 0.3)	(0.6, 0.6)	(0.9, 0.2)
	$\mathbb{F}_3$	(0.9, 0.2)	(0.9, 0.2)	(0.3, 0.8)	(0.3, 0.8)	(0.6, 0.6)	(0.3, 0.8)
	$\mathbb{F}_4$	(0.8, 0.3)	(0.7, 0.5)	(0.9, 0.2)	(0.6, 0.6)	(0.8, 0.3)	(0.3, 0.8)
e_4	$\mathbb{F}_1$	(0.9, 0.2)	(0.7, 0.5)	(0.3, 0.8)	(0.7, 0.5)	(0.9, 0.2)	(0.8, 0.3)
	$\mathbb{F}_2$	(0.3, 0.8)	(0.3, 0.8)	(0.7, 0.5)	(0.9, 0.2)	(0.8, 0.3)	(0.8, 0.3)
	$\mathbb{F}_3$	(0.9, 0.2)	(0.8, 0.3)	(0.3, 0.8)	(0.5, 0.7)	(0.6, 0.6)	(0.3, 0.8)
	$\mathbb{F}_4$	(0.8, 0.3)	(0.6, 0.6)	(0.8, 0.3)	(0.6, 0.6)	(0.9, 0.2)	(0.6, 0.6)

Step 2. Four individual assessments are integrated with the use of FFWPHM operator ($\xi = \zeta = 3$) to obtain the collective decision matrix $\widetilde{R} = \left(\widetilde{h}_{ij} \right)_{4 \times 6}$, as summarized in Table 5.

Table 5. Collective decision matrix $\widetilde{R}$.

	Λ_1	Λ_2	Λ_3
$\mathbb{F}_1$	(0.8548, 0.2530)	(0.7926, 0.3503)	(0.5371, 0.7259)
$\mathbb{F}_2$	(0.5200, 0.7350)	(0.4058, 0.7929)	(0.7926, 0.3503)
$\mathbb{F}_3$	(0.8696, 0.2564)	(0.8655, 0.2570)	(0.4027, 0.7794)
$\mathbb{F}_4$	(0.8151, 0.3291)	(0.7946, 0.3963)	(0.8446, 0.2798)

	Λ_4	Λ_5	Λ_6
$\mathbb{F}_1$	(0.6009, 0.6741)	(0.8655, 0.2545)	(0.7960, 0.3454)
$\mathbb{F}_2$	(0.8092, 0.3933)	(0.6887, 0.5574)	(0.8748, 0.2570)
$\mathbb{F}_3$	(0.4238, 0.7869)	(0.6256, 0.5797)	(0.4027, 0.7794)
$\mathbb{F}_4$	(0.6021, 0.6201)	(0.8034, 0.4115)	(0.4948, 0.7478)

Step 3. Make the decision matrix normalized. Since Λ_4 is a cost-type attribute, it is transformed into a benefit-type attribute using Equation (14). Thus, the normalized decision matrix $R = \left(h_{ij} \right)_{4 \times 6}$ has been obtained and represented in Table 6.

Table 6. Normalized decision matrix R.

	Λ_1	Λ_2	Λ_3
$\mathbb{F}_1$	(0.8548, 0.2530)	(0.7926, 0.3503)	(0.5371, 0.7259)
$\mathbb{F}_2$	(0.5200, 0.7350)	(0.4058, 0.7929)	(0.7926, 0.3503)
$\mathbb{F}_3$	(0.8696, 0.2564)	(0.8655, 0.2570)	(0.4027, 0.7794)
$\mathbb{F}_4$	(0.8151, 0.3291)	(0.7946, 0.3963)	(0.8446, 0.2798)

	Λ_4	Λ_5	Λ_6
$\mathbb{F}_1$	(0.6741, 0.6009)	(0.8655, 0.2545)	(0.7960, 0.3454)
$\mathbb{F}_2$	(0.3933, 0.8092)	(0.6887, 0.5574)	(0.8748, 0.2570)
$\mathbb{F}_3$	(0.7869, 0.4238)	(0.6256, 0.5797)	(0.4027, 0.7794
$\mathbb{F}_4$	(0.6201, 0.6021)	(0.8034, 0.4115)	(0.4948, 0.7478)

Step 4. Compute the average solution matrix $AV = \left[AV_j \right]_{1\times 6}$.

$$\left[AV_j \right]_{1\times 6} = \left\langle \begin{array}{l} (0.8070, 0.3539), (0.7746, 0.4101), (0.7178, 0.4852), \\ (0.6618, 0.5935), (0.7710, 0.4289), (0.7333, 0.4769) \end{array} \right\rangle$$

Step 5. Compute the PDA and NDA, as listed in Tables 7 and 8.

Table 7. PDA.

	Λ_1	Λ_2	Λ_3	Λ_4	Λ_5	Λ_6
$\mathbb{F}_1$	0.2642	0.1497	0.0000	0.1067	0.6653	0.6201
$\mathbb{F}_2$	0.0000	0.0000	0.7800	0.0000	0.0000	1.2824
$\mathbb{F}_3$	0.3313	0.5955	0.0000	4.0925	0.0000	0.0000
$\mathbb{F}_4$	0.0511	0.1102	1.2710	0.0000	0.1831	0.0000

Table 8. NDA.

	Λ_1	Λ_2	Λ_3	Λ_4	Λ_5	Λ_6
$\mathbb{F}_1$	0.0000	0.0000	1.8905	0.0000	0.0000	0.0000
$\mathbb{F}_2$	1.5330	2.0905	0.0000	6.8093	0.5956	0.0000
$\mathbb{F}_3$	0.0000	0.0000	2.5965	0.0000	0.8682	2.4275
$\mathbb{F}_4$	0.0000	0.0000	0.0000	0.7508	0.0000	2.0389

Step 6. First, using the BWM suggested in this study, the subjective weights $\varphi_i(i = 1, \ldots, 6)$ of the attributes were determined. Following the advice of the expert panel, the greatest and worst characteristics were Λ_1 and Λ_4, respectively. The preferences for the finest and worst attributes in relation to other attributes were ascertained using FF information to obtain *FFBO* and *FFOW*:

$$FFBO = ((0.5, 0.5), (0.85, 0.25), (0.92, 0.15), (0.9, 0), (0.91, 0.2), (0.93, 0.5))$$

$$FFOW = ((0.9, 0), (0.95, 0.2), (0.85, 0.25), (0.5, 0.5), (0.92, 0.2), (0.95, 0.15)).$$

Calculate the entropy value of each FFN using Equation (7):

$$EBO = (1.000, 0.8579, 0.6323, 0.7176, 0.6776, 0.6014)$$

$$EWO = (0.7176, 0.4597, 0.8579, 1.0000, 0.6324, 0.4597)$$

A linear model of the problem is constructed as follows:

$$\min x_1$$
$$s.t \begin{cases} |\varphi_1 - (\varphi_1 + \varphi_2) \times 0.8579| \le x_1 \\ |\varphi_1 - (\varphi_1 + \varphi_3) \times 0.6323| \le x_1 \\ |\varphi_1 - (\varphi_1 + \varphi_4) \times 0.7176| \le x_1 \\ |\varphi_1 - (\varphi_1 + \varphi_5) \times 0.6776| \le x_1 \\ |\varphi_1 - (\varphi_1 + \varphi_6) \times 0.6014| \le x_1 \\ |\varphi_2 - (\varphi_2 + \varphi_4) \times 0.4597| \le x_1 \\ |\varphi_3 - (\varphi_3 + \varphi_4) \times 0.8579| \le x_1 \\ |\varphi_5 - (\varphi_5 + \varphi_4) \times 0.6324| \le x_1 \\ |\varphi_6 - (\varphi_6 + \varphi_4) \times 0.4597| \le x_1 \\ \varphi_1 + \varphi_2 + \varphi_3 + \varphi_4 + \varphi_5 + \varphi_6 = 1 \\ \varphi_1, \varphi_2, \varphi_3, \varphi_4, \varphi_5, \varphi_6 \ge 0 \end{cases}$$

The subjective weights were then calculated using the LINGO 18.0 software.

Second, the EWM mentioned in Section 5.1 was implemented to calculate the objective weights. Finally, the integrated weights were calculated using Equation (12). Attribute weights of different types were presented in Table 9.

Table 9. Attribute weights of different types.

Method	Λ_1	Λ_2	Λ_3	Λ_4	Λ_5	Λ_6
Subjective weights φ	0.288	0.090	0.223	0.078	0.189	0.132
Objective weights ϕ	0.249	0.201	0.143	0.075	0.142	0.190
Integrated weights ω	0.400	0.100	0.178	0.033	0.150	0.140

Step 7. Using the integrated weights from Table 9 as a guide, aggregate the PDA and NDA to obtain SP_i and SN_i, and then normalize them to get NSP_i and NSN_i. Finally, the final evaluation value AS_i was calculated and the alternatives were ranked. Table 10 provides a summary of the findings.

Table 10. Calculation results and ranking under integrated weights.

	SP_i	SN_i	NSP_i	NSN_i	AS_i	Ranking
$\mathbb{F}_1$	0.3104	0.3384	0.9620	0.7000	0.8310	1
$\mathbb{F}_2$	0.3192	1.1280	0.9890	0.0000	0.4945	4
$\mathbb{F}_3$	0.3227	0.0948	1.0000	0.1712	0.5856	3
$\mathbb{F}_4$	0.2864	0.3095	0.8875	0.7256	0.8066	2

As shown in Table 10, the alternatives are ranked $\mathbb{F}_1 \succ \mathbb{F}_4 \succ \mathbb{F}_3 \succ \mathbb{F}_2$ and $\mathbb{F}_1$ is the optimal alternative.

6.1. Sensitivity Analysis

Attribute weights hold an important position in evaluation and decision results, and effective GSS can be aided by reasonable attribute weights. Subjective weights based on the BWM, objective weights based on the EWM, and integrated weights were then calculated in this study. Here, the impact of EDAS based on different weight types on GSS was analyzed. The results calculated using the subjective and objective weights are listed in Tables 11 and 12, respectively.

Table 11. Calculation results and ranking under subjective weights.

	SP_i	SN_i	NSP_i	NSN_i	AS_i	Ranking
$\mathbb{F}_1$	0.3055	0.4216	0.6524	0.6689	0.6607	2
$\mathbb{F}_2$	0.3432	1.2733	0.7330	0.0000	0.3665	4
$\mathbb{F}_3$	0.4682	1.0635	1.0000	0.1648	0.5824	3
$\mathbb{F}_4$	0.3427	0.3277	0.7319	0.7427	0.7373	1

Table 12. Calculation results and ranking under objective weights.

	SP_i	SN_i	NSP_i	NSN_i	AS_i	Ranking
$\mathbb{F}_1$	0.3115	0.2703	0.6118	0.8103	0.7111	1
$\mathbb{F}_2$	0.2808	1.4252	0.5516	0.0000	0.2758	4
$\mathbb{F}_3$	0.5091	1.8558	1.0000	0.3995	0.6998	2
$\mathbb{F}_4$	0.2512	0.3252	0.4935	0.7717	0.6326	3

The evaluation values and change trends of each alternative under different weight types are shown in Figure 2.

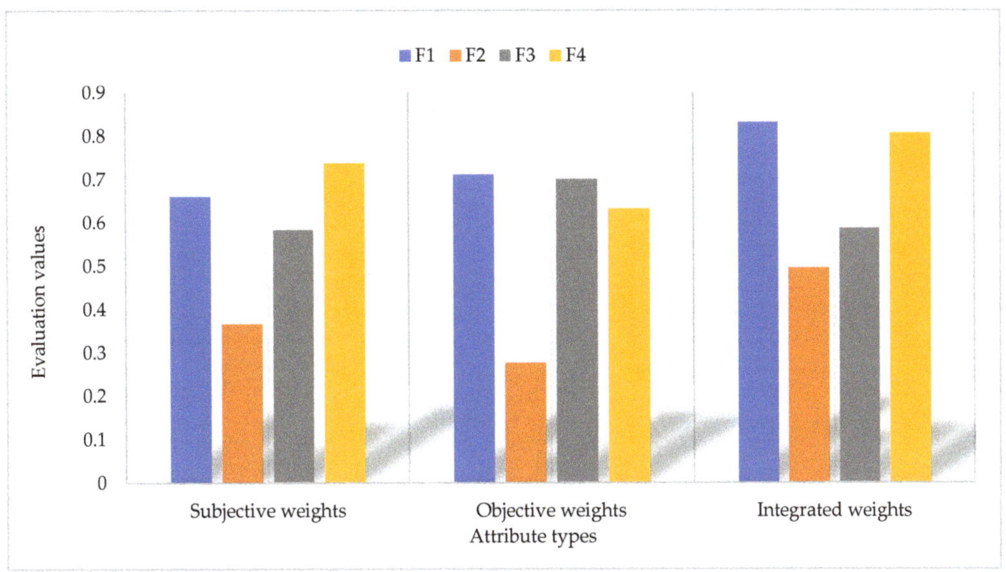

Figure 2. Evaluation values of each alternative under different types of weights.

From Tables 11 and 12 and Figure 2, we can see that the alternative under the subjective weight is ranked $\mathbb{F}_4 \succ \mathbb{F}_1 \succ \mathbb{F}_3 \succ \mathbb{F}_2$, the ranking list under the objective weight is $\mathbb{F}_1 \succ \mathbb{F}_3 \succ \mathbb{F}_4 \succ \mathbb{F}_2$, and the $\mathbb{F}_1 \succ \mathbb{F}_4 \succ \mathbb{F}_3 \succ \mathbb{F}_2$ is under the integrated method. On the basis of these ordered lists, we can see that when the attribute weights are different, the ranking of the schemes differs. There is a slight difference between the objective and integrated weights of the alternative ranking, whereas the subjective and integrated weights are quite different. Therefore, it is impossible to disregard the availability of objective weights while evaluating and choosing green suppliers. Since the objective weight generally rests with objective data, we should focus on both the subjective awareness of evaluation experts and the objectivity of the original data in the evaluation process. Integrating subjective and objective weights leads to more rational decisions.

The decision's effectiveness is directly correlated with the change in parameters. We will conduct sensitivity analyses on several parameters involved in the integrated EDAS group decision model below. In the above numerical study, when aggregating expert evaluation information using the proposed FFWPHM operator, only case $\xi = \zeta = 3$ is considered, which cannot comprehensively demonstrate the stability of the FFWPHM operator and EDAS in evaluating green suppliers. Subsequently, we discuss the selection of the ideal solution for the green suppliers in relation to the adjustment of the parameters ξ and ζ in the FFWPHM operator.

The proposed FFWPHM operator includes two variables, ξ and ζ, thus when those values change, the way in which expert assessment data is integrated will likewise change. Due to the different decision matrices obtained by the aggregation of operators, the integrated weights based on aggregated data also change accordingly. The changes to the optimal GSS achieved by different parameters in ξ and ζ remain to be discussed. We set ξ and ζ as different real numbers and fused the rigorous assessment data. Table 13 lists the ranking outcomes using the EDAS and FFWPHM operator under different parameter combinations.

Table 13. Calculation results under different ξ and ζ.

	AS_1	AS_2	AS_3	AS_4	Ranking
FFWPHM$^{(2,2)}$ − EDAS	0.7856	0.5000	0.4786	0.8019	$\mathbb{F}_4 \succ \mathbb{F}_1 \succ \mathbb{F}_2 \succ \mathbb{F}_3$
FFWPHM$^{(2.5,3)}$ − EDAS	0.8348	0.5033	0.5840	0.8152	$\mathbb{F}_1 \succ \mathbb{F}_4 \succ \mathbb{F}_3 \succ \mathbb{F}_2$
FFWPHM$^{(3,3.5)}$ − EDAS	0.8007	0.4551	0.5986	0.7804	$\mathbb{F}_1 \succ \mathbb{F}_4 \succ \mathbb{F}_3 \succ \mathbb{F}_2$
FFWPHM$^{(3.5,4)}$ − EDAS	0.7722	0.4186	0.6079	0.7558	$\mathbb{F}_1 \succ \mathbb{F}_4 \succ \mathbb{F}_3 \succ \mathbb{F}_2$
FFWPHM$^{(4,4.5)}$ − EDAS	0.7393	0.3778	0.6217	0.7266	$\mathbb{F}_1 \succ \mathbb{F}_4 \succ \mathbb{F}_3 \succ \mathbb{F}_2$
FFWPHM$^{(4.5,5)}$ − EDAS	0.7081	0.3380	0.6383	0.6974	$\mathbb{F}_1 \succ \mathbb{F}_4 \succ \mathbb{F}_3 \succ \mathbb{F}_2$
FFWPHM$^{(5,5)}$ − EDAS	0.6968	0.3228	0.6445	0.6862	$\mathbb{F}_1 \succ \mathbb{F}_4 \succ \mathbb{F}_3 \succ \mathbb{F}_2$
FFWPHM$^{(5.5,5.5)}$ − EDAS	0.6668	0.2822	0.6672	0.6582	$\mathbb{F}_3 \succ \mathbb{F}_1 \succ \mathbb{F}_4 \succ \mathbb{F}_2$
FFWPHM$^{(6,6)}$ − EDAS	0.6374	0.2446	0.6913	0.6309	$\mathbb{F}_3 \succ \mathbb{F}_1 \succ \mathbb{F}_4 \succ \mathbb{F}_2$

From Table 13, it is noted that while ξ and ζ change between $[2.5, 5]$, the scheme's sort remains consistent $\mathbb{F}_1 \succ \mathbb{F}_4 \succ \mathbb{F}_3 \succ \mathbb{F}_2$, which is the same as the result when $\xi = \zeta = 3$. This means that the proposed integrated EDAS decision model in the FF environment has a certain stability under different values of ξ and ζ. It can also be observed that the ranking of the solutions alters as ξ and ζ are beyond the above mentioned range. However, the poorest option is always the alternative $\mathbb{F}_2$, while the best choice changes from $\mathbb{F}_1$ to $\mathbb{F}_3$. In other words, the operator can consistently aggregate decision information while remaining adaptable. To adapt to various decision situations, DMs can adjust the parameters in response to their risk preferences.

6.2. Comparative Analysis

Some current representative decision mechanisms, specifically TOPSIS [17], WASPAS [26], VIKOR [27], and ARAS [27], are adopted for comparative analysis to further verify the viability and applicability of the proposed framework in the FF environment. To guarantee the uniformity of the results, under the situation of $\xi = \zeta = 3$, the weight vector $(0.288, 0.090, 0.223, 0.078, 0.189, 0.132)^T$ computed in this work is incorporated into the calculation procedure of each approach. Tables 14 and 15 display the main computing results and rankings produced by various decision techniques.

Table 14. Main computing results and ranking under EDAS, VIKOR, and ARAS.

	Proposed Integrated EDAS		VIKOR [27]				ARAS [27]		
	AS_i	Ranking	S_i	R_i	Q_i	Ranking	$sc(\mathbb{F}_i)$	K_i	Ranking
$\mathbb{F}_1$	0.6607	2	0.2938	0.1834	0.1588	3	0.4487	0.7358	1
$\mathbb{F}_2$	0.3665	4	0.6155	0.2880	1.0000	1	0.2067	0.3389	4
$\mathbb{F}_3$	0.5824	3	0.5440	0.2230	0.7102	2	0.3141	0.5150	3
$\mathbb{F}_4$	0.7373	1	0.3005	0.1189	0.0121	4	0.4268	0.6998	2

Table 15. Main computing results and rankings under TOPSIS and WASPAS.

	TOPSIS [17]				WASPAS [26]			
	$D(\mathbb{F}_i, \mathbb{F}^+)$	$D(\mathbb{F}_i, \mathbb{F}^-)$	$\varsigma(\mathbb{F}_i)$	Ranking	$\mathbb{C}_i^{(1)}$	$\mathbb{C}_i^{(2)}$	$\mathbb{C}_i$	Ranking
$\mathbb{F}_1$	0.1803	0.2249	−0.0510	2	0.5380	0.4526	0.4953	1
$\mathbb{F}_2$	0.2013	0.2159	−0.2103	3	0.3977	0.2817	0.3397	4
$\mathbb{F}_3$	0.2140	0.2046	−0.3322	4	0.4580	0.2852	0.3716	3
$\mathbb{F}_4$	0.1750	0.2297	0.0000	1	0.5273	0.4613	0.4943	2

As observed by Tables 14 and 15, not only do the computing values computed by the established model differ from the current methods, but also the final green supplier ranking produced is significantly varied. Only the optimal solution obtained by [17] is compatible with the suggested model and $\mathbb{F}_4$ is the best green supplier. Different core ideas of each decision method cause variations in ranking.

By modifying the relationship between the weighted summation and the product model, WASPAS can increase decision accuracy. The core idea of the ARAS approach for choosing the appropriate solution is to identify the weighted decision matrix's optimal function for calculating the utility level of the decision object. The proximity to optimal solutions, both positive and negative, is emphasized by TOPSIS and VIKOR. However, TOPSIS only adds up the distances between negative and positive ideal alternatives, while VIKOR additionally takes into account the relative significance of these distances. The distinguishing characteristic of EDAS is that it assesses options by figuring out their distance from the typical option. With EDAS, the extremely positive and negative ideal solutions can be converted into the average solution, which has a more realistic meaning.

According to the study presented above and the thorough comparison shown in Table 16, compared to the created model, the following drawbacks of the other decision approaches exist:

Table 16. A comprehensive comparison of different approaches.

	Established Model	[17]	[26]	[27]	[27]
Ranking method	EDAS	TOPSIS	WASPAS	VIKOR	ARAS
Decision process	Group	Single	Group	Single	Single
Multiple aggregation strategies	Yes	No	No	No	No
DMs' weights	Assumed	No	Computed	No	No
Criteria weights	Integrated	Assumed	Objective	Assumed	Assumed
Parameters involved	Yes	No	Yes	Yes	No

(1) In regard to the ranking approach, it is not suitable to utilize the closeness degree formula that was finally employed for ranking in [17] when an alternative to being considered is a positive ideal solution. The concept of superior and inferior solutions is transformed by EDAS into a compromise idea, which significantly improves the influence of extreme values on the decision outcome. The FF weighted average (FFWA) operator engaged in research [26,27] may result in information loss and even rank inability when membership or non-membership is equal to zero in the FF environment.

(2) Only simple decision-making environments are covered by [17,27]. Due to insufficient information and poor consideration, a single DM might not be capable of making appropriate decisions. Meanwhile, the introduction of the FFWA operator into the MAGDM by [26] may cause incorrect initial assessment information aggregation. The decision-making model proposed assumes the participation of numerous DMs, and the choice results generated by the group of DMs with their collective wisdom are more practical to implement.

(3) All other approaches engaged in the comparison only focus on the objective data and consider the objective weights of attributes in their investigations but neglect the subjective judgment of DMs, which is a main drawback. Subjective weights are rather realistic and aid in lowering the bias of the results. The integrated weighting technique constructed can measure the importance of attributes more comprehensively and also addresses the unscientific effects brought on by too strong subjective psychology in the calculation.

7. Conclusions

This study constructed a comprehensive series of index systems for GSS, and an innovative MAGDM strategy for selecting the ideal green supplier was created. The proposed model used the FFWPHM operator to aggregate the FF information of the reviewing experts, which overcomes the obstacles of data correlation and incompleteness. A novel BWM based on FF entropy was combined with the EWM to compute the integrated weight of each attribute, which respects the subjective judgment of DMs and relies on objective data. Finally, EDAS was used to evaluate the options, and the numerical analysis shows that option 1 is the most reasonable green supplier. The findings of the sensitivity analysis show that the sequence of alternatives remained the same when the parameters were altered within a certain range, thus demonstrating the robustness of the proposed model. The results of the comparative analysis highlight the limitations of other methods and illustrate the strong applicability of the suggested approach.

This study provides management suggestions for companies and suppliers. In the context of green development nowadays, suppliers should properly reconcile environmental protection with economic efficiency. From the case study, we find that DMs place a high priority on environmental management. So suppliers can improve their competitiveness by upgrading their environmental management capabilities. Enterprises should also create social value while improving their profits. It costs labor and material resources to implement green technology and product innovation in a short period of time, but these investments are beneficial to boost the core competitiveness of enterprises from a development standpoint.

Although our proposed hybrid model can provide applied value for GSS, there are still limitations in the research. Green supplier evaluation involves various indicators in multiple dimensions; however, the constructed index system does not cover all the sub-criteria. Future research may expand or add some other related indicators to build a more scientific and comprehensive index system. This paper discusses the ideal situation where the DMs' weight is known; however, the social interactions between experts can be complex, and the situations where the DMs' weight is unknown exist in the actual decision process. In future research, we can introduce social networks into the calculation of expert weights.

Author Contributions: Writing—original draft preparation, S.Z. and W.C.; methodology, J.G. and E.Z. All authors have read and agreed to the published version of the manuscript.

Funding: This work is supported by the Social Sciences Planning Projects of Zhejiang (21QNYC11ZD).

Data Availability Statement: Not applicable.

Appendix A

Here we will briefly review the fundamentals of FFSs, specifically the basic concepts, operations, and comparison rules between FFNs. The basic concepts of HM and PA operators are reviewed to enable laying of the groundwork for suggesting new operators.

Definition A1 ([17]). *Let X be a non-empty set. An FFS is presented as follows:*

$$F = \{\langle x_j, u_F(x_j), v_F(x_j)\rangle | x_j \in X\}, \tag{A1}$$

where $u : X \to [0,1]$ is the membership function $u_F(x_j)\,(0 \le u_F(x_j) \le 1)$ and $v : X \to [0,1]$ is the non-membership function $v_F(x_j)\,(0 \le v_F(x_j) \le 1)$. For $x_j \in X$, it satisfies the condition $0 \le (u_F(x_j))^3 + (v_F(x_j))^3 \le 1$. If $\pi_F(x_j) = \sqrt[3]{1 - (u_F(x_j))^3 - (v_F(x_j))^3}$, then $\pi_F(x_j)$ is defined to be the indeterminacy of the set F. For clarity and brevity, $F = (u_F, v_F)$ denotes an FFN.

Definition A2 ([17]). *If $\lambda > 0$ exists, let $\mathfrak{A}_1 = (u_1, v_1)$ and $\mathfrak{A}_2 = (u_2, v_2)$ be two FFNs. The algorithms between FFNs are as follows:*

(1) $\mathfrak{A}_1^c = (v_1, u_1)$;

(2) $\mathfrak{A}_1 \oplus \mathfrak{A}_2 = \left(\sqrt[3]{u_1^3 + u_2^3 - u_1^3 u_2^3}, v_1 v_2\right)$;

(3) $\mathfrak{A}_1 \otimes \mathfrak{A}_2 = \left(u_1 u_1, \sqrt[3]{v_1^3 + v_2^3 - v_1^3 v_2^3}\right)$;

(4) $\lambda \mathfrak{A}_1 = \left(\sqrt[3]{1 - (1 - u_1^3)^\lambda}, (v_1)^\lambda\right)$;

(5) $\quad \mathfrak{A}_1^\lambda = \left((u_1)^\lambda, \sqrt[3]{1 - (1 - v_1^3)^\lambda} \right).$

Definition A3 ([17]). *Let* $\mathfrak{A} = (u_{\mathfrak{A}}, v_{\mathfrak{A}})$ *be an FFN; its score and accuracy functions are determined as follows:*

$$O(\mathfrak{A}) = (u_{\mathfrak{A}})^3 - (v_{\mathfrak{A}})^3, \tag{A2}$$

$$\Theta(\mathfrak{A}) = (u_{\mathfrak{A}})^3 + (v_{\mathfrak{A}})^3, \tag{A3}$$

where $O(F) \in [-1, 1]$ *and* $\Theta(F) \in [0, 1]$.

Definition A4 ([17]). *Let* $\mathfrak{A}_1 = (u_1, v_1)$ *and* $\mathfrak{A}_2 = (u_2, v_2)$ *be two FFNs; then,*
(1) *if* $O(\mathfrak{A}_1) > O(\mathfrak{A}_2)$, *then* $\mathfrak{A}_1 > \mathfrak{A}_2$;
(2) *if* $O(\mathfrak{A}_1) = O(\mathfrak{A}_2)$, *then,*

  (a) *if* $\Theta(\mathfrak{A}_1) > \Theta(\mathfrak{A}_2)$, *then* $\mathfrak{A}_1 > \mathfrak{A}_2$;
  (b) *if* $\Theta(\mathfrak{A}_1) = \Theta(\mathfrak{A}_2)$, *then* $\mathfrak{A}_1 = \mathfrak{A}_2$.

Definition A5 ([50]). *Let* $\mathfrak{A}_1 = (u_1, v_1)$ *and* $\mathfrak{A}_2 = (u_2, v_2)$ *be two FFNs; then, the standard Hamming distance between* $\mathfrak{A}_1$ *and* $\mathfrak{A}_2$ *is described below:*

$$d(\mathfrak{A}_1, \mathfrak{A}_2) = \frac{1}{2}\left(\left| u_1^3 - u_2^3 \right| + \left| v_1^3 - v_2^3 \right| + \left| \pi_1^3 - \pi_2^3 \right| \right). \tag{A4}$$

Definition A6 ([29]). *Let* $\mathfrak{A}_i \geq 0 (i = 1, 2, \ldots, \lambda)$ *be the set of real values, where* $\xi, \zeta \geq 0$. *Then,*

$$HM(\mathfrak{A}_1, \mathfrak{A}_2, \ldots, \mathfrak{A}_\lambda) = \left(\frac{2}{\lambda(\lambda + 1)} \sum_{i=1, j=i}^{\lambda} \mathfrak{A}_i^\xi \mathfrak{A}_j^\zeta \right)^{\frac{1}{\xi + \zeta}} \tag{A5}$$

is called the HM operator.

Definition A7 ([28]). *Let* $\mathfrak{A}_i \geq 0 (i = 1, 2, \ldots, \lambda)$ *be the set of real numbers. Then the PA operator is defined as:*

$$PA(\mathfrak{A}_1, \mathfrak{A}_2, \ldots, \mathfrak{A}_\lambda) = \sum_{i=1}^{\lambda} \frac{1 + \sigma(\mathfrak{A}_i)}{\sum\limits_{t=1}^{\lambda} (1 + \sigma(\mathfrak{A}_t))}, \tag{A6}$$

where $\Omega(\mathfrak{A}_i, \mathfrak{A}_j)$ *denotes the support of* $\mathfrak{A}_i$ *and* $\mathfrak{A}_j$, *satisfying the properties mentioned below:*
(1) $\Omega(\mathfrak{A}_i, \mathfrak{A}_j) \in [0, 1]$;
(2) $\Omega(\mathfrak{A}_i, \mathfrak{A}_j) = \Omega(\mathfrak{A}_j, \mathfrak{A}_i)$;
(3) *If* $\left| \mathfrak{A}_i - \mathfrak{A}_j \right| \leq \left| \mathfrak{A}_k - \mathfrak{A}_l \right|$, *then* $\Omega(\mathfrak{A}_i, \mathfrak{A}_j) \geq \Omega(\mathfrak{A}_k, \mathfrak{A}_l)$.

References

1. Wang, D.W.; Li, J. Coastal haze pollution, economic and financial performance, and sustainable transformation in coastal cities. *J. Coast. Res.* **2020**, *109*, 1–7. [CrossRef]
2. Ahmed, Z.; Asghar, M.M.; Malik, M.N.; Nawaz, K. Moving towards a sustainable environment: The dynamic linkage between natural resources, human capital, urbanization, economic growth, and ecological footprint in China. *Resour. Policy* **2020**, *67*, 101677. [CrossRef]
3. Zhang, K.M.; Wen, Z.G. Review and challenges of policies of environmental protection and sustainable development in China. *J. Environ. Manag.* **2008**, *88*, 1249–1261. [CrossRef] [PubMed]
4. Tseng, C.H.; Chang, K.H.; Chen, H.W. Strategic orientation, environmental innovation capability, and environmental sustainability performance: The case of Taiwanese suppliers. *Sustainability* **2019**, *11*, 1127. [CrossRef]
5. Cabral, I.; Grilo, A.; Cruz-Machado, V. A decision-making model for lean, agile, resilient and green supply chain management. *Int. J. Prod. Res.* **2012**, *50*, 4830–4845. [CrossRef]

6. Hervani, A.A.; Helms, M.M.; Sarkis, J. Performance measurement for green supply chain management. *Benchmarking Int. J.* **2005**, *12*, 330–353. [CrossRef]
7. Gupta, H.; Barua, M.K. Supplier selection among SMEs on the basis of their green innovation ability using BWM and fuzzy TOPSIS. *J. Clean. Prod.* **2017**, *152*, 242–258. [CrossRef]
8. Rajesh, R.; Ravi, V. Supplier selection in resilient supply chains: A grey relational analysis approach. *J. Clean. Prod.* **2015**, *86*, 343–359. [CrossRef]
9. Zhu, W.; He, Y. Green product design in supply chains under competition. *Eur. J. Oper. Res.* **2017**, *258*, 165–180. [CrossRef]
10. Boran, F.E.; Genç, S.; Kurt, M.; Akay, D. A multi-criteria intuitionistic fuzzy group decision making for supplier selection with TOPSIS method. *Expert Syst. Appl.* **2009**, *36*, 11363–11368. [CrossRef]
11. Govindan, K.; Rajendran, S.; Sarkis, J.; Murugesan, P. Multi criteria decision making approaches for green supplier evaluation and selection: A literature review. *J. Clean. Prod.* **2015**, *98*, 66–83. [CrossRef]
12. Banaeian, N.; Mobli, H.; Fahimnia, B.; Nielsen, I.E.; Omid, M. Green supplier selection using fuzzy group decision making methods: A case study from the agri-food industry. *Comput. Oper. Res.* **2018**, *89*, 337–347. [CrossRef]
13. Kilic, H.S.; Yalcin, A.S. Modified two-phase fuzzy goal programming integrated with IF-TOPSIS for green supplier selection. *Appl. Soft Comput.* **2020**, *93*, 106371. [CrossRef]
14. Zhou, F.; Chen, T.Y. An integrated multicriteria group decision-making approach for green supplier selection under Pythagorean fuzzy scenarios. *IEEE Access* **2020**, *8*, 165216–165231. [CrossRef]
15. Atanassov, K.T. Intuitionistic fuzzy sets. *Fuzzy Sets Syst.* **1986**, *20*, 1–137. [CrossRef]
16. Yager, R.R. Pythagorean membership grades in multicriteria decision making. *IEEE Trans. Fuzzy Syst.* **2013**, *22*, 958–965. [CrossRef]
17. Senapati, T.; Yager, R.R. Fermatean fuzzy sets. *J. Ambient Intell. Hum. Comput.* **2020**, *11*, 663–674. [CrossRef]
18. Nguyen, N.B.T.; Lin, G.H.; Dang, T.T. A Two phase integrated fuzzy decision-making framework for green supplier selection in the coffee bean supply chain. *Mathematics* **2021**, *9*, 1923. [CrossRef]
19. Tavana, M.; Shaabani, A.; Santos-Arteaga, F.J.; Valaei, N. An integrated fuzzy sustainable supplier evaluation and selection framework for green supply chains in reverse logistics. *Environ. Sci. Pollut. Res.* **2021**, *28*, 53953–53982. [CrossRef]
20. Fazlollahtabar, H.; Kazemitash, N. Green supplier selection based on the information system performance evaluation using the integrated Best-Worst Method. *Facta Univ. Ser. Mech. Eng.* **2021**, *19*, 345–360. [CrossRef]
21. Wang, C.N.; Nguyen, T.L.; Dang, T.T. Two-Stage Fuzzy MCDM for Green Supplier Selection in Steel Industry. *Intell. Autom. Soft Comput.* **2022**, *33*, 1245–1260. [CrossRef]
22. Çalık, A. A novel Pythagorean fuzzy AHP and fuzzy TOPSIS methodology for green supplier selection in the Industry 4.0 era. *Soft Comput.* **2021**, *25*, 2253–2265. [CrossRef]
23. Zeng, S.Z.; Pan, Y.; Jin, H.H. Online Teaching Quality Evaluation of Business Statistics Course Utilizing Fermatean Fuzzy Analytical Hierarchy Process with Aggregation Operator. *Systems* **2022**, *10*, 63. [CrossRef]
24. Wei, D.M.; Meng, D.; Rong, Y.; Liu, Y.; Garg, H.; Pamucar, D. Fermatean Fuzzy Schweizer–Sklar Operators and BWM-Entropy-Based Combined Compromise Solution Approach: An Application to Green Supplier Selection. *Entropy* **2022**, *24*, 776. [CrossRef] [PubMed]
25. Tan, J.D.; Liu, Y.; Senapati, T.; Grag, H.; Rong, Y. An extended MABAC method based on prospect theory with unknown weight information under Fermatean fuzzy environment for risk investment assessment in B&R. *J. Ambient Intell. Hum. Comput.* **2022**, 1–30. [CrossRef]
26. Mishra, A.R.; Rani, P. Multi-criteria healthcare waste disposal location selection based on Fermatean fuzzy WASPAS method. *Complex Intell. Syst.* **2021**, *7*, 2469–2484. [CrossRef]
27. Gül, S. Fermatean fuzzy set extensions of SAW, ARAS, and VIKOR with applications in COVID-19 testing laboratory selection problem. *Expert Syst.* **2021**, *38*, e12769. [CrossRef]
28. Yager, R.R. The power average operator. *IEEE Trans. Syst. Man Cybern. Part A Syst. Hum.* **2001**, *31*, 724–731. [CrossRef]
29. Beliakov, G.; Pradera, A.; Calvo, T. *Aggregation Functions: A Guide for Practitioners*; Springer: Berlin/Heidelberg, Germany, 2007; Volume 221.
30. Shi, M.H.; Yang, F.; Xiao, Y. Intuitionistic fuzzy power geometric Heronian mean operators and their application to multiple attribute decision making. *J. Intell. Fuzzy Syst.* **2019**, *37*, 2651–2669. [CrossRef]
31. Liu, P.D.; Mahmood, T.; Khan, Q. Group decision making based on power Heronian aggregation operators under linguistic neutrosophic environment. *Int. J. Fuzzy Syst.* **2018**, *20*, 970–985. [CrossRef]
32. Rezaei, J. Best-worst multi-criteria decision-making method. *Omega* **2015**, *53*, 49–57. [CrossRef]
33. Liu, S.; Zhang, J.X.; Niu, B.; Liu, L.; He, X.J. A novel hybrid multi-criteria group decision-making approach with intuitionistic fuzzy sets to design reverse supply chains for COVID-19 medical waste recycling channels. *Comput. Ind. Eng.* **2022**, *169*, 108228. [CrossRef] [PubMed]
34. Ma, Y.F.; Zhao, Y.Y.; Wang, X.Y.; Feng, C.Y.; Zhou, X.Y.; Lev, B. Integrated BWM-Entropy weighting and MULTIMOORA method with probabilistic linguistic information for the evaluation of Waste Recycling Apps. *Appl. Intell.* **2023**, *53*, 813–816. [CrossRef]
35. Feng, D.; Fu, X.Y.; Jiang, S.F.; Jing, L.T. Conceptual Solution Decision Based on Rough Sets and Shapley Value for Product-Service System: Customer Value-Economic Objective Trade-Off Perspective. *Appl. Sci.* **2021**, *11*, 11001. [CrossRef]

36. Krishankumar, R.; Arun, K.; Kumar, A.; Rani, P.; Ravichandran, K.S.; Gandomi, A.H. Double-hierarchy hesitant fuzzy linguistic information-based framework for green supplier selection with partial weight information. *Neural Comput. Appl.* **2021**, *33*, 14837–14859. [CrossRef]
37. Xiong, L.; Zhong, S.Q.; Liu, S.; Zhang, X.; Li, Y.F. An approach for resilient-green supplier selection based on WASPAS, BWM, and TOPSIS under intuitionistic fuzzy sets. *Math. Probl. Eng.* **2020**, *2020*. [CrossRef]
38. Liu, P.; Pan, Q.; Xu, H.X.; Zhu, B.Y. An Extended QUALIFLEX Method with Comprehensive Weight for Green Supplier Selection in Normal q-Rung Orthopair Fuzzy Environment. *Int. J. Fuzzy Syst.* **2022**, *24*, 2174–2202. [CrossRef]
39. Baki, R. An Integrated Multi-criteria Structural Equation Model for Green Supplier Selection. *Int. J. Precis. Eng. Manuf. Green Technol.* **2022**, *9*, 1063–1076. [CrossRef]
40. Zhang, S.Q.; Wei, G.W.; Gao, H.; Wei, C.; Wei, Y. EDAS method for multiple criteria group decision making with picture fuzzy information and its application to green suppliers selections. *Technol. Econ. Dev. Econ.* **2019**, *25*, 1123–1138. [CrossRef]
41. Zhang, N.; Su, W.H.; Zhang, C.H.; Zeng, S.Z. Evaluation and selection model of community group purchase platform based on WEPLPA-CPT-EDAS method. *Comput. Ind. Eng.* **2022**, *172*, 108573. [CrossRef]
42. Mishra, A.R.; Rani, P.; Pandey, K. Fermatean fuzzy CRITIC-EDAS approach for the selection of sustainable third-party reverse logistics providers using improved generalized score function. *J. Ambient Intell. Hum. Comput.* **2022**, *13*, 295–311. [CrossRef] [PubMed]
43. He, Y.; Wei, G.W.; Chen, X.D. Taxonomy-based multiple attribute group decision making method with probabilistic uncertain linguistic information and its application in supplier selection. *J. Intell. Fuzzy Syst.* **2021**, *41*, 3237–3250. [CrossRef]
44. Xu, D.S.; Cui, X.X.; Xian, H.X. An extended EDAS method with a single-valued complex neutrosophic set and its application in green supplier selection. *Mathematics* **2020**, *8*, 282. [CrossRef]
45. Wu, Y.N.; Xu, C.B.; Huang, Y.; Li, X.Y. Green supplier selection of electric vehicle charging based on Choquet integral and type-2 fuzzy uncertainty. *Soft Comput.* **2020**, *24*, 3781–3795. [CrossRef]
46. Kang, X.; Xu, X.J.; Yang, Z.L. Evaluation and selection of green suppliers for papermaking enterprises using the interval basic probability assignment-based intuitionistic fuzzy set. *Complex Intell. Syst.* **2022**, *8*, 4187–4203. [CrossRef]
47. Gegovska, T.; Koker, R.; Caka, T. Green supplier selection using fuzzy multiple-criteria decision-making methods and artificial neural networks. *Comput. Intell. Neurosci.* **2020**, *2020*. [CrossRef]
48. Krishankumar, R.; Gowtham, Y.; Ahmed, I.; Ravichandran, K.S.; Kar, S. Solving green supplier selection problem using q-rung orthopair fuzzy-based decision framework with unknown weight information. *Appl. Soft Comput.* **2020**, *94*, 106431. [CrossRef]
49. Deng, Z.; Wang, J.Y. Evidential Fermatean fuzzy multicriteria decision-making based on Fermatean fuzzy entropy. *Int. J. Intell. Syst.* **2021**, *36*, 5866–5886. [CrossRef]
50. Liu, P.D.; Chen, S.M.; Wang, P. Multiple-attribute group decision-making based on q-rung orthopair fuzzy power maclaurin symmetric mean operators. *IEEE Trans. Syst. Man Cybern. Syst.* **2018**, *50*, 3741–3756. [CrossRef]

Article

Simulation of Manufacturing Scenarios' Ambidexterity Green Technological Innovation Driven by Inter-Firm Social Networks: Based on a Multi-Objective Model

Xuan Wei [1], Hongyu Wu [2], Zaoli Yang [3,*], Chunjia Han [4] and Bing Xu [5]

1   School of Statistics and Mathematics, Shandong University of Finance and Economics, Jinan 250001, China
2   Business School, Qingdao University, Qingdao 266000, China
3   School of Economics and Management, Beijing University of Technology, Beijing 100000, China
4   School of Business, Economics & Informatics, Birkbeck, University of London, London WC1E 6BT, UK
5   Edinburgh Business School, Heriot-Watt University, Edinburgh TD1 3HE, UK
*   Correspondence: yangzaoli@bjut.edu.cn

Abstract: The mechanism of the impact of inter-firm social networks on innovation capabilities has attracted much research from both theoretical and empirical perspectives. However, as a special emerged and developing complex production system, how the scenario factors affect the relationship between these variables has not yet been analyzed. This study identified several scenario factors which can affect the firm's technological innovation capabilities. Take the manufacturing scenario in China as an example, combined with the need for firms' ambidexterity innovation and green innovation capability, a multi-objective simulation model is constructed. Past empirical analysis results on the relationship between inter-firm social network factors and innovation capabilities are used in the model. In addition, a numerical analysis was conducted using data from the Chinese auto manufacturing industry. The results of the simulation model led to several optimization strategies for firms that are in a dilemma of development in the manufacturing scenario.

Keywords: manufacturing scenario; inter-firm social network; ambidexterity and green innovation capability; simulation

Citation: Wei, X.; Wu, H.; Yang, Z.; Han, C.; Xu, B. Simulation of Manufacturing Scenarios' Ambidexterity Green Technological Innovation Driven by Inter-Firm Social Networks: Based on a Multi-Objective Model. *Systems* **2023**, *11*, 39. https://doi.org/10.3390/systems11010039

Academic Editor: Mitsuru Kodama

Received: 18 November 2022
Revised: 3 January 2023
Accepted: 6 January 2023
Published: 10 January 2023

1. Introduction

This research focuses on how to combine scenario factors with previous technological innovation theories to solve manufacturing firms' development problems. It has been widely accepted that technological innovation is a decisive factor in the development of a country, a region, or an organization to win strategic advantages, competitive advantages, or profit sources [1]. Environmental sound technological innovation has become one of the hottest topics in recent years in manufacturing industries [2]. Leading and driving development through green technological innovation has become an urgent requirement for manufacturing firms' development [3].

Meanwhile, the development of any firms, industries, and different countries is not isolated but is connected by a variety of complex related networks [4]. Under the scenario of global climate change, green technological innovation is the objective of each manufacturing firm for future development on the base of the development of complex related networks [5]. Previous research has conducted much research on the relationship between inter-firm social networks and innovation capabilities from both theoretical and empirical aspects [6]. Related theories include structural hole theory [7], weak correlation theory [8], network centrality theory [9], network effect theory [10], open innovation theory [11], and complex network theory [12].

The cross-over studies of the inter-firm social network theory and green technological innovation, which consider the influence of scenario factors, can not only provide a deep understanding of the mechanism of internal and external factors affecting technological innovation, but also can shed light on the impact of inter-firm social networks on manufacturing firms' performance. In addition, from the perspective of practice, the inter-firm social network structure and the coordination of internal and external resources could be optimized. However, there is still a lack of analysis for the linking of the scenario factors and the relationship between inter-firm social networks and different types of green technological innovation.

The main objectives of this study are as follows. First, this study aims to grasp the key impact factors of the scenario that affect the manufacturing firms' technological innovation and to sort out the existing research on the summarizing of the concept of manufacturing scenario as an example. Second, this study aims to analyze the relationship between the key scenario factors faced by manufacturing industries, inter-firm social networks, and firms' ambidexterity of technological innovation. Third, based on the results of previous studies and data we collected, a numerical simulation analysis is conducted to find out the optimal solution of the relationship between inter-firm social networks and innovation capability in the Chinese auto manufacturing industry.

The possible contribution of this research is as follows. First, through a comprehensive refinement of key scenario factors, this study summarized three aspects of scenario factors faced by manufacturing industries: institutional scenario, economic scenario, and cultural and environmental scenario. Second, a multi-objective programming simulation model, which considers profit maximizing, inter-firm social network benefits, equilibrium of the ambidexterity innovation, and green technological innovation, is established. Third, the results of this study provide a theoretical basis for manufacturing firms to make their social network strategies to achieve green technological innovation and for the government to achieve policy goals by affecting inter-firm social networks.

The content of the remainder of this study is arranged as follows: Section 2 analyzes the meaning of scenario factors faced by manufacturing industries, inter-firm social networks, and ambidexterity green technological innovation through literature reviews. Section 3 describes the research design. Section 4 gives the mathematical and technical route of the simulation model. Section 5 describes the results of the numerical simulation. Section 6 discusses the results and Section 7 summarizes the conclusion of this study.

2. Literature Review

2.1. Manufacturing Scenario Factors

Previous studies have analyzed the connotation of scenario factors from different perspectives. We summarized all the related literature about scenarios faced by manufacturing industries in studies of technological innovation which we can get. From the view of the content of the scenario, it can be divided into the institutional scenario, economic scenario, and cultural environmental scenario.

First, researchers from the perspective of institutional scenario believe that the manufacturing scenario has the following characteristics: (1) There is an environmental protection system that emphasizes energy conservation, emission reduction, energy tax, and carbon tax in terms of the environmental system [13]; (2) From the perspective of the institutional environment, manufacturing industries have characteristics of the incompleteness of policies, inter-regional institutional heterogeneous, and institutional uncertainty [14]; (3) The role of government is a dilemma for the development of manufacturing industries [15].

Second, from the perspective of the economic scenario, the research shows that the main characteristics of the global economic scenario faced by manufacturing industries are as follows: (1) The manufacturing industries is in a transition period from traditional production systems to a new system with more high-technologies and more environmentally sound [16]. In this process, there is a relatively large gap between developing and developed countries in firms' technological innovation capabilities [17]; (2) The organizational network in the manufacturing scenario is unbounded in the supply chain cooperation [18].

With the deep integration of global competition, the green technological transition is the main challenge for the continued development of economic development [19].

Third, from the perspective of cultural and environmental scenarios, the main characteristic of the current cultural and environmental scenario faced by manufacturing industries are (1) Information technology is fully infiltrated, and informatization is the core of the technological system changes [20]; (2) Cultural integration and exchange are conducive to the innovation and development of the manufacturing industry [21].

The direction of future expected scenarios faced by manufacturing industries mainly includes: The requirements for (1) ambidexterity development of organizations in the manufacturing scenario [22]; (2) improving network capabilities [23]; (3) global integration [24]; (4) business model innovation [25]; (5) innovative entrepreneurship and the social responsibility of manufacturing firms [26]; (6) the energy efficiency improvement and low carbonization [27]; and (7) open innovation [28].

2.2. Inter-Firm Social Networks

The concept of the social network was first created in the study of sociology [8]. Researchers treat various subjects (such as a single person, firm, or other organization) as players [7]. Due to the relationship of contracts or cooperation between firms, they can exchange information and resources so that there is a wide network of relationships between firms. Each firm's network constitutes the firm's third type of capital, which is social capital [29]. A whole firm's social network with structural characteristics, cognitive characteristics, and relationship characteristics is also formed by all firm's networks [7]. From the broadest perspective, since a single firm is both a node in the social network and a set of multiple natural individuals, which also can be nodes in the social network, hence, the broad meaning of the firm social network includes the relationship of intra-firm and inter-firm. Both social networks have made certain progress in current research [6]. To reduce the complexity of the study, this study uses a narrow concept of the social network, that is, the inter-firm social network. The network between firms and firms, the network between firms and universities or scientific research institutions, the network between firms and intermediaries, and the network between firms and customers are the main constituent units and structures of this narrowly defined inter-firm social network [30].

Although the same emphasis is placed on interrelationships between organizations, there are differences in the concepts of clusters and inter-firm social networks. The concept of clusters mainly comes from economic literature [31] especially related to spatial geography. The concept of social networks mainly comes from sociological literature [32]. The clusters emphasized the returns to scale within a region [33]. The social network emphasized the type of linkage and its benefits. The nature of the cluster is the agglomeration and "open membership", which means substituting the formal structures or strong long-term relations with the local resource pool [34]. However, there is nothing inherently spatial about the social network because of its "path dependence" for one actor [6]. In this study, the relationship that brings benefits to the firm is not limited to the association caused by geographical location, but also the structure of its connection, so we use social networks instead of clusters.

2.3. Ambidexterity Green Technological Innovation

After the 'Earth Summit', which was held in Rio de Janeiro, Brazil in 1992, a series of notions related to a green and sustainable economy attracted attention from both scholars and practitioners [35], such as bio-economy [36], industrial ecology [37], circular economy [38], and cleaner production [39]. In this study, we define green technological innovation as technological innovation, which is environmentally sound and can be used to support the transition toward sustainability [2]. The factors that drive the firm's green technological innovation include both internal and external factors [40]. The scene outside the firm provides a firm with a platform and environment for cooperation, competition, or supply and demand relationships with customers, suppliers, distributors, manufacturers, universities, or intermediaries [41]. When embedding in social networks, on the one hand, firms can collect market intelligence information, which provides the basis for the direction of innovation of the products or services [42]. On the other hand, the firm can make use of not-here inventions or the extroverted marketization of idle technology through open innovation. Through the influence of these two aspects, firms can absorb and integrate external knowledge and resources, which is conducive to the improvement of green technological innovation [43].

Specifically, the scenario of institutional, economic, and cultural environmental faced by manufacturing firms have a certain impact on the results of social network characteristics and green technological innovation, e.g., Zhou, et al. [44] found that the government social capital of TMT is positively related to firms' innovation performance and firms' network prestige plays a mediating role in this relationship. Mansell [45] found that research concerning changes in the techno-economic paradigm cautioned that assessments of these changes needed to go beyond market dynamics to examine social, cultural, and political issues.

It can be found that previous research on the firm's green technological innovation in the manicuring scenario has analyzed the use of internal and external knowledge and resources, as well as the openness and exploratory nature of innovation from their respective perspectives. Therefore, the ambidexterity green technological innovation of firms is a critical perspective that studied the key factors to innovation in the manufacturing scenario [46]. Therefore, the ambidexterity-balanced and green innovation capability is the future pursuit of the manufacturing scenario in the field of technological innovation. The research on the law of a firm's social network and its impact on the ambidexterity green innovation capability has significance in both theoretical and practical aspects.

3. Research Design

Due to the lack of consideration of scenario factors, previous research on the relationship between manufacturing inter-firm social networks and green technological innovation capabilities is usually aimed at various factors inner firm. The typical research paradigm is to regard the firm's social network as the cause variable and the firm's technological innovation as the result variable. Then, the other properties are used as intermediate variables in these research models [9]. Related research methods of previous research usually include parametric methods and non-parametric methods. Parametric methods mainly include econometric models [47], stochastic frontier analysis [48], and multiple regression analysis [49]. Non-parametric methods mainly include game-theory-based analysis [50], DEA analysis [51], and qualitative research [52].

One of the contributions of this study is that the manufacturing scenario factors' consideration in the relationship between inter-firm social networks and ambidexterity green technological innovation. The research object goes beyond the boundaries of the firm. Hence that the inter-firm social network is not only the main independent variable but also an intermediate variable between scenario factors and ambidexterity green technological innovation. In addition, in the relationship system of related variables, not only the ambidexterity green technological innovation of the firm is the object of optimization, but also the firm's social network. Moreover, scenario factors put forward directional requirements for the optimization of the two factors. These characteristics require that the research design of this study must satisfy not only the logical relationship between the variables in the model but also the synergy of optimization of the variables. Therefore, this study adopts a multi-objective planning simulation method combined with qualitative analysis methods. The specific research framework and implementation route are shown in Figure 1.

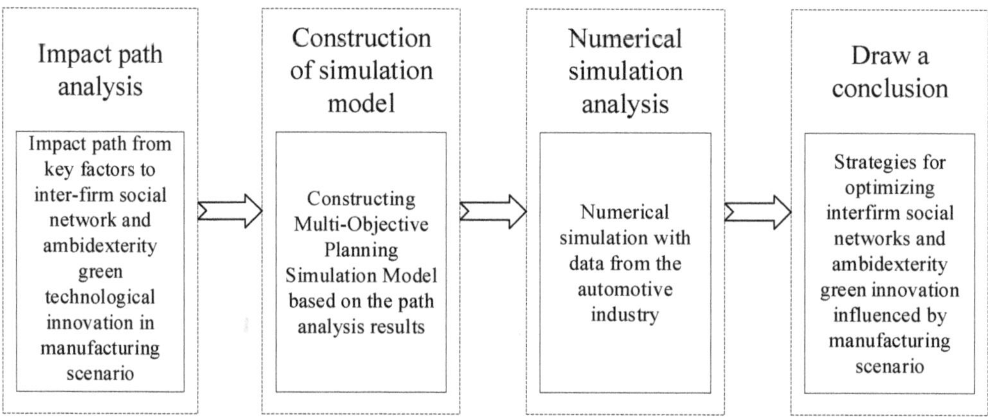

Figure 1. Technology roadmap.

To address the question of the optimization strategy of the relationship between inter-firm social networks and the firm's ambidexterity green technological innovation in the manufacturing scenario, this research designed the following research framework and technical route (as shown in Figure 1).

First, based on a review of past research, the key factors that affect ambidexterity green technological innovation in the manufacturing scenario were clarified, and a diagram of the path of influence was drawn. This stage mainly adopts the method of subjective qualitative data analysis, that is, sorting and coding the related research in the past, and deriving the influence path relationship mechanism between various variables [53].

Second, based on the results of path analysis, this study built a simulation model. Since there are many requirements for the development of the firm's innovation capabilities, this study built a multi-objective planning simulation model. The establishment of the goal system is based on the economic output of ambidexterity green technological innovation. Referring to past technological innovation models, we regard maximizing benefits as the basic objective. Then, combining the relationship between the scenario factors, the inter-firm social network, and the firm's innovation capability, we established the ambidexterity innovation objective, the objective of minimizing environmental pollution, and the social network benefit objective.

Third, based on the measurement results of the relationship between the influencing factors in the past research and the real data of the auto industry companies in China, specific numerical simulations are carried out. We collected and used the results of previous research on the relationship between variables in the Chinese auto industry as simulation data so that the experiment fits the manufacturing scenario.

Fourth, this study concludes and puts forward strategic recommendations to optimize inter-firm social networks and the firm's ambidexterity green technological innovation under the influence of key factors in the manufacturing scenario.

4. Path Analysis and Simulation Model Setting

4.1. Inter-Firm Social Network Benefit Objective

4.1.1. Path Analysis

Manufacturing scenario factors of the cultural environment require firms to pursue more favorable social network positions. First, the production system of modern manufacturing has always emphasized cooperation and exchange domestically and overseas through collaboration between industry, universities, and research institutes and supply chain collaboration [54]. This system has achieved significant performance. Therefore, the knowledge environment, knowledge sharing level, and relationship network construction are important situational factors that influence the technological innovation of manufacturing firms [55]. Second, the increasing attention to absorptive capacity, social network theory, and open innovation theory shows the importance of organizational cross-border cooperation behavior for technological innovation [52]. The formation of innovation in the manufacturing scenario is highly dependent on the relationship between the organization and other organizations, such as governments, suppliers, customers, and universities [47]. Therefore, the construction of the national innovation system will affect the level of innovation capabilities of manufacturing enterprises.

4.1.2. Objective Function

In a manufacturing scenario, the knowledge-sharing environment, relational network, and national innovation system make the social network benefits generated able to be divided into direct social network benefits and indirect social network benefits. Further, the level of social network benefits is affected by the absorptive capacity. Therefore, there are the following objective functions:

$$\max N_t^{social\ network\ benefit} = \sum_{t=0}^{y} \beta_i^d D_{t-i} + \sum_{i=0}^{y} \beta_i^b B_{t-i} \tag{1}$$

where the variables in capital letters all represent n-th order column vectors, t represents the period in which the variable in. At the t-th period, $N_t^{social\ network\ benefit}$ represents the social network efficiency of firms; D_{t-i} represents the scale of the direct link of the firm at the i-th time-lag order; β_i^d represents the coefficient of the influence of the direct social network scale on the total social network benefit at the i-th time-lag order. B_{t-i} refers to the scale of the indirect link of the firm at the i-th time-lag order; β_i^b represents the coefficients of the influence of the indirect social network scale on the total social network benefit at the i-th time-lag order. The objective of limiting conditions include:

$$\begin{aligned} D_t &= \alpha_1 C_t^{resource\ input\ for\ social\ network} \\ B_t &= \alpha_2 C_t^{resource\ input\ for\ social\ network} \end{aligned} \tag{2}$$

where α_1 and α_2 represent the influence level of the input for the social network on the scale of the direct social network and indirect social network, respectively. $C_t^{resource\ input\ for\ social\ network}$ represents the available resources of the firm for the development of social networks during t-th period.

4.2. Balance of the Ambidexterity Technological Innovation Objective

4.2.1. Path Analysis

Manufacturing scenario factors of economics require firms to pursue a balance of ambidexterity and technological innovation. The development of manufacturing firms' technological innovation has certain characteristics. On the one hand, technological innovation development in manufacturing industries relies on learning from the experience of other players, which means there is a norm of imitating innovation [56]. On the other hand, the development of high-tech industries and emerging industries emphasizes the disruptive innovation strategy [57]. These scenario factors have formed the current situation of the co-existence of incremental innovation and breakthrough innovation [41]. Therefore, manufacturing technological innovation has the objective of the balance of ambidexterity innovation. The resources must be balanced between the levels of two types of innovation [22].

4.2.2. Objective Function

This study sets the ambidexterity innovation as product innovation and production technology innovation. At the same time, according to past research, indirect social network effects mainly affect breakthrough innovation, and direct social network effects mainly affect progressive innovation. The objective of the balanced ambidexterity innovation could be expressed by the following:

$$\begin{aligned} \max A_t^{product\ technology\ innovation} &= \alpha_3 N_t^D C_{t-1}^{input\ for\ the\ product\ technology\ innovation} \int \alpha_4 dt \\ \max A_t^{product\ innovation} &= \alpha_5 \alpha_6 N_t^B C_{t-1}^{input\ for\ the\ product\ innovation} \end{aligned} \tag{3}$$

where $A_t^{product\ technology\ innovation}$ and $A_t^{product\ innovation}$ represent the benefit of production technological innovation and product innovation at time t, respectively. $C_{t-1}^{input\ for\ the\ product\ technology\ innovation}$ and $C_{t-1}^{input\ for\ the\ product\ innovation}$ represent the input of production technological innovation and product innovation at the i-th lag order of time t, respectively. α_3 and α_4 represent the patents increasing because of the production technological innovation and its price effect, respectively. α_5 and α_6 represent the patents increasing because of product innovation and its yield effect, respectively.

4.3. Objective of Minimizing Environmental Pollution

4.3.1. Path Analysis

Manufacturing scenario factors of the institution require firms to pursue the objective of minimizing environmental pollution. First, the world is currently facing various severe environmental problems. An environmentally sustainable system that pulls and pushes the green technological innovation of firms in manufacturing is necessary [5]. Second, as many countries' governments increasingly emphasize ecological and environmental civilization, society and the market are increasingly encouraging the firm's behavior of green innovation and social responsibility innovation [16]. Therefore, the cultural environment of the firms' ethical environment will also affect the manufacturing industries' innovative behavior.

4.3.2. Objective Function

Therefore, in the ambidexterity context of policy pressure and social ethical incentives, manufacturing enterprises have an environmental pollution minimization objective:

$$\min C_t^{cost\ of\ pollution} = \alpha_{7(t-1)} Q_t - \alpha_8 \alpha_9 C_{t-1}^{input\ for\ control\ pollution} \tag{4}$$

where $C_t^{cost\ of\ pollution}$ represents the actual emission cost of the firm during the t-th period. $\alpha_{7(t-1)}$ represents the pollution cost per unit product at the 1 order lag of time t. Q_t represents the output of the firm. α_8 and α_9 represent the number of patents increased per unit of emission reduction input and the emission cost reduction per patent can save, respectively.

4.4. The Objective Function of Profit Maximization

4.4.1. Path Analysis

In the research on technological innovation, product orientation is accepted by most researchers [58]. In the socialist market economic system, innovation emphasizes market-oriented competition. Hence that product orientation and market orientation are also important guiding factors for the development of innovation capabilities of manufacturing firms [59].

4.4.2. Objective Function

In this situation, manufacturing firms have the objective of maximizing profits. This study sets the income of a firm as the product of output and product price. The above analysis reflects that the firm's input includes product R&D investment, production technology innovation investment, and emission reduction technology investment. In addition, the firm's cost should also include the production cost per unit of product taxes payable, so the objective function of profit maximization is:

$$\max\pi_t = R_t - C_t^{product} - C_t^{input\ for\ product\ technology\ innovation} \\ -C_t^{input\ for\ product\ innovation} - C_t^{input\ for\ pollution\ discharge\ rights} - \\ C_t^{input\ for\ pollution\ control\ technology} \tag{5}$$

where R_t represents the firm's income. π_t means the revenue of the firm. At the same time, since the input in period t often depends on the income of the previous period, hence that this study set the following equation:

$$R_t = (A_t^{product\ technology\ innovation} + Q_t)* \\ \max(A_t^{product\ innovation}, p_t) \\ C_t^{product} = \alpha_{10}R_t \\ C_t^{input\ for\ product\ technology\ innovation} = \alpha_{11}R_t \\ C_t^{input\ for\ product\ innovation} = \alpha_{12}R_t \\ C_t^{input\ for\ pollution\ control\ technology} = \alpha_{13}R_t \\ \alpha_{11} + \alpha_{12} = 1 \\ \alpha_{11} = \frac{N_t^D}{N_t^D + N_t^B} \tag{6}$$

where α_{10}, α_{11}, α_{12}, and α_{13} represent the ratio of the cost of production, the input of product technology innovation, the input of product innovation, and input of pollution control technologies in total income in the $t - 1$ period.

5. Numerical Simulation Based on the Chinese Automobile Industry

5.1. Data Description and Initial Values

This study selects the firms' data of the automobile industry among Chinese listed companies for the simulation. In the automobile manufacturing category (by the WIND industry classification) listed on the Chinese A-share market, we found that there are 16 firms whose main products in the past three years are completed vehicle businesses. These firms including Jiangling Motors Co., Changan Automobile Co., FAW Car Co., FAW Xiali Automobile Co., BYD Co., Dongfeng Motor Co., SAIC Motor Co., FOTON Co., Jianghuai Automobile Co., King Long Motor Co., Sokon Industry Group Stock Co., Guangzhou Automobile Group Co., Great Wall Motor Co., and Lifan Industry Co. The data of these firms are used as the basis for numerical simulation to set the initial values. At the same time, three companies, FAW Xiali Automobile Co., Dongfeng Motor Co., and Lifan Industry Co., whose output and sales have been declining in 2016–2018 were selected as the key analysis objects. The measurement methods and collected data of each variable are shown in Table 1.

5.2. Numerical Simulation Results

5.2.1. Firms' Profit Evolution without R&D

Some Chinese auto companies are currently facing both sales and price decline pressure. In the case of the price and sales volume decreasing with the average decline rate of 2015–2016 and operating costs remaining unchanged without innovation input changing, the simulation flowchart is shown in Figure A1. The simulation results are shown in Figure 2. Figure 2 shows that the declining trend of FAW Xiali, Dongfeng Motor, and Lifan in 10 years is more obvious, and the companies face huge problems in sustainable development.

5.2.2. Incremental Innovation Capability Enhanced

To reverse the decline, based on the analysis in the above chapters, consider the firm using a certain proportion of R&D investment to improve its product technology innovation capability. First, based on the industry's average R&D investment level (2.4118 percent of income), the number of patents increased per 100 million yuan of investment (100, past research results), and the average sales growth per patent (500, according to the motor industry data calculation), and the sales increase value that each patent may bring is used as the sensitivity analysis variable. The simulation flowchart is shown in Figure A2. As shown in Figure 3, when the sales increase of each patent is 100, the profits of the three companies have improved in the short term, but they still cannot recover the decline in the long term.

Table 1. Initial values and data sources of variables.

Var	Calculation Method	Data Source
Q_t	Car Sales Volume of each company in 2016–2018	FS
p_t	$\dfrac{\textit{Total Vehicle Sales of Each Enterprise in 2018}}{\textit{Automobiles Sales Volume in 2018}}$	FS
$C_t^{social\ network}$	The business reputation of each company at t year	FS
α_1, α_2	$\alpha_1 = \dfrac{\textit{Average Annual Increase in Degree Centrality}}{\textit{Average Annual Increase in Business Reputation}}$ $\alpha_2 = \dfrac{\textit{Average Annual Increase in Betweenness Centrality}}{\textit{Average Annual Increase in Business Reputation}}$	FS
β_i^d, β_i^b	The coefficient of the effect of degree centrality and betweenness centrality in i-th order time lag on the social network benefit	PL
α_3, α_5	$\alpha_3 = \alpha_5 = \dfrac{\textit{Average Annual Increase in Patents}}{\textit{Average Annual Increase in R\&D Expense}}$	FS
α_4, α_6	$\alpha_4 = \dfrac{\textit{Average Annual Change in Price}}{\textit{Average Annual Increase in Patents}}$ $\alpha_6 = \dfrac{\textit{Average Annual Change in Yield}}{\textit{Average Annual Increase in Patents}}$	FS
$\alpha_{7(t-1)}$	$\dfrac{\textit{Expense on Pollution Emission}}{\textit{Yield}_{t-1}}$	FS
α_9	$\dfrac{\textit{Average Decrease of Unit Production Cost of Pollution Emision}}{\textit{Average Annual Increase on Patents}}$	FS
$\alpha_{10}, \alpha_{11}, \alpha_{12},$ and α_{13}	$\alpha_{10} = cost\ of\ production/total\ income_{t-1}$ $\alpha_{11} = \alpha_{12} = R\&D\ Expense/total\ income_{t-1}$ $\alpha_{13} = Cost\ on\ Pollution\ Emission\ of\ Each\ Production$	FS

Note: FS: Financial Statements; PL: previous literature.

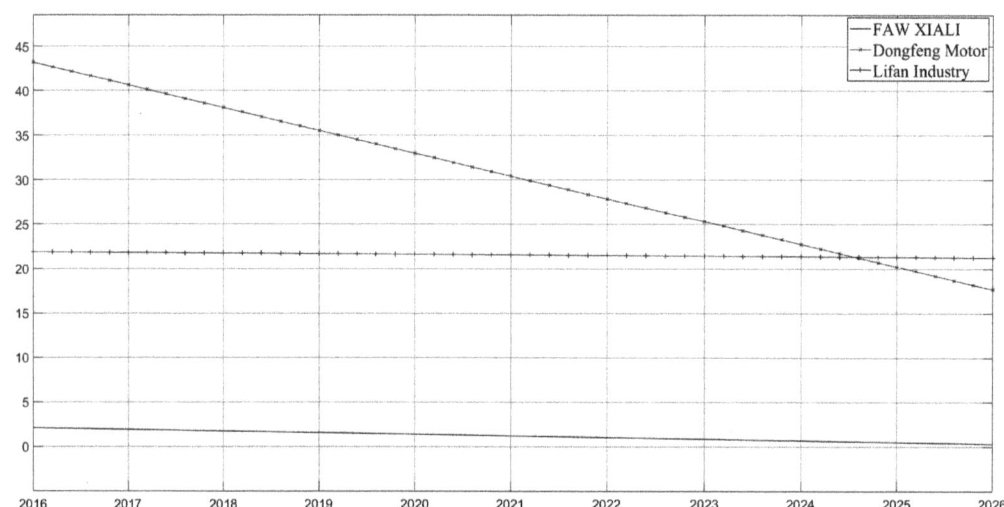

Figure 2. Basic situation corporate profit changes.

5.2.3. Product Innovation Capability Enhanced

To solve the problem of long-term decline, consider that the enterprise will invest a certain amount of R&D expenses to develop new products. Similarly, set the initial value of the industry's average R&D investment level, the increasing number of patents per 100-million-yuan investment, and the average price increase brought by each patent as 2.4118 percentage of income, 100 and 22.099 million CNY (calculated based on data from the automotive industry). The simulation flowchart is shown in Figure A3. The simulation results are shown in Figure 4. It can be seen that when companies are pursuing product innovation capabilities, for Dongfeng Motor and Lifan Industry co., firms' profits will decline in the short term, but will rise after reaching the lowest point. However, for FAW Xiali, due to its obvious downward trend in recent years, it is difficult to turn the crisis into a safe situation within the predictable range under the current industry's average product innovation capability input.

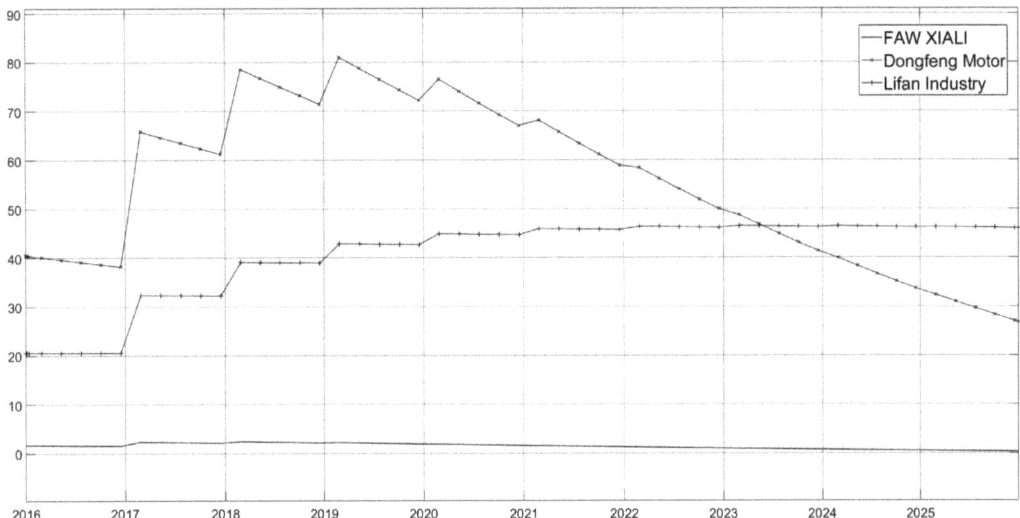

Figure 3. Changes in profit of incremental innovation capability enhanced.

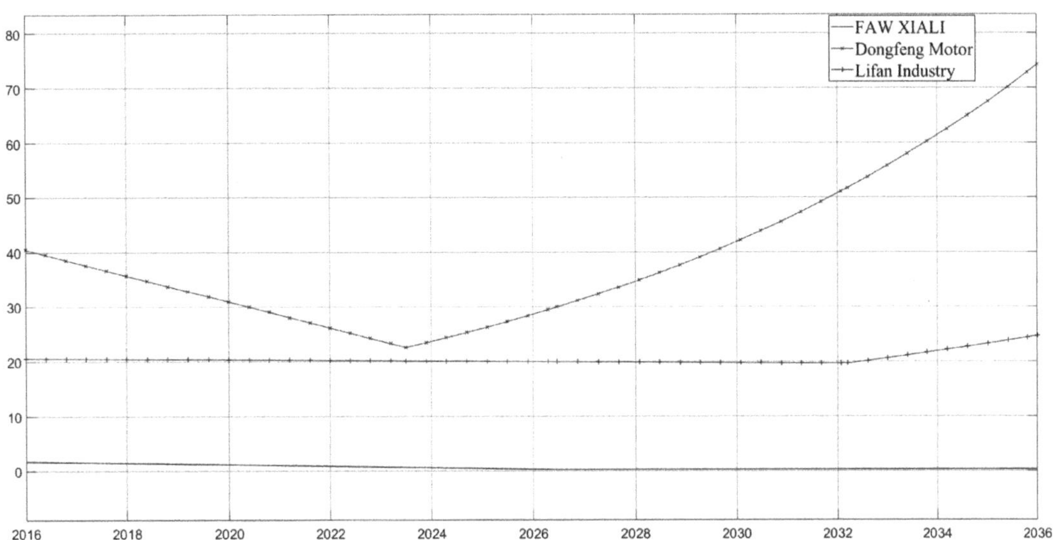

Figure 4. Profit changes considering product innovation enhanced.

5.2.4. Considering Social Networks Effect and Ambidexterity Innovation

Based on previous research, the increase in business reputation and intangible assets can be used as a driving force for the increase of inter-firm social networks, the degree of centrality influences production technology innovation capability, and the betweenness centrality influences product innovation capability. According to the calculation of the automobile industry data, the average increase rate of business reputation and intangible assets is 0.172279, the degree centrality increase from a 100-million-yuan business reputation is 1.697282, and the betweenness centrality increase per 100 million yuan is -0.00055. The lag effect of the degree and betweenness centrality is derived from the calculation results in past studies. The simulation flowchart is shown in Figure A4. The results are shown in Figures 5 and 6. As can be seen from Figure 6, under the current average level of social network structure in the industry, automotive industry companies are more affected by indirect networks, corresponding to the current overall downward trend in the automotive industry, companies are looking for breakthrough technological changes and innovations. According to the simulation analysis of this study, if reasonable R&D investment can be made according to the industry average, as shown in Figure 5, with the increase in the proportion of breakthrough innovation investment, Dongfeng Motor and Lifan will have a manager's performance under the current technical framework. The process of rising first and then descending, and then undergoing a new period of rapid growth due to product upgrades. However, FAW Xiali's investment in R&D at the current industry average level still cannot restore the decline.

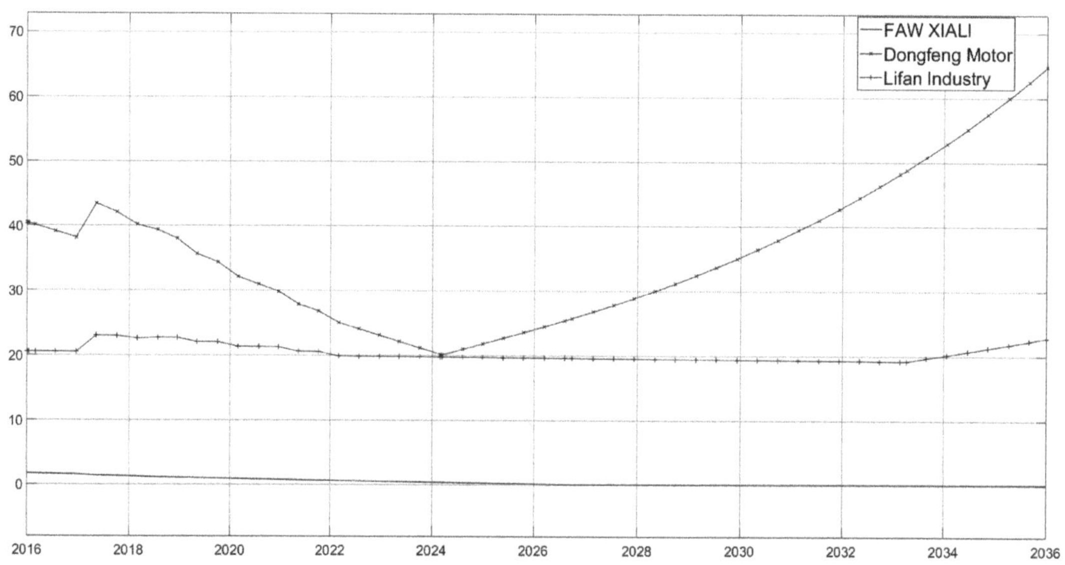

Figure 5. Profit change of ambidexterity innovation with the social network effect.

5.2.5. Consider the Environmental Regulation's Effect

According to "People's Republic of China Environmental Protection Tax Law implementation regulations" annexed "Environmental protection tax item tax table" calculates the average pollution tax per car is 114 CNY. According to the calculated results by Shi et al. [60], the sewage cost (185 CNY) can be saved for each additional patent. Wang and Qian [61] showed that in addition to environmental efficiency and economic efficiency of environmental tax deductions, corporate environmental investment also has social benefits for stakeholders, such as consumers, residents, investors, and creditors. Ye Tong's (2018) research showed that investment in emission-reduction technology has a significant impact on corporate goodwill. The simulation flowchart is shown in Figure A5. The results are shown in Figures 7 and 8. As can be seen from the comparison between Figures 7 and 8, when a company makes a certain amount of emission reduction investment, on the one hand, the innovation brought by the emission reduction investment can reduce emissions, thereby reducing the corresponding emission costs, on the other hand, because with the increase of goodwill and intangible assets, emission reduction investment can affect performance through the enterprise's social network effect. Eventually, the company's emission costs have dropped significantly (Figure 7), while the company's profits will not decline significantly (Figure 8).

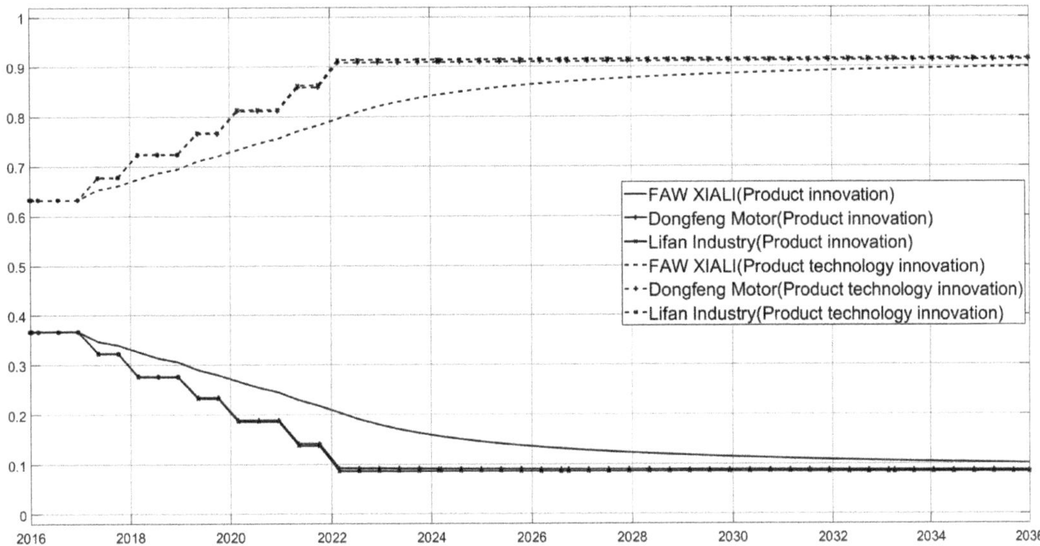

Figure 6. Ambidexterity innovation changes with the social network effect.

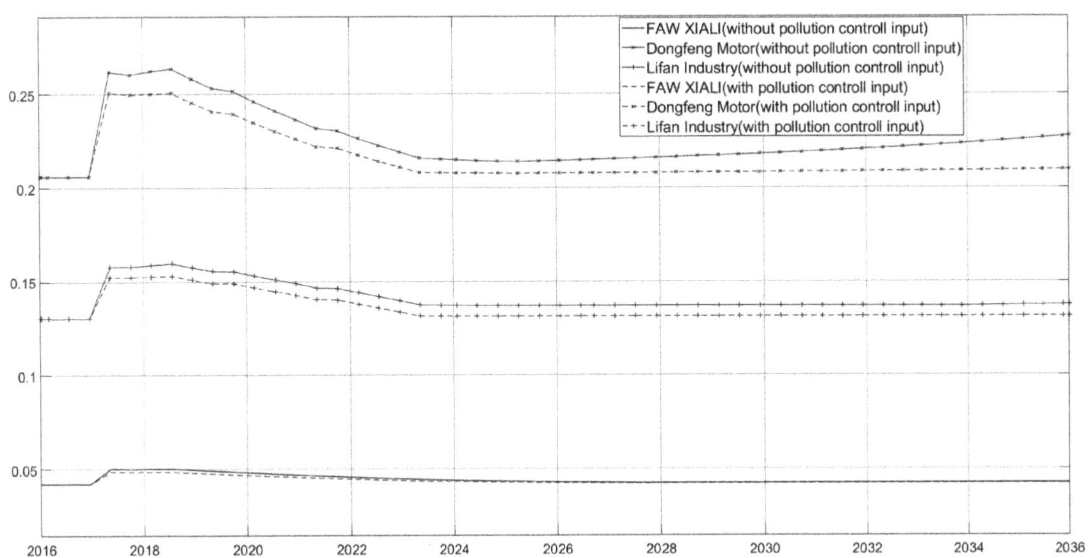

Figure 7. Comparison of emission taxes with a different pollution control input.

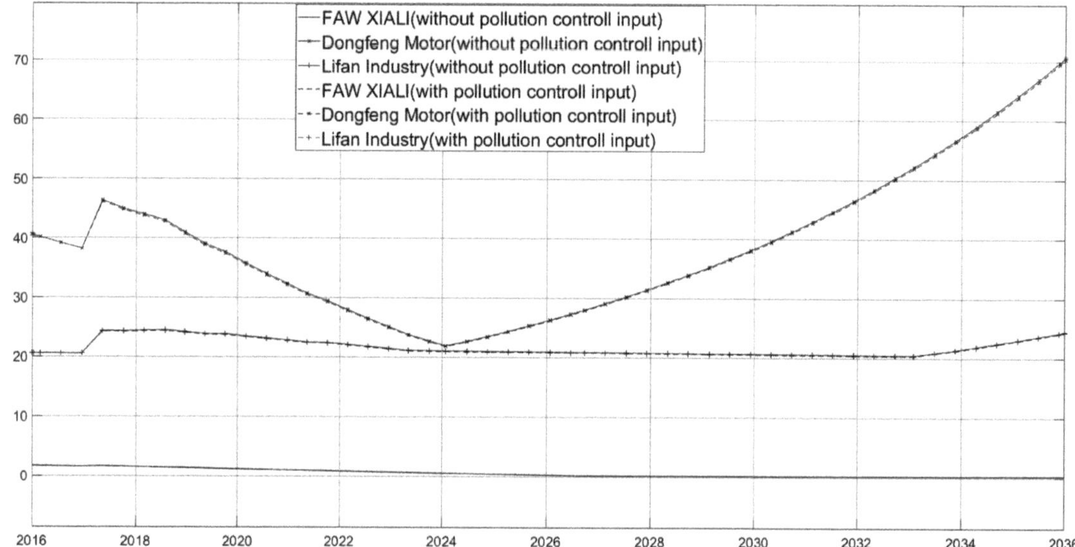

Figure 8. Comparison of profit with a different pollution control input.

6. Discussion

Based on the combination of scenario analysis theory, social network theory, and technological innovation theory, this research establishes a mechanism model of scenario factors and inter-firm social networks affecting ambidexterity green technological innovation. Real data are used to carry out a simulation based on a multi-objective programming model. By comparing with previous literature, the results of this study can draw the following theoretical and practical implications.

6.1. Relationship with the Existing Literature

Previous studies have summarized a variety of theoretical factors that have a direct impact on firms' green technology innovation. Such as firms' internal R&D funds and personnel, and external search and absorptive capacity [62]. However, the impact of scenario factors, such as environmental changes, is usually regarded as an uncontrollable exogenous variable [63]. This study argues that these scenario factors can be systematically classified and internalized as the specific goals of firms' technological innovation. The inter-firm social network plays an important mediating role in this theoretical system. An important premise behind this is that previous research has found that firms can actively adjust their social network resources in order to achieve the goal of technological innovation [6].

6.2. Contributions and Suggestions

This study presents a theoretical mechanism with multi factors including scenario factors, inter-firm social networks, and ambidexterity green technological innovation. Simulation modeling and numerical case analysis help this research to discover the firms' multi-objective operation system under this theoretical framework. In this system, the economic transformation, institutional change, and cultural environment requirements make firms produce the strategic goal of maximizing income through balanced ambidexterity green technology innovation and the investment of firms in their social networks conducive to the accomplishment of this strategic goal. The implications of these results can be classified into the following perspectives.

First, for firms and innovators, the simulation model and results of this study show that the strategic goals of firms are influenced by scenario factors. The impact of different scenario factors means that firms should establish different goals. There is a systematic relationship between these goals. The lack of goals based on the judgment of the scenario factors may lead to the failure of the firm's innovation. A reasonable investment in inter-firms' social networks is an important guarantee for establishing correct goals, discovering innovation opportunities, and obtaining innovation resources. Therefore, firms should fully consider the influencing factors and transmission mechanisms in different scenarios, to realize the construction of the social network and the development of innovation abilities that are beneficial to themselves. For example, in the global manufacturing network, China and Chinese firms have a special position. There are many different forms of innovation in China, such as imitation innovation, improved innovation, and independent innovation. Additionally, the whole world is facing a huge problem of environmentally sustainable development. These scenario factors put forward higher requirements on the dynamic management capabilities and network dynamic capabilities of firms in China. Specifically, the simulation analysis of this study shows that the investment of firms in emission reduction is not only conducive to saving their emission costs, but also conducive to firms' reputation.

Second, for the government and policymakers, the suggestion is to promote the construction of a social network platform that is widely participated by all subjects in society. So, the national innovation system is developing towards green, product-oriented, and multi-sectoral cooperation. The simulation paths and results of this study support the view that the inter-firm social network is a type of resource that can be rewarded in varying proportions through active management. Further, the cost reduction and income increase of this resource are not enough only through the operation of the firm itself. The establishment of a more effective social network also requires the government to build a more complete digital technology facility and supply chain development platform, and make the knowledge provided by public R&D more closely integrated with industrial development.

7. Conclusions

The simulation results of this study show that the interaction of the three scenario factors of institutional, economic, and cultural environment in the manufacturing industries results in the need for multi-objective development of the firm.

First, the market-oriented driven force and the social-network-based technological innovation strategies provide a fierce market competition environment and opportunities for innovation for manufacturing firms. In the fierce market competition, the firm's R&D investment determines the firm's profit, and different directions of R&D investment have different impacts on firms' short-term and long-term operations. These orientations include ambidexterity reforms of technology, incremental innovation aimed at reducing costs and improving production efficiency, and breakthrough innovation aimed at changing the current production structure. Incremental innovation is conducive to the short-term growth of firms, while breakthrough innovation is conducive to the long-term growth of firms. Firms must review the situation and determine whether they should pursue short-term growth or long-term breakthrough in the current market scenario.

Second, the cultivation of knowledge-sharing is emphasized by manufacturing industries. The inter-firm social network is important to the development of firms. Therefore, the construction of the national innovation system makes it necessary for firms to pursue their social network benefits. The simulation results of this study show that inter-firm social network is an effective tool for firms to make decisions when reviewing the situation and the situation. The local network's growth is conducive to the improvement of the firm's incremental innovation, while the indirect global network centrality is conducive to the improvement of the firm's breakthrough innovation. However, firms are also limited by the local centrality and global centrality of their social networks to a certain extent, so they should pay attention to their social network development planning to maintain the balance of ambidexterity innovation.

Third, there is currently a need for manufacturing firms to reduce environmental pollution by promoting both government environmental policies and social advocacy for corporate responsibility. The simulation results of this study show that the joint efforts of government, society, and firms can achieve a win-win situation for both enterprise and environmental benefits. Firms' R&D investment in reducing emissions can effectively reduce pollutant emissions and discharge costs [64]. However, firms may face the problem of input costs exceeding cost savings. Therefore, the encouragement and promotion of government and society are necessary for emission reduction activities by firms. The increase in the goodwill of firms and the social network effect could alleviate this problem and even benefit the firms.

However, this study also has the following limitations and corresponding future research plans. First, in the simulation model of this research, only one specific social network among enterprises is set. There are many types of inter-firm social network links, such as links with the customer, links with suppliers, and links with competitors. Different types and strengths of social network links may have different effects on the realization of innovation goals. Therefore, future research needs to extend the model to different types of social network links. Second, the investment benefits of green innovation and social network effects are idealized in the simulation model of this study. In practice, there is a relatively large risk in innovation activities [65]. Previous studies have shown that proper social network embedding is beneficial to reduce this risk, but it cannot eliminate this risk. How to better play the role of inter-firm social networks depends on further research.

Author Contributions: Conceptualization, X.W. and Z.Y.; Methodology, X.W.; Software, H.W. and C.H.; Validation, X.W., H.W., Z.Y., C.H. and B.X.; Formal analysis, X.W.; Investigation, X.W. and B.X.; Resources, Z.Y.; Data curation, H.W., C.H. and B.X.; Writing–original draft, X.W.; Writing–review and editing, X.W. and Z.Y.; Visualization, H.W.; Supervision, Z.Y.; Project administration, Z.Y.; Funding acquisition, X.W. All authors have read and agreed to the published version of the manuscript.

Funding: This research was funded by National Natural Science Foundation of China, grant number 72102121.

Data Availability Statement: The data presented in this study are available on request from the corresponding author.

Appendix A

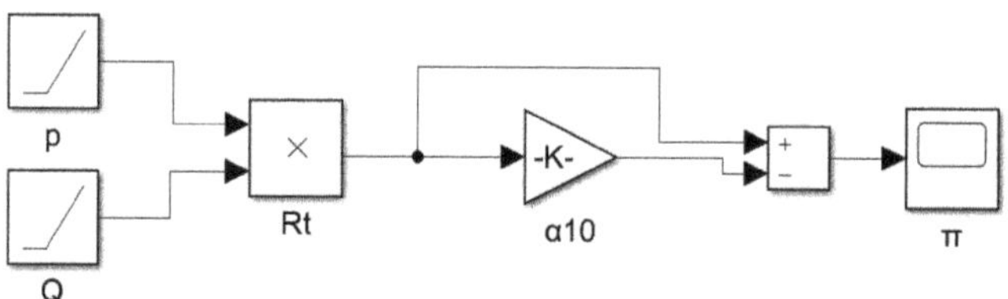

Figure A1. Simulation flowchart of basic situation.

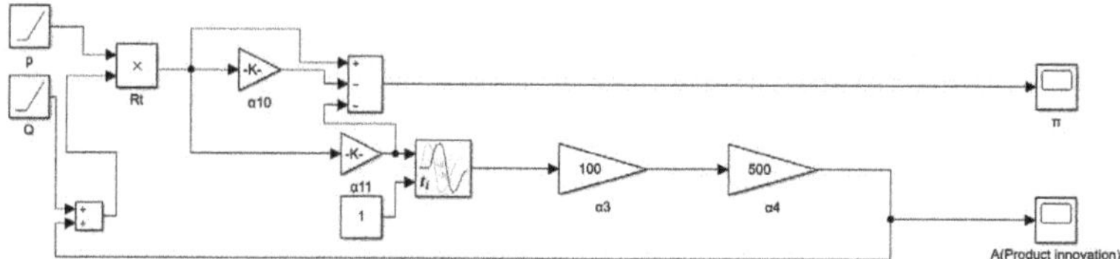

Figure A2. Simulation flowchart of incremental innovation capability enhanced.

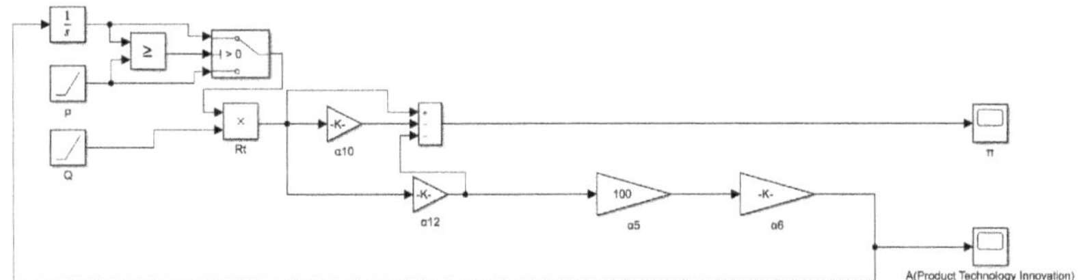

Figure A3. Simulation flow chart of product innovation enhanced.

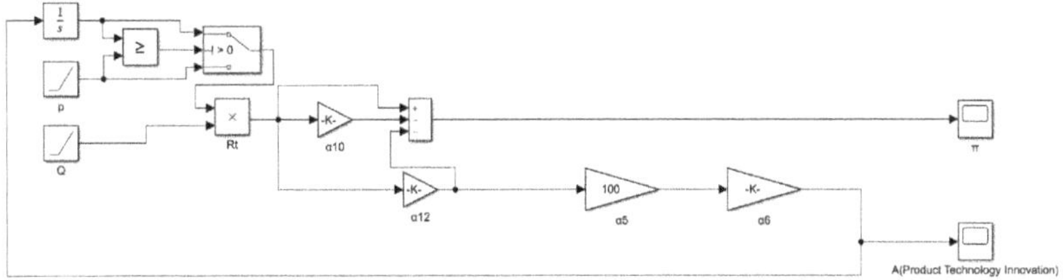

Figure A4. Simulation flow chart of social network effect and ambidexterity innovation.

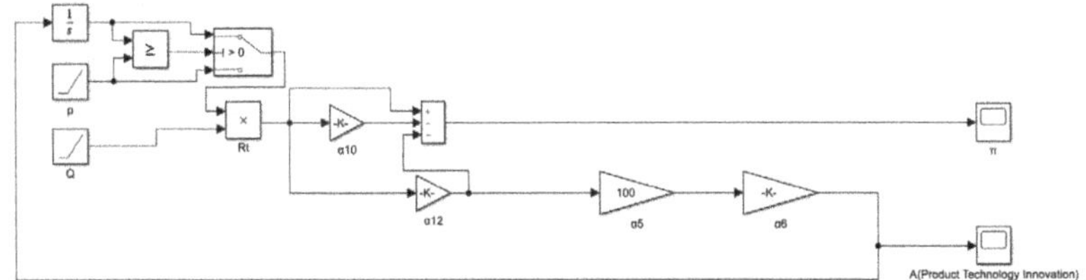

Figure A5. Simulation flowchart of environmental regulation effect.

References

1. Malacina, I.; Teplov, R. Supply chain innovation research: A bibliometric network analysis and literature review. *Int. J. Prod. Econ.* **2022**, *251*, 108540. [CrossRef]
2. Wei, X.; Liu, R.; Chen, W. How the COVID-19 Pandemic Impacts Green Inventions: Evidence from a Quasi-Natural Experiment in China. *Sustainability* **2022**, *14*, 10385. [CrossRef]
3. Yang, Z.; Shang, W.-L.; Zhang, H.; Garg, H.; Han, C. Assessing the green distribution transformer manufacturing process using a cloud-based q-rung orthopair fuzzy multi-criteria framework. *Appl. Energy* **2022**, *311*, 118687. [CrossRef]
4. Binz, C.; Truffer, B.; Li, L.; Shi, Y.; Lu, Y. Conceptualizing leapfrogging with spatially coupled innovation systems: The case of onsite wastewater treatment in China. *Technol. Forecast. Soc. Chang.* **2012**, *79*, 155–171. [CrossRef]
5. Dosi, G.; Soete, L. On the syndemic nature of crises: A Freeman perspective. *Res. Policy* **2022**, *51*, 104393. [CrossRef] [PubMed]
6. Chen, W.; Wei, X.; Liu, R. Influence of Interfirm Social Networks on Technological Innovation and its Time Lag Effect: A Meta-Analysis. *IEEE Access* **2019**, *7*, 167019–167031. [CrossRef]
7. Burt, R.S. *Structural Holes: The Social Structure of Competition*; Harvard University Press: Cambridge, MA, USA, 1992.
8. Granovetter, M.S. The strength of weak ties. *Am. J. Sociol.* **1973**, *78*, 1360–1380. [CrossRef]
9. Ahuja, G. Collaboration networks, structural holes, and innovation: A longitudinal study. *Adm. Sci. Q.* **2000**, *45*, 425–455. [CrossRef]
10. Uzzi, B. The Sources and Consequences of Embeddedness for the Economic Performance of Organizations: The Network Effect. *Am. Sociol. Rev.* **1996**, *61*, 674. [CrossRef]
11. Singh, S.K.; Gupta, S.; Busso, D.; Kamboj, S. Top management knowledge value, knowledge sharing practices, open innovation and organizational performance. *J. Bus. Res.* **2021**, *128*, 788–798. [CrossRef]
12. Wang, C.; Hu, Q. Knowledge sharing in supply chain networks: Effects of collaborative innovation activities and capability on innovation performance. *Technovation* **2020**, *94–95*, 102010. [CrossRef]
13. Zhang, Y.; Hu, H.; Zhu, G.; You, D. The impact of environmental regulation on enterprises' green innovation under the constraint of external financing: Evidence from China's industrial firms. *Environ. Sci. Pollut. Res.* **2022**, *29*, 1–22. [CrossRef]

14. Tarraco, E.L.; Borini, F.M.; Bernardes, R.C.; Navarrete, S.D.S. The Differentiated Impact of the Institutional Environment on Eco-Innovation and Green Manufacturing Strategies: A Comparative Analysis Between Emerging and Developed Countries. *IEEE Trans. Eng. Manag.* **2021**, *68*, 1–12. [CrossRef]

15. Javeed, S.A.; Latief, R.; Jiang, T.; Ong, T.S.; Tang, Y. How environmental regulations and corporate social responsibility affect the firm innovation with the moderating role of Chief executive officer (CEO) power and ownership concentration? *J. Clean. Prod.* **2021**, *308*, 127212. [CrossRef]

16. Stern, N.; Valero, A. Innovation, growth and the transition to net-zero emissions. *Res. Policy* **2021**, *50*, 104293. [CrossRef]

17. Hu, D.; Jiao, J.; Tang, Y.; Xu, Y.; Zha, J. How global value chain participation affects green technology innovation processes: A moderated mediation model. *Technol. Soc.* **2022**, *68*, 101916. [CrossRef]

18. Honorato, C.; de Melo, F.C.L. Industry 4.0: The Case-Study of a Global Supply Chain Company. *J. Ind. Integr. Manag.* **2022**, *7*, 1–18. [CrossRef]

19. Lv, X.; Lu, X. Green Growth or Gray Growth: Measuring Green Growth Efficiency of the Manufacturing Industry in China. *Systems* **2022**, *10*, 255. [CrossRef]

20. Lei, Y.; Guo, Y.; Zhang, Y.; Cheung, W. Information technology and service diversification: A cross-level study in different innovation environments. *Inf. Manag.* **2021**, *58*, 103432. [CrossRef]

21. Gu, Y.; Hu, L.; Hou, C. Leveraging diverse ecosystem partners for innovation: The roles of regional innovation environment and partnership heterogeneity. *Econ. Res. -Ekon. Istraz.* **2022**, *35*, 1–20. [CrossRef]

22. Jacob, J.; Mei, M.-Q.; Gunawan, T.; Duysters, G. Ambidexterity and innovation in cluster SMEs: Evidence from Indonesian manufacturing. *Ind. Innov.* **2022**, *29*, 948–968. [CrossRef]

23. Bhatti, S.H.; Ahmed, A.; Ferraris, A.; Hussain, W.; Wamba, S.F. Big data analytics capabilities and MSME innovation and performance: A double mediation model of digital platform and network capabilities. *Ann. Oper. Res.* **2022**, *318*, 1–24. [CrossRef]

24. Wei, J.; Wu, A.; Peng, X. A Study on China's Strategic Management: The Problem of Circumstances and the Theoretical Front-line. *Manag. World* **2014**, *20*, 167–171.

25. Jin, Y.; Ji, S.; Liu, L.; Wang, W. Business model innovation canvas: A visual business model innovation model. *Eur. J. Innov. Manag.* **2021**, *25*, 1469–1493. [CrossRef]

26. Luo, W.; Zhang, C.; Li, M. The influence of corporate social responsibilities on sustainable financial performance: Mediating role of shared vision capabilities and moderating role of entrepreneurship. *Corp. Soc. Responsib. Environ. Manag.* **2022**, *29*, 1266–1282. [CrossRef]

27. Zhai, D.; Zhao, X.; Bai, Y.; Wu, D. Effective Evaluation of Green and High-Quality Development Capabilities of Enterprises Using Machine Learning Combined with Genetic Algorithm Optimization. *Systems* **2022**, *10*, 128. [CrossRef]

28. Gao, L.; Ma, W. Open innovation: Connotation, framework and Chinese context. *Manag. World* **2014**, *6*, 157–169.

29. Dahiyat, S.E.; Khasawneh, S.M.; Bontis, N.; Al-Dahiyat, M. Intellectual capital stocks and flows: Examining the mediating roles of social capital and knowledge transfer. *VINE J. Inf. Knowl. Manag. Syst.* **2021**, *53*, 11–42. [CrossRef]

30. Martin-Rios, C.; Erhardt, N.L.; Manev, I.M. Interfirm collaboration for knowledge resources interaction among small innovative firms. *J. Bus. Res.* **2022**, *153*, 206–215. [CrossRef]

31. Lis, A.M. Development of proximity in cluster organizations. *Entrep. Sustain. Issues* **2020**, *8*, 116–132. [CrossRef]

32. Hu, C.; Yang, H.; Yin, S. Insight into the Balancing Effect of a Digital Green Innovation (DGI) Network to Improve the Performance of DGI for Industry 5.0: Roles of Digital Empowerment and Green Organization Flexibility. *Systems* **2022**, *10*, 97. [CrossRef]

33. Petrenko, Y.; Vechkinzova, E.; Antonov, V. Transition from the industrial clusters to the smart specialization: A case study. *Insights Reg. Dev.* **2019**, *1*, 118–128. [CrossRef] [PubMed]

34. Gordon, I.R.; McCann, P. Industrial clusters: Complexes, agglomeration and/or social networks? *Urban Stud.* **2000**, *37*, 513–532. [CrossRef]

35. Loiseau, E.; Saikku, L.; Antikainen, R.; Droste, N.; Hansjürgens, B.; Pitkänen, K.; Leskinen, P.; Kuikman, P.; Thomsen, M. Green economy and related concepts: An overview. *J. Clean. Prod.* **2016**, *139*, 361–371. [CrossRef]

36. Donner, M.; de Vries, H. How to innovate business models for a circular bio-economy? *Bus. Strateg. Environ.* **2021**, *30*, 1932–1947. [CrossRef]

37. Awan, U. Industrial Ecology in Support of Sustainable Development Goals. In *Responsible Consumption and Production*; Springer: Cham, Switzerland, 2020; pp. 370–380.

38. Mazzoni, F. Circular economy and eco-innovation in Italian industrial clusters. Best practices from Prato textile cluster. *Insights Reg. Dev.* **2020**, *2*, 661–676. [CrossRef]

39. Amjad, M.S.; Rafique, M.Z.; Khan, M.A. Leveraging Optimized and Cleaner Production through Industry 4.0. *Sustain. Prod. Consum.* **2021**, *26*, 859–871. [CrossRef]

40. Wang, M.; Wang, H. Knowledge search and innovation performance: The mediating role of absorptive capacity. *Oper. Manag. Res.* **2022**, *15*, 1060–1070. [CrossRef]

41. Wei, X.; Chen, W.; Lin, C. Evolution, Classification, and Relationships of the Theories of Innovation Capabilities. *Q. J. Manag.* **2019**, *13*, 112–157.

42. Randhawa, K.; Wilden, R.; Gudergan, S. How to innovate toward an ambidextrous business model? The role of dynamic capabilities and market orientation. *J. Bus. Res.* **2021**, *130*, 618–634. [CrossRef]

43. Qin, X. The Impact of Interregional Collaboration on Multistage R&D Productivity and Their Interregional Gaps in Chinese Provinces. *Mathematics* **2022**, *10*, 1310. [CrossRef]
44. Zhou, Y.; Zhu, H.; Zhu, L.; Liu, G.; Zou, Y. Be close to government and academy: TMT social capital, network prestige and firm's innovation performance. *Chin. Manag. Stud.* 2022, *ahead-of-print*. [CrossRef]
45. Mansell, R. Adjusting to the digital: Societal outcomes and consequences. *Res. Policy* **2021**, *50*, 104296. [CrossRef]
46. Cancela, B.L.; Coelho, A.; Neves, M.E.D. Greening the business: How ambidextrous companies succeed in green innovation through to sustainable development. *Bus. Strat. Environ.* **2022**, *31*, 1–15. [CrossRef]
47. Vega-Jurado, J.; Gutiérrez-Gracia, A.; Fernández-De-Lucio, I. Does external knowledge sourcing matter for innovation? Evidence from the Spanish manufacturing industry. *Ind. Corp. Chang.* **2009**, *18*, 637–670. [CrossRef]
48. Wang, D.; Wan, K.; Yang, J. Measurement and evolution of eco-efficiency of coal industry ecosystem in China. *J. Clean. Prod.* **2019**, *209*, 803–818. [CrossRef]
49. Ahuja, G. The duality of collaboration: Inducements and opportunities in the formation of interfirm linkages. *Strateg. Manag. J.* **2000**, *21*, 317–343. [CrossRef]
50. Liu, W.; Yang, J. The Evolutionary Game Theoretic Analysis for Sustainable Cooperation Relationship of Collaborative Innovation Network in Strategic Emerging Industries. *Sustainability* **2018**, *10*, 4585. [CrossRef]
51. Sueyoshi, T.; Goto, M. Weak and strong disposability vs. natural and managerial disposability in DEA environmental assessment: Comparison between Japanese electric power industry and manufacturing industries. *Energy Econ.* **2012**, *34*, 686–699. [CrossRef]
52. West, J.; Bogers, M. Leveraging External Sources of Innovation: A Review of Research on Open Innovation. *J. Prod. Innov. Manag.* **2014**, *31*, 814–831. [CrossRef]
53. Miles, M.B.; Huberman, A.M. *Qualitative Data Analysis: An Expanded Sourcebook*; Sage Publications, Inc.: Thousand Oaks, CA, USA, 1994.
54. Lundvall, B.-A. *National Systems of Innovation: Towards a Theory of Innovation and Interactive Learning*; Pinter Publishers: London, UK, 1992; pp. 318–330.
55. Xie, X.; Fang, L.; Zeng, S.; Huo, J. How does knowledge inertia affect firms product innovation? *J. Bus. Res.* **2016**, *69*, 1615–1620. [CrossRef]
56. Li, X.; Zhou, Y.; Xue, L.; Huang, L. Roadmapping for industrial emergence and innovation gaps to catch-up: A patent-based analysis of OLED industry in China. *Int. J. Technol. Manag.* **2016**, *72*, 105–143. [CrossRef]
57. Si, S.; Chen, H. A literature review of disruptive innovation: What it is, how it works and where it goes. *J. Eng. Technol. Manag.* **2020**, *56*, 101568. [CrossRef]
58. Schmookler, J. Invention and economic growth. In *Invention and Economic Growth*; Harvard University Press: Cambridge, MA, USA, 2013.
59. Parrilli, M.D.; Balavac, M.; Radicic, D. Business innovation modes and their impact on innovation outputs: Regional variations and the nature of innovation across EU regions. *Res. Policy* **2020**, *49*, 104047. [CrossRef] [PubMed]
60. Shi, D.; Ding, H.; Wei, P.; Liu, J. Can smart city construction reduce environmental pollution. *China Ind. Econ.* **2018**, *6*, 117–135.
61. Wang, F.; Qian, R. Study on Enterprise Efficiency of Environmental Protection Investment in China—Based on the Data of 704 A-Share Listed Companies from 2010 to 2014. *Collect. Essays Financ. Econ.* **2017**, *32*, 43–52.
62. He, X.; Xia, M.; Li, X.; Lin, H.; Xie, Z. How Innovation Ecosystem Synergy Degree Influences Technology Innovation Performance—Evidence from China's High-Tech Industry. *Systems* **2022**, *10*, 124. [CrossRef]
63. Epicoco, M. Technological Revolutions and Economic Development: Endogenous and Exogenous Fluctuations. *J. Knowl. Econ.* **2021**, *12*, 1437–1461. [CrossRef]
64. Chen, X.; Liu, Y.; Mcelroy, M. Transition towards carbon-neutral electrical systems for China: Challenges and perspectives. *Front. Eeg. Manag.* **2022**, *9*, 504–508. [CrossRef]
65. Bai, S.; Bi, X.; Han, C.; Zhou, Q.; Shang, W.-L.; Yang, M.; Wang, L.; Ieromonachou, P.; He, H. Evaluating R&D efficiency of China's listed lithium battery enterprises. *Front. Eeg. Manag.* **2022**, *9*, 473–485.

Article

A Gaussian-Shaped Fuzzy Inference System for Multi-Source Fuzzy Data

Yun Zhang [1] and Chaoxia Qin [2,*]

[1] College of Finance and Economics, Sichuan International Studies University, Chongqing 400031, China
[2] College of Computer Science, Sichuan University, Chengdu 610065, China
* Correspondence: qinchaoxia2@stu.scu.edu.cn

Abstract: Fuzzy control theory has been extensively used in the construction of complex fuzzy inference systems. However, we argue that existing fuzzy control technologies focus mainly on the single-source fuzzy information system, disregarding the complementary nature of multi-source data. In this paper, we develop a novel Gaussian-shaped Fuzzy Inference System (GFIS) driven by multi-source fuzzy data. To this end, we first propose an interval-value normalization method to address the heterogeneity of multi-source fuzzy data. The contribution of our interval-value normalization method involves mapping heterogeneous fuzzy data to a unified distribution space by adjusting the mean and variance of data from each information source. As a result of combining the normalized descriptions from various sources for an object, we can obtain a fused representation of that object. We then derive an adaptive Gaussian-shaped membership function based on the addition law of the Gaussian distribution. GFIS uses it to dynamically granulate fusion inputs and to design inference rules. This proposed membership function has the advantage of being able to adapt to changing information sources. Finally, we integrate the normalization method and adaptive membership function to the Takagi–Sugeno (T–S) model and present a modified fuzzy inference framework. Applying our methodology to four datasets, we confirm that the data do lend support to the theory implying the improved performance and effectiveness.

Keywords: multi-source fuzzy data; normalization method; membership function; Gaussian-shaped fuzzy inference

Citation: Zhang, Y.; Qin, C. A Gaussian-Shaped Fuzzy Inference System for Multi-Source Fuzzy Data. *Systems* **2022**, *10*, 258. https://doi.org/10.3390/systems10060258

Academic Editors: Zaoli Yang, Yuchen Li and Ibrahim Kucukkoc

Received: 30 October 2022
Accepted: 13 December 2022
Published: 15 December 2022

Publisher's Note: MDPI stays neutral with regard to jurisdictional claims in published maps and institutional affiliations.

1. Introduction

Traditional fuzzy control models have achieved remarkable success in making inferences for single-source fuzzy information systems. They have been increasingly applied to intelligent driving [1–3], intelligent medical [4–6], intelligent factories [7–9], and various other fields. It is believed that information fusion enhances the descriptive ability of objects through the use of data redundancy and complementarity [10–12]. A Fuzzy Inference System (FIS) could be improved by combining information fusion theory with fuzzy control theory to enable the system to perform better in terms of control and decision-making tasks than before. Motivated by this, we examine a fuzzy inference model that is driven by multi-source fuzzy data and look at how multi-source fuzzy data affect the precision of the inference model.

The fuzzy set (FS) and the rough set (RS) are two commonly used methods for describing fuzzy data or knowledge in a different way. As far back as 1965, scholar Zade was the first to propose the fuzzy set theory [13]. Its essence is to mine the decision-making value of fuzzy data by constructing membership functions and performing fuzzy set operations. Zadeh's fuzzy set is modeled as:

$$A = A(x_1)/x_1 + A(x_2)/x_2 + \cdots + A(x_n)/x_n, \tag{1}$$

where $A(x_i)$ represents the membership degree of x_1 in set A, and "+" represents the component connection symbol in a fuzzy set; see [13]. Rough set theory was first proposed by Pawlak in 1982 [14]. The core idea is to use equivalence relation R to construct the equivalence class of object x in universe U. The approximation set composed of equivalent objects is called the lower approximation set, denoted as:

$$\underline{R}(x) = \{x \in U | [x]_R \subseteq U\}, \tag{2}$$

see [15]. In contrast, the approximation set composed of equivalent objects and similar objects is called the upper approximation set, denoted as:

$$\overline{R(x)} = \{x \in U | [x]_R \cap U \neq \varnothing\}, \tag{3}$$

see [15]. A key aspect of using fuzzy set theory to describe fuzzy knowledge lies in modeling membership functions [16]. There are four candidates for membership functions: bell-shaped curve function [17], S-shaped curve function [18], Z-shaped curve function [19], and π-shaped curve function [20]. Theoretically, the bell-shaped curve function is most widely used because it follows the law of large numbers and the central limit theorem [21,22].

Unfortunately, bell-shaped curve functions do not directly work for fuzzy systems involving multiple sources, for two reasons. First, it requires the function input to be a continuous and accurate real number. Second, it requires data from different sources to be analyzed using the same metrics so that comparability of data can be ensured. Our solution to these two problems is to use interval values to represent fuzzy data (such as measurement errors, degrees, spatial and temporal distances, emotions, etc.) and then to normalize heterogeneous fuzzy data to exact real numbers using interval normalization. Upon normalization, these real numbers will all be subjected to the same quantization factor (mean and variance). Additionally, we propose an adaptive membership function model that considers the dynamics of a multi-source environment, such as different information sources joining and leaving the system. With the aid of the membership function model, not only are we able to represent the overall distribution of data from multiple sources, but we are also able to adjust the membership value of the input variables to accommodate dynamic changes in the data. Combining the normalization method with the adaptive membership function, we develop a Gaussian-shaped Fuzzy Inference System (GFIS) driven by multi-source fuzzy data.

Our article makes two contributions to this literature. The first contribution is to propose an interval normalization method that is based on the normalization idea of Z-scores [23], and describe the meaning of fuzzy data by the standard deviations of each data value from the mean. The interval normalization method can be formalized as:

$$N_x = \frac{0.5 \times (\underline{x} + \overline{x}) - \mu}{\sigma}, \tag{4}$$

where $x = [\underline{x}, \overline{x}]$ is an interval. The greatest challenge and innovation of our method lies in calculating mean μ and variance σ for fuzzy data. In this work, we present formal models for mean and variance of fuzzy data, and we develop an approach for normalizing fuzzy data (interval value) with these formal models. The interval-value representation and normalization of fuzzy data are the premise of designing the fuzzy model of input variables in GFIS.

The second contribution of this article consists of deriving a Gaussian-shaped membership function for the input variable in GFIS. This function can be used to analyze normal-distributed data, and we denote it as:

$$\mu(x) = f(N(x|0, m)), \tag{5}$$

where m is the number of information sources. A significant advantage of the proposed membership function lies in its ability to be adapted to the change of information sources. As a general rule, membership functions based on Gaussian distributions require dynamic mean and variance parameters to be specified as hyperparameters. When function variables are normalized in advance, we can ensure that the mean value of function variables is always 0 and that variance increases linearly with the number of information sources available. Therefore, we do not have to incur large computational costs to obtain the hyperparameters of Gaussian-shaped membership functions. Additionally, a Gaussian-shaped membership function can handle dynamic changes in information sources effectively. It is important to note that the GFIS, which comprises the normalization method and Gaussian-shaped membership function, is thus applicable to multi-source fuzzy environments.

The rest of the article is organized as follows: Section 2 provides a selected literature overview. Section 3 introduces preliminaries. Section 4 outlines the methodology proposed. Section 5 details the experimental design and Section 6 reports the experimental results. In Section 7, we present the conclusions.

2. Related Work

Due to its smoothness and ability to reflect people's thinking characteristics, the Gaussian membership function has been widely used. Below, we provide a brief description of several Gaussian membership function methods and their applications in the literature. It is worth mentioning that our method (among the three alternatives discussed) is the only one that takes into account both data heterogeneity and fuzzy data fusion.

Hosseini et al. presented a Gaussian interval type-2 membership function and applied it to the classification of lung CAD (Computer Aided Design) [24]. Type-2 interval membership functions extend type-1 interval membership functions. In the type-2 fuzzy system, the membership of the type-1 fuzzy set is also characterized by a fuzzy set, which can improve the ability to deal with uncertainties. However, there are additional theoretical and computational challenges associated with the Gaussian interval type-2 membership function.

In Kong et al.'s study, a Gaussian membership function combined with a neural network model was designed to help diagnose automobile faults [25]. The system has been proven to perform better in terms of reasoning accuracy than either a network model or a fuzzy inference model alone. However, it relies on a single source of information to derive fuzzy inferences from multivariable data.

Li et al. proposed a Gaussian fuzzy inference model for multi-source homogeneous data [26]. The model uses three indicators (center representation, population, and density function) to describe single-source information and adopts a mixed Gaussian model to represent multi-source fusion information. This model does not address the heterogeneity and fuzziness of multi-source data. In addition, the model fails to account for differences between variables in a multidimensional dataset in terms of magnitude and measurement standards.

In previous research, different fuzzy inference models and membership functions were proposed for various practical applications. We propose a Gaussian fuzzy control inference model to solve the fuzzy inference problem in a multi-source fuzzy environment. Our model can improve medical CAD diagnostic accuracy by fusing X-ray images from different institutions. For automobile fault diagnosis, our model can describe each diagnosis parameter with fuzzy data. By further modeling these fuzzy parameters with fuzzy sets, we enhance fuzzy systems' ability to cope with uncertainty. The fuzzy normalization model for multi-source data fusion enables us to map multivariate data to a dimensionless distribution space to ensure that the data metrics are aligned.

3. Preliminary

The Fuzzy Inference System (FIS), also known as a fuzzy system, is a software application that utilizes fuzzy set theory and fuzzy inference technology to process fuzzy information. In order to illustrate FIS's application scenarios, let us take car following as an example. Due to the fact that the driver's behavior is fuzzy and uncertain during the process of controlling the car, it can be difficult to accurately describe the driver's behavior. When a car follows another car, it is necessary to maintain a safe distance in order to ensure the safety of drivers. FIS can be used to control the distance between a car and the one it is following. Specifically, based on the driver's experience, the fuzzy rules of FIS for car following are summarized as follows. When a driver believes that the relative distance is far greater than the safe distance and the relative speed is fast, he or she accelerates appropriately. This will make the difference between the relative and safe distances as small as possible.

The classical FIS consists of five basic components: the definition of inputs and outputs; the construction of fuzzification strategies; the construction of knowledge bases; the design of fuzzy inference mechanisms; and the defuzzification of output. See Figure 1 for details.

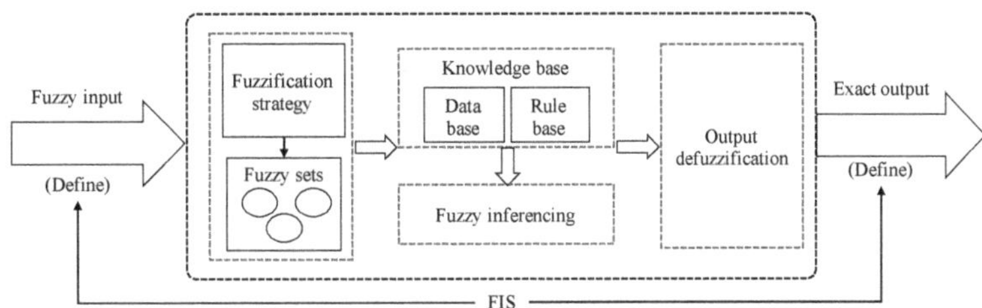

Figure 1. The architecture of a fuzzy inference system (FIS). The architecture of FIS describes the mapping process from a given input to an output. The process consists of five parts: defining input and output, formulating a fuzzification strategy, building a knowledge base, designing fuzzy inference mechanism, and defuzzification of output. See the main text for details.

(1) The definition of inputs and outputs

In an FIS, the inputs and outputs correspond to the observation variables and operation variables, respectively. Definition of inputs and outputs includes determining parameters, variable numbers, data formats, etc. An FIS that has one input variable is called a single variable FIS and an FIS that has more than one input variable is called a multivariable FIS. FIS driven by multi-source fuzzy data encounters challenges in normalizing heterogeneous fuzzy data due to the fact that the traditional method of normalizing data fails in this situation.

(2) The construction of fuzzification strategies

Fuzzification is the process of assigning each input variable to a fuzzy set with a certain degree of membership. An input variable can be either an exact value or fuzzy data with noise [27]. It is therefore necessary to consider the format of input variables when developing a strategy for fuzzification. In particular, fuzzy data are typically presented in discrete nominal formats or in aggregate interval-value formats, which makes mathematical fitting of membership functions challenging. In this paper, we use interval normalization to convert heterogeneous fuzzy data into continuous and exact values. We also use a math function to derive the membership function for fuzzy data.

(3) The construction of knowledge bases

In a knowledge base, there are two parts: a database and a rule base, respectively [28]. Among the features in the database are membership functions, scale transformation factors, and fuzzy set variables. The rule base contains some fuzzy control conditions and fuzzy logic relationships.

(4) The design of fuzzy inference mechanisms

Fuzzy inference uses fuzzy control conditions and fuzzy logics to predict the future status of operating variables. This is the core of an FIS. In an FIS, syllogisms [29–31] are commonly used to make inferences, which can be expressed as follows:

Truth: IF x_1 is $A_1, \cdots$, and x_n is A_n, THEN y is B.

Condition: x_1 is A_1', $\cdots$, and x_n is A_n'.

The inference result: y is B'.

According to the FIS, truth is represented by fuzzy implication relations, denoted as $A \xrightarrow{f(x)} B$. The inference result is derived from the combination of fuzzy conditions and fuzzy logics.

(5) The defuzzification of output

In general, the result derived from the FIS is a fuzzy value or a set, which must be deblurred to identify a clear control signal or a decision output. Most commonly used defuzzification methods include the maximum membership [32], the weighted average [33], and the center of gravity [34].

Maximum Membership Given k FIS submodels, the output of each FIS submodel is y_i with the membership degree of $\mu(x_i)$. The final output of the FIS model is given by:

$$y = y_i,\tag{6}$$

where

$$i = \arg\max_i\{\mu(x_1), \mu(x_2), \ldots, \mu(x_k)\},\tag{7}$$

see [32].

Weighted Average Given k FIS submodels, the output of each FIS submodel is y_i with the membership degree of $\mu(x_i)$. The final output of the FIS model is:

$$y = \frac{\sum_{i=1}^{i=k} y_i \times \mu(x_i)}{k},\tag{8}$$

as k is the number of sub FIS models, see [33].

Center of Gravity Given k FIS submodels, the output of each FIS submodel is y_i with the membership degree of $\mu(x_i)$. The final output of the FIS model is:

$$y = \frac{\sum_{i=1}^{i=k}(\mu(x_i) \times y_i)}{\sum_{i=1}^{i=k} \mu(x_i)},\tag{9}$$

as k is the number of FIS submodels, see [34].

4. The Methodology

4.1. Normalization of Heterogeneous Fuzzy Data

We begin this section by defining mean and variance in the interval-value universe by referring to mean and variance of real numbers.

Definition 1 (interval-value mean). *Let* $X = \{x_1, \cdots, x_i, \cdots, x_n\}$ *denote the universe* X *of intervals, where* $x_i = [\underline{x_i}, \overline{x_i}]$ *represents the i-th interval. The interval-value mean is the sum of the average values of all intervals divided by the number of intervals, denoted as:*

$$\mu = \frac{\sum_{i=1}^{i=n} 0.5 \times (\underline{x_i} + \overline{x_i})}{n}.\tag{10}$$

Definition 2 (interval-value variance). *Let* $X = \{x_1, \cdots, x_i, \cdots, x_n\}$ *denote the universe* X *of intervals, where* $x_i = [\underline{x_i}, \overline{x_i}]$ *represents the i-th interval. The interval-value variance is the weighted square sum of the deviation degree between each interval-value mean and the overall mean of the interval-value universe, denoted as:*

$$\sigma^2 = \frac{\sum_{i=1}^{i=n}\left(0.5 \times (\underline{x_i} + \overline{x_i}) - \mu\right)^2}{n}\tag{11}$$

with $\frac{1}{n}$ as the weight.

In the next step, we develop the normalization model for multi-source fuzzy data with interval-value format using the z-score normalization method [23] in the real number universe. Interval-value normalization is the process of mapping heterogeneous fuzzy data to the same data distribution space with a mean of 0 and a variance of 1. By transforming the data metrics of various platforms or organizations, interval-value normalization facilitates the analysis of multi-source fuzzy data. It is shown in Equation (12) that the core for normalizing interval values is μ and σ, where μ is the mean and σ is the standard deviation of the interval-value universe. The interval-value normalization method is applicable to the case where the maximum and minimum values in the universe are unknown:

$$N_{x_i} = \frac{0.5 \times \left(\underline{x_i} + \overline{x_i}\right) - \mu}{\sigma}. \tag{12}$$

Given a k-dimensional interval value x_i in which components (dimensions) are mutually independent, x_i is written as $\left[(\underline{x_{i1}}, \underline{x_{i2}}, \cdots, \underline{x_{ik}}), (\overline{x_{i1}}, \overline{x_{i2}}, \cdots, \overline{x_{ik}})\right]$. At this point, the mean of the interval-value universe is a k-dimensional vector, denoted as:

$$\mu = \left(\frac{\sum_{i=1}^{i=n} 0.5 \times \left(\underline{x_{i1}} + \overline{x_{i1}}\right)}{n}, \cdots, \frac{\sum_{i=1}^{i=n} 0.5 \times \left(\underline{x_{ik}} + \overline{x_{ik}}\right)}{n}\right). \tag{13}$$

The variance of interval-value universe is a diagonal matrix, given by Equation (14), where σ_i^2 is the variance of the i-th dimension. All non-diagonal elements of matrix σ^2 are zero since the different dimensions of the data are independent. The normalization result corresponding to interval x_i is also a k-dimensional vector, given by Equation (15):

$$\sigma^2 = \begin{bmatrix} \sigma_1^2 & \cdots & 0 \\ \vdots & \ddots & \vdots \\ 0 & \cdots & \sigma_k^2 \end{bmatrix}. \tag{14}$$

$$N_{x_i} = \left(\frac{0.5 \times \left(\underline{x_{i1}} + \overline{x_{i1}}\right) - \mu_1}{\sigma_1}, \cdots, \frac{0.5 \times \left(\underline{x_{ik}} + \overline{x_{ik}}\right) - \mu_k}{\sigma_k}\right). \tag{15}$$

On the basis of the above-mentioned notions, we propose an algorithm for normalizing multidimensional intervals (Algorithm 1). Algorithm 1 is divided into three program loops: (1) looping $n \times k$ times to obtain the mean of each dimension of the interval-value universe (see line 1–6), whose time complexity is $O(n \times k)$; (2) looping $n \times k$ times to obtain the standard deviation of each dimension of the interval-value universe (see line 7–12), whose time complexity is $O(n \times k)$; and (3) looping $n \times k$ times to obtain the normalized value of each dimension of each interval object (see lines 13–17), whose time complexity is also $O(n \times k)$. Hence, the total time complexity of Algorithm 1 is $O(3 \times n \times k)$.

Algorithm 1: Interval-value normalization

---

input : A set of intervals $X = \{r_1, r_2, \cdots, r_n\}$.
output: A set of real numbers

1 // Each dimension. **for** $i \leftarrow 1$ **to** k **do**
2   // Each interval. **for** $j \leftarrow 1$ **to** n **do**
3      $sum[i]+ = 0.5 \times (x_{ji} + x_{ji})$
4   **end**
5    $\mu[i] = sum[i]/n$
6 **end**
7 // Each dimension. **for** $i \leftarrow 1$ **to** k **do**
8   // Each interval. **for** $j \leftarrow 1$ **to** n **do**
9      $E[i] = \left(0.5 \times \left(x_{ji} + \overline{x_{ji}}\right) - \mu[i]\right)^2$
10   **end**
11    $\sigma[i] = \sqrt{\frac{E[i]}{n}}$
12 **end**
13 // Each interval. **for** $i \leftarrow 1$ **to** n **do**
14   // Each dimension. **for** $j \leftarrow 1$ **to** k **do**
15      $N_{x_{ij}} = \frac{0.5 \times \left(x_{ij} + \overline{x_{ij}}\right) - \mu_j}{\sigma_j}$
16   **end**
17 **end**
18 **return** N_X

4.2. Membership Function Modeling

The method above normalizes a random fuzzy dataset to a random real number set N_X with a mean of 0 and a variance of 1. It has now been possible to resolve the heterogeneity and fuzziness of multi-source data. Therefore, we can fit the membership function using the distribution function of the data. According to the definition of normal distribution, the distribution law of a one-dimensional random number can be expressed by a standard univariate normal distribution with $N(N_x|0,1) = \frac{1}{\sqrt{2\pi}} \times e^{\frac{-N_x^2}{2}}$. Multidimensional random real numbers correspond to a standard multivariate normal distribution, given by:

$$N(N_x|0, E) = \frac{1}{\left(\sqrt{2\pi}\right)^k} \times e^{\frac{-N_x^T \cdot N_x}{2}}. \tag{16}$$

when $k = 1$, $N(N_x|0, E)$ is equivalent to $N(N_x|0,1)$. We thus use $N(N_x|0, E)$ to uniformly identify the distribution law of normalized data. The integration of standard normal distribution is used to determine the probability that the corresponding value (fuzzy data) will occur. Based on this, we construct a membership function for fuzzy data x, which can be denoted as:

$$\mu(N_x) = \frac{\int_{-\infty}^{x} N(N_x|0, E)dx - \int_{-\infty}^{min} N(N_x|0, E)dx}{\int_{-\infty}^{max} N(N_x|0, E)dx - \int_{-\infty}^{min} N(N_x|0, E)dx}, \tag{17}$$

where $\int_{-\infty}^{max} N(N_x|0, E)dx - \int_{-\infty}^{min} N(N_x|0, E)dx$ is the total probability of fuzzy set $[min, max]$, and $\mu(N_x)$ is the membership degree of the normalized value N_x or the corresponding fuzzy data x.

Considering the dynamic nature of the multi-source environment, we design an adaptive membership function model to increase FIS's adaptability. According to the addition law of normal distribution [35], namely:

$$N(x_1|u_1, \sigma_1^2) + N(x_2|u_2, \sigma_2^2) = N(x_1 + x_2|u_1 + u_2, \sigma_1^2 + \sigma_2^2), \tag{18}$$

the data from several same data distribution spaces can be added and the sum still meets the normal distribution. In the event that we consider the data from the *i*-th information source as belonging to a standard normal distribution $N(N_{x_i}|0, E)$, the fusion data from m independent information sources are still grouped according to the normal distribution $N(N_x|0, \sum_{i=1}^{i=m} E_i)$, where $N_x = \sum_{i=1}^{i=m} N_{x_i}$ is the fusion data of multi-source normalized data. Accordingly, the adaptive membership function of fuzzy data x is modeled as:

$$\mu(N_x) = \frac{\int_{-\infty}^{x} N(N_x|0, E \times m)dx - \int_{-\infty}^{min} N(N_x|0, E \times m)dx}{\int_{-\infty}^{max} N(N_x|0, E \times m)dx - \int_{-\infty}^{min} N(N_x|0, E \times m)dx}, \tag{19}$$

where the *min* and *max* are the minimum value and the maximum value in the fusion dataset, respectively.

According to Equation (19), as long as the law of large numbers and the central limit theorem are satisfied, we can continuously calculate the membership value of fusion data as its variance changes linearly with the number of information sources.

4.3. Integration into T–S Model

As shown in Figure 2, the T–S model [36] is a fuzzy inference model that divides the global nonlinear reasoning system into several simple linear reasoning systems. The outputs of multiple subsystems are fused into a final decision result. Instead of the T–S model, an alternative model may be used, such as the Mamdani model. The T–S model is chosen for its simplicity of design. Both the IF and Then parts of the Mamdani model are ambiguous. Therefore, the Mamdani model needs to rely on prior knowledge to design reliable Then rules. In contrast, the output variable of the T–S model is a precise constant or a linear function, and its degree of automation is higher. The working principle of the multi-input T–S model is as follows:

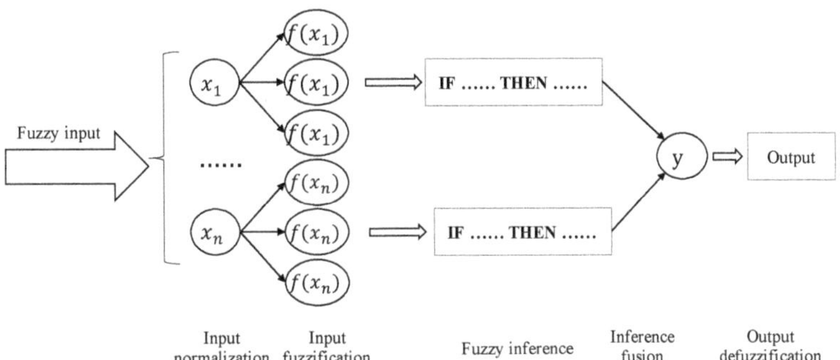

Figure 2. Sample diagram of a T–S model. A T–S model is a nonlinear system characterized by a set of "IF–THEN" fuzzy rules. Each rule indicates a subsystem, and the entire T–S model is a linear combination of all these subsystems. See the text for details.

(1) As a result of normalizing and fuzzifying the input, the input is mapped to the fuzzy set of the input universe, which corresponds to a membership function $f(x)$.
(2) It is important to select the right fuzzy rules and inference methods in order to derive results from sub-T–S models.
(3) The results of all sub-T–S models are merged to yield a total fuzzy output;
(4) The fuzzy output is deblurred to arrive at the final decision.

Example 1. *Suppose that two input fuzzy sets are known as* $S_1 = \{x|x \in [1,2]\}$ *with the membership function:*

$$f(x_1) = \frac{\int_{-\infty}^{x} N(x|0,1)dx - \int_{-\infty}^{1} N(x|0,1)dx}{\int_{-\infty}^{2} N(x|0,1)dx - \int_{-\infty}^{1} N(x|0,1)dx}, \tag{20}$$

and $S_2 = \{x|x \in [-1,1]\}$ *with the membership function:*

$$f(x_2) = \frac{\int_{-\infty}^{x} N(x|0,1)dx - \int_{-\infty}^{-1} N(x|0,1)dx}{\int_{-\infty}^{1} N(x|0,1)dx - \int_{-\infty}^{-1} N(x|0,1)dx}. \tag{21}$$

There are three T–S fuzzy rules:

(1) *R1: if $x_1 \in S_1$, then $y_1 = x_1$;*
(2) *R2: if $x_2 \in S_2$, then $y_2 = 2 \times x_2$;*
(3) *R3: if $x_1 \in S_1$ and $x_2 \in S_2$, then $y_3 = x_1 + x_2$.*

When $x_1 = 1.5$ and $x_2 = 1$ are observed, we obtain:

(1) *R1: $f(x_1) = 0.68$ and $y_1 = 1.5$;*
(2) *R2: $f(x_2) = 1$ and $y_2 = 2$;*
(3) *R3: $f(x_1) = 0.68$ and $f(x_2) = 1$, and $y_3 = 1.5 + 1 = 2.5$.*

We first use the direct product method [37] to calculate the weight of each sub-T–S model: $w_1 = f(x_1) = 0.68$, $w_2 = f(x_2) = 1$, and $w_3 = f(x_1) \times f(x_2) = 0.68$. Then, calculate the total output according to the weighted average:

$$y = \frac{(w_1 \times y_1 + w_2 \times y_2 + w_3 \times y_3)}{count(y_i)} = 1.57. \tag{22}$$

Finally, it is necessary to explain the total output and return a defuzzification decision.

5. Experiments

5.1. Data Description

To illustrate our method, three datasets are retrieved from the University of California Irvine (UCL) Machine Learning Repository to conduct the effectiveness and efficiency analysis. Table 1 summarizes the details of the three datasets.

Table 1. Descriptions of datasets.

No.	Datasets	Objects	Continuous Attributes	Classes	Abbreviations
1	Wine [38]	178	13	3	Wine
2	User Knowledge Modeling [39]	403	5	4	User
3	Climate Model Simulation [40]	540	18	2	Climate

The Wine dataset [38] consists of three grape varieties with 13 chemical components. Thirteen chemical components are taken as the 13 dimensions of an input parameter (x_i) and three grape varieties are taken as the three categories of an output parameter (y_i). GFIS is used for inferencing the category of grape based on the 13 kinds of chemical components. The fuzzy rule is that, if $N_{x_i} \in S_i$, then $y_i = N_{x_i}$ with the weight $\mu(N_{x_i})$, $i \in [1,13]$. Our proposed membership function is used to calculate the weight of each sub-T–S model output. Finally, we perform the K-means algorithm to cluster and explain the final inference results. L1-Distance [41] is used as the distance formula. By calculating the clustering precision, we assess the effectiveness of the proposed normalization method and the membership function. The more precise the clustering, the more effective the proposed methods. The User dataset [39] consists of four user varieties with five kinds of study data. Five kinds of study data are taken as the five dimensions of an input parameter and four user varieties are taken as the four categories of an output parameter. The Climate dataset [40] consists of two kinds of climate varieties with 18 kinds of climate data. Eighteen climate data are taken as the 18 dimensions of an input parameter and two kinds of climate varieties are taken as the two categories of an output parameter. To fully verify the effectiveness of the proposed methods, the same fuzzy inference program is conducted on the two datasets, respectively. Due to the lack of a public database containing multi-source fuzzy data, we refer to [42,43] to preprocess the original datasets in order to obtain target datasets:

Step 1: Let $X = \{x_1, x_2, \cdots, x_n\}$ denote an original dataset. Construct an interval-valued dataset, denoted as:

$$X' = \{\left[\underline{x'_1}, \overline{x'_1}\right], \cdots, \left[\underline{x'_n}, \overline{x'_n}\right]\}, \tag{23}$$

where

$$x'_i = [\underline{x'_i}, \overline{x'_i}] = [x_i - \alpha\sigma, x_i + \alpha\sigma], \tag{24}$$

σ is the standard deviation of the i-th attribute in the same class and and α is a noise condition factor. In the benchmark analysis, we set $\alpha = 1$

Step 2: Let the number of information sources be m. Construct a multi-source interval-valued dataset by copying each piece of data m times.

Step 3: Get a random number r from a normal distribution $N(0, 0.1)$. If $r > 0$,

$$x'_i = [\underline{x'_i} \times (1 - r), \overline{x'_i} \times (1 + r)], \tag{25}$$

otherwise

$$x'_i = [\underline{x'_i} \times (1 + r), \overline{x'_i} \times (1 - r)]. \tag{26}$$

Take a test dataset with two attributes and two categories as an example.

The original test data are: category 1 $\{460, 0.2; 550, 1.3\}$ and category 2 $\{580, 4.0; 570, 3.5\}$. The mean and standard deviation of the first attribute of Category 1 are $\frac{460+550}{2} = 505$ and

$$\sigma_{11} = \sqrt{\frac{(460 - 505)^2 + (550 - 505)^2}{2 - 1}} = 63.64. \tag{27}$$

The mean and standard deviation of the first attribute of Category 2 are $\frac{0.2+1.3}{2} = 0.75$ and

$$\sigma_{12} = \sqrt{\frac{(0.2 - 0.75)^2 + (1.3 - 0.75)^2}{2 - 1}} = 0.78. \tag{28}$$

Similarly, the standard deviations of the two attributes of Category 2 are $\sigma_{21} = 70.71$ and $\sigma_{22} = 0.35$.

Let $\alpha = 1$. The interval value data set is: Category 1 $\{396.36 \sim 523.64, -0.58 \sim 0.98; 486.36 \sim 613.64, 0.52 \sim 2.08\}$ and Category 2 $\{509.29 \sim 650.71, 3.65 \sim 4.35; 499.29 \sim 640.71, 3.15 \sim 3.85\}$.

In the end, we can construct a multi-source dataset by following steps 2 and 3.

5.2. Experimental Settings

We performed all experiments using Pycharm on MacOS 12.1 with an Intel Core i7 2.6GHz processor and 16 GB RAM. Three different experiments were conducted to illustrate the impact of multi-source fuzzy data on fuzzy inference accuracy. Three kinds of experiments are described briefly below:

(1) Experiment 1 is conducted on the original dataset. The original dataset refers to the dataset which has not been processed to remove noise (fuzzy processing).
(2) Experiment 2 is conducted on the normalized dataset. The normalized dataset is obtained by fuzzifying (noise addition) and normalizing the original dataset.
(3) Experiment 3 is conducted on the fusion dataset. Fusion data are obtained by summing normalized data from different information sources, denoted as:

$$N_{x_i} = \sum_{j=1}^{m} N_{x_{ij}}, \tag{29}$$

where N_{x_i} is the i-th fusion data and $N_{x_{ij}}$ is the i-th normalized data from the j-th information source.

5.3. Performance Measurement

In this subsection, we evaluate the effectiveness and efficiency of the proposed methodology. In our study, the dataset goes through four stages of state: original data, interval-valued data, normalized data, and fusion data. Since the input parameters of GFIS must be exact numerical values, we only perform GFIS inference on the original, normalized, and fusion datasets. For the purpose of defuzzying and interpreting the inference results, the K-means algorithm is used to cluster the inference results, with each cluster indicating a decision category. Clustering results are compared with the actual decision categories to obtain the clustering precision, which reflects the effectiveness of the proposed normalization method and the membership function. The proposed method becomes more effective as clustering precision increases.

In a multi-source environment, GFIS inference on normalized data are performed independently for each information source, and the weighted average is used to indicate the final GFIS inference. However, the difference is that the GFIS inference of fusion data is to fuse the data of all information sources first and then perform GFIS inference on the fusion data. To facilitate memory, the GFIS inference experiment conducted on normalized data is called Non-fused GFIS, while the GFIS inference experiment on fusion data is called Fused GFIS. In Non-fused GFIS, the clustering precision of inference results is denoted as:

$$p = \sum_{i=1}^{No.\ of\ IS} \left(\frac{\sum_{j=1}^{count\ of\ clusters\ in\ each\ IS} \frac{object\ No.\ with\ correct\ classification\ of\ each\ cluster\ in\ each\ IS}{object\ No.\ of\ each\ cluster\ in\ each\ IS}}{count\ of\ clusters\ in\ each\ IS} \right) / No.\ of\ IS, \tag{30}$$

where IS stands for information sources. In Fused GFIS, the clustering precision of inference results is expressed as:

$$p = \frac{\sum_{i=1}^{count\ of\ clusters\ on\ the\ fusion\ dataset} \frac{number\ of\ objects\ with\ correct\ classification\ in\ each\ cluster}{number\ of\ all\ objects\ in\ each\ cluster}}{count\ of\ clusters\ on\ the\ fusion\ dataset}. \tag{31}$$

We have used K-means clustering to quantify the precision of inference results for different GFIS models. We are now ready to test the operating efficiency of the normalization method, the fusion method, and two types of GFIS models. Additionally to accuracy, operating efficiency is an equally crucial metric for evaluating models or methods, as it tells us how much it costs to operate them. A high level of operating efficiency indicates a high level of availability and feasibility.

Our fusion method is based on the addition law of normal distribution; see Equation (18). Through the transformation of data distribution space, we guarantee that the normalized data from each information source satisfy the standard normal distribution $N(N_x|0, E)$ and the fusion data from m independent information sources meet the normal distribution $N\left(\sum_{i=1}^{i=m} N_{x_i}|0, \sum_{i=1}^{i=m} E_i\right)$. We need to calculate the sum of each object with respect to the sources of information as well as the variance of the fusion data according to the normal distribution formula. Thus, the proposed fusion method has a total time cost equal to the sum of the two calculation steps.

6. Results

6.1. Effectiveness Analysis Results

Table 2 reports the clustering precision achieved using GFIS on three different datasets. In general, the higher the precision, the better the inference ability of GFIS becomes and, therefore, the more effective the proposed method will be. For the dataset Wine, three experiments have an accuracy of 90.69%, 86.32%, and 90.14%, respectively. GFIS precision in three experiments is 52.88%, 43.24%, and 44.90% for the User dataset, respectively. As shown in the last row of Table 2, the precision of GFIS is 55.62%, 55.52%, and 66.13%

for the Climate dataset, respectively. As presented in Figure 3, we can draw three conclusions. First, data normalization will reduce the clustering precision of inference results (0.18–18.23%). This is because normalization will scale the distance of the original data and add some noise. Second, data fusion can improve the clustering precision of inference results (3.84–19.11%). The reason for this is that data fusion can help eliminate part of the errors caused by different sources of information. Third, the clustering precision of inference results is related to datasets. The closer the dataset is to the normal distribution, the higher the clustering precision is.

Table 2. Clustering precision of GFIS inference on different datasets.

Dataset	Precision		
	Original data	Normalized data	Fusion data
Wine	90.69%	86.32%	90.14%
User	52.88%	43.24%	44.90%
Climate	55.62%	55.52%	66.13%

Notes: The "original data" refer to Experiment 1 performed on the original dataset. The "normalized data" refer to Experiment 2 conducted on the normalized dataset. The "Fusion data" refer to Experiment 3 conducted on the fusion dataset. In total, 20 information sources are available, and the number of data objects is what counts.

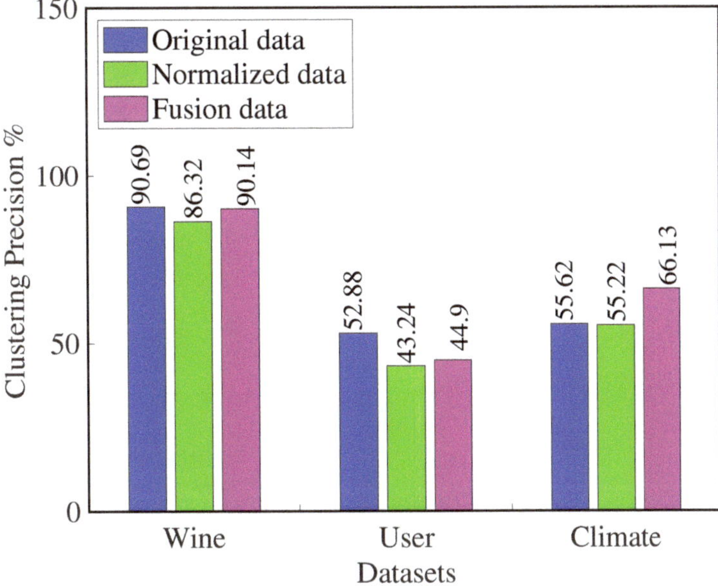

Figure 3. Clustering precision of GFIS inference on different datasets. The normalization of data will reduce the clustering precision of the inference results (0.18–18.23%) because normalization scales the distance between the original data and adds some noise to the results. Data fusion increases the clustering precision by 3.84–19.11% due to its ability to eliminate part of the errors from diverse information sources.

In Table 3 , we have shown the clustering precision of inference results generated by two different GFIS models across three datasets as IS quantity changes. The bold value in Table 3 indicates the better of the two GFIS models. The precision data in Table 3 reveal that the Fused GFIS is better than the Non-fused GFIS under all the test conditions with respect to the number of IS. Results of the test verify the effectiveness and superiority of the hypothesis. That is, information fusion technology combined with fuzzy control theory enhances the decision-making capability of FIS. Figure 4 demonstrates how the clustering precision changes with the increase or decrease of information sources. As illustrated in Figure 4, whether it is the Fused GFIS model or the Non-fused GFIS model, clustering precision does not appear to be related to the number of information sources. This is because, although information fusion can improve the accuracy of clustering, adding more information sources will also bring more noise.

Table 3. Clustering precision of GFIS inference with different numbers of information sources.

| Dataset | Number of information sources (IS) | | | | | | | | | |
| | 10 | | 20 | | 30 | | 40 | | 50 | |
	Non-Fused GFIS	Fused GFIS	Non-Fused GFIS	Fused GFIS	Non-Fused GFIS	Fused GFIS	Non-Fused GFIS	Fused GFIS	Non-Fused GFIS	Fused GFIS
Wine	81.13%	**90.14%**	86.32%	**90.14%**	87.80%	**91.08%**	88.51%	**90.14%**	89.18%	**90.61%**
User	41.02%	**46.36%**	43.24%	**46.73%**	43.94%	**47.67%**	43.28%	**47.67%**	43.21%	**50.59%**
Climate	54.61%	**54.85%**	55.52%	**66.13%**	54.67%	**59.71%**	57.60%	**58.90%**	54.96%	**58.80%**

Notes: "Non-fused GFIS" refers to GFIS that is driven by normalized data from independent sources. "Fused GFIS" refers to the GFIS driven by fusion data from multiple information sources.

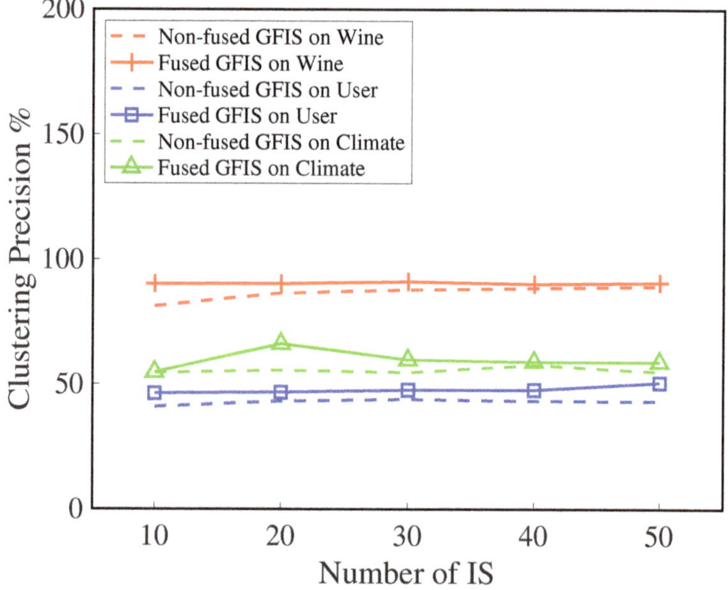

Figure 4. Clustering precision of GFIS inference with different numbers of information sources (IS). When it comes to the number of IS, the clustering precision of Fused GFIS is better than that of Non-fused GFIS under all test conditions. In either model, the clustering precision does not appear to be related to the number of information sources.

6.2. Efficiency Analysis Results

Table 4 reports the time cost associated with the normalization of heterogeneous fuzzy data. Test results show that the time cost of data normalization varies with datasets and the number of data objects. Apart from that, under all test conditions, the time cost of data normalization is controlled at the second level, which is generally acceptable. By using linear fitting, Figure 5 demonstrates the linear relationship between the time cost of data normalization and the number of data objects. Figure 5 illustrates that all datasets demonstrate a positive linear relationship between the normalization time cost and the number of data objects, which is consistent with the time complexity $O(3 \times n \times k)$ in the normalization algorithm (see Algorithm 1 for details).

Table 4. Time cost of data normalization for heterogeneous fuzzy data with different numbers of data objects (seconds).

Dataset	Number of Data Objects			
	50	100	150	200
Wine	9.10	18.56	27.80	36.93
User	3.67	7.19	10.28	13.82
Climate	13.00	25.92	39.25	51.35

Notes: Experiments were conducted on interval-valued data. The number of IS is 20.

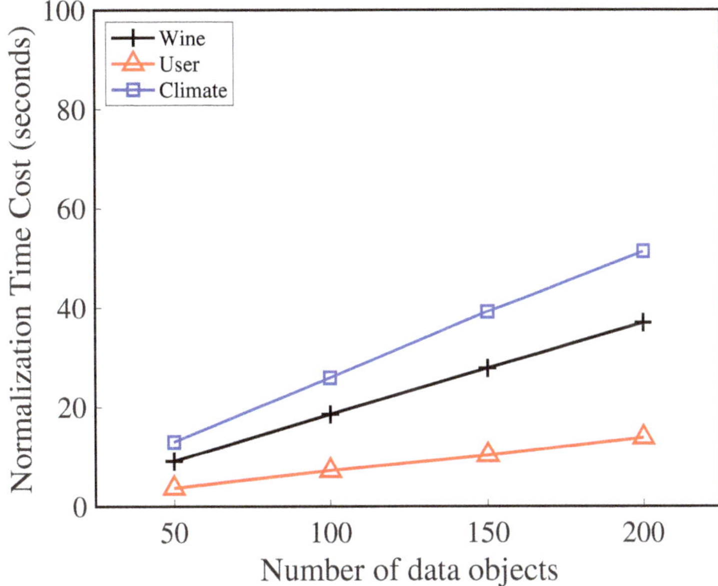

Figure 5. Time cost of data normalization for heterogeneous fuzzy data with different numbers of data objects. For all datasets, the normalization time cost has a positive linear relationship with the number of data objects, which is consistent with the time complexity $O(3 \times n \times k)$ in the normalization algorithm (see Algorithm 1 for details).

Table 5 shows the time cost of fusion of normalized data. The non-fused GFIS models incur no time burden, as opposed to the GFIS models that undergo data fusion. Figure 6 displays how the fusion time cost varies based on the number of IS in multi-source normalized data. Results from all datasets show a positive linear relationship between fusion time cost and IS number. The reason for this is that our fusion method requires us to calculate the sum of each data object's multi-source normalized value $\sum_{i=1}^{i=m} N_{x_i}$ and the sum of variances from all information sources $\sum_{i=1}^{i=m} E_i$. In this regard, the positive correlation between the time complexity of the fusion method and the number of IS can be formulated as $O(T) \propto m$, where $O(T)$ is the time cost of fusing an information source and m is the number of information sources.

Table 5. Time cost of normalized data fusion with different numbers of IS (seconds).

Dataset	Number of Information Sources (IS)				
	10	20	30	40	50
Wine	7.35	14.05	21.62	28.84	36.59
User	6.29	12.49	18.56	24.19	30.21
Climate	30.89	59.61	87.51	117.33	149.28

Notes: Experiments were conducted on the normalized dataset. The number of data objects is the actual value.

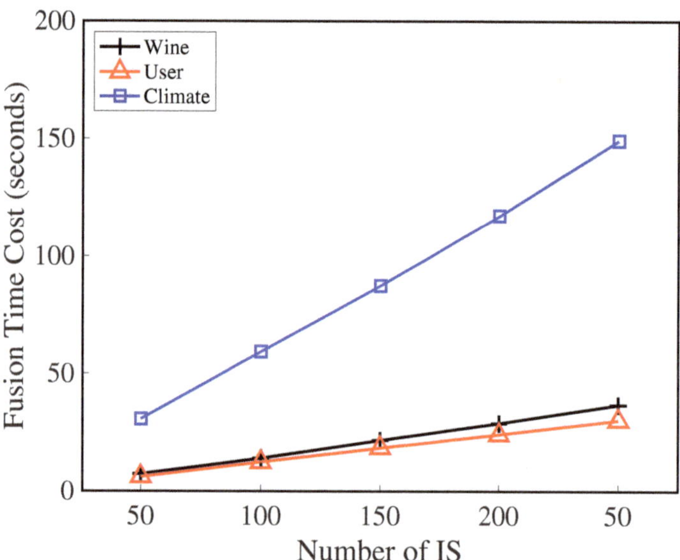

Figure 6. Time cost of normalized data fusion with different numbers of IS. For all datasets, the fusion time cost and the number of IS are positively correlated. Using the core formula of the fusion method, if we assume that the time cost of fusing one information source is $O(T)$, and m is the number of sources, we obtain $O(T) \ltimes m$ as the total time cost.

Table 6 presents a summary of the time cost associated with GFIS inferences. Compared with Non-fused GFIS, the inference time cost of Fused GFIS additionally includes the time cost of data fusion (see Table 5). However, the inference time cost of Fused GFIS is still lower than that of Non-fused GFIS, demonstrating the effectiveness and superiority of Fused GFIS. Figure 7 displays the time trend of GFIS inference with IS increasing or decreasing. The results show that, for both GFIS models, the inference time increases linearly with IS number for all three data sets. A fused GFIS, however, has a significantly lower time cost than a non-fused one. A major reason for this is the improved adaptability of the membership function in the GFIS model, which enables GFIS to properly handle incremental data with fuzzy inference.

Table 6. Time cost of GFIS inference with different numbers of IS (seconds).

Dataset	Number of information sources (IS)									
	10		20		30		40		50	
	Non-Fused GFIS	Fused GFIS	Non-Fused GFIS	Fused GFIS	Non-Fused GFIS	Fused GFIS	Non-Fused GFIS	Fused GFIS	Non-Fused GFIS	Fused GFIS
Wine	27.92	15.34	55.39	29.07	89.34	43.67	114.49	57.95	138.72	72.80
User	24.84	14.11	49.83	26.00	74.06	37.64	101.84	49.89	120.53	61.89
Climate	120.63	67.48	204.25	119.56	295.02	170.53	383.36	224.94	478.23	280.56

Notes: Experiments were conducted on the non-fused and fused datasets. The number of data objects is the actual value.

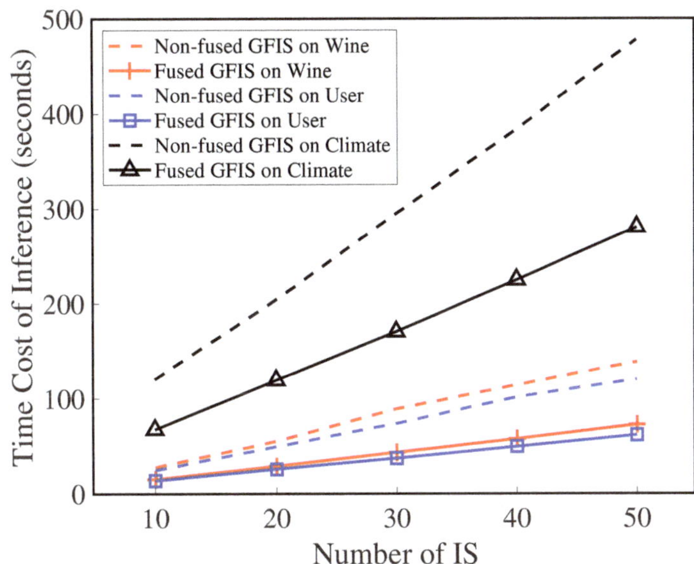

Figure 7. Time cost of GFIS inference with different numbers of IS. It has been found that the inference time cost for both two GFIS models increases linearly with the number of IS for all datasets. Fused GFIS, however, has a much lower time cost than non-fused GFIS, demonstrating the effectiveness and adaptability of the proposed membership function.

6.3. Sensitivity Analysis Results

In this subsection, we perform two robustness checks. In particular, we perform some additional tests regarding data noise and analyze the sensitivity with respect to different dataset choices.

The baseline tests run on three datasets. To demonstrate that the results are not driven by a specific dataset choice, another dataset, Iris [44], is also considered. There are three classes in the dataset, each with 50 objects, and each class represents a type of iris plant. We use the shape of the flower to identify the category of flowers. Each dataset is successively filled with $\pm\sigma$, $\pm2\sigma$, $\pm3\sigma$, $\pm4\sigma$ and $\pm5\sigma$ of noise to determine the GFIS sensitivity to data noise. We test whether different noises influence the accuracy of inference for the Non-fused GFIS and the Fused GFIS.

Table 7 reports the experimental results. It is evident from the experimental results that data noise does not seem to change the benchmark conclusion of our experiment. In addition, we find that the amount of noise has no significant impact on the accuracy of each model (see Figure 8). This is because we use the mean value of fuzzy data to offset the effect of noise.

Moreover, we notice that, although the Fused and Non-fused GFIS on Iris, Wine and User are close to each other, the Fused and Non-fused GFIS on Climate seem far from each other. Table 7 shows that the optimization ratio of Fused GFIS to Non-fused GFIS is stable within 3% and 6% for datasets Iris, Wine, and User. However, for dataset Climate, the optimization ratio of Fused GFIS to Non-fused GFIS varies between 3% and 20%. This is due to the fact that the dataset Climate is unbalanced, with category 1 to category 2 ratios of 1:11. As a result of this, the GFIS model is moderately sensitive to class 1 on a smaller scale, and overall inference accuracy is affected to some extent.

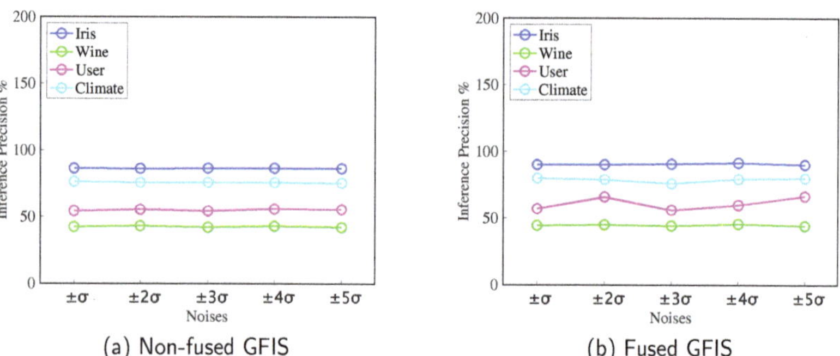

(a) Non-fused GFIS   (b) Fused GFIS

Figure 8. Influence of noise on the accuracy of different GFIS models. (**a**) non-fused GFIS; (**b**) fused GFIS.

Table 7. Inference precision of different GFIS models with the noise of each dataset changed.

Datasets	Noise	Non-Fused GFIS	Fused GFIS	Optimization
	σ	86.46	90.14	4.25%
	2σ	86.32	90.14	4.43%
Wine	3σ	86.59	90.61	4.64%
	4σ	86.57	91.49	5.68%
	5σ	86.39	90.14	4.34%
	σ	42.48	44.43	4.59%
	2σ	43.24	44.9	3.84%
User	3σ	42.17	44.19	4.79%
	4σ	43.06	45.48	5.62%
	5σ	42.14	44.12	4.70%
	σ	54.37	57.13	5.08%
	2σ	55.52	66.13	19.11%
Climate	3σ	54.35	56.24	3.48%
	4σ	55.82	59.88	7.27%
	5σ	55.54	66.53	19.79%
	σ	76.5	80	4.58%
	2σ	75.97	79	3.99%
Iris	3σ	76.03	79.33	4.34%
	4σ	75.93	79.33	4.48%
	5σ	75.6	80	5.82%

Notes: "Non-fused GFIS" refers to GFIS that is driven by single-source data. "Fused GFIS" refers to the GFIS driven by fusion data from multiple information sources. "Optimization" is the ratio: (Fused GFIS-Non-fused GFIS)/Non-fused GFIS

7. Conclusions

In this article, we develop a new Gaussian-shaped fuzzy inference system that is suitable for multi-source fuzzy environments. To achieve this goal, our first step is to normalize multi-source fuzzy data to remove the heterogeneity of multi-source data. In order to obtain the fused data of an object, we sum the normalized descriptions of it from different information sources. We then propose an adaptive membership function for the fusion data, which provides the basis for granulating the input for GFIS and designing its inference rules. We also propose a novel fuzzy inference framework by integrating the normalization method and adaptive membership function with the T–S model.

We conducted extensive experiments on three benchmark datasets to evaluate the effectiveness of the proposed methods. Three main conclusions can be drawn from the experimental results. First, the normalization of interval-value data can slightly reduce the clustering accuracy of the original data since it scales the distances and adds some noise. Second, the Fused GFIS model's inference precision is significantly higher than that of the Non-fused GFIS model. This is due to the fact that data fusion can remove some of the errors from different sources of information. Third, with an adaptive membership function, the proposed GFIS can handle fuzzy inferences of incremental data more efficiently.

There are some limitations to the proposed GFIS, which helps us identify future research directions. First, all data variables should be independent. Second, all data variables have to follow the law of large numbers and the central limit theorem. Third, when applying GFIS, it is necessary to select appropriate fuzzy rules and inference logic.

Author Contributions: Conceptualization, Y.Z. and C.Q.; methodology, C.Q.; software, C.Q.; validation, Y.Z. and C.Q.; formal analysis, Y.Z.; investigation, Y.Z.; resources, Y.Z.; writing—original draft preparation, Y.Z. and C.Q.; writing—review and editing, Y.Z. and C.Q.; visualization, C.Q.; supervision, Y.Z.; funding acquisition, Y.Z. All authors have read and agreed to the published version of the manuscript.

Funding: This research was supported by the Science and Technology Research Program of Chongqing Municipal Education Commission (Grant No. KJQN202200910).

Institutional Review Board Statement: Not applicable.

Informed Consent Statement: Not applicable.

Data Availability Statement: The data that support the findings of this study are openly available in https://archive.ics.uci.edu/ml/datasets/wine, https://archive.ics.uci.edu/ml/datasets/User+Knowledge+Modeling,https://archive.ics.uci.edu/ml/datasets/climate+model+simulation+crashes, and https://archive.ics.uci.edu/ml/datasets/iris (accessed on 15 September 2022).

Acknowledgments: We would like to thank Guo Bing, Dai Cheng, Lu Junyu, Su Hong, Shen Yan, and Zhang Zhen for their helpful and constructive discussions and comments.

Conflicts of Interest: The authors declare no conflict of interest. The funders had no role in the design of the study; in the collection, analyses, or interpretation of data; in the writing of the manuscript, or in the decision to publish the results.

References

1. Pae, D.S.; Choi, I.H.; Kang, T.K.; Lim, M.T. Vehicle detection framework for challenging lighting driving environment based on feature fusion method using adaptive neuro-fuzzy inference system. *Int. J. Adv. Robot. Syst.* **2018**, *15*, 1729881418770545. [CrossRef]
2. Bylykbashi, K.; Qafzezi, E.; Ikeda, M.; Matsuo, K.; Barolli, L.; Fuzzy-based Driver Monitoring System (FDMS): Implementation of two intelligent FDMSs and a testbed for safe driving in VANETs. *Future Gener. Comput. Syst.* **2020**, *105*, 665–674. [CrossRef]
3. Hussain, S.; Kim, Y.-S.; Thakur, S.; Breslin, J.G.; Optimization of waiting time for electric vehicles using a fuzzy inference system. *IEEE Trans. Intell. Transp. Syst.* **2022**, *23*, 15396–15407. [CrossRef]
4. Wu, J.; Hu, R.; Li, M.; Liu, S.; Zhang, X.; He, J.; Chen, J.; Li, X. Diagnosis of sleep disorders in traditional Chinese medicine based on adaptive neuro-fuzzy inference system. *Biomed. Signal Process. Control* **2021**, *70*, 102942. [CrossRef]
5. Colella, Y.; Valente, A.S.; Rossano, L.; Trunfio, T.A.; Fiorillo, A.; Improta, G. A fuzzy inference system for the assessment of indoor air quality in an operating room to prevent surgical site infection. *Int. J. Environ. Res. Public Health* **2022**, *19*, 3533. [CrossRef]
6. Singh, P.; Kaur, A.; Batth, R.S.; Kaur, S.; Gianini, G. Multi-disease big data analysis using beetle swarm optimization and an adaptive neuro-fuzzy inference system. *Neural Comput. Appl.* **2021**, *33*, 10403–10414. [CrossRef]
7. Paul, S.K.; Chowdhury, P.; Ahsan, K.; Ali, S.M.; Kabir, G. An advanced decision-making model for evaluating manufacturing plant locations using fuzzy inference system. *Expert Syst. Appl.* **2022**, *191*, 116378. [CrossRef]
8. Weldcherkos, T.; Salau, A.O.; Ashagrie, A. Modeling and design of an automatic generation control for hydropower plants using Neuro-Fuzzy controller. *Energy Rep.* **2021**, *7*, 6626–6637. [CrossRef]

9. Geramian, A.; Abraham, A. Customer classification: A Mamdani fuzzy inference system standpoint for modifying the failure mode and effect analysis based three-dimensional approach. *Expert Syst. Appl.* **2021**, *186*, 115753. [CrossRef]
10. Beres, E.; Adve, R. Selection cooperation in multi-source cooperative networks. *IEEE Trans. Wirel. Commun.* **2008**, *187*, 104831. [CrossRef]
11. Cvetek, D.; Muštra, M.; Jelušić, N.; Tišljarić, L. A survey of methods and technologies for congestion estimation based on multisource data fusion. *Appl. Sci.* **2021**, *11*, 2306. [CrossRef]
12. Chen, F.; Yuan, Z.; Huang, Y. Multi-source data fusion for aspect-level sentiment classification. *Knowl.-Based Syst.* **2020**, *187*, 104831. [CrossRef]
13. Zade, L. Fuzzy sets. *Inf. Control* **1965**, *8*, 338–353. [CrossRef]
14. Pawlak, Z. Rough sets. *Int. J. Comput. Inf. Sci.* **1982**, *11*, 341–356. [CrossRef]
15. Kumar, P.; Krishna, P.R.; Bapi, R.S.; De S.K. Clustering using similarity upper approximation. In Proceedings of 2006 IEEE International Conference on Fuzzy Systems, Vancouver, BC, Canada, 16–21 July 2006; pp. 839–844.
16. Ali, O.A.M.; Ali, A.Y.; Sumait, B.S. Comparison between the effects of different types of membership functions on fuzzy logic controller performance. *Int. J.* **2015**, *76*, 76–83.
17. Maturo, F.; Fortuna, F. Bell-shaped fuzzy numbers associated with the normal curve. In *Topics on Methodological and Applied Statistical Inference*; Springer, Cham, Switzerland, 2016; pp. 131–144.
18. Chang, C.T. An approximation approach for representing S-shaped membership functions. *IEEE Trans. Fuzzy Syst.* **2010**, *18*, 412–424.
19. Zhu, A.X.; Yang, L.; Li, B.; Qin, C.; Pei, T.; Liu, B. Construction of membership functions for predictive soil mapping under fuzzy logic. *Geoderma* **2010**, *155*, 164–174. [CrossRef]
20. Mandal, S.N.; Choudhury, J.P.; Chaudhuri, S.R.B. In search of suitable fuzzy membership function in prediction of time series data. *Int. J. Comput. Sci. Issues* **2012**, *9*, 293–302.
21. Jenish, N.; Prucha, I.R. Central limit theorems and uniform laws of large numbers for arrays of random fields. *J. Econom.* **2009**, *150*, 86–98. [CrossRef]
22. SYa, S.; Melkumova, L.E. Normality assumption in statistical data analysis. In *CEUR Workshop Proceedings*; Annecy, France, 2016; pp. 763–768. Available online: https://ceur-ws.org/Vol-1638/Paper90.pdf (accessed on 15 September 2022).
23. Fei, N.; Gao, Y.; Lu, Z.; Xiang, T. Z-score normalization, hubness, and few-shot learning. In Proceedings of the IEEE/CVF International Conference on Computer Vision, Montreal, QC, Canada, 10–17 October 2021; pp. 142–151.
24. Hosseini, R.; Qanadli, S.D.; Barman, S.; Mazinani, M.; Ellis, T.; Dehmeshki, J. An automatic approach for learning and tuning Gaussian interval type-2 fuzzy membership functions applied to lung CAD classification system. *IEEE Trans. Fuzzy Syst.* **2011**, *20*, 224–234. [CrossRef]
25. Kong, L.; Zhu, S.; Wang, Z. Feature subset selection-based fault diagnoses for automobile engine. In Proceedings of the 2011 Fourth International Symposium on Computational Intelligence and Design, Washington, DC, USA, 28–30 October 2011; pp. 367–370.
26. Li, Z.; He, T.; Cao, L.; Wu, T.; McCauley, P.; Balas, V.E.; Shi, F. Multi-source information fusion model in rule-based Gaussian-shaped fuzzy control inference system incorporating Gaussian density function *J. Intell. Fuzzy Syst.* **2015**, *29*, 2335–2344. [CrossRef]
27. Deng, Y.; Ren, Z.; Kong, Y.; Bao, F.; Dai, Q. A hierarchical fused fuzzy deep neural network for data classification. *IEEE Trans. Fuzzy Syst.* **2016**, *25*, 1006–1012. [CrossRef]
28. Xue, D.; Yadav, S.; Norrie, D.H. Knowledge base and database representation for intelligent concurrent design. *Comput.-Aided Des.* **1999**, *31*, 131–145. [CrossRef]
29. Wang, X.; Ruan, D.; Kerre, E.E. Fuzzy inference and fuzzy control. In *Mathematics of Fuzziness—Basic Issues*; Springer: Berlin/Heidelberg, Germany, 2009; pp. 189–205.
30. Vemuri, N.R. Investigations of fuzzy implications satisfying generalized hypothetical syllogism. *Fuzzy Sets Syst.* **2017**, *323*, 117–137. [CrossRef]
31. Zadeh, L.A. Syllogistic reasoning in fuzzy logic and its application to usuality and reasoning with dispositions. *IEEE Trans. Syst. Man, Cybern.* **1985**, *1985*, 754–763. [CrossRef]
32. Zhao, X.; Liu, Y.; He, X. Fault diagnosis of gas turbine based on fuzzy matrix and the principle of maximum membership degree. *Energy Procedia* **2012**, *16*, 1448–1454. [CrossRef]
33. Liou, T.S.; Wang, M.J.J. Fuzzy weighted average: An improved algorithm. *Fuzzy Sets Syst.* **1992**, *49*, 307–315. [CrossRef]
34. Van Broekhoven, E.; De Baets, B. Fast and accurate center of gravity defuzzification of fuzzy system outputs defined on trapezoidal fuzzy partitions. *Fuzzy Sets Syst.* **2006**, *157*, 904–918. [CrossRef]
35. Lemons, D.S. An introduction to stochastic processes in physics. *Am. J. Phys.* **2003**, *71*, 191. [CrossRef]
36. Johansen, T.A.; Shorten, R.; Murray-Smith, R. On the interpretation and identification of dynamic Takagi–Sugeno fuzzy models. *IEEE Trans. Fuzzy Syst.* **2000**, *8*, 297–313. [CrossRef]
37. Weisstein, E.W. Direct product. From *MathWorld—A Wolfram Web Resource*. 2006. Available online: https://mathworld.wolfram.com/DirectProduct.html (accessed on 15 September 2022).
38. Aeberhard, S.; Coomans, D.; de Vel, O. Comparative analysis of statistical pattern recognition methods in high dimensional settings. *Pattern Recogn.* **1994**, *27*, 1065–1077. [CrossRef]

39. Kahraman, H.T.; Sagiroglu, S.; Colak, I. Developing intuitive knowledge classifier and modeling of users' domain dependent data in web. *Knowl. Based Syst.* **2013**, *37*, 283–295. [CrossRef]
40. Lucas, D.D.; Klein, R.; Tannahill, J.; Ivanova, D.; Brandon, S.; Domyancic, D.; Zhang, Y. Failure analysis of parameter-induced simulation crashes in climate models. *Geosci. Model Dev. Discuss* **2013**, *6*, 585–623. [CrossRef]
41. Kashima, H.; Hu, J.; Ray, B.; Singh, M. K-means clustering of proportional data using L1 distance. In Proceedings of the 2008 19th International Conference on Pattern Recognition, Tampa, FL, USA, 8–11 December 2013; Volume 6, pp. 585–623.
42. Leung, Y.; Fischer, M.M.; Wu, W.Z.; Mi, J.S. A rough set approach for the discovery of classification rules in interval-valued information systems. *Int. J. Approx. Reason* **2008**, *47*, 233–246. [CrossRef]
43. Huang, Y.; Li, T.; Luo, C.; Fujita, H.; Horng, S.J. Dynamic Fusion of Multisource Interval-Valued Data by Fuzzy Granulation. *IEEE Trans. Fuzzy Syst.* **2018**, *26*, 3403–3417. [CrossRef]
44. Dasarathy, B.V. Nosing Around the Neighborhood: A New System Structure and Classification Rule for Recognition in Partially Exposed Environments. *IEEE Trans. Pattern Anal. Mach. Intell.* **1980**, *2*, 67–71. [CrossRef]

 *systems*

Article

Parallel Learning of Dynamics in Complex Systems

Xueqin Huang [1], Xianqiang Zhu [1,*], Xiang Xu [1], Qianzhen Zhang [1] and Ailin Liang [1,2]

[1] Science and Technology on Information Systems Engineering Laboratory, National University of Defense Technology, Changsha 410073, China
[2] College of Economics and Management, Nanjing University of Aeronautics and Astronautics, Nanjing 210016, China
* Correspondence: zhuxianqiang@nudt.edu.cn

Abstract: Dynamics always exist in complex systems. Graphs (complex networks) are a mathematical form for describing a complex system abstractly. Dynamics can be learned efficiently from the structure and dynamics state of a graph. Learning the dynamics in graphs plays an important role in predicting and controlling complex systems. Most of the methods for learning dynamics in graphs run slowly in large graphs. The complexity of the large graph's structure and its nonlinear dynamics aggravate this problem. To overcome these difficulties, we propose a general framework with two novel methods in this paper, the Dynamics-METIS (D-METIS) and the Partitioned Graph Neural Dynamics Learner (PGNDL). The general framework combines D-METIS and PGNDL to perform tasks for large graphs. D-METIS is a new algorithm that can partition a large graph into multiple subgraphs. D-METIS innovatively considers the dynamic changes in the graph. PGNDL is a new parallel model that consists of ordinary differential equation systems and graph neural networks (GNNs). It can quickly learn the dynamics of subgraphs in parallel. In this framework, D-METIS provides PGNDL with partitioned subgraphs, and PGNDL can solve the tasks of interpolation and extrapolation prediction. We exhibit the universality and superiority of our framework on four kinds of graphs with three kinds of dynamics through an experiment.

Keywords: complex systems; graph partition; graph neural networks; ordinary differential equation; dynamics

Citation: Huang, X.; Zhu, X.; Xu, X.; Zhang, Q.; Liang, A. Parallel Learning of Dynamics in Complex Systems. *Systems* **2022**, *10*, 259. https://doi.org/10.3390/systems10060259

Academic Editors: Zaoli Yang, Yuchen Li and Ibrahim Kucukkoc

Received: 22 November 2022
Accepted: 12 December 2022
Published: 15 December 2022

Publisher's Note: MDPI stays neutral with regard to jurisdictional claims in published maps and institutional affiliations.

1. Introduction

Complex networks or graphs are ubiquitous in life, and each individual is a node or vertex in many kinds of graphs. It is very important to know what complex networks are and how they affect us. Many systematic problems can be constructed as the mathematical tool of a 'Graph' to carry out research and solve many practical problems, such as the global outbreak of the WannaCry computer blackmail virus [1], the COVID-19 global pandemic [2], the rapid spread of monkeypox [3], and the spread of rumors in social networks [4]. All of these can be modeled as a graph or a complex network model. For these complex systems, constructing effective graph models is helpful for better predictions and control. Specifically, graphs help us to prevent the spread of epidemics, block the spread of computer viruses, crack down on terrorist networks [5], improve the robustness of power grids, strengthen public opinion monitoring, and so on. Dynamics is a mainstream approach for studying the dynamic processes of vertexes in a graph. There are now many studies on network dynamics [6], and the existing dynamics models of graphs are still worthy of further study. Dynamics makes the solution to the dynamic process of the network easily explainable. Nonlinear dynamics models have been widely studied and applied in different fields, including applied mathematics [7,8], statistical physics [9], and engineering [10]. Some networks' evolution mechanisms are known at the beginning of their establishment, but the real world is so complex that the potential dynamics of a large number of complex networks are unknown. It is difficult to construct complex models of these unknown differential equations. Dynamics modeling on a graph also becomes more challenging when considering the unknown elements of dynamics and the large scale of the complex system itself.

Fortunately, in the era of big data, many complex network systems have produced a large amount of available data in the process of dynamic development. When seeking a model for the data, we can learn its dynamics on a graph with a combination of ODEs and GNNs. In addition, after the complex system is abstracted into a graph, its complex network structure, large-scale edges and vertexes, and complex dynamics processes form a series of NP-complete problems [11,12]. This results in the poor performance of many models and algorithms on the graph. However, there is a better method, namely, graph partition. This process evenly divides the large-scale graph into a series of subgraphs to adapt to distributed applications. Therefore, the learning process of the dynamics on a graph can be accelerated by graph partition. Based on this, we proposed a model framework for graph partition to accelerate the graph neural dynamics learning process. This method skillfully combines the dynamic process on the graph with the fast graph-partition algorithm METIS and realizes a large-scale graph-partition method considering a dynamic process. After the large graph is divided evenly, the dynamics of each subgraph can further be learned in parallel by combining GNNs [13–16] and differential equations. This helps us to recognize, predict, and control a complex system more quickly and accurately.

Our work can be used for two tasks in a general framework. One is partitioning a large graph with network dynamics for parallel tasks downstream. Another is learning the unequal time interval states of subgraph dynamics for interpolation and extrapolation predictions. In task one, this model is more accurate and faster than the usual spectral clustering; the execution efficiency is very high, i.e., one to two orders of magnitude faster than the common partition algorithm; and a graph with millions of vertexes can be divided into 256 classes in a few seconds. In task two, the model can learn unequal interval (continuous-time) dynamics in graphs. It obtains more accurate results than most graphs and dynamics, and it is more than twice as fast as other models.

Overall, the main contributions of this paper are as follows:

(1)   A novel algorithm: We propose a novel algorithm for graph partition, namely, Dynamics-METIS (D-METIS). D-METIS can partition a large graph into multiple subgraphs, and it innovatively considers two balances of the subgraphs, i.e., the balance of vertexes and the balance of cumulative dynamic changes.

(2)   A novel model: This novel model is called the Partitioned Graph Neural Dynamics Learner (PGNDL). The PGNDL is a parallel model that combines ordinary differential equation systems and GNN. Thus, it can quickly learn the dynamics of large graphs. It can also learn unequal interval (continuous-time) dynamics on any graph.

(3)   More efficient parallel general framework: The experimental results show that our framework completed the tasks on various graphs faster than the most well-known framework, NDCN [17], with at least twice the efficiency.

(4)   More accurate in regression tasks: The PGNDL (D-METIS) performs accurately on various dynamics and networks.

The main purpose of this paper is to accelerate the dynamics learning process on large graphs, apply the graph-partition algorithm to cut large graphs into needed subgraphs, and then implement the neural dynamics learning model in parallel in each subgraph. Compared to the existing method of graph partition, our model can learn more complex dynamics on larger graphs and achieve faster, more accurate, and more interpretable results. Our D-METIS considers not only the balance of the number of nodes in each subgraph but also the balance of the degree of dynamic changes in each subgraph. Additionally, our PGNDL model is different from other existing GNNs [14–17]. It is a parallel learning method, which can reduce the complexity of each thread and improve the computational efficiency.

To illustrate the proposed framework, we review knowledge of graph partition and GNNs with ODE (Section 2). Then, we define the methods and framework's terminology and its algorithms (Section 3). Following this, we demonstrate the framework on different graphs with different dynamics using 24 datasets consisting of 400 vertexes and 2000 vertexes (Section 4). Finally, we summarize this work (Section 5).

2. Related Work

2.1. Graph Partition

Graph partitioning involves evenly dividing a large graph into a series of subgraphs. This means subgraphs can be executed in parallel. If the current subgraph needs information from other subgraphs, partitioning must consider information transfer. The quality of the graph partition affects the storage cost of each machine and the communication cost among machines. According to the memory cost of partition, it can be divided into offline and streaming partition algorithms. For large-scale graph data, streaming partition is particularly important when the memory of a single machine cannot meet the requirements of the partition algorithm [18]. The partition method of graph data can be divided into vertex partitioning or edge-cut partitioning. For graph data with power law distribution, some vertexes may have many edges; if we run the vertex partitioning, many edges will be missing, and the edge load will be uneven, but edge partition can deal with this kind of problem [19]. The two goals of graph partition are load balancing (reducing storage costs) and minimizing cuts (reducing communication costs). At the same time, the two goals of optimization are balanced graph partitioning.

As you can see, graph partition is an NP-hard problem [19]. In normal circumstances, the relaxation is to optimize load balancing and ensure minimum cuts as much as possible. We can define a Graph as $G = (V, E)$, which means graph G has $|V|$ vertexes and $|E|$ edges.

The edge partitioning is shown as follows:

$$\max_{i \in [1,k]} |E_i| \leq \frac{(1 + \alpha)|E|}{k} \tag{1}$$

$$RF_v = \sum_{i-1}^{k} \frac{V(E_i)}{|V|} \tag{2}$$

In Formula (1), the balanced graph-partitioning problem is defined as creating k disjoint sets of vertices (partitions); $|E|$ means the number of edges in graph G; $V(E_i)$ means a set of vertexes representing the association of all edges E_i in a subgraph; and the parameter α controls the balance rate. In Formula (2) RF_v represents the replication factor of the vertex and measures the number of vertex cuts. Regarding edge partition, linear deterministic greedy partitioning (LDG) [18] considers using a greedy algorithm to put neighbor vertexes together during partition to reduce edge cutting and ensure the vertex load balance of each subgraph. Compared with LDG, Fennel's [20] scoring function has relaxed the constraint on the number of vertices in the subgraph from multiplication to subtraction. METIS [21] is a hierarchical partitioning algorithm. The core idea is to reduce the size of the original graph via continuously sparsely merging vertexes and edges for a given original graph structure and then, to a certain extent, segmenting the reduced graph structure. Finally, the small, segmented graph is restored to the original graph structure to ensure the balance of each subgraph. METIS adopts the heavy edge-matching strategy in the sparse phase. When dividing the reduced subgraph, it randomly initializes a node to conduct a width first search to obtain the subgraph with the minimum tangent edge and then maps it to the original graph structure. Because METIS needs to traverse and scale the entire graph structure, it is inefficient when segmenting large-scale graphs with large memory consumption.

Additionally, the edge partitioning is as follows:

$$\max_{i \in [1,k]} |V_i| \leq \frac{(1 + \alpha)|V|}{k} \tag{3}$$

where $|V_i|$ means the number of vertexes in each subgraph, representing the load balancing of the vertex; and the parameter α controls the balance rate. Regarding edge partition, neighbor expansion (NE) [19] edge partition also considers the locality of neighbors; for the boundary vertex, it selects a candidate vertex whose neighbor is the closest to the outside of the boundary, which can ensure the maximization of the assigned neighbor edge to ensure the minimum node-repetition rate. Degree-Based Hashing (DBH) [22] divides the vertex allocation edge by judging the degree information of the vertex. For power-law graphs, the locality of low-degree vertexes is easy to maintain. Meanwhile, for high-degree vertexes, it is impossible to allocate all edges on a subgraph because there are too many vertexes associated. Thus, the algorithm tries to maintain the locality of low-degree vertexes as much as possible. Additionally, the generalizable approximate graph-partitioning (GAP) framework [23] is a vertex-partition algorithm based on GNN.

2.2. GNNs with ODE

This is a new way to combine Ordinary Differential Equations (ODE) [17,24–26] and GNNs to learn the non-linear and high-dimensional dynamics of graphs. Neural Dynamics on Complex Networks (NDCN) [17] is a successful network class. NDCN captures the instantaneous rate of change of vertex states by differential equation systems and GNNs instead of mapping through a discrete number of layers forward. It integrates GNN layers in continuous time rather than discrete depth. The continuous-time dynamics on a graph can describe by a differential equation system, such as Formula (4).

$$\frac{dX(t)}{dt} = f(X, G, W, t) \tag{4}$$

where $X(t) \in \mathbb{R}^{v \times d}$ represents the state of a dynamic system consisting of v-linked vertexes at time $t \in [0, \infty)$, and $X(0)$ is the initial state of this system at time $t = 0$. Each vertex is characterized by v dimensional features. $G = (V, E)$ is the network structure capturing how vertexes interact with each other. $W(t)$ are parameters that control how the system evolves. The function f governs the instantaneous rate of change of dynamics on the graph. We can obtain the NDCN model as follows:

$$\underset{W(t), \Theta(T)}{\arg\min} \quad \mathcal{L} = \int_0^T \mathcal{R}(X, G, W, t)dt + \mathcal{S}(Y(X(T), \Theta)) \tag{5}$$

$$subject\ to \qquad X_h(t) = f_e(X(t), W_e)$$
$$\frac{dX_h(t)}{dt} = f(X_h, G, W_h, t) \tag{6}$$
$$X(t) = f_d(X_h(t), W_d)$$

where $\int_0^T \mathcal{R}(X, G, W, t)dt$ is the 'running' loss of the continuous-time dynamics on graphs from $t = 0$ to T, and $\mathcal{S}(Y(X(T), \Theta))$ is the 'terminal' loss at time T. Vertexes can have various semantic labels encoded by one-hot encoding $Y(X(T), \Theta)$, and Θ represents the parameters of this classification function. The first constraint transforms $X(t)$ into hidden space $X_h(t)$ through encoding function f_e, and $X(0) = X_0$. The second constraint is the governing dynamics in the hidden space. The third constraint decodes the hidden signal back to the original space with a decoding function.

The neural structures of the model are illustrated in Figure 1.

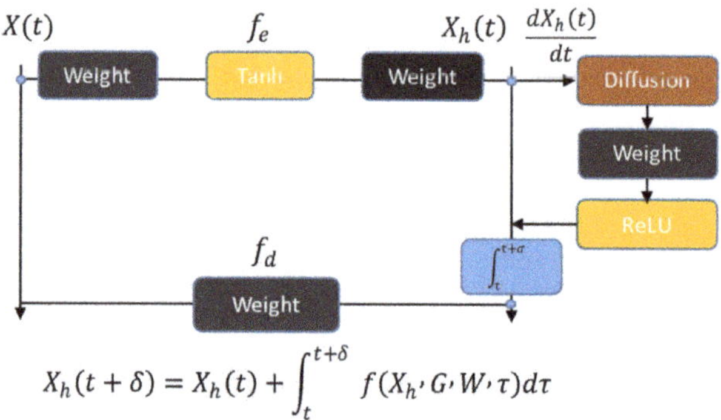

$$X_h(t + \delta) = X_h(t) + \int_{t}^{t+\delta} f(X_h, G, W, \tau)d\tau$$

Figure 1. Illustration of an NDCN instance.

We can see that the input of NDCN is the node state $X(t)$ at time t, while the output is the node state after time $(t + \delta)$. The NDCN will first map the input $X(t)$ into the hidden space using $X_h(t)$ to represent $X(t)$ after the hidden layer. The hidden layer is an encoding function f_e. Additionally, the dynamics of influence and information diffusion between nodes are modeled in the hidden space, which is completed through the GNN. By integrating the dynamic process $X_h(t)$ from t to $t + \delta$, we can obtain the output state. Then, we use the decoder f_d to obtain the status of nodes in the original space. Moreover, f_e and f_d are flexible as any deep neural structure (including the linear weighting layer and activation function).

From the point of view of the dynamic system, continuous depth can be interpreted as continuous physical time, and the output of any hidden GNN layer at time t is the instantaneous network dynamics process of conventional neural network models. Additionally, a unified framework for automated interactions and dynamics discovery (AIDD) was proposed [27]. It is based on the more rigorous mathematical form of Markov dynamics and local network interaction to express the problem. Additionally, it provides a unified objective function based on logarithmic likelihood. This kind of model has been applied to many fields, such as climate studies [28], rumor detection [29], and healthcare [30].

3. Methodology

Cutting a graph with dynamic characteristics is a new problem. Additionally, finding a graph-partition method that can match the downstream applications and how to dynamically reconstruct each subgraph are two difficulties inherent in this problem. The following methods and frameworks are proposed to resolve this problem.

In this section, we propose the Dynamics METIS (D-METIS) algorithm to solve the first difficulty of the problem. For the second difficulty, we use the data-driven method based on GNNs to obtain the dynamics of subgraphs cut by D-METIS, called the Partitioned Graph Neural Dynamics Learner (PGNDL), and we give its application ways to resolve the regression task of interpolation prediction and extrapolation prediction. Finally, we elaborate on the general framework.

3.1. Dynamics METIS

We proposed a novel method named Dynamics METIS (D-METIS) for cutting large graphs while considering the dynamics in the graph. There are three given rules when designing D-METIS:

1. The graph structure damage caused by partitioning should be minimized;
2. The structure of each subgraph should be evenly distributed to facilitate the synchronization and parallel assessment of downstream tasks;
3. The distribution of the dynamics state change degree of each subgraph should be even and convenient for downstream application and analysis.

Thus, our D-METIS method can compress the dynamics states for more efficient graph-partitioning tasks and also take advantage of the information on state changes of vertexes, which takes the partitioning task closer to reality.

3.1.1. METIS Algorithm

METIS is a multilevel k-way partition algorithm [21]. The graph $G = (V, E)$ is first coarsened into a small-scale graph, which contains a small number of points. Then, the k-path is divided for the coarsened graph. At this time, the number of points is greatly reduced due to the coarsening, so a much smaller graph needs to be divided now, and the time complexity is greatly reduced. After the partition, each subgraph is refined step-by-step until the number of original points is restored to obtain the original graph. The three steps of coarsening, dividing, and refining can be seen graphically in Figure 2.

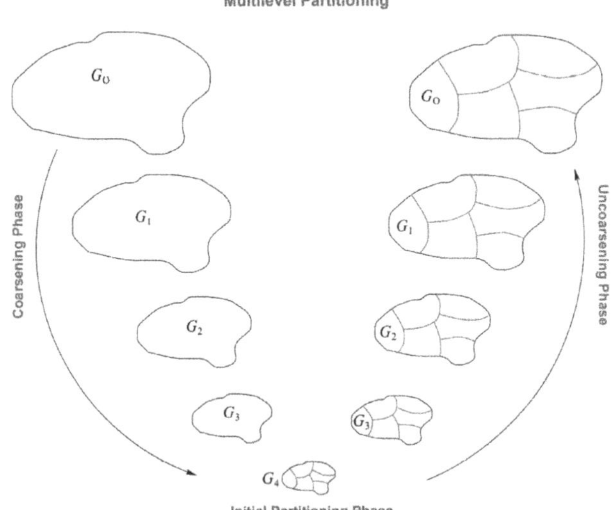

Figure 2. The three steps of multilevel k-way graph partitioning. G_0 is the input, which is also the finest graph, G_{i+1} is the second most coarse graph of G_i, and G_4 is the coarsest graph.

3.1.2. METIS for a Graph with Dynamics

The traditional METIS algorithm is applied to some static graphs without dynamics. However, there are many graphs with dynamics in the real world, so we need a new method for partitioning with dynamics. We enabled the METIS algorithm to be suitable for a graph with dynamics by designing the dynamic process compression strategy of vertexes. First, we give a case of dynamic process compression on a graph, as shown in Figure 3; in Figure 3a, we give every vertex five states; and in Figure 3b, we sequentially compute the sum of dynamic changes for each vertex as the new weight for vertexes. For example, the vertex numbered 1 has five states: $2{\to}3{\to}5{\to}7{\to}9$, and is then compressed into a weight = 7. This compression method will be mathematically detailed later.

Let $G = (V, E)$ denote an undirected graph consisting of a vertex set V and an edge set E. A pair of vertexes makes up an edge (i.e., $e = \{v_1, v_2\}$ where $v_1, v_2 \in V$). The number of vertexes in the graph is denoted as $n = |V|$, and the number of edges is denoted by $m = |E|$.

Additionally, vertexes can have a group of weights $v_i(t)$. where $i \in (1, 2, 3, \ldots, n)$, $v_i \in V$, $t \in [0, T]$. If there are no weights specified on vertexes, they are assumed to be one.

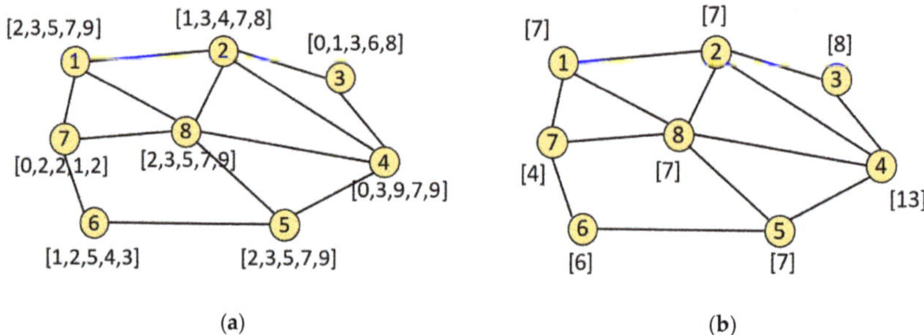

(a) (b)

Figure 3. Dynamic process compression on a graph. (**a**) Original graph with dynamics process; (**b**) Compression result of a graph with dynamics.

To facilitate the partitioning of graphs with dynamics, the total dynamic change of vertexes is counted as the new unique weight of vertexes denoted by w_i and is calculated as Formula (7).

$$w_i = \sum_{t=0}^{T} |v_i(t) - v_i(t+1)| \tag{7}$$

Therefore, referring to the original steps of the METIS algorithm [21], the steps of our D-METIS algorithm are as follows:

1.   Obtaining graph data;
2.   Compressing dynamics process information;
3.   Converting the adjacency matrix to Compressed Sparse Row (CSR) format;
4.   Coarsening;
5.   Initial partitioning;
6.   Refinement.

This is a model for large graph partition with its dynamic changes. The balanced graph-partitioning problem is defined as creating C disjoint sets of partitions, $V = V_1 \cup V_2 \ldots \cup V_C$, with the constraint that the sum of the weights in any given set does not exceed some threshold ε, which is greater than the average weight of a set V_c.

$$C \frac{\max_c |V_c|}{|V|} \leq 1 + \varepsilon \tag{8}$$

D-METIS's objective is to minimize the weight of inter-partition edge *edgecuts* while not exceeding the balance constraint.

$$edgecuts = \sum_{c=1}^{C} \sum_{v \in V_c} \sum_{u \in \Gamma(v), u \notin V_c} \theta\{v, u\} \tag{9}$$

$$\sum_{c=1}^{C} G_c + relink(edgecuts) = G \tag{10}$$

where v, $u \in V$, $c \in \{1, 2, 3, \ldots, C\}$, $V_c \subseteq V$. The links between vertexes and the vertexes in each V_c form a subgraph G_c. *relink(edgecuts)* is a function for relinking all the cut edges between subgraphs; it turns the subgraphs into the original large graph G.

The specific value of C depends on downstream task requirements. A larger value can be taken for faster results, and a smaller value can be used for improved accuracy. As we determined after many experiments, $C = |V|/l, l \in (100, 400)$, $C \geq 2$ is best.

3.2. Partitioned Graph Neural Dynamics Learner

The Partitioned Graph Neural Dynamics Learner (PGNDL) is a model of learning dynamics on partitioned graphs. It is a parallel model proposed specifically for large graphs. First of all, we used a differential equation system, as presented in Formula (11), to describe the dynamics on subgraphs cut by D-METIS.

$$\frac{dX_c(t)}{dt} = f_c(X_c, G_c, W_c, t) \tag{11}$$

where $X_c(t) \in \mathbb{R}^{v \times d}$ represents the state of a dynamic system consisting of v-linked vertexes in the subgraph G_c at time $t \in [0, T]$. $W_c(t)$ are parameters that control how the system evolves. The function f_c governs the instantaneous rate of change of dynamics on the subgraph G_c.

Such a problem can be seen as an optimal control problem so that the goal becomes to learn the best control parameters $W_c(t)$ for Formula (11). Unlike the traditional optimal control modeling, we modeled Formula (10) using GNN. After integrating Formula (11) over continuous time, the graph neural ODE model was proposed, illustrated as Formula (12).

$$X_c(t) = X_c(0) + \int_0^t f_c(X_c, G_c, W_c, \tau)d\tau \tag{12}$$

Formula (11) can have continuous layers with a real number t depth corresponding to the continuous-time dynamics on subgraph G_c. Thus, it can also be interpreted as a continuous-time GNN; the concept of continuous-time GNN was elaborated on in [17]. To increase the express ability of this model, we can encode the subgraph signals $X_c(t)$ from the original space to hidden space signals $X_{c,h}(t)$ so that this model can learn the dynamics better in a hidden space.

Then, a general model of graph neural dynamics learner can be denoted as follows:

$$\underset{W(t)}{\operatorname{argmin}}\mathcal{L} = \sum_{c=1}^{C} \left(\int_0^T \mathcal{R}(X_c, G_c, W_c, t)dt \right) \tag{13}$$

Subject to

$$X_{c,h}(t) = f_e(X_c(t), W_{c,e}) \tag{14}$$

$$\frac{dX_{c,h}(t)}{dt} = f(X_{c,h}, G_c, W_{c,h}, t) \tag{15}$$

$$X_c(t) = f_d(X_{c,h}(t), W_{c,d}) \tag{16}$$

where the objective formula, Formula (13), means the total loss of the continuous-time dynamics on subgraphs $\{G_1, G_2, G_3, \ldots\ldots\ldots, G_C\}$ from $t = 0$ to $t = T$. The constraint formula, Formula (14), transforms $X_c(t)$ into hidden space $X_{c,h}(t)$, and f_e is the encoding function. The constraint formula, Formula (15), is the governing dynamics in the hidden space by f. The constraint formula, Formula (16), decodes the hidden signal back to the original space by decoding function f_d. Additionally, f_e, f, and f_d are flexible to be any deep neural structures.

The PGNDL can learn differential equation systems to predict unequal interval states of the vertex, which means the PGNDL can learn the continuous-time dynamics on a graph (or subgraphs) at an arbitrary physical time t. The arbitrary physical times mean is unequally sampled with different observational time intervals. Additionally, there are two situations: when $t < T$ and t is not in $\{t, t+a, t+2a, \ldots\}$ (a is the value of equal sampling interval), it can be called interpolation prediction; when $t > T$, it is called extrapolation prediction.

We used ℓ1-norm loss as the loss function of the continuous-time dynamics on subgraphs and adopted two fully connected neural layers with a nonlinear hidden layer as f_e. A GCN model with a simplified diffusion operator Φ denotes the instantaneous rate of dynamics changes on subgraphs in the hidden space, and f_d is a linear decoding function to obtain the original signal for regression tasks. In addition, since our model uses a parallel learning mechanism on multiple subgraphs to minimize the total prediction error, it is easy to think of summing the errors of each subgraph to obtain the objective function (17). Thus, the model is:

$$\underset{W_*,b_*}{argmin}\mathcal{L} = \sum_{c=1}^{C} \left(\int_0^T |X_c(t) - \widehat{X_c(t)}| dt \right) \tag{17}$$

Subject to

$$X_{c,h}(t) = tanh(X_c(t)W_{c,e} + b_{c,e})W_0 + b_0 \tag{18}$$

$$\frac{dX_{c,h}(t)}{dt} = ReLU(\Phi X_{c,h}(t)W_c + b_c) \tag{19}$$

$$X_c(t) = X_{c,h}(t)W_{c,d} + b_{c,d} \tag{20}$$

where $\widehat{X_c(t)} \in \mathbb{R}^{n \times d}$ is the supervised dynamic information available at time stamp t. $|\cdot|$ denotes l_1-norm loss between $X_c(t)$ and $\widehat{X_c(t)}$ at time $t \in [0, T]$. Φ is the normalized graph Laplacian, $\Phi = D^{-\frac{1}{2}}(D - A)D^{-\frac{1}{2}}$, where $A \in \mathbb{R}^{n \times n}$ is the adjacency matrix of the network and $D \in \mathbb{R}^{n \times n}$ is the corresponding node degree matrix of subgraph c. $W_c \in \mathbb{R}^{d_e \times d_e}$ and $b_c \in \mathbb{R}^{n \times d_e}$ are shared parameters in subgraph c. $W_{c,e} \in R^{d \times d_e}$ and $W_0 \in R^{d_e \times d_e}$ are the matrices in linear layers for encoding, and $W_d \in \mathbb{R}^{d_e \times d}$ are for decoding. $b_{c,e}, b_0, b_c,$ and $b_{c,d}$ are the biases at the corresponding layers. Additionally, we designed the graph neural differential equation system as (19) to learn the network dynamics in a data-driven way. Thus, we can obtain $X_c(t)$, which means the states of subgraph c at time t.

3.3. General Framework

In this part, we summarize the general parallel framework of the work in a flowchart, as shown in Figure 4.

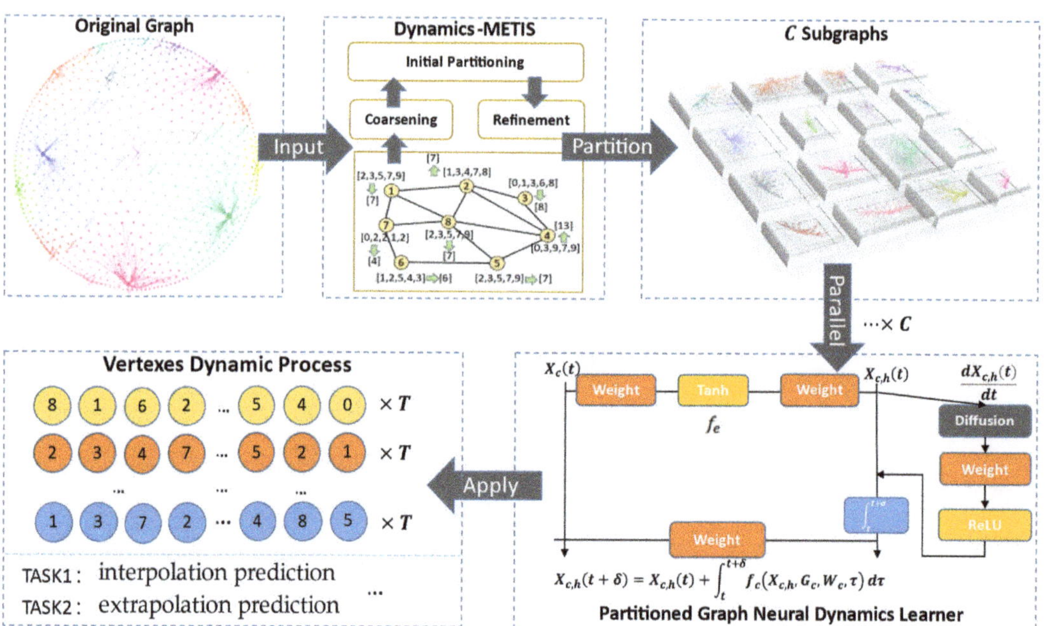

Figure 4. General framework.

In this framework, the original graph is taken as input data and contains the adjacency matrix $A_{i,j}$, and dynamic states of each vertex. We need to convert $A_{i,j}$ into the CSR format, then input $CSR(A_{i,j})$ and dynamic states of vertexes into the D-METIS algorithm. In the D-METIS algorithm, the total changes of states of each vertex on the original graph are first calculated as a compressed representation of the vertex dynamics changes. Additionally, coarsening, initial partitioning, and refinement are executed just like METIS [21]. After dividing the original large graph into C subgraphs, we can use PGNDL on C subgraphs in parallel to learn the dynamics of each subgraph. Then, the dynamic states of any vertex at a continuous time can be predicted according to the actual demands and tasks.

4. Experiments

4.1. Setup

In setup, four classes of graphs and three dynamics models were used to generate the simulation data. All the experiments were conducted with 11th Gen Intel (R) CPU @ 2.30 GHz with 32 GB of RAM. To ensure the generality of the results, each dataset was executed 10 times to obtain the average value.

4.1.1. Datasets

We chose the following graphs as our experimental datasets to verify the effectiveness of the model and framework:

- Random graph proposed by Erdós and Rényi [31];
- Power-law graph proposed by Albert Barabási [32];
- Community graph proposed by Santo Fortunato [33];
- Small World graph proposed by Watts and Strogatz [34].

Therefore, we obtained 4 graphs with 400 vertexes and 4 graphs with 2000 vertexes using the 4 classes of network models. Additionally, we set the initial value $X(0)$ the same for all the experiments, and thus, different dynamics were only due to their different dynamics' rules and networks' structures. We generated these graphs using the python package 'networkx2.0'. The specific generation parameters are shown in Appendix A (open source code in Github).

As we can see in Table 1, there are four classes of graphs, where $|V|$ and $|E|$ mean the numbers of vertexes and edges in different graphs, respectively.

Table 1. Statistics for four simulated datasets.

Graphs	$|V|$	$|E|$
Random	400	8050
	2000	200,160
Power Law	400	1975
	2000	9975
Community	400	1201
	2000	159,866
Small World	400	6308
	2000	5976

4.1.2. Dynamics Simulation on the Graph

The following three continuous-time network dynamics were used for dynamic simulation on the graph, where $\vec{x_i}(t) \in \mathbb{R}^{d \times 1}$ is the d dimensional features of vertex i at time t,

$$X(t) = \left[\ldots, \vec{x_i}(t), \ldots \right]^T \in \mathbb{R}^{n \times d}.$$

- Mutualistic interaction dynamics [35].
- This is a dynamic among species in ecology, and its equation is

$$\frac{d\vec{x_i}(t)}{dt} = b_i + \vec{x_i}\left(1 - \frac{\vec{x_i}}{k_i}\right)\left(\frac{\vec{x_i}}{c_i} - 1\right) + \sum_{j=1}^{n} A_{i,j}\frac{\vec{x_i}\vec{x_j}}{d_i + e_i\vec{x_i} + h_j\vec{x_j}} \tag{21}$$

- The operations there between vectors are element-wise. The mutualistic differential equation systems capture the abundance $\vec{x_i}(t)$ of species i, consisting of incoming migration term b_i, logistic growth with population capacity k_i and Allee effect with cold-start threshold c_i, and the mutualistic interaction term with interaction network A.
- Gene-regulatory dynamics [36].
- This can be described by an equation as follows:

$$\frac{d\vec{x_i}(t)}{dt} = -b_i\vec{x_i}f + \sum_{j=1}^{n} A_{i,j}\frac{\vec{x}_j^h}{\vec{x}_j^h + 1} \tag{22}$$

- where the first term models degradation when $f = 1$ or dimerization when $f = 2$, and the second term captures genetic activation tuned by the Hill coefficient h.
- SIS dynamics [37,38].
- S (Susceptible), a susceptible person, refers to a healthy person who lacks immunity and is vulnerable to infection after contact with an infected person. I (Infectious), the patient, refers to the infectious patient, and the infection can be transmitted to S and changed into I; R (Recovered) refers to a person with immunity after recovery. If the disease is a lifelong immune infectious disease, a person cannot be changed into S or I again. If the immune period is limited, a person can be changed into S again and then be reinfected. The mathematical expression is

$$N^* \, di(t)/dt = \lambda^* \, s(t)^* \, N^* i(t) - \mu^* \, N^* \, i(t) \tag{23}$$

- The total number of people is N. At time t, the ratio of various groups to the total number of people is, respectively, recorded as $s(t)$ and $i(t)$, and the number of various groups is $S(t)$ and $I(t)$. When $t = 0$ at the initial time, the initial ratio of the number of different types of people is $s0$ and $i0$. The average number of susceptible persons effectively contacted by each patient at each time point is λ, and the ratio of the number of cured patients to the total number of patients at each time point is μ.

4.2. Balance Analysis of D-METIS

We analyzed the graph-partition effect of the D-METIS algorithm from two aspects: the balance of the vertex number in every subgraph and the dynamic cumulative change of subgraphs.

4.2.1. Dynamic Cumulative Change of Subgraphs

As far as we know, this work is the first to consider how to segment a graph with dynamic processes. To analyze the dynamic balance effect of the D-METIS algorithm on each segmented subgraph, we compared three graph-partition methods, namely, random partition, METIS, and D-METIS, which are all graph-partition tasks for 400-node power-law networks. The partition results were positioned from 3 subgraphs to 11 subgraphs to verify the stability of our D-METIS algorithm through various cutting degrees.

As shown in Figure 5, the abscissa is the number of cut subgraphs; the ordinate is the number of nodes/dynamics-changes in subgraphs. The red lines are our D-METIS, which ensures the stability of dynamic cumulative change in each task. METIS performs poorly, and random partition performs worst.

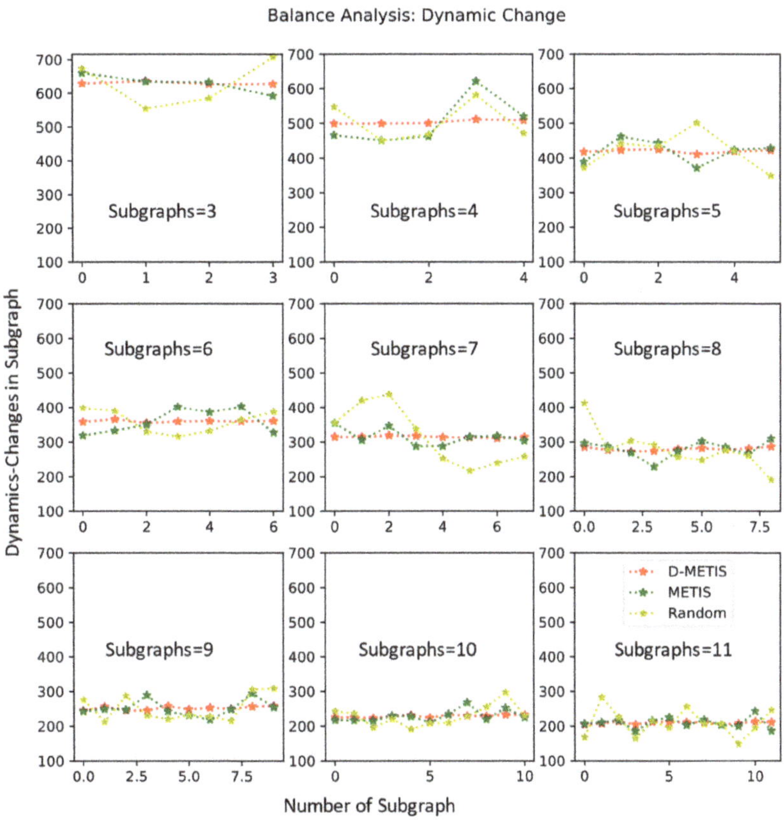

Figure 5. Balance analysis: dynamic cumulative change of subgraphs.

4.2.2. Vertex Distribution of Subgraphs

Additionally, we compared the three graph-partition methods to analyze the vertex distribution balance effect of the D-METIS algorithm on each subgraph; as seen in Figure 6, D-METIS performs much better than random partition and a little worse than METIS, as D-METIS should consider the constraint of dynamics balance. Despite this, D-METIS still splits the large graph evenly into multiple subgraphs. This is enough to balance the running time of downstream parallel tasks.

4.3. Learning Graph Dynamics with Unequal Interval Sampling

The PGNDL can be used for learning graph dynamics with unequal interval sampling and can complete interpolation/extrapolation prediction tasks using the 'dopri5' method with time-step 1 in the forward-integration process. To verify the progressiveness of the model, we compared and analyzed the NDCN [17] model, the PGNDL (with METIS), and our PGNDL (with D-METIS). In the PGNDL (with METIS) and PGNDL (with D-METIS), we ran the PGNDL (with METIS) and PGNDL (with D-METIS) in parallel on the subgraphs from the original graphs generated by three dynamic models and four graph types mentioned above. The NDCN is a single-threaded model used as our baseline. The three models used parameters of the same scale. Additionally, we repeated this experiment 10 times in the same way in each large graph; each repetition had 1000 iterations. We analyzed the fixed vertex numbers of 400 and 2000 and took the average $\ell 1$ loss of 10 runs as the final result for comparison. According to the same specifications, we also analyzed the efficiency of our models.

Using the $\ell 1$ loss results of the above three models, we analyzed the effect of graph dynamics learning with METIS and D-METIS algorithms in various graphs and their dynamics models. The experimental results proved the effectiveness of D-METIS. In terms of the calculation run time, the advantages of our model are obvious.

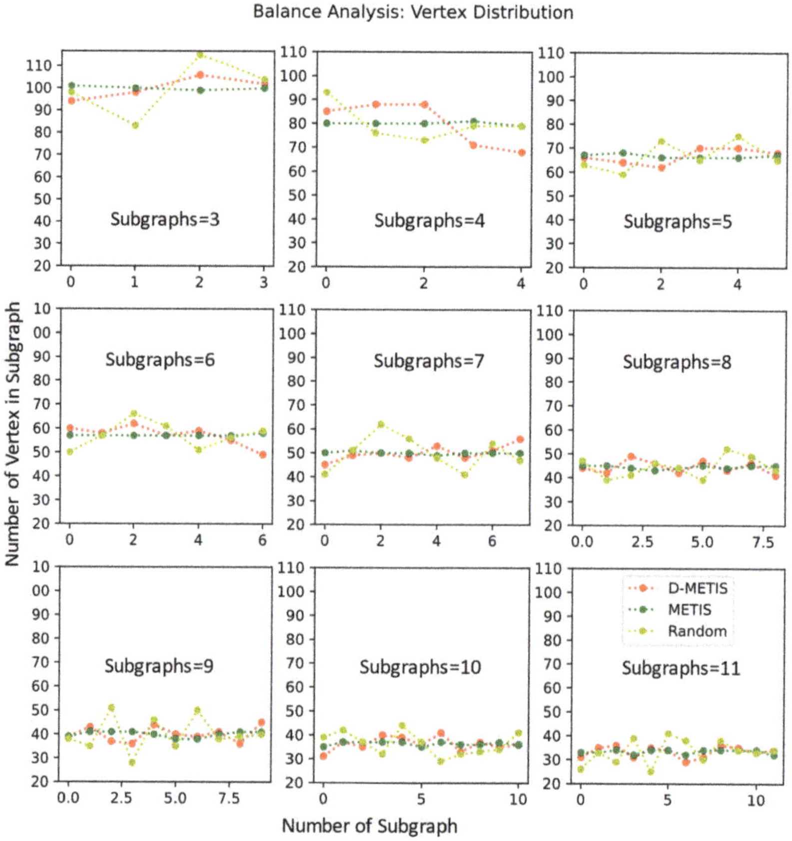

Figure 6. Balance analysis: vertex distribution of subgraphs.

4.3.1. Interpolation-Prediction Task

We irregularly sampled 120 snapshots of the $[0, T]$ dynamics $\left\{ X_c\left(\hat{t_1}\right), \ldots, X_c\left(\hat{t_{120}}\right) \mid 0 \leq t_1 < \ldots < t_{120} \leq T \right\}$; the intervals between $t_1 \ldots t_{120}$ were random and different. Then, we picked 80 snapshots randomly from the top 100 as the training set and used the remained 20 snapshots in the top 100 to test the interpolation-prediction task.

The experimental results for the NDCN, PGNDL (METIS), and our PGNDL (D-METIS) are shown in Table 2 (for n = 400) and Table 3 (for n = 2000).

As can be seen in Tables 2 and 3, our PGNDL (D-METIS) model attained 17 ↑ and 1—symbols when n = 400 and 15 ↑ when n = 2000; this means our model performs better in most cases. Additionally, we noticed that the PGNDL performs best if the dynamics is SIS Dynamics. Similarly, we also found that when the class of the graph was Community or Small World, using our model for the interpolation-prediction task was a better choice.

Table 2. Accuracy of the interpolation-prediction task. The original graph size is $A_{400 \times 400}$, and cut every graph into 4 subgraphs.

Model	Dynamics	Random	Power Law	Community	Small World
NDCN	SIS Dynamics	0.023	0.287	0.025	0.037
	Mutualistic interaction	0.472	0.341	0.831	0.436
	Gene Regulation	1.951	0.719	2.529	1.053
PGNDL (METIS)	SIS Dynamics	0.005	0.291	0.012	0.033
	Mutualistic interaction	0.503	0.437	0.523	0.393
	Gene Regulation	3.451	1.534	2.671	0.891
PGNDL (D-METIS)	SIS Dynamics	0.004 ↑↑	0.273 ↑↑	0.011 ↑↑	0.033 ↑-
	Mutualistic interaction	0.460 ↑↑	0.486 ↓↓	0.457 ↑↑	0.407 ↑↓
	Gene Regulation	2.780 ↓↑	1.568 ↓↓	2.456 ↑↑	0.849 ↑↑

The following is an explanation of some symbols used in the tables: ↑↑ means the marked result is better than that of the NDCN and PGNDL (METIS). ↑↓ means the marked result is better than that of the NDCN but worse than the PGNDL (METIS). ↓↓ means the marked result is worse than that of the NDCN and PGNDL (METIS). ↓↑ means the marked result is worse than that of the NDCN but better than the PGNDL (METIS). -↑ means the marked result is equal to that of the NDCN and better than the PGNDL (METIS). ↑- means the marked result is better than that of the NDCN and equal to the PGNDL (METIS). -↓ means the marked result is equal to that of the NDCN but worse than the PGNDL (METIS). ↓- means the marked result is worse than that of the NDCN and equal to the PGNDL (METIS). The symbol definitions above also apply to Tables 3–5.

Table 3. Accuracy of the interpolation-prediction task. The original graph size is $A_{2000 \times 2000}$, and we cut every graph into 8 subgraphs.

Model	Dynamics	Random	Power Law	Community	Small World
NDCN	SIS Dynamics	0.028	0.137	0.024	0.111
	Mutualistic interaction	0.538	0.368	1.098	0.482
	Gene Regulation	11.150	1.248	24.090	1.110
PGNDL (METIS)	SIS Dynamics	0.024	0.024	0.014	0.043
	Mutualistic interaction	0.644	0.538	0.697	0.446
	Gene Regulation	11.582	2.085	47.890	2.892
PGNDL (D-METIS)	SIS Dynamics	0.008 ↑↑	0.022 ↑↑	0.005 ↑↑	0.037 ↑↑
	Mutualistic interaction	0.664 ↓↓	0.635 ↓↓	0.735 ↑↓	0.431 ↑↑
	Gene Regulation	10.141 ↓↑	1.548 ↓↑	35.250 ↓↑	2.030 ↓↑

Table 4. Accuracy of the extrapolation-prediction task. The original graph size is $A_{400 \times 400}$, and we cut every graph into 4 subgraphs.

Model	Dynamics	Random	Power Law	Community	Small World
NDCN	SIS Dynamics	0.017	0.021	0.008	0.021
	Mutualistic interaction	0.223	0.245	0.434	0.227
	Gene Regulation	2.287	0.371	3.070	0.870
PGNDL (METIS)	SIS Dynamics	0.009	0.019	0.014	0.024
	Mutualistic interaction	0.516	0.390	0.512	0.300
	Gene Regulation	3.592	1.312	2.501	0.994
PGNDL (D-METIS)	SIS Dynamics	0.005 ↑↑	0.020 ↑↓	0.006 ↑↑	0.019 ↑↑
	Mutualistic interaction	0.420 ↓↑	0.360 ↓↑	0.220 ↑↑	0.210 ↑↑
	Gene Regulation	2.860 ↓↑	1.907 ↓↓	2.597 ↑↓	2.490 ↓↓

Table 5. Accuracy of the extrapolation-prediction task. The original graph size is $A_{2000 \times 2000}$, and we cut every graph into 8 subgraphs.

Model	Dynamics	Random	Power Law	Community	Small World
NDCN	SIS Dynamics	0.004	0.021	0.004	0.019
	Mutualistic interaction	0.103	0.299	0.493	0.194
	Gene Regulation	15.710	0.548	25.940	1.258
PGNDL(METIS)	SIS Dynamics	0.004	0.024	0.004	0.023
	Mutualistic interaction	0.644	0.538	0.686	0.346
	Gene Regulation	11.582	2.025	54.130	3.271
PGNDL (D-METIS)	SIS Dynamics	**0.004 - -**	**0.020 ↑↑**	**0.003 ↑↑**	**0.019 -↑**
	Mutualistic interaction	**0.634 ↓↑**	**0.530 ↓↑**	**0.487 ↑↑**	**0.337 ↓↑**
	Gene Regulation	**10.611 ↑↓**	**2.034 ↓↓**	**32.640 ↓↑**	**2.625 ↓↑**

4.3.2. Extrapolation-Prediction Task

Different from the interpolation-prediction task, extrapolation prediction requires 80 snapshots in the top 100 as the training set and 20 snapshots of the tail 20 for testing.

The results of the extrapolation-prediction task are shown in Tables 4 and 5. We attained 16 ↑ when n = 400 and 13 ↑ and 2—when n = 2000. This result is similar to the result of the interpolation-prediction task and shows that our model is very suitable for SIS Dynamics and Community graphs.

In addition, we also found that the results of almost all PGNDL models based on D-METIS are more accurate than those using only METIS. This shows that D-METIS plays a positive role in helping the model learn dynamics in the graph.

4.3.3. Complexity and Time-Consumption Analysis

First, we compared the space complexity of NDCN and PGNDL:

$$\frac{O_{PGNDL}\left(|V|^2 \cdot Para_{GNN}\right)}{O_{NDCN}\left(\left(\frac{|V|}{|C|}\right)^2 \cdot Para_{GNN}\right)} = \frac{1}{|C|} \tag{24}$$

where $Para_{GNN}$ is the parameters in the GNN's neural network structure to be optimized; thus, the space complexity of our PGNDL is $\frac{1}{|C|}$ of the NDCN.

Additionally, since our PGNDL is a parallel implementation of the NDCN, the time complexity is consistent with the NDCN.

To further analyze the actual efficiency of the PGNDL, it is necessary to conduct a detailed analysis of the runtime, which is beneficial for summarizing engineering experience for practice. The PGNDL can complete the extrapolation- and interpolation-prediction tasks simultaneously due to the same training process. Therefore, we can estimate the time consumption of two tasks at once. In other words, the time consumption of these two tasks is consistent. We recorded the above experimental runtime of extrapolation prediction and interpolation prediction.

The running-time statistics of each model are as follows.

It can be seen in Table 6 that our PGNDL (D-METIS) is the fastest model in every class of graph or dynamics; it is **2 to 3 times faster** than NDCD when n = 400.

Table 6. Time consumption. The original graph size: $A_{400 \times 400}$; number of subgraphs is 4.

Model	Dynamics	Random	Power Law	Community	Small World
NDCN	SIS Dynamics	66.8	67.6	64.1	67.4
	Mutualistic interaction	74.8	75.8	74.6	73.2
	Gene Regulation	82.5	78.2	83.4	74.9
PGNDL (METIS)	SIS Dynamics	35.4	30.2	37.6	29.5
	Mutualistic interaction	35.5	30.4	37.8	29.8
	Gene Regulation	35.4	30.2	37.9	29.9
PGNDL (D-METIS)	SIS Dynamics	**34.2**	**28.5**	**36.8**	**28.2**
	Mutualistic interaction	**33.4**	**28.7**	**37.2**	**28.4**
	Gene Regulation	**33.9**	**28.5**	**37.1**	**28.4**

When n = 2000, as seen in the results of Table 7, our model PGNDL with D-METIS is **2 to 4 times** faster than NDCD.

Table 7. Time consumption. The original graph size: $A_{2000 \times 2000}$; number of subgraphs is 8.

Model	Dynamics	Random	Power Law	Community	Small World
NDCN	SIS Dynamics	207.8	234.7	223.6	179.4
	Mutualistic interaction	198.8	240.7	260.8	276.1
	Gene Regulation	342.9	238.5	432.3	286.8
PGNDL (METIS)	SIS Dynamics	118.3	77.3	145.3	77.0
	Mutualistic interaction	119.2	80.1	146.6	77.5
	Gene Regulation	118.6	79.7	147.8	77.4
PGNDL (D-METIS)	SIS Dynamics	**115.7**	**76.7**	**146.2**	**75.1**
	Mutualistic interaction	**116.9**	**77.8**	**149.2**	**76.2**
	Gene Regulation	**116.1**	**77.3**	**148.5**	**77.5**

In addition, by comparing 400 vertexes with 2000 vertexes, we can infer that as the number of vertexes increases, the number of cut subgraphs will increase, and the acceleration effect will be more significant.

5. Conclusions

This work proposed a general parallel framework that contains a graph-partition accelerated graph neural dynamics learning model called the PGNDL and a novel graph-partition algorithm entitled D-METIS for graphs with dynamics. The PGNDL can learn unequal interval sampled dynamics states in graphs. Different from other graph-learning methods, our model has an appropriate graph-partition mechanism that reduces the graph size and then uses parallel learning for subgraphs; benefiting from this, our model is more than twice as fast as others. Furthermore, we obtained more accurate results on some mainstream network types and dynamics. We used the PGNDL to learn unequal time interval states of subgraphs dynamics for interpolation prediction and extrapolation prediction in parallel. Based on the PGNDL, four graph networks and three dynamic processes were tested and analyzed in the experimental part. We found that the graph dynamics learning of the graph-partition parallel acceleration method was faster than other methods by at least 200%, and it is very suitable for SIS dynamics and Community graphs; in these cases, our model performed accurately and efficiently. In the future, we will try to apply D-METIS and PGNDL to the prevention and control of infectious diseases in large communities and the prediction of interest transfer in large communities. Additionally, we will explore its limitations and wider application scope.

Author Contributions: Conceptualization, X.H. and A.L.; Methodology, X.H.; Software, X.H., X.X. and A.L.; Validation, X.H.; Formal analysis, X.H. and Q.Z.; Investigation, X.Z.; Resources, X.Z., X.X. and Q.Z.; Data curation, X.Z., X.X. and A.L.; Writing—original draft, X.Z.; Supervision, X.Z., X.X., Q.Z. and A.L.; Project administration, X.Z., Q.Z. and A.L.; Funding acquisition, X.Z. All authors have read and agreed to the published version of the manuscript.

Funding: This research was founded by the National Natural Science Foundation of China, grant number 61273322; the Changsha Science and Technology Bureau, grant number KQ2009009; and the Huxiang Youth Talent Support Program, grant number 2021RC3076.

Institutional Review Board Statement: Not applicable.

Informed Consent Statement: Not applicable.

Data Availability Statement: See in Appendix A.

Conflicts of Interest: The authors declare no conflict of interest.

Appendix A

https://github.com/Huangbuffer/PGNDL (accessed on 15 November 2022). The codes and parameters are open-sourced at the link above. The datasets can be generated in Dynamics-METIS.py.

References

1. Christensen, K.K.; Liebetrau, T. A new role for 'the public'? Exploring cyber security controversies in the case of WannaCry. *Intell. Natl. Secur.* **2019**, *34*, 395–408. [CrossRef]
2. Akaev, A.; Zvyagintsev, A.I.; Sarygulov, A.; Devezas, T.; Tick, A.; Ichkitidze, Y. Growth Recovery and COVID-19 Pandemic Model: Comparative Analysis for Selected Emerging Economies. *Mathematics* **2022**, *10*, 3654. [CrossRef]
3. Abdelhamid, A.A.; El-Kenawy, E.-S.M.; Khodadadi, N.; Mirjalili, S.; Khafaga, D.S.; Alharbi, A.H.; Ibrahim, A.; Eid, M.M.; Saber, M. Classification of Monkeypox Images Based on Transfer Learning and the Al-Biruni Earth Radius Optimization Algorithm. *Mathematics* **2022**, *10*, 3614. [CrossRef]
4. Doerr, B.; Fouz, M.; Friedrich, T. Why rumors spread so quickly in social networks. *Commun. ACM* **2012**, *55*, 70–75. [CrossRef]
5. Fan, C.; Zeng, L.; Sun, Y.; Liu, Y.-Y. Finding key players in complex networks through deep reinforcement learning. *Nat. Mach. Intell.* **2020**, *2*, 317–324. [CrossRef]
6. Tanaka, G.; Morino, K.; Aihara, K. Dynamical robustness in complex networks: The crucial role of low-degree nodes. *Sci. Rep.* **2012**, *2*, 232. [CrossRef]
7. Acharya, A. An action for nonlinear dislocation dynamics. *J. Mech. Phys. Solids* **2022**, *161*, 104811. [CrossRef]
8. Lyu, J.; Liu, F.; Ren, Y. Fuzzy identification of nonlinear dynamic system based on selection of important input variables. *J. Syst. Eng. Electron.* **2022**, *33*, 737–747. [CrossRef]
9. Newman, M.E.; Barabási, A.L.E.; Watts, D.J. *The Structure and Dynamics of Networks*; Princeton University Press: Princeton, NJ, USA, 2011; Volume 12.
10. Lü, J.; Wen, G.; Lu, R.; Wang, Y.; Zhang, S. Networked Knowledge and Complex Networks: An Engineering View. *IEEE/CAA J. Autom. Sin.* **2022**, *9*, 1366–1383. [CrossRef]
11. Wood, R.K. Deterministic network interdiction. *Math. Comput. Model.* **1993**, *17*, 1–18. [CrossRef]
12. Phillips, C.A. The network inhibition problem. In Proceedings of the Twenty-fifth Annual ACM Symposium on Theory of Computing, San Diego, CA, USA, 16–18 May 1993; pp. 776–785.
13. Brockschmidt, M. GNN-FiLM: Graph Neural Networks with Feature-wise Linear Modulation. In Proceedings of the 37th International Conference on Machine Learning, PMLR 119, Virtual, 13–18 July 2020; pp. 1144–1152.
14. Narayan, A.; Roe, P.H.O. Learning graph dynamics using deep neural networks. *Ifac-Papersonline* **2018**, *51*, 433–438. [CrossRef]
15. Seo, Y.; Defferrard, M.; VanderGheynst, P.; Bresson, X. Structured sequence modeling with graph convolutional recurrent networks. In Proceedings of the International Conference on Neural Information, Siem Reap, Cambodia, 13–16 December 2018; Springer: Cham, Switzerland, 2018; pp. 362–373.
16. Ma, S.; Liu, J.; Zuo, X. Survey on Graph Neural Network. *J. Comput. Res. Dev.* **2022**, *59*, 47–80.
17. Zang, C.; Wang, F. Neural Dynamics on Complex Networks. In Proceedings of the 26th ACM SIGKDD International Conference on Knowledge Discovery & Data Mining, ACM, Virtual Event, 6–10 July 2020; pp. 892–902.
18. Stanton, I.; Kliot, G. Streaming graph partitioning for large distributed graphs. In Proceedings of the 18th ACM SIGKDD International Conference on Knowledge Discovery and Data Mining, KDD '12, New York, NY, USA, 12–16 August 2012; pp. 1222–1230.

19. Zhang, C.; Wei, F.; Liu, Q.; Tang, Z.G.; Li, Z. Graph edge partitioning via neighborhood heuristic. In Proceedings of the 23rd ACM SIGKDD International Conference on Knowledge Discovery and Data Mining, Halifax, NS, Canada, 13–17 August 2017; Volume 8, pp. 605–614.
20. Tsourakakis, C.; Gkantsidis, C.; Radunovic, B.; Vojnovic, M. Fennel: Streaming graph partitioning for massive scale graphs. In Proceedings of the 7th ACM International Conference on Web Search and Data Mining, ACM, New York City, NY, USA, 24–28 February 2014; pp. 333–342.
21. Karypis, G.; Kumar, V. A fast and high quality multilevel scheme for partitioning irregular graphs. *SIAM J. Sci. Comput.* **1998**, *20*, 359–392. [CrossRef]
22. Xie, C.; Yan, L.; Li, W.J.; Zhang, Z. Distributed Power-Law Graph Computing: Theoretical and Empirical Analysis. In Proceedings of the Advances in Neural Information Processing Systems, Montreal, QC, Canada, 8–13 December 2014; pp. 1673–1681.
23. Nazi, A.; Hang, W.; Goldie, A.; Ravi, S.; Mirhoseini, A. Gap: Generalizable approximate graph partitioning framework. *arXiv* **2019**, arXiv:1903.00614.
24. Craig, T. A Treatise on Linear Differential Equations. *Nature* **1890**, *41*, 508–509.
25. Shampine, L.F. *Computer Methods for Ordinary Differential Equations and Differential-Algebraic Equations (Book Review)*; SIAM Review: Philadelphia, PA, USA, 1999; Volume 41, pp. 400–401.
26. Chen, R.T.Q.; Rubanova, Y.; Bettencourt, J.; Duvenaud, D.K. Neural Ordinary Differential Equations. *Adv. Neural Inf. Process. Syst.* **2018**, *31*, 6571–6583.
27. Zhang, Y.; Guo, Y.; Zhang, Z.; Chen, M.; Wang, S.; Zhang, J. Universal framework for reconstructing complex networks and node dynamics from discrete or continuous dynamics data. *Phys. Rev. E* **2022**, *106*, 034315. [CrossRef]
28. Yu, D.; Zhou, Y.; Zhang, S.; Liu, C. Heterogeneous Graph Convolutional Network-Based Dynamic Rumor Detection on Social Media. *Complexity* **2022**, *2022*, 8393736. [CrossRef]
29. Hwang, J.; Choi, J.; Choi, H.; Lee, K.; Lee, D.; Park, N. Climate Modeling with Neural Diffusion Equations. In Proceedings of the 2021 IEEE International Conference on Data Mining (ICDM), Auckland, New Zealand, 7–10 December 2021.
30. Wang, F.; Cui, P.; Pei, J.; Song, Y.; Zang, C. Recent Advances on Graph Analytics and Its Applications in Healthcare. In Proceedings of the KDD '20: 26th ACM SIGKDD International Conference on Knowledge Discovery & Data Mining, Virtual Event, 6–10 July 2020.
31. Erdős, P.; Rényi, A. On the evolution of random graphs. *Publ. Math. Inst. Hung. Acad. Sci.* **1960**, *5*, 17–60.
32. Barabási, A.-L.; Albert, R. Emergence of scaling in random networks. *Science* **1999**, *286*, 509–512. [CrossRef] [PubMed]
33. Fortunato, S. Community detection in graphs. *Phys. Rep.* **2009**, *486*, 75–174. [CrossRef]
34. Watts, D.J.; Strogatz, S.H. Collective dynamics of 'small world' networks. *Nature* **1998**, *393*, 440–442. [CrossRef] [PubMed]
35. Gao, J.; Barzel, B.; Barabási, A.-L. Author Correction: Universal resilience patterns in complex networks. *Nature* **2019**, *568*, E5. [CrossRef]
36. Alon, U. *An Introduction to Systems Biology: Design Principles of Biological Circuits*; Chapman & Hall/CRC: London, UK, 2007; 320p, ISBN 1584886420. GBP 30.99.
37. Tong, Y.; Ahn, I.; Lin, Z. The impact factors of the risk index and diffusive dynamics of a SIS free boundary model. *Infect. Dis. Model.* **2022**, *7*, 605–624. [CrossRef] [PubMed]
38. Jing, X.; Liu, G.; Jin, Z. Stochastic dynamics of an SIS epidemic on networks. *J. Math. Biol.* **2022**, *84*, 50. [CrossRef]

Article

Novel Hybrid MPSI–MARA Decision-Making Model for Support System Selection in an Underground Mine

Miloš Gligorić [1,*], Zoran Gligorić [1], Suzana Lutovac [1], Milanka Negovanović [1] and Zlatko Langović [2]

[1] Faculty of Mining and Geology, University of Belgrade, Đušina 7, 11000 Belgrade, Serbia
[2] Faculty of Hotel Management and Tourism, University of Kragujevac, 36210 Vrnjačka Banja, Serbia
* Correspondence: milos.gligoric@rgf.bg.ac.rs

Abstract: An underground mine is a very complex production system within the mining industry. Building up the underground mine development system is closely related to the installation of support needed to provide the stability of mine openings. The selection of the type of support system is recognized as a very hard problem and multi-criteria decision making can be a very useful tool to solve it. In this paper we developed a methodology that helps mining engineers to select the appropriate support system with respect to geological conditions and technological requirements. Accordingly, we present a novel hybrid model that integrates the two following decision-making components. First, this study suggests a new approach for calculating the weights of criteria in an objective way named the Modified Preference Selection Index (MPSI) method. Second, the Magnitude of the Area for the Ranking of Alternatives (MARA) method is proposed as a novel multi-criteria decision-making technique for establishing the final rank of alternatives. The model is tested on a hypothetical example. Comparative analysis confirms that the new proposed MPSI–MARA model is a very useful and effective tool for solving different MCDM problems.

Keywords: objective weights of criteria; MPSI method; multi-criteria decision-making method; MARA method; support system selection; underground mining

Citation: Gligorić, M.; Gligorić, Z.; Lutovac, S.; Negovanović, M.; Langović, Z. Novel Hybrid MPSI–MARA Decision-Making Model for Support System Selection in an Underground Mine. *Systems* **2022**, *10*, 248. https://doi.org/10.3390/systems10060248

Academic Editors: Zaoli Yang, Yuchen Li and Ibrahim Kucukkoc

Received: 24 October 2022
Accepted: 1 December 2022
Published: 9 December 2022

Publisher's Note: MDPI stays neutral with regard to jurisdictional claims in published maps and institutional affiliations.

1. Introduction

Supporting is one of the most important parts of every underground mine project that needs special attention. It represents the mining activity that is primarily directed at preventing a disastrous mine roof fall. As well as securing roof stability, monitoring and controlling deformations along the underground drifts are extremely important characteristics of every support system technology. Depending on the rock types and geological conditions that govern the deposits, supporting is implemented to a greater or lesser extent. Underground coal mines require extensive utilization of supporting systems, while the application of support in polymetallic deposits is less required.

Since the ore deposits are inclined to surface and are almost exhausted using surface mine technologies, it is necessary to change the exploitation technology by using underground mining technology. The same case applies to coal deposits that are excavated using underground mining methods. Having in mind that the coal is sedimentary rock contained of accompanying soft rocks with low strength, it means that underground objects in coal deposits should be secured from self-destruction. Because of these complex geological conditions, every underground coal mine requires the application of certain support system technologies. Realizing the stability of the underground structures, providing safe workplace conditions for personnel and creating, as much as possible, a longer useful life are essential functions of every support system in an underground coal mine.

In recent years, multi-criteria decision-making (MCDM) methods have experienced increasing application in solving many problems in different fields of research. Economics, mathematics and engineering are just some of the scientific areas where the application of MCDM methods play a vital role in creating a reliable model in the decision-making process. In accordance with that, an enormous number of MCDM methods have been developed for solving such complex problems. With the rapid progress of new MCDM methods, methods for defining the weights of criteria were developed at the same time. It means that the optimal solution largely depends on the weights of criteria. Considering the huge impact on the final results in decision-making process, criteria weight determination is one of the most significant problems in MCDM models.

The MPSI method represents a novel method for determining the weight coefficients in the different MCDM problems in an objective way. This method is based on the traditional Preference Selection Index (PSI) method creating an improved and modified version called the Modified Preference Selection Index (MPSI) method. This modification is quite slight and is only directed at eliminating a certain step from the original PSI method. This minimal modification upgrades the final values of the weight coefficients and makes them closer to values obtained by other objective weighting methods.

The MARA method presents a new multi-criteria decision-making method providing support to decision makers for solving different problems in numerous spheres. This method is based on Magnitude of the Area for the Ranking of Alternatives named the MARA method. The two main functions under the optimal alternative and each alternative are established as the major concepts of this proposed method. Calculating the area under the optimal and each alternative has a pivotal part to play in computing the magnitude of the area that represents the core of this study. The area below the alternative is calculated by definite integration of the linear function over an interval of 0–1. The proposed methodology is easy to understand and is a low time-consumption approach.

The aim of this study is to provide rational assistance to underground mining engineers through the process of support system selection in underground mines. This MCDM model is based on integrating the MPSI method and MARA method as two novel approaches. The MPSI method represents a new technique for evaluating the objective weights of criteria. The MARA method is the new proposed method applied for the ranking of alternatives (support systems in underground mines).

The paper is organized as follows. Firstly, Section 1 provides the Introduction and aim of this paper while the Literature Review is described in Section 2. Then, Section 3 represents a detailed description of the new objective method named MPSI that is used for determining the criteria weights. The novel multi-criteria decision-making method called the MARA method is expressed in Section 4. A comprehensive illustration of the hybrid MPSI–MARA model is demonstrated through the Numerical Example in Section 5, while the validation of the proposed model is verified through the Comparative Analysis in Section 6. Finally, Section 7 deals with the discussion of the proposed methods while concluding considerations and further research are explained in Section 8.

2. Literature Review

Some authors have investigated and estimated different approaches and combined versions of support systems in underground mines as well as in tunnels. Kang [1] analysed the deformations and damage features of coal mine roadways under complex geological and geotechnical characteristics. Additionally, roadway support techniques such as the application of rock bolting, steel arches, grouting and combined support design in coal mines in different countries were analysed. Several case studies have analysed typical support system technologies. In order to better understand the behaviour of support system techniques in coal mines, Mark and Barczak [2] introduced elementary instructions of supporting procedure for mining engineers. The key factors that affect the mine structure such as rock strength, roof span and forces applied to the mine roof were explained. In addition, other significant factors that have an influence on a coal mine roof such as the load-carrying capacity, installation time and quality as well as the ability of the support system to secure skin control were described. Elrawy et al. [3] illustrated the stability evaluation of the underground tunnel in a nickel mine considering of various collapse assessment criteria such as strength of the rock mass, the extent of failure zone and rock mass deformation. They simulated different variants of rock support systems through several cases to select the best one. Putra [4] described the basic concepts of the underground support elements and their selection based on an empirical approach and finite element method. Canbulat and Merwe [5] developed an optimum roof support system methodology based on a stochastic technique and probabilistic approach to input data.

As well as these conventional approaches for supporting in mines, the application of powered roof support mechanisms known as longwall systems is growing rapidly especially in deep coal mines. In the global coal mining industry, a longwall system is recognized as a vital structure for providing protection and safety conditions during the underground mining process. There are many authors who have analysed, investigated, monitored and selected the most efficient longwall system for roof support in underground conditions [6–10].

Numerous researchers have observed and investigated the support design system in underground conditions as a multi-criteria decision-making concept. Rafiee et al. [11] applied the traditional Technique for Order Preference by Similarity to Ideal Solution (TOPSIS) method to select the most efficient support system in a Naien water transporting tunnel. They created an initial decision-making matrix considering four alternatives (models of support systems) with respect to five criteria such as the probability of failure, safety factor, cost, applicability factor and displacement. Jalalifear et al. [12] presented an optimum rock bolt system selection for supporting the underground facilities based on a hybrid Analytical Hierarchy Process (AHP)–Entropy–TOPSIS approach. AHP and Entropy methods are used to compute the weights of the total of twenty-six criteria, while the TOPSIS methodology is applied to choose the most preferred rock bolt system design of which there are a total of nine. Shaffiee Haghshenas et al. [13] developed an MCDM model to estimate the most suitable tunnel support system. This methodology is established as a combined version of the FDAHP (Fuzzy Delphi Analytic Hierarchy Process) and ELECTRE (Elimination and Choice Expressing Reality) procedures. Five tunnel support systems with respect to six criteria were used to rank the supporting system and examined through numerical calculation in a case study. Oraee et al. [14] proposed an approach for tunnel support system selection based on the AHP method. Ten alternatives (support systems) and seven criteria such as vertical displacement at point 1, point 2 and point 3, horizontal displacement at point 3, support system costs, support system performance and safety factors are involved in the model. Oraee and Bakhtavar [15] studied the optimum tunnel support system selection using the integration of the AHP, TOPSIS and Preference Ranking Organization METHod for Enrichment Evaluations (PROMETHEE) techniques. To validate the capability of the utilized methods for support system selection, a case study from [14] was implemented. Tajvidi Asr et al. [16] introduced the support system selection in the Beheshtabad tunnel by applying the Simple Additive Weighting (SAW), TOPSIS and Linear Assignment (LA) methods. Six variants of support systems as alternatives with respect to six criteria (cost, safety factor, applicability, installation time, displacement and mechanization) were included in the MCDM process for final ranking. Yavuz et al. [17] presented an optimum support design for a main haulage road in the Western Lignite Corporation (WLC), Tuncbilek, in Turkiye based on the AHP method.

The main and very important part of every MCDM method is assessing the weights of criteria. There are many methods for determining the weights of criteria. All of these methods can be divided into three groups: subjective weighting methods, objective weighting methods and combined weighting methods. Subjective methods for criteria weight determination are based on estimation and opinion by the decision maker. These methods represent the subjective preference by the decision maker of the input data that directly influence the final choice in decision making. There are many subjective weighting methods that have been widely used in various areas such as the Full Consistency Method (FUCOM) [18,19], Step-Wise weight Assessment Ratio Analysis (SWARA) [20,21], PIvot Pairwise RElative Criteria Importance Assessment (PIPRECIA) [22,23], Decision-MAking

Trial and Evaluation Laboratory (DEMATEL) [24,25], KEmeny Median Indicator Ranks Accordance (KEMIRA) [26], WEight Balancing Indicator Ranks Accordance (WEBIRA) [27], Simple Multi-Attribute Rating Technique (SMART) [28] and many other methods. Objective methods for criteria weight determination, unlike subjective methods, are based on information from the "raw" input data using mathematical tools for obtaining the criteria weights. These methods exclude any subjective favour and judgement of the decision maker that makes this approach more acceptable and more suitable. Many researchers have applied objective weighting methods in different scientific fields such as Entropy [29,30], CRiteria Importance Through Intercriteria Correlation (CRITIC) [31,32], Standard Deviation (SD) [33,34], Mean Weight (MW) [35,36], Coefficient of Variation (CV) [37], MEthod based on the Removal Effects of Criteria (MEREC) [38,39], LOgarithmic Percentage Change-driven Objective Weighting (LOPCOW) [40], Criteria Impact LOSs (CILOS) [41,42], Integrated Determination Of CRiteria Weight (IDOCRIW) [41,42] and many other methods. Combined methods for criteria weight determination represent the hybrid of subjective and objective methods. These methods take the positive characteristics of the other methods, subjective and objective, and integrate them into one model. There are many authors who have utilized the combined weighting methods [43–46].

Many authors have applied the PSI method in different scientific disciplines to investigate MCDM problems. Moreover, many authors have used the PSI method combined with other methods under uncertain environments for ranking the alternatives. This method is simple and easy to understand for the decision maker and widely applied in various scientific fields for solving multi-criteria decision-making (MCDM) problems such as: the design stage of the production system life cycle [47], the machining optimization problem [48], the fuel buses selection problem [49], the ranking of the performance factors of a flexible manufacturing system [50], warehouse location for a supermarket [51], natural thermal insulation material selection for the external walls [52], the potential tourism selection of countries [53], the problem of determining laser cutting process conditions [54], machine performance evaluation [55], the problem of human resource management [56], etc. From the very extensive literature review, there are no methods dealing with modified or improved approaches of the PSI method. It was one of the key reasons and main motivations to investigate the original PSI method and to develop a modified application of the PSI method.

In addition to numerous studies implementing the conventional and relatively new MCDM techniques [57–60], there are many authors who have examined and established an enormous number of MCDM methods that cover uncertain environments creating improved and extended versions of existing MCDM methods [61–65].

3. MPSI Method

In order to calculate the objective weights of criteria, we created a new approach for determining the weight coefficients. We applied well-known Preference Selection Index (PSI) method and made a slight modification of that method. Preference Selection Index (PSI) method was first developed by Maniya and Bhatt [66] as a new type of decision-making method for solving multi-criteria decision-making problems.

MPSI method is based on the degree of the oscillation, i.e., variation in the preference value for each criterion. That variation actually presents the distance between normalized value and mean value per criterion and is expressed by using the Euclidean distance. MPSI method is characterized as a very simple and easy to understand approach for defining the objective weights of criteria. Moreover, this newly developed method is not highly time consuming when calculating the weight coefficients. This makes the MPSI method a greatly flexible and applicable method for solving various different MCDM problems.

This new method is composed of the following steps:

Step 1. Construct the initial decision-making matrix as follows:

$$(A/C) = \left[x_{ij}\right]_{mxn} = \begin{bmatrix} A/C & C_1 & C_2 & \cdots & C_n \\ A_1 & x_{11} & x_{12} & \cdots & x_{1n} \\ A_2 & x_{21} & x_{22} & \cdots & x_{2n} \\ \vdots & \vdots & \vdots & \ddots & \vdots \\ A_m & x_{m1} & x_{m2} & \cdots & x_{mn} \end{bmatrix} \tag{1}$$

where:

$A_1, A_2, \ldots, A_m$ represents the vector of corresponding alternatives,

$C_1, C_2, \ldots, C_n$ represents the vector of corresponding criteria,

x_{ij} represents the evaluation of the alternative i with respect to criterion j,

m is the number of alternatives,

n is the number of criteria.

Step 2. Form the normalized decision-making matrix R as follows:

Depending on criterion tendency, a simple linear normalization technique is used to transform a different input data value into compatible scale, i.e., unity interval $[0, 1]$.

For beneficial (maximization) criteria:

$$r_{ij} = \frac{x_{ij}}{\underset{i=1,2,\ldots,m}{max\ x_{ij}}} \tag{2}$$

for non-beneficial (minimization) criteria:

$$r_{ij} = \frac{\underset{i=1,2,\ldots,m}{min\ x_{ij}}}{x_{ij}} \tag{3}$$

The normalized decision-making matrix R is formed as:

$$R(A/C) = \left[r_{ij}\right]_{mxn} = \begin{bmatrix} A/C & C_1 & C_2 & \cdots & C_n \\ A_1 & r_{11} & r_{12} & \cdots & r_{1n} \\ A_2 & r_{21} & r_{22} & \cdots & r_{2n} \\ \vdots & \vdots & \vdots & \ddots & \vdots \\ A_m & r_{m1} & r_{m2} & \cdots & r_{mn} \end{bmatrix} \tag{4}$$

where:

r_{ij} represents the normalized value of the corresponding criterion, $0 < r_{ij} < 1$.

Step 3. Calculate the mean value v_j of the normalized evaluations of criterion j and it is calculated with following equation:

$$v_j = \frac{1}{m} \sum_{i=1}^{m} r_{ij} \tag{5}$$

Step 4. Calculate the preference variation value p_j as follows:

$$p_j = \sum_{i=1}^{m} \left(r_{ij} - v_j\right)^2 \tag{6}$$

Step 5. The criteria weights w_j are determined using the following equation:

$$w_j = \frac{p_j}{\sum_{j=1}^{n} p_j} \tag{7}$$

4. MARA Method

Every decision-making problem consists of two main elements such as alternatives and criteria. Based on these elements, decision maker forms an initial decision-making matrix to choose the best possible alternative as follows:

$$
D = \left[x_{ij}\right]_{m \times n} =
\begin{bmatrix}
A/C & C_1 & C_2 & \cdots & C_n \\
A_1 & x_{11} & x_{12} & \cdots & x_{1j} \\
A_2 & x_{21} & x_{22} & \cdots & x_{2j} \\
\vdots & \vdots & \vdots & \ddots & \vdots \\
A_m & x_{m1} & x_{m2} & \cdots & x_{mn}
\end{bmatrix}
\tag{8}
$$

where:

$A = [A_1, A_2, \ldots, A_m]$ is a given set of alternatives,
$C = [C_1, C_2, \ldots, C_n]$ is a given set of criteria,
m is the total number of alternatives,
n is the total number of criteria,
$\left[x_{ij}\right]_{m \times n}$ is an assessment of alternative A_i with respect to a set of criteria.
The procedure of the MARA method is composed of the following steps:
Step 1. Normalization of input data
In this step, the same procedure of normalization of input data is applied as described in Step 2 of the Modified Preference Selection Index (MPSI) method.
Step 2. Weighted normalization
Weighted normalization is based on product of the criterion weight w_j with corresponding normalized value r_{ij} in the following way:

$$
g_{ij} = w_j r_{ij}, \ \forall i \in [1, 2, \ldots, m], \ \forall j \in [1, 2, \ldots, n]
\tag{9}
$$

The result of weighted normalization is weighted normalized matrix shown as:

$$
G = \left[g_{ij}\right]_{m \times n} =
\begin{bmatrix}
A/C & C_1 & C_2 & \cdots & C_n \\
A_1 & g_{11} & g_{12} & \cdots & g_{1j} \\
A_2 & g_{21} & g_{22} & \cdots & g_{2j} \\
\vdots & \vdots & \vdots & \ddots & \vdots \\
A_m & g_{m1} & g_{m2} & \cdots & g_{mn}
\end{bmatrix}
\tag{10}
$$

Step 3. Optimal alternative determination
Optimal alternative consists of elements that are determined in the following way:

$$
s_j = max\left(g_{ij} | 1 \leq j \leq n\right), \ \forall i \in [1, 2, \ldots, m]
\tag{11}
$$

The final set of the optimal alternative is shown as:

$$
S = \left\{s_1, s_2, \ldots, s_j\right\}, \ j = 1, 2, \ldots, n
\tag{12}
$$

Step 4. Decomposition of the optimal alternative
Decomposition of the optimal alternative represents the division of the optimal alternative into two subsets or two portions. The set S can be illustrated as the union of the two subsets:

$$
S = S^{max} \cup S^{min}
\tag{13}
$$

If k characterizes the total number of benefit criteria, then $l = n - k$ denotes the total number of cost criteria. Accordingly, the optimal alternative is determined as follows:

$$
S = \left\{s_1, s_2, \ldots, s_k\right\} \cup \left\{s_1, s_2, \ldots, s_l\right\}; k + l = j
\tag{14}
$$

Step 5. Decomposition of each alternative

As is described in the Step 4, decomposition of each alternative is defined in a similar way:

$$T_i = T_i^{max} \cup T_i^{min}, \ \forall i \in [1, 2, \ldots, m] \tag{15}$$

$$T_i = \{t_{i1}, t_{i2}, \ldots, t_{ik}\} \cup \{t_{i1}, t_{i2}, \ldots, t_{il}\}, \ \forall i \in [1, 2, \ldots, m] \tag{16}$$

Step 6. Intensity of the element

For the optimal alternative, intensity of the element can be calculated as:

$$S_k = s_1 + s_2 + \ldots + s_k \tag{17}$$

$$S_l = s_1 + s_2 + \ldots + s_l \tag{18}$$

The intensity for each alternative is computed in the same way as for the optimal alternative:

$$T_{ik} = t_{i1} + t_{i2} + \ldots + t_{ik}, \ \forall i \in [1, 2, .., m] \tag{19}$$

$$T_{il} = t_{i1} + t_{i2} + \ldots + t_{il}, \ \forall i \in [1, 2, .., m] \tag{20}$$

Step 7. Magnitude of the Area for the Ranking of Alternatives (MARA)

The proposed method is based on the creation of the following two main linear functions. The first function takes into account the optimal alternative, and it is created through the following two points $(0, S_k)$ and $(1, S_l)$. Function is of the linear form as follows:

$$f^{opt}(S_k, S_l) = \frac{S_l - S_k}{1 - 0}(x - S_k) + S_k = (S_l - S_k)x + S_k \tag{21}$$

Analogically, we can create the second function considering the *i*th alternative by the following way:

$$f^i(T_{ik}, T_{il}) = \frac{T_{il} - T_{ik}}{1 - 0}(x - T_{ik}) + T_{ik} = (T_{il} - T_{ik})x + T_{ik} \tag{22}$$

Chart of main functions defined by Equations (21) and (22) is represented in Figure 1.

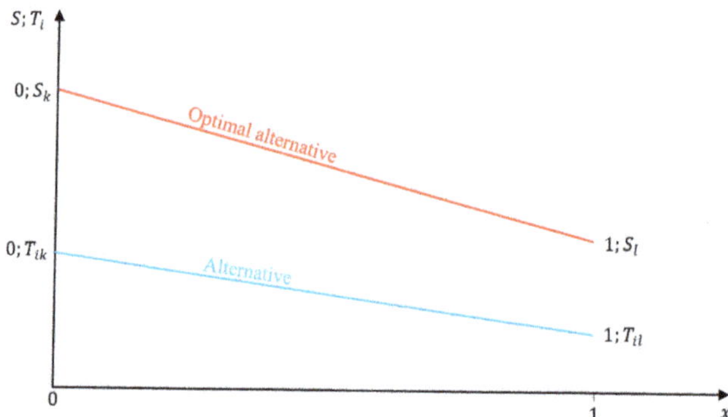

Figure 1. Function of the optimal and *i*th alternative.

Area under the optimal alternative is calculated as:

$$F^{opt} = \int_0^1 f^{opt}(S_k, S_l)dx = \int_0^1 ((S_l - S_k)x + S_k)dx = \frac{S_l - S_k}{2} + S_k \tag{23}$$

Area under the *i*th alternative is computed by the following method:

$$F^i = \int_0^1 f^i(T_{ik}, T_{il})dx = \int_0^1 ((T_{il} - T_{ik})x + T_{ik})dx = \frac{T_{il} - T_{ik}}{2} + T_{ik}; \forall i \in [1, 2, \ldots, m] \tag{24}$$

Magnitude of the Area of the ith alternative is represented by the following equation:

$$M_i = \int_0^1 f^{opt}(S_k, S_l)dx - \int_0^1 f^i(T_{ik}, T_{il})dx; \forall i \in [1, 2, \ldots, m] \tag{25}$$

Final ranking of the alternatives is determined according to ascending order of M_i. Figure 2 shows a detailed flow chart of novel hybrid MPSI–MARA model.

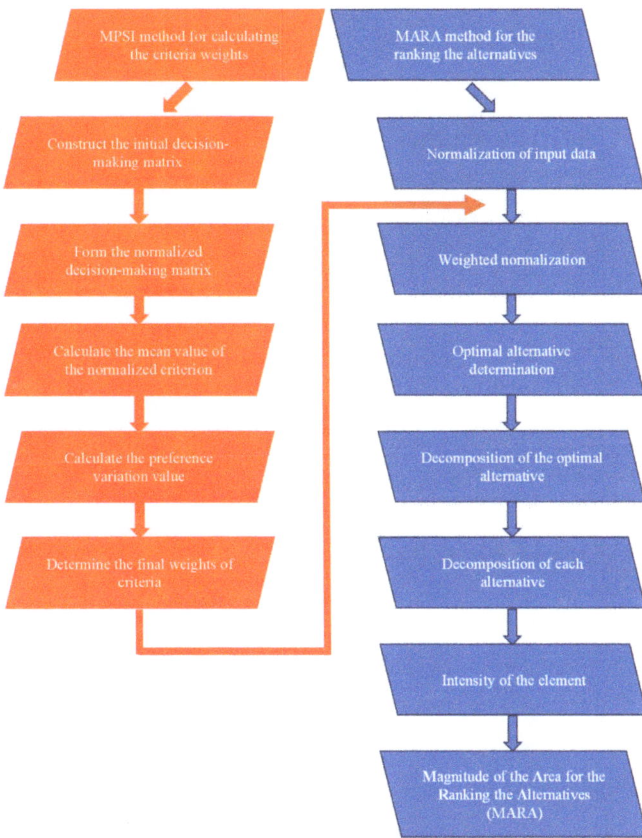

Figure 2. Hybrid MPSI–MARA model framework.

5. Numerical Example

As the underground support system is one of the key factors in every underground mine, mining engineers must dedicate special attention to its selection. Underground coal mine management is faced with the challenge of selecting the underground support system. The coal deposit has a reserve of approximately 10 million tons of coal. The coal deposit is divided into two zones. Because the first zone of around 6 million tons of coal is almost exhausted, the second zone of approximately 4 million tons should be excavated in the near future. Underground mine engineers have created a mine project where approximately 1200 metres of drifts should be supported. Shotcrete, anchors, wired mesh, steel sets, timber and combination are some of the possible variants of underground support systems that could be used in this coal deposit. It should be noted that the problem is hypothetical.

As we mentioned earlier, mining engineers should select one of the combined support systems that are usually applied. Four alternatives (underground support systems) according to the four criteria are evaluated for this problem. A detailed description of the alternatives and criteria is illustrated below.

5.1. Description of Alternatives

A1—Shotcrete 7–10 cm + anchors + wired mesh

This support system is composed of three basic parts: shotcrete, anchors and wired mesh. Shotcrete is the main component of this system and has the most important role in this support system. It is applied in the form of a layer of a certain thickness on the roof and walls of the underground facilities. Anchors and wired mesh represent the additional structural reinforcement of this system. They are designed to increase the strength and toughness of shotcrete and to prevent the collapse of the underground facilities. Figure 3 shows a graphical visualization of this alternative.

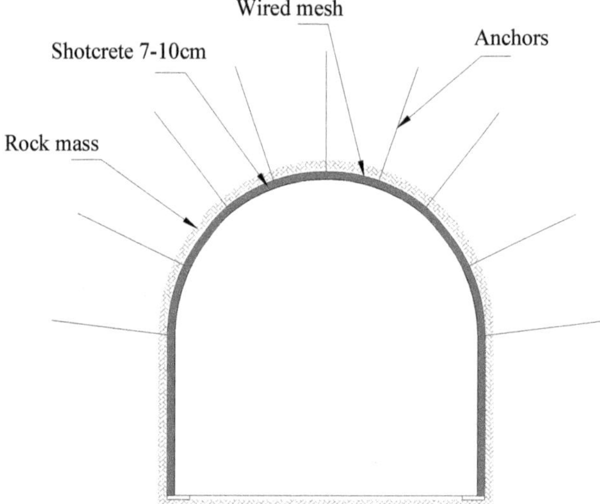

Figure 3. Illustration of A1.

A2—Shotcrete 10–12 cm + anchors + wired mesh + steel frame

This support system is very similar to the previous system (A1). There are only two small differences between these two support systems. The first small difference refers to the thickness of the layer during the "shotcreting", while the second difference is reflected in the fact that this system uses a steel frame as additional structural reinforcement. The steel frame is a very significant part of this system, which further increases its support load capacity. In this way, the safety factor of this system is also increased. The only disadvantage of this system is the very long installation time per meter of length. This alternative is graphically represented in Figure 4.

A3—Steel arches (steel frame + timber)

This support system consists of a steel frame and timber. The steel frame is formed in the shape of an arch and placed in a full drift profile. Timber is used to fill the space between the steel frame and the rock mass mostly in the roof of the underground facility. Since this system is able to support potentially dangerous zones in an underground mine, the required installation time is quite long. Moreover, this system is characterized by high installation costs per meter of length. In Figure 5, steel arches are graphically described.

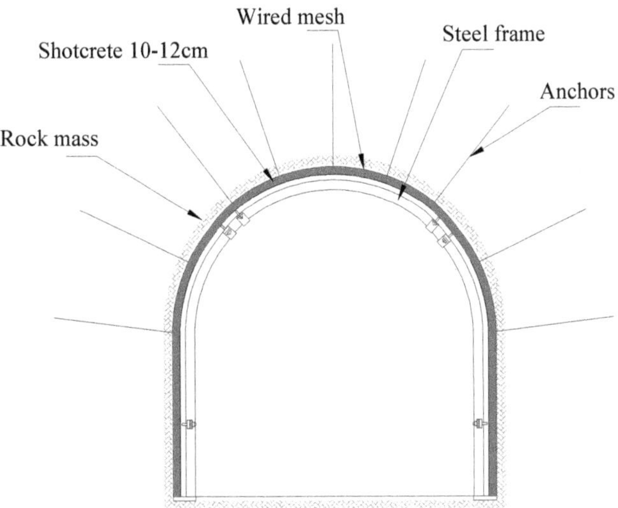

Figure 4. Illustration of A2.

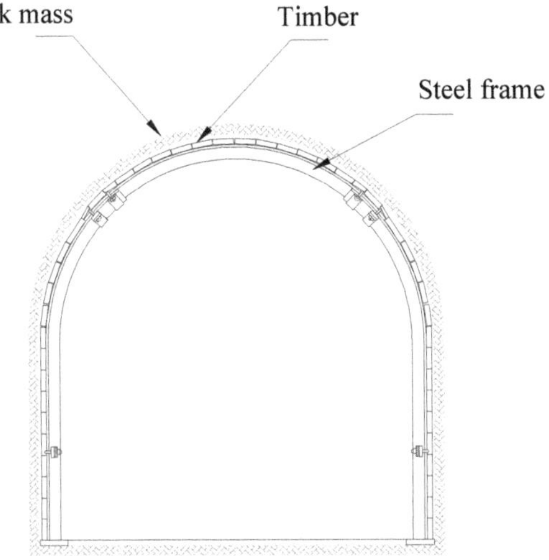

Figure 5. Illustration of A3.

A4—Steel frame + wired mesh + AT anchors

This support system is designed of a combination of a steel frame, wired mesh and AT anchors. The steel frame provides a secure and stable support predisposed to long-term supporting. The steel frame also ensures a high level of support after significant deformation. The wired mesh is the additional structural reinforcement that protects the employees from small fragments of the mined rock mass and rock samples from the roof. AT (Advanced Technology) anchors represent the system of anchors incorporated into the rock mass. AT anchors are characterized by their quick and easy installation, decreasing the number of employees needed for the development of the drifts and additionally increasing the stability of the underground facility. Combining these three components of support, the load capacity of support is extremely increased. The negative side of this system is the very high installation cost per meter of length. This alternative is clearly illustrated in Figure 6.

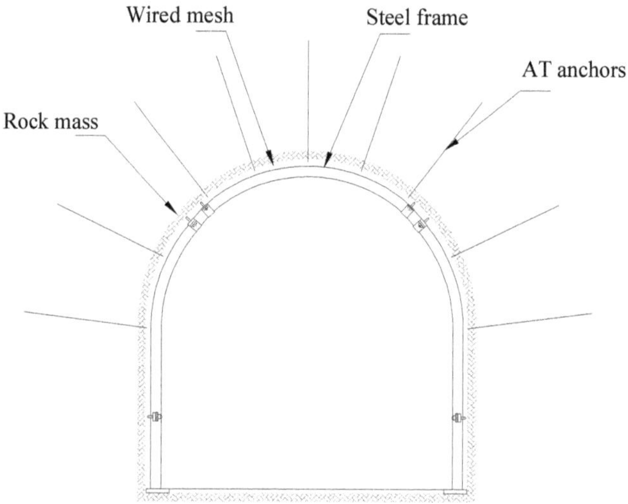

Figure 6. Illustration of A4.

5.2. Description of Criteria

C1—Installation costs

Installation costs are one of the most significant parameters during the process of support selection in underground mines. It represents the costs that are needed to install the appropriate support system and maintenance costs. Additionally, it includes many types of costs such as capital costs, operating costs, direct and indirect costs, etc. The installation costs are expressed in dollars per meter of length. This criterion should be defined as minimum.

C2—Support load capacity

Support load capacity is a crucial attribute on the basis of which the support system will be selected. It represents the key role in the process of the optimal support system selection in underground mines. In the first row, the support load capacity depends on the ratio between the rock mass and the support system. It means that the stresses and deformations in the rock mass caused by the development of underground facilities have a great impact on the support system. Accordingly, the support load capacity is illustrated as the resistance of the support system to underground pressure. It is expressed in kilopascals. This criterion is characterized as maximum.

C3—Safety factor

Safety factor is a very dominant indicator of the quality of each support system. It indicates how the proposed support system is able to provide stable and secure work conditions in underground mines. It represents a target value of every support system that should be a greater than the target, ensuring the maximum reliability of the system. This criterion should be defined as maximum.

C4—Installation time

Installation time is one of the most important parameters that has a high impact on the support system selection. It is characterized as the time that is needed to install the sections of the support system per meter of length. It has a huge influence on the moment when the exploitation of the ore deposit will be started. Moreover, the final production plan and the final financial results of the project largely depend on this criterion. It is expressed in minutes per meter of length. This criterion tends to be minimum.

The initial decision-making matrix with input data is represented in Table 1. In addition, the tendency of each criterion is shown in Table 1.

Table 1. Initial decision-making matrix with input data.

Alternative/ Criterion	C1	C2	C3	C4
	min	max	max	min
A1	500	300	1.85	70
A2	700	750	1.95	120
A3	800	200	1.75	80
A4	900	600	2.00	60

Table 2 shows the normalized values of the input data for constructing the normalized decision-making matrix.

Table 2. The normalized decision-making matrix.

Alternative/ Criterion	C1	C2	C3	C4
	min	max	max	min
A1	1.0000	0.4000	0.9250	0.8571
A2	0.7143	1.0000	0.9750	0.5000
A3	0.6250	0.2667	0.8750	0.7500
A4	0.5556	0.8000	1.0000	1.0000

The Modified Preference Selection Index (MPSI) method is applied for assessing the weights of criteria. The weight coefficient of each criterion is represented in Table 3 and Figure 7.

Table 3. The weights of criteria by MPSI method.

Weight/ Criterion	C1	C2	C3	C4
	w_1	w_2	w_3	w_4
w	0.1885	0.5763	0.0152	0.2200

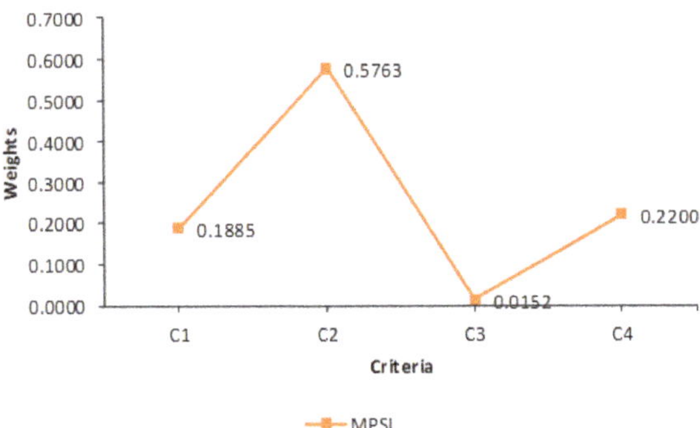

Figure 7. The weights of criteria obtained by MPSI method.

The weighted normalized decision-making matrix is presented in Table 4.

Table 4. The weighted normalized decision-making matrix.

Alternative/ Criterion	C1	C2	C3	C4
	min	max	max	min
A1	0.1885	0.2305	0.0140	0.1886
A2	0.1346	0.5763	0.0148	0.1100
A3	0.1178	0.1537	0.0133	0.1650
A4	0.1047	0.4611	0.0152	0.2200

Using Equation (11), each element of the optimal alternative is determined. The results are shown in Table 5.

Table 5. Optimal alternative determination.

Optimal Alternative/ Criterion	C1	C2	C3	C4
	min	max	max	min
	s_1	s_2	s_3	s_4
S	0.1885	0.5763	0.0152	0.2200

In Table 6, decomposition of the optimal alternative is represented.

Table 6. Decomposition of the optimal alternative.

Optimal Alternative/ Criterion	C1	C2	C3	C4
	min	max	max	min
	s_1	s_2	s_3	s_4
S^{max}		0.5763	0.0152	
S^{min}	0.1885			0.2200

The decomposition of alternatives is presented in Table 7.

Table 7. Decomposition of alternatives.

Alternative/ Criterion		C1	C2	C3	C4
		min	max	max	min
		t_1	t_2	t_3	t_4
A1	T_1^{max}		0.2305	0.0140	
	T_1^{min}	0.1885			0.1886
A2	T_2^{max}		0.5763	0.0148	
	T_2^{min}	0.1346			0.1100
A3	T_3^{max}		0.1537	0.0133	
	T_3^{min}	0.1178			0.1650
A4	T_4^{max}		0.4611	0.0152	
	T_4^{min}	0.1047			0.2200

The intensity of the optimal alternative and alternatives is computed by Equations (17)–(20). The results are shown in Table 8.

Table 8. The intensity of the optimal alternative and alternatives.

Alternative	max S_k T_{ik}	min S_l T_{il}
S	0.5915	0.4085
A1	0.2446	0.3770
A2	0.5911	0.2446
A3	0.1670	0.2828
A4	0.4763	0.3247

Using Equations (21)–(24), we calculated the area under the optimal alternative and alternatives. A brief description of the calculation process is represented for alternative 2 as follows:

$$F^2 = \int_0^1 f^2(0.5911,\ 0.2446)dx = \int_0^1 ((0.2446 - 0.5911)x + 0.5911)dx = \frac{0.2446 - 0.5911}{2} + 0.5911 = 0.4179 \tag{26}$$

The same procedure is valid to the other alternatives.

In Table 9, the values of the area under the optimal alternative and alternatives are shown.

Table 9. The area under optimal alternative and alternatives.

Alternative	Area	Values
Optimal Alternative	F^{opt}	0.5000
A1	F^1	0.3108
A2	F^2	0.4179
A3	F^3	0.2249
A4	F^4	0.4005

The Magnitude of the Area of the Alternative is computed by Equation (25). For example, the Magnitude of the Area of alternative 2 is calculated as follows:

$$M_2 = \int_0^1 f^{opt}(0.5915,\ 0.4085)dx - \int_0^1 f^2(0.5911,\ 0.2446)dx = 0.5000 - 0.4179 = 0.0821 \tag{27}$$

The same procedure is valid to other alternatives.

Table 10 and Figure 8 show the Magnitude of the Area of the Alternatives and final ranking of the alternatives, which is determined in ascending order of M_i.

Table 10. Magnitude of the Area of Alternatives and final ranking of the alternatives.

Alternative	Magnitude of the Area of Alternative M_i	Values	Rank
A1	M_1	0.1892	3
A2	M_2	0.0821	1
A3	M_3	0.2751	4
A4	M_4	0.0995	2

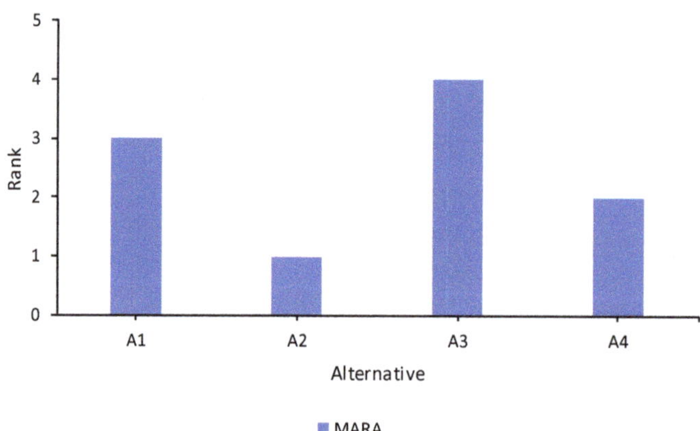

Figure 8. Final rank of the MARA method.

6. Comparative Analysis

Firstly, we compare the MPSI method with traditional methods for determining the objective weights of criteria, including the Entropy method [67], the MEthod based on the Removal Effects of Criteria (MEREC) method [68], the Coefficient of Variation (CV) method [69] and the CRITIC method [70]. The Entropy, Coefficient of Variation and CRITIC methods use a linear sum normalization procedure, while the MPSI and MEREC methods use a linear max-min normalization technique to evaluate the weights of criteria. The weights of the criteria calculated by each objective weighting method are shown in Table 11 and Figure 9.

Table 11. The weights of criteria calculated by each objective weighting method.

Method/Weight	w_1	w_2	w_3	w_4
Entropy	0.1216	0.6694	0.0073	0.2017
MEREC	0.1662	0.4958	0.0473	0.2908
CV *	0.2019	0.4746	0.0503	0.2732
CRITIC	0.2702	0.3249	0.0451	0.3598
MPSI (proposed)	0.1885	0.5763	0.0152	0.2200

* CV—Coefficient of Variation.

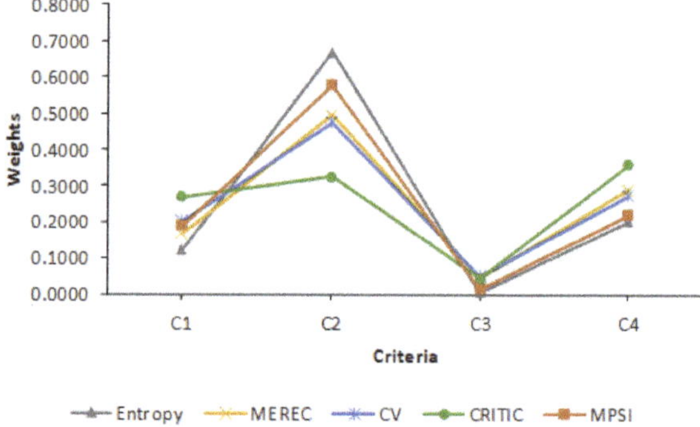

Figure 9. The weights of criteria obtained by applied methods.

The Spearman's rank correlation coefficient was used to evaluate the performance of the applied objective weighting methods for defining the weights of criteria. The correlation coefficients are represented in Table 12.

Table 12. Correlation coefficients of the methods for defining the weights of criteria.

Correlation Coefficient	Entropy	MEREC	CV *	CRITIC	MPSI (Proposed)
Entropy		0.9636	0.9610	0.5936	0.9894
MEREC	0.9636		0.9935	0.7711	0.9732
CV *	0.9610	0.9935		0.7928	0.9837
CRITIC	0.5936	0.7711	0.7928		0.6806
MPSI (proposed)	0.9894	0.9732	0.9837	0.6806	
Average	0.8769	0.9253	0.9327	0.7095	0.9067

* CV—Coefficient of Variation.

From these obtained results, it is clearly visible that the new MPSI method gives a high degree of correlation with all traditional objective weighting methods. The MPSI method has the strongest correlation with the Entropy method with 0.9894. Additionally, a very high correlation is evident with the Coefficient of Variation (CV) method (0.9837) and MEREC method (0.9732). The correlation coefficient with the CRITIC method is lower than the others with 0.6806 but still high. The average Spearman's rank correlation coefficient with 0.9067 confirms that the MPSI method stands side by side with the other applied objective weighting methods.

The next phase in the comparative analysis is a comparison procedure of the MARA method with well-known and traditional multi-criteria decision-making methods such as the Technique for Order of Preference by Similarity to Ideal Solution (TOPSIS) method [71], COmplex PRoportional ASsessment (COPRAS) method [72], Simple Additive Weighting (SAW) method [73] and Multi-Attributive Border Approximation area Comparison (MABAC) method [74]. It should be noted that the weights obtained by the MPSI method are used to compute the final rank of alternatives. The final rank of alternatives by the applied MCDM methods is shown in Table 13 and Figure 10.

Table 13. Final rank of alternatives by applied MCDM methods.

	TOPSIS	Rank	COPRAS	Rank	SAW	Rank	MABAC	Rank	MARA (Proposed)	Rank
A1	0.2723	3	0.2254	3	0.6216	3	0.0325	3	0.1892	3
A2	0.8034	1	0.3184	1	0.8358	1	0.2325	1	0.0821	1
A3	0.1370	4	0.1622	4	0.4498	4	−0.2564	4	0.2751	4
A4	0.7099	2	0.2940	2	0.8010	2	0.2042	2	0.0995	2

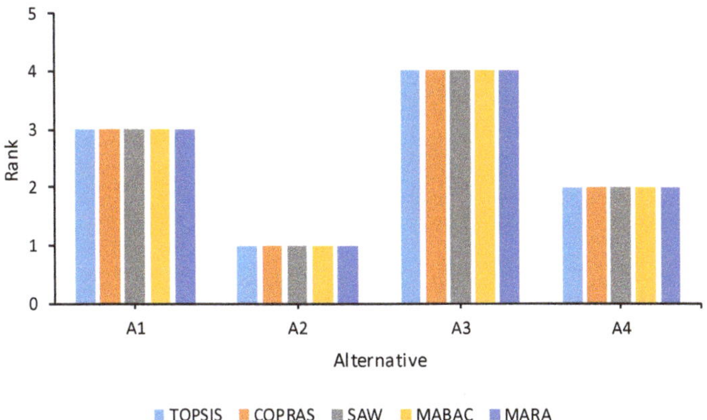

Figure 10. Chart of the final rank of alternatives.

7. Discussion

In this study, the four possible support systems are determined according to the four adopted attributes. From an extensive numerical and comparative analysis, the importance of the weights of attributes as well as the detailed procedure of ranking the alternatives are clarified. Corresponding to the MPSI method, criterion C2 (support load capacity) with 0.5763 is the most dominant attribute. It is followed by criterion C4 (installation time) with 0.2200, then criterion C1 (installation costs) with 0.1885 and criterion C3 (safety factor) with 0.0152. Regarding the MARA method's findings, alternative A2 (shotcrete 10–12 cm + anchors + wired mesh + steel frame) occupies the top place with its exceptional performances of supporting. It is followed by alternative A4 (steel frame + wired mesh + AT anchors), then A1 (shotcrete 7–10 cm + anchors + wired mesh) and the last one is A3 (steel arches (steel frame + timber)).

The problem of the support system selection plays a very important role in the optimization process of any underground mine. In order to achieve a positive and profitable final mine production plan, the support system has quite a high impact on the creation of an acceptable technological and economic model. For this purpose, we developed this new hybrid model that helps mining engineers to select the optimal support system. The model is tested on a hypothetical example and shown to be a powerful tool for solving this common problem in an underground coal mine. This new methodology is verified and validated through a comparative analysis with classical MCDM methods, and the results show that the developed model is extremely acceptable and reliable.

This hybrid MPSI–MARA decision-making model introduces a novel methodology to select the optimal support system in an underground mine. This integrated MCDM procedure, which associates two primary phases, the weighting process and the ranking process, provides an effective and flexible tool for solving such complex MCDM problems. Applicability and flexibility to real-life situations, a relatively short time for calculation and simplicity are just some of the positive characteristics that are identified in the developed decision algorithm.

8. Conclusions

Having in mind that the support system selection in underground mines occupies a significant part of every mine project, it must be subjected to a powerful tool for it to be solved. The installation costs, support load capacity, safety factors and installation time are identified as the most dominant attributes for analysing and optimizing the optimal support system selection.

In this paper, we developed a novel hybrid MPSI–MARA approach to select the best support system in underground mines. The MPSI method represents a novel procedure for calculating the weights of the criteria in the different MCDM problems in an objective way. The MARA method is developed as a new method for ranking the alternatives based on the creation of two main linear functions.

The main contribution of this paper is reflected in the fact that two methods that belong to MCDM optimization problems are developed. The first method refers to assigning the weights of criteria in an objective way, while the second method represents a novel approach for the ranking of the alternatives. This innovative model, verified by a great relationship with traditional MCDM methods, contributes to the significant development of numerous scientific disciplines related to MCDM activities such as operational research, artificial intelligence, optimization systems, etc.

The proposed paper offers enormous help for decision makers to overcome this critical problem in the process of underground mine development. Based on this work, mining engineers and mining companies can create a mine project in a much better and easier way. Moreover, they can predict the profitability of the mine production plan considering the installation costs and installation time of each support system. Given the study's findings, it is clearly visible that the support load capacity represents the most important factor for the support system selection in an underground mine. Since the support load capacity of the support system A2 has the highest value compared to other support systems, there is great potential for it to be selected as the best. Alternative 2 (A2), i.e., the support system composed of shotcrete 10–12 cm + anchors + wired mesh + steel frame, is selected as the best one from the set of possible support systems. The applied decision-making methods produced the same rank order of alternatives. The MARA method is characterized by a high degree of correlation with the compared MCDM methods. The results show that the proposed method is suitable for solving any other MCDM problems. Further research can be directed on the inclusion of an uncertain environment that describes the behaviour of the input data. This would be of great importance in order to obtain a much more reliable and effective final rank of alternatives. Moreover, future studies can focus on incorporating several MCDM methods into the process of the combined MPSI–MARA model. In that way, our new hybrid model would be upgraded and able to solve other complex problems in various research areas.

Author Contributions: Conceptualization, M.G. and Z.G.; methodology, M.G. and Z.G.; validation, S.L., M.N. and Z.L.; formal analysis, S.L. and M.N.; investigation, M.G.; data curation, M.G.; writing—original draft preparation, M.G. and Z.G.; writing—review and editing, M.G. and Z.G.; visualization, S.L., M.N. and Z.L.; supervision, Z.G. All authors have read and agreed to the published version of the manuscript.

Funding: This research received no external funding.

Institutional Review Board Statement: Not applicable.

Informed Consent Statement: Not applicable.

Data Availability Statement: Not applicable.

Conflicts of Interest: The authors declare no conflict of interest.

References

1. Kang, H. Support technologies for deep and complex roadways in underground coal mines: A review. *Int. J. Coal Sci. Technol.* **2014**, *1*, 261–277. [CrossRef]
2. Mark, C.; Barczak, T.M. Fundamentals of coal mine roof support. New Technology for Coal Mine Roof Support. *Proc. NIOSH Open Ind. Brief.* **2000**, *9453*, 23–42.
3. Abdellah, W.; Abdelhaffez, G.; Saleem, H. Stability assessment of underground openings using different rock support systems. *Rud. Zb.* **2020**, *35*, 49–63.
4. Putra, R.A.M. Underground Support System Determination: A Literature Review. *Undergr. Support Syst. Determ. A Lit. Rev.* **2021**, *83*, 14.
5. Canbulat, I.; Van der Merwe, J.N. Design of optimum roof support systems in South African collieries using a probabilistic design approach. *J. S. Afr. Inst. Min. Metall.* **2009**, *109*, 71–88.
6. Klenowski, G.; McNamara, P. Development of support systems for longwall mining in the Bowen Basin, Central Queensland. In Proceedings of the 2020 Coal Operators' Conference, Wollongong, Australia, 12–14 February 2020; University of Wollongon Mining Engineering: Wollongong, Australia, 2020.
7. Szurgacz, D.; Zhironkin, S.; Cehlár, M.; Vöth, S.; Spearing, S.; Liqiang, M. A step-by-step procedure for tests and assessment of the automatic operation of a powered roof support. *Energies* **2021**, *14*, 697. [CrossRef]
8. Prusek, S.; Rajwa, S.; Wrana, A.; Krzemień, A. Assessment of roof fall risk in longwall coal mines. *Int. J. Min. Reclam. Environ.* **2017**, *31*, 558–574. [CrossRef]
9. Wang, D.; Zeng, X.; Wang, G.; Li, R. Adaptability Analysis of Four-Leg Hydraulic Support with Large Mining Height under Impact Dynamic Load. *Shock Vib.* **2022**, *2022*, 2168871. [CrossRef]
10. Yetkin, M.E.; Simsir, F.; Ozfirat, M.K.; Yenice, H.; Ozfirat, P.M. A fuzzy approach to selecting roof supports in longwall mining. *S. Afr. J. Ind. Eng.* **2016**, *27*, 162–177. [CrossRef]
11. Rafiee, R.; Ataei, M.; Kamali, M. Probabilistic stability analysis of Naien water transporting tunnel and selection of support system using TOPSIS approach. *Sci. Res. Essays* **2011**, *6*, 4442–4454.
12. Jalalifar, H.; Behaadini, M.; Bazzazi, A.A. The optimum rock bolt support system selection by using AHP-Entropy-TOPSIS method. *Ariel* **2009**, *149*, 40–120.
13. Shaffiee Haghshenas, S.; Mikaeil, R.; Abdollahi Kamran, M.; Shaffiee Haghshenas, S.; Hosseinzadeh Gharehgheshlagh, H. Selecting the Suitable Tunnel Supporting System Using an Integrated Decision Support System, (Case Study: Dolaei Tunnel of Touyserkan, Iran). *J. Soft Comput. Civ. Eng.* **2019**, *3*, 51–66.

14. Oraee, K.; Hosseini, N.; Gholinejad, M. A New Approach for Determination of Tunnel Supporting System Using Analytical Hierarchy Process (AHP). In Proceedings of the 2009 Coal Operators' Conference, Wollongong, Australia, 12–13 February 2009; The AusIMM Illawarra Branch: Wollongong, Australia, 2009; pp. 78–89.

15. Hosseini, N.; Oraee, K.; Gholinejad, M. Selection of tunnel support system by using multi criteria decision-making tools. In Proceedings of the 29th International Conference on Ground Control in Mining, Morgantown, WV, USA, 27–29 July 2010; Department of Mining Engineering, College of Engineering and Mineral Resources, West Virginia University: Morgantown, WV, USA, 2010; pp. 1–9.

16. Tajvidi Asr, E.; Hayaty, M.; Rafiee, R.; Ataie, M.; Jalali, S.E. Selection of optimum tunnel support system using aggregated ranking of SAW, TOPSIS and LA methods. *Int. J. Appl. Oper. Res.* **2015**, *5*, 49–63.

17. Yavuz, M.; Iphar, M.; Once, G. The optimum support design selection by using AHP method for the main haulage road in WLC Tuncbilek colliery. *Tunn. Undergr. Space Technol.* **2008**, *23*, 111–119. [CrossRef]

18. Pamučar, D.; Stević, Ž.; Sremac, S. A New Model for Determining Weight Coefficients of Criteria in MCDM Models: Full Consistency Method (FUCOM). *Symmetry* **2018**, *10*, 393. [CrossRef]

19. Badi, I.; Muhammad, L.J.; Abubakar, M.; Bakır, M. Measuring Sustainability Performance Indicators Using FUCOM-MARCOS Methods. *Oper. Res. Eng. Sci. Theory Appl.* **2022**, *5*, 99–116. [CrossRef]

20. Das, P.P.; Chakraborty, S. SWARA-CoCoSo method-based parametric optimization of green dry milling processes. *J. Eng. Appl. Sci.* **2022**, *69*, 35. [CrossRef]

21. Singh, R.K.; Modgil, S. Supplier selection using SWARA and WASPAS—A case study of Indian cement industry. *Meas. Bus. Excell.* **2020**, *24*, 243–265. [CrossRef]

22. Stanujkic, D.; Zavadskas, E.K.; Karabasevic, D.; Smarandache, F.; Turskis, Z. The use of the pivot pairwise relative criteria importance assessment method for determining the weights of criteria. *Rom. J. Econ. Forecast.* **2017**, *20*, 116–133.

23. Jauković-Jocić, K.; Karabašević, D.; Jocić, G. The use of the PIPRECIA method for assessing the quality of e-learning materials. *Ekonomika* **2020**, *66*, 37–45. [CrossRef]

24. Sharma, M.; Joshi, S.; Kumar, A. Assessing enablers of e-waste management in circular economy using DEMATEL method: An Indian perspective. *Environ. Sci. Pollut. Res.* **2020**, *27*, 13325–13338. [CrossRef] [PubMed]

25. Dwijendra, N.K.A.; Akhmadeev, R.; Tumanov, D.; Kosov, M.; Shoar, S.; Banaitis, A. Modeling social impacts of high-rise residential buildings during the post-occupancy phase using dematel method: A case study. *Buildings* **2021**, *11*, 504. [CrossRef]

26. Krylovas, A.; Zavadskas, E.K.; Kosareva, N.; Dadelo, S. New KEMIRA method for determining criteria priority and weights in solving MCDM problem. *Int. J. Inf. Technol. Decis. Mak.* **2014**, *13*, 1119–1133. [CrossRef]

27. Krylovas, A.; Kosareva, N.; Zavadskas, E.K. WEBIRA-comparative analysis of weight balancing method. *Int. J. Comput. Commun. Control* **2018**, *12*, 238–253. [CrossRef]

28. Oktavianti, E.; Komala, N.; Nugrahani, F. Simple multi attribute rating technique (SMART) method on employee promotions. *J. Phys. Conf. Ser.* **2019**, *1193*, 012028. [CrossRef]

29. Li, X.; Wang, K.; Liu, L.; Xin, J.; Yang, H.; Gao, C. Application of the entropy weight and TOPSIS method in safety evaluation of coal mines. *Procedia Eng.* **2011**, *26*, 2085–2091. [CrossRef]

30. Zhu, Y.; Tian, D.; Yan, F. Effectiveness of Entropy Weight Method in Decision-Making. *Math. Probl. Eng.* **2020**, *2020*, 3564835. [CrossRef]

31. Mukhametzyanov, I. Specific character of objective methods for determining weights of criteria in MCDM problems: Entropy, CRITIC and SD. *Decis. Mak. Appl. Manag. Eng.* **2021**, *4*, 76–105. [CrossRef]

32. Singh, T. Optimum design based on fabricated natural fiber reinforced automotive brake friction composites using hybrid CRITIC-MEW approach. *J. Mater. Res. Technol.* **2021**, *14*, 81–92. [CrossRef]

33. Alhabo, M.; Zhang, L. Multi-criteria handover using modified weighted TOPSIS methods for heterogeneous networks. *IEEE Access* **2018**, *6*, 40547–40558. [CrossRef]

34. Nguyen, P.-H.; Tsai, J.-F.; Nguyen, V.-T.; Vu, D.-D.; Dao, T.-K. A Decision Support Model for Financial Performance Evaluation of Listed Companies in The Vietnamese Retailing Industry. *J. Asian Financ. Econ. Bus.* **2020**, *7*, 1005–1015. [CrossRef]

35. Şahin, M. A comprehensive analysis of weighting and multicriteria methods in the context of sustainable energy. *Int. J. Environ. Sci. Technol.* **2021**, *18*, 1591–1616. [CrossRef] [PubMed]

36. Sałabun, W.; Wątróbski, J.; Shekhovtsov, A. Are MCDA Methods Benchmarkable? A Comparative Study of TOPSIS, VIKOR, COPRAS, and PROMETHEE II Methods. *Symmetry* **2020**, *12*, 1549. [CrossRef]

37. El-Santawy, M.F.; Ahmed, A.N. CV-VIKOR: A new approach for allocating weights in multi-criteria decision making problems. *Life Sci. J.* **2012**, *9*, 5875–5877.

38. Ecer, F.; Aycin, E. Novel Comprehensive MEREC Weighting-Based Score Aggregation Model for Measuring Innovation Performance: The Case of G7 Countries. *Informatica* **2022**, 1–31. [CrossRef]

39. Shanmugasundar, G.; Sapkota, G.; Čep, R.; Kalita, K. Application of MEREC in Multi-Criteria Selection of Optimal Spray-Painting Robot. *Processes* **2022**, *10*, 1172. [CrossRef]

40. Ecer, F.; Pamucar, D. A novel LOPCOW-DOBI multi-criteria sustainability performance assessment methodology: An application in developing country banking sector. *Omega* **2022**, *112*, 102690. [CrossRef]

41. Trinkūnienė, E.; Podvezko, V.; Zavadskas, E.K.; Jokšienė, I.; Vinogradova, I.; Trinkūnas, V. Evaluation of quality assurance in contractor contracts by multi-attribute decision-making methods. *Econ. Res.-Ekon. Istraživanja* **2017**, *30*, 1152–1180. [CrossRef]

42. Čereška, A.; Zavadskas, E.K.; Bucinskas, V.; Podvezko, V.; Sutinys, E. Analysis of steel wire rope diagnostic data applying multi-criteria methods. *Appl. Sci.* **2018**, *8*, 260. [CrossRef]

43. Odu, G.O. Weighting methods for multi-criteria decision making technique. *J. Appl. Sci. Environ. Manag.* **2019**, *23*, 1449–1457. [CrossRef]

44. Wang, Y.; Zhao, N.; Jing, H.; Meng, B.; Yin, X. A Novel Model of the Ideal Point Method Coupled with Objective and Subjective Weighting Method for Evaluation of Surrounding Rock Stability. *Math. Probl. Eng.* **2016**, *2016*, 8935156. [CrossRef]

45. Pan, X.; Liu, H.; Huan, J.; Sui, Y.; Hong, H. Allocation model of carbon emission permits for the electric power industry with a combination subjective and objective weighting approach. *Energies* **2020**, *13*, 706. [CrossRef]

46. Mahmoody Vanolya, N.; Jelokhani-Niaraki, M. The use of subjective–objective weights in GIS-based multi-criteria decision analysis for flood hazard assessment: A case study in Mazandaran, Iran. *GeoJournal* **2021**, *86*, 379–398. [CrossRef]

47. Attri, R.; Grover, S. Application of preference selection index method for decision making over the design stage of production system life cycle. *J. King Saud Univ.-Eng. Sci.* **2015**, *27*, 207–216. [CrossRef]

48. Petković, D.; Madić, M.; Radovanović, M.; Gečevska, V. Application of The Performance Selection Index Method For Solving Machining MCDM Problems. *Facta Univ. Ser. Mech. Eng.* **2017**, *15*, 97–106. [CrossRef]

49. Vahdani, B.; Zandieh, M.; Tavakkoli-Moghaddam, R. Two novel FMCDM methods for alternative-fuel buses selection. *Appl. Math. Model.* **2011**, *35*, 1396–1412. [CrossRef]

50. Jain, V. Application of combined MADM methods as MOORA and PSI for ranking of FMS performance factors. *Benchmarking Int. J.* **2018**, *25*, 1903–1920. [CrossRef]

51. Ulutaş, A.; Balo, F.; Sua, L.; Demir, E.; Topal, A.; Jakovljević, V. A new integrated grey MCDM model: Case of warehouse location selection. *Facta Univ. Ser. Mech. Eng.* **2021**, *19*, 515. [CrossRef]

52. Ulutaş, A.; Balo, F.; Sua, L.; Karabasevic, D.; Stanujkic, D.; Popovic, G. Selection of insulation materials with PSI-CRITIC based CoCoSo method. *Rev. La Construcción* **2021**, *20*, 382–392. [CrossRef]

53. Stanujkic, M.; Stanujkic, D.; Karabasevic, D.; Sava, C.; Popovic, G. Comparison of Tourism Potentials Using Preference Selection Index Method. *QUAESTUS Multidiscip. Res. J.* **2020**, 177–187.

54. Madic, M.; Antucheviciene, J.; Radovanovic, M.; Petkovic, D. Determination of laser cutting process conditions using the preference selection index method. *Opt. Laser Technol.* **2017**, *89*, 214–220. [CrossRef]

55. Sari, E.B. Measuring The performances of the machines via Preference Selection Index (PSI) method and comparing them with values of Overall Equipment Efficiency (OEE). *Izmir J. Econ.* **2019**, *34*, 573–581.

56. Vahdani, B.; Mousavi, S.M.; Ebrahimnejad, S. Soft computing-based preference selection index method for human resource management. *J. Intell. Fuzzy Syst.* **2014**, *26*, 393–403. [CrossRef]

57. Urošević, K.; Gligorić, Z.; Miljanović, I.; Beljić, Č.; Gligorić, M. Novel Methods in Multiple Criteria Decision-Making Process (MCRAT and RAPS)—Application in the Mining Industry. *Mathematics* **2021**, *9*, 1980. [CrossRef]

58. Thanh, N.V. Designing a MCDM Model for Selection of an Optimal ERP Software in Organization. *Systems* **2022**, *10*, 95. [CrossRef]

59. Stević, Ž.; Tanackov, I.; Puška, A.; Jovanov, G.; Vasiljević, J.; Lojaničić, D. Development of Modified SERVQUAL–MCDM Model for Quality Determination in Reverse Logistics. *Sustainability* **2021**, *13*, 5734. [CrossRef]

60. Wang, C.N.; Tsai, H.T.; Ho, T.P.; Nguyen, V.T.; Huang, Y.F. Multi-criteria decision making (MCDM) model for supplier evaluation and selection for oil production projects in Vietnam. *Processes* **2020**, *8*, 134. [CrossRef]

61. Ashraf, A.; Ullah, K.; Hussain, A.; Bari, M. Interval-Valued Picture Fuzzy Maclaurin Symmetric Mean Operator with application in Multiple Attribute Decision-Making. *Rep. Mech. Eng.* **2022**, *3*, 301–317. [CrossRef]

62. Arora, H.; Naithani, A. Significance of TOPSIS approach to MADM in computing exponential divergence measures for pythagorean fuzzy sets. *Decis. Mak. Appl. Manag. Eng.* **2022**, *5*, 246–263. [CrossRef]

63. Bairagi, B. A novel MCDM model for warehouse location selection in supply chain management. *Decis. Mak. Appl. Manag. Eng.* **2022**, *5*, 194–207. [CrossRef]

64. Riaz, M.; Athar Farid, H.M. Picture fuzzy aggregation approach with application to third-party logistic provider selection process. *Rep. Mech. Eng.* **2022**, *3*, 318–327. [CrossRef]

65. Wang, X.; Zhang, C.; Deng, J.; Su, C.; Gao, Z. Analysis of factors influencing miners' unsafe behaviors in intelligent mines using a novel hybrid MCDM model. *Int. J. Environ. Res. Public Health* **2022**, *19*, 7368. [CrossRef] [PubMed]

66. Maniya, K.; Bhatt, M.G. A selection of material using a novel type decision-making method: Preference selection index method. *Mater. Des.* **2010**, *31*, 1785–1789. [CrossRef]

67. Dehdasht, G.; Ferwati, M.S.; Zin, R.M.; Abidin, N.Z. A hybrid approach using entropy and TOPSIS to select key drivers for a successful and sustainable lean construction implementation. *PLoS ONE* **2020**, *15*, e0228746. [CrossRef]

68. Keshavarz-Ghorabaee, M.; Amiri, M.; Zavadskas, E.K.; Turskis, Z.; Antucheviciene, J. Determination of objective weights using a new method based on the removal effects of criteria (MEREC). *Symmetry* **2021**, *13*, 525. [CrossRef]

69. Xu, Y.; Lai, K.K.; Leung, W.K.J. A consensus-based decision model for assessing the health systems. *PLoS ONE* **2020**, *15*, e0237892. [CrossRef]

70. Vujičić, M.D.; Papić, M.Z.; Blagojević, M.D. Comparative analysis of objective techniques for criteria weighing in two MCDM methods on example of an air conditioner selection. *Tehnika* **2017**, *72*, 422–429. [CrossRef]

71. Karim, R.; Karmaker, C.L. Machine selection by AHP and TOPSIS methods. *Am. J. Ind. Eng.* **2016**, *4*, 7–13.

72. Balali, A.; Valipour, A.; Edwards, R.; Moehler, R. Ranking effective risks on human resources threats in natural gas supply projects using ANP-COPRAS method: Case study of Shiraz. *Reliab. Eng. Syst. Saf.* **2021**, *208*, 107442. [CrossRef]
73. Ibrahim, A.; Surya, R.A. The Implementation of Simple Additive Weighting (SAW) Method in Decision Support System for the Best School Selection in Jambi. *J. Phys. Conf. Ser.* **2019**, *1338*, 012054. [CrossRef]
74. Pamučar, D.; Ćirović, G. The selection of transport and handling resources in logistics centers using Multi-Attributive Border Approximation area Comparison (MABAC). *Expert Syst. Appl.* **2015**, *42*, 3016–3028. [CrossRef]

 *systems*

Article

Type-1 Robotic Assembly Line Balancing Problem That Considers Energy Consumption and Cross-Station Design

Yuanying Chi [1], Zhaoxuan Qiao [1], Yuchen Li [1,*], Mingyu Li [2] and Yang Zou [2]

[1] School of Economics and Management, Beijing University of Technology, Beijing 100124, China
[2] Beijing Benz Automotive Co., Ltd., Beijing 100176, China
* Correspondence: liyuchen@bjut.edu.cn

Abstract: Robotic assembly lines are widely applied to mass production because of their adaptability and versatility. As we know, using robots will lead to energy-consumption and pollution problems, which has been a hot-button topic in recent years. In this paper, we consider an assembly line balancing problem with minimizing the number of workstations as the primary objective and minimizing energy consumption as the secondary objective. Further, we propose a novel mixed integer linear programming (MILP) model considering a realistic production process design—cross-station task, which is an important contribution of our paper. The "cross-station task" design has already been applied to practice but rarely studied academically in type-1 RALBP. A simulated annealing algorithm is developed, which incorporates a restart mechanism and an improvement strategy. Computational tests demonstrate that the proposed algorithm is superior to two other classic algorithms, which are the particle swarm algorithm and late acceptance hill-climbing algorithm.

Keywords: robotic assembly line balancing; energy consumption; cross-station task; mixed integer liner programming; simulated annealing

Citation: Chi, Y.; Qiao, Z.; Li, Y.; Li, M.; Zou, Y. Type-1 Robotic Assembly Line Balancing Problem That Considers Energy Consumption and Cross-Station Design. *Systems* **2022**, *10*, 218. https://doi.org/10.3390/systems10060218

Academic Editor: William T. Scherer

Received: 30 September 2022
Accepted: 4 November 2022
Published: 15 November 2022

Publisher's Note: MDPI stays neutral with regard to jurisdictional claims in published maps and institutional affiliations.

1. Introduction

Assembly lines are widely utilized in mass production, such as automobiles and household appliances. Robots have gained appeal with line designers as a result of advances in manufacturing technology and the quest for production efficiency. Assembly line with robots is referred to as the robotic assembly line. Compared with human workers, robots can process tasks quickly and accurately without having to worry about undue fatigue. Further, a robot with multiple arms is also more adaptable and capable of processing a variety of tasks [1]. However, the utilization of robots will consume energy and create pollution problems. Fysikopoulos et al. [2] pointed out that approximately 9–12% of the cost of manufacturing a car is spent on energy, and for every 20% reduction in energy consumption, the final manufacturing cost will drop by roughly 2–2.4%. Therefore, reducing energy consumption plays a vital role in protecting the environment and boosting business competitiveness.

The assembly line balancing problem (ALBP) was first raised by Bryton [3], and refers to assigning tasks to workstations to achieve the optimization objective. The ALBP can be divided into four types according to different inputs and objectives, i.e., minimizing the number of workstations with a given cycle time is called type-1 ALBP. Minimizing the cycle time given the number of workstations is called type-2 ALBP. If the number of workstations and cycle time are unknown simultaneously, maximizing the line efficiency is labeled type-E ALBP. Aiming to find a feasible balancing solution when both are given is type-F ALBP.

The ALBP with robots is called the robotic assembly line balancing problem (RALBP), which was first proposed by Rubinovitz and Bukchin [4], and is a significant problem in the industrial sector. Since different types of robots have different processing task times and energy consumption per unit of time, the allocation of robots also needs to be considered in RALBP. In this paper, we consider minimizing the number of workstations with a given cycle time, i.e., type-1 RALBP, to be the primary objective and minimizing energy consumption based on the optimal number of workstations to be the secondary objective. Further, we introduce the "cross-station task" design, which is quite popular in practice but is rarely considered in academic settings [5], to improve production efficiency in the RALBP, and propose a novel mixed integer linear mathematical model for this problem. A simulated annealing algorithm encapsulating a restart mechanism and an improvement rule are developed to solve the problem.

The type-1 ALBP has been widely studied in the literature. Kilincci [6] presented a heuristic algorithm based on the Petri net approach to solve the type-1 simple ALBP. Manavizadeh et al. [7] solved a U-shaped balancing type-1 problem with different types of workers and designed an alert system based on the optimal number of stations to balance workload. A simulated annealing algorithm (SA) was used to solve this problem. Li et al. [8] investigated 14 meta-heuristics for type-1 two-sided ALBP and two new decoding schemes with a reduced search space developed. Comprehensive experiments have shown that the improved iterated greedy algorithm is the most efficient in solving the benchmark problems. Li et al. [9] investigated type-1 assembly line balancing considering uncertain task time. An algorithm based on the branch and bound remember algorithm was developed to solve this problem, and the effectiveness of the algorithm was demonstrated. Li et al. [10] incorporated uncertain task time attributes in type-1 U-shaped ALBP. They proposed an algorithm based on the branch and bound remember algorithm to solve this problem. Zhang [11] proposed an immune genetic algorithm (IGA) which aimed to minimize the number of workstations as well as the workload. Baskar and Anthony Xavior [12] investigated a few heuristic algorithms based on slope indices, which is a method of assigning tasks to stations, to solve the type-1 simple ALBP. Pınarbaşı and Alakaş [13] formulated a constraint programming (CP) model for type-1 ALBP considering assignment restrictions. The author compared the results of different models and showed that CP is the best one. Huang et al. [14] considered a mixed-model two-sided ALBP that aimed at minimizing the number of mated-stations. A combinatorial Benders-decomposition-based exact algorithm was used to solve the proposed problem. The computational tests showed that this algorithm can obtain exact results on large-sized problem instances.

Ever since Rubinovitz and Bukchin [4] came up with RALBP and proposed an efficient heuristic to solve it [15], it has become a prevalent research direction for ALBP. Hong and Cho [16] solved the type-1 RALBP considering assembly cost. A simulated annealing algorithm was utilized as the optimization tool. Gao et al. [17] proposed an innovative genetic algorithm (GA) hybridized with local search to solve the type-2 RALBP. Five local search procedures were developed to enhance the search ability of GA. Li et al. [1] designed a cuckoo search method through different neighborhood generation methods to solve the two-sided RALBP. The computational tests showed that the proposed algorithm outperformed other meta-heuristics. Janardhanan et al. [18] considered sequence-dependent setup times for RALBP to minimize the cycle time. They proposed a migrating birds optimization algorithm (MBO) and demonstrated the effectiveness of the MBO. Sun and Wang [19] developed a hybrid algorithm that combines the branch-and-bound (B&B) and estimation of the distribution algorithm to minimize the cycle time on the robotic assembly line. Aslan [20] investigated an two-sided RALBP with sequence-dependent setup times, and a variable neighborhood search (VNS) algorithm was utilized to solve this problem.

Other features of RALBP have also been studied. Michels et al. [21] studied spot welding robotic assembly lines based on an automotive company located in Brazil. A mixed-integer linear programming (MILP) model was developed. Pereira et al. [22] solved cost-oriented RALBP (cRALBP), taking into account that different types of robots have different costs. Rabbani et al. [23] investigated four-sided human–robot collaborations on ALBP, where the tasks are performed on the left, right, above, and beneath sides. An augmented multi-objective particle swarm optimization was used to solve the model. Koltai et al. [24]

analyzed the short- and long-term effects of adding robots to human ALs for the operation of the line. The use of the models was demonstrated using a case study involving a power inverter. Lahrichi et al. [25] investigated two variants of type-2 RALBP with sequence dependence. The first variant is given different types of robots, each of which can be arbitrarily assigned to multiple stations. Another variant is that given a group of robots, each robot can only be assigned to one station.

Further, the use of robots on assembly line creates energy consumption problems; some scholars treat energy consumption as one of the optimization objectives. Mukund Nilakantan et al. [26] proposed models with dual objectives to minimize the cycle time and total energy consumption simultaneously. The particle swarm optimization was used to solve this problem. Nilakantan et al. [27] proposed a multi-objective co-evolutionary algorithm to solve the energy-related RALBP. Zhang et al. [28] investigated a U-shaped RALBP and developed a multiobjective mixed-integer non-linear model to optimize carbon emissions. Hybrid Pareto–grey wolf optimization (HPGWO) was designed, and its effectiveness was demonstrated. Zhou and Wu [29] aimed to optimize the total energy consumption and a productivity-related objective simultaneously in RALBP. A novel algorithm based on a well-known enhanced decomposition-based multi-objective algorithm (MOEA/D) was designed to solve this problem. Zhang et al. [30] investigated mixed-model U-shaped RALBP and proposed a hybrid multi-objective dragonfly algorithm (HMODA) to achieve the goals of energy saving and efficiency. Belkharroubi and Yahyaoui [31] minimized energy consumption on a mixed-model RALBP. A cuckoo search algorithm, which was based on the memory principle, was developed to tackle this problem. The authors tested its effectiveness by comparing other algorithms.

By analyzing the previous research works, we can conclude that the RALBP had been studied extensively from various angles. However, the mathematical models are always nonlinear, which is not desired for a commercial solver specialized in solving mixed-integer problems, e.g., CPLEX. We did not find any literature about the type-1 RALBP with the energy consumption objective. Further, the multi-functional robots and their implied application to the assembly line regarding the cross-station design are absent.

The contribution of our paper is as follows: (1) To our best knowledge, we did not find the studies of the multi-objective optimized type-1 RALBP considering energy consumption. Thus, this work fills the research gap in RALBP. (2) We introduce the "cross-station task" design, which has already been applied in practice but rarely studied academically, into type-1 RALBP for the first time. (3) We leverage and modify the simulated annealing algorithm for solving this problem, where incorporates an improvement mechanism of exact algorithms.

The remainder of this paper is organized as follows. Section 2 describes the cross-station task design in detail, and a simple example is presented. We propose a MILP model considering the cross-station task design in Section 3. A simulated annealing algorithm is designed for the problem in Section 4. Computational results are shown in Section 5. Section 6 concludes the paper.

2. RALBP-CS Design

In this section, we introduce the cross-station task design to reduce the idle time of each station as much as possible on the assembly line. This idea is similar to certain other studies. Grzechca and Foulds [32] relaxed the assumption that a task cannot be split among two or more stations, i.e., a task can be split into multiple subtasks, then changed the priority graph for research. Nanda and Scher [33] relaxed the assumption and studied overlapping workstations, where a task can be processed by a pair of workstations simultaneously.

In the cross-station task design, a task could be processed at three stations, which is a more realistic design than the previous two designs in some manufacturing systems. The cross-station task design is a practical application [5]. The three stations include the current station and its front and rear stations, if they exist. To achieve this, one station can "borrow" ("lend") its cycle time from (to) its front or rear stations. If this task is assigned to the very front part of the current station, it could be processed in advance at the front station. That is, the current station borrows its cycle time from the front station. If this task is assigned to the very rear part of the current station, it could be processed with a delay at the rear station. That is, the current station borrows its cycle time from the rear station. For example, in Figure 1, task 2 is originally assigned to station 2. Since station 2 has borrowed a portion of the cycle time of station 1, task 2 can start being processed in advance at station 1. Likewise, task 4 is originally assigned to station 3. Since station 2 has lent a portion of its cycle time to station 3, task 4 can start being processed in advance at station 2. There are two points worth noting: (1) Tasks can only be assigned to one station and one robot, but they can be processed when the work-in-process (WIP) is at adjacent stations. (2) A robot is multi-functional with different robotic arms, as mentioned in Section 1. For example, robot 3 is processing task 4 with one robotic arm and operating task 6 with another arm.

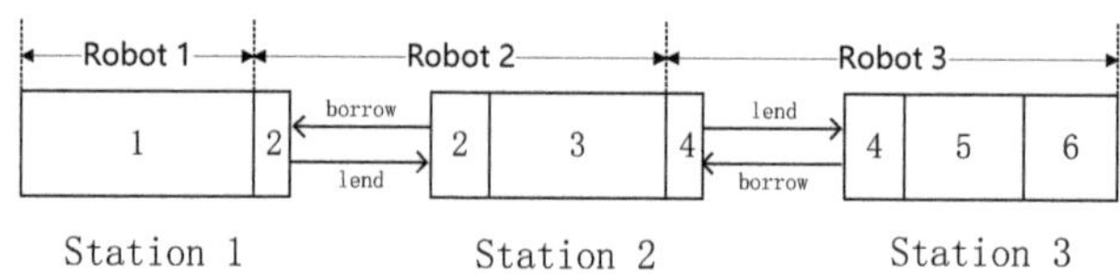

Figure 1. Assembly line with cross-station task design.

Then, a simple example is presented.

Example 1. *To intuitively recognize the advantages of this design, suppose there is only one task sequence and a type of robot, which is shown in Figure 2. The task number is stored in the node, and the number outside the code displays the time the robot takes to process the task. There is a given cycle time of 11.*

Task assignment is shown in Figure 3. The shaded part represents idle time. Without the RALBP-CS design, tasks 1, 2, and 3 are assigned to station 1; tasks 4, 5 and 6 are assigned to station 2; and tasks 7, and 8 are assigned to station 3. Thus, there are 3 stations installed, and some idle time is incurred on the line. If the RALBP-CS design is applied to the line, tasks 1, 2, 3, and 4 are assigned to station 1, which borrows 1 unit of the cycle time of station 2. Then, tasks 5, 6, 7, and 8 are assigned to station 2. The number of stations that are installed is two (one less than the line without considering RALBP-CS design), and there is no idle time at all on line.

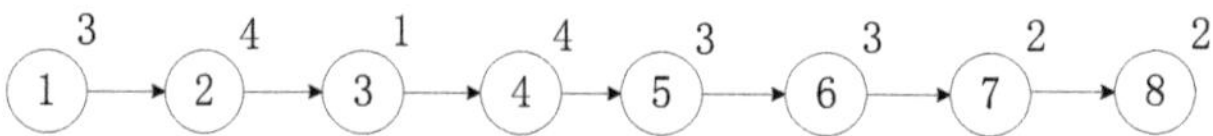

Figure 2. Priority relationship for example 1.

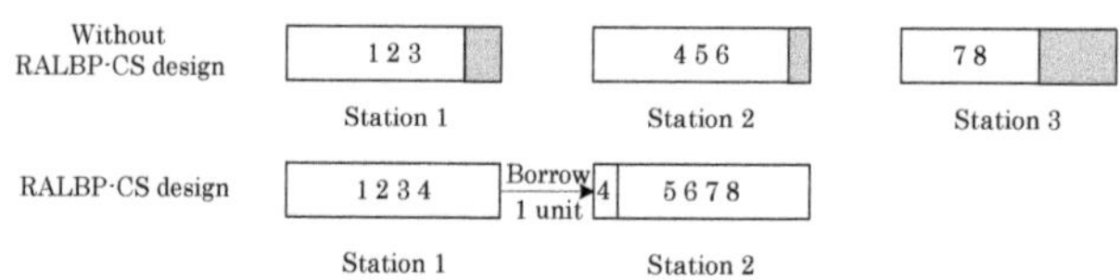

Figure 3. Task assignment for example 1.

3. Mathematical Modeling

In this section, the MILP mathematical model is formulated. The primary goal is to minimize the number of workstations. The secondary objective is to minimize total energy consumption. The decision variables are summarized at Table 1.

Table 1. Notations.

Parameters	Descriptions
n	Total number of tasks
R	Total number of robot types
J	The maximum number of workstations that are allowed to be opened
m	The number of workstations that are actually installed
i, h	Index of tasks, $i, h = 1, 2, \cdots n$
s, j	Index of stations. $s, j = 1, 2, \cdots J$
r	Index of robots. $r = 1, 2, \cdots R$
c	Cycle time
ω	Weight coefficient
Rte_s	The remaining capacity at station s
TEC	Total energy consumption
TEC_{max}	The upper bound of total energy consumption
OEC_r	Operation energy consumption of the robot r per time unit
SEC_r	Standby energy consumption of the robot r per time unit
t_{ir}	The task $i's$ processing time by robot r
$Pr(i)$	Set of direct predecessors of task i
γ	Maximum amount of time that can be borrowed from one station by another
ϕ	A large positive number

Decision variables	Descriptions
x_{ij}	1, if task i is assigned to workstation j ; 0, otherwise
y_{rj}	1, if robot r is allocated to workstation j ; 0, otherwise
z_{irj}	1, if task i and robot r are assigned to workstation j ; 0, otherwise
$d_{j,s}$	1, if station j borrows time from station s; 0, otherwise
$q_{j,s}$	A positive value shows the amount of time station j borrows from station s; 0, otherwise

3.1. Assumptions

The basic assumptions underlying the problem are as follows.

- One single product is manufactured on the assembly line.
- Robots are multi-functional with numerous arms that can handle different tasks simultaneously.
- The precedence relations between the tasks are given previously.
- A task can only be assigned to one station and one robot.
- Each station can only borrow time from its adjacent stations.
- The task-processing time is dependent on the type of robot assigned to it.
- Each robot can be assigned to any station and can process any task.

The notations are presented in Table 1 and used throughout the paper.

3.2. Formulation

In the model, TEC represents the total energy consumption. TEC_{max} represents the upper bound of total energy consumption, which is a fixed parameter by taking the maximum operating and idle energy consumption of the robots for each task, and then summing them. Thus, TEC/TEC_{max} is in the range $[0, 1]$. Now, the mathematical model is presented.

$$\min \quad Obj = m + TEC/TEC_{max}, \tag{1}$$

$$s.t. \quad m = \sum_{r=1}^{R} \sum_{j=1}^{J} y_{rj}, \tag{2}$$

$$TEC = OEC + SEC, \tag{3}$$

$$OEC = \sum_{j=1}^{J} \sum_{r=1}^{R} \sum_{i=1}^{n} OPC_r z_{irj} t_{ir}, \tag{4}$$

$$SEC = \sum_{j=1}^{J} \sum_{r=1}^{R} g_{rj} SPC_r, \tag{5}$$

$$g_{rj} \leq c + q_{j,j+1} + q_{j,j-1} - q_{j+1,j} - q_{j-1,j} - \sum_{r=1}^{R} \sum_{i=1}^{n} z_{irj} t_{ir} + (1 - y_{rj})\psi, \forall j = 1, \cdots, J, \tag{6}$$

$$g_{rj} \geq c + q_{j,j+1} + q_{j,j-1} - q_{j+1,j} - q_{j-1,j} - \sum_{r=1}^{R} \sum_{i=1}^{n} z_{irj} t_{ir} - (1 - y_{rj})\psi, \forall j = 1, \cdots, J, \tag{7}$$

$$\sum_{r=1}^{R} y_{rj} \leq 1, \forall j = 1, \cdots, J, \tag{8}$$

$$\sum_{r=1}^{R} y_{rj} \geq x_{ij}, \forall j = 1, \cdots, J, \forall i = 1, \cdots, n, \tag{9}$$

$$\sum_{r=1}^{R} y_{rj} \leq \sum_{i=1}^{n} x_{ij}, \forall j = 1, \cdots, J, \tag{10}$$

$$\sum_{r=1}^{R} y_{rj} \geq \sum_{r=1}^{R} y_{rj+1}, \forall j = 1, \cdots, J-1, \tag{11}$$

$$\sum_{j=1}^{J} x_{ij} = 1, \forall i = 1, \cdots, n, \tag{12}$$

$$\sum_{i=1}^{n} j x_{hj} \leq \sum_{i=1}^{n} j x_{ij}, \forall h \in p(i), \tag{13}$$

$$x_{ij} + y_{rj} \leq z_{irj} + 1, \forall i = 1, \cdots, n, \forall r = 1, \cdots, R, \forall j = 1, \cdots, J, \tag{14}$$

$$\left(1 - x_{ij}\right) + y_{rj} \leq \left(1 - z_{irj}\right) + 1, \forall i = 1, \cdots, n, \forall r = 1, \cdots, R, \forall j = 1, \cdots, J, \tag{15}$$

$$x_{ij} + \left(1 - y_{rj}\right) \leq \left(1 - z_{irj}\right) + 1, \forall i = 1, \cdots, n, \forall r = 1, \cdots, R, \forall j = 1, \cdots, J, \tag{16}$$

$$\left(1 - x_{ij}\right) + \left(1 - y_{rj}\right) \leq \left(1 - z_{irj}\right) + 1, \forall i = 1, \cdots, n, \forall r = 1, \cdots, R, \forall j = 1, \cdots, J, \tag{17}$$

$$q_{j,s} \leq \gamma, \forall j = 1, \cdots, J, \forall s = 1, \cdots, J, \tag{18}$$

$$d_{j,j+1} + d_{j+1,j} \leq 1, \forall j = 1, \cdots, J-1, \tag{19}$$

$$\varphi d_{j,j+1} \geq q_{j,j+1}, \forall j = 1, \cdots, J-1, \tag{20}$$

$$\varphi d_{j+1,j} \geq q_{j+1,j}, \forall j = 1, \cdots, J-1, \tag{21}$$

$$q_{j+1,j} \geq 0.1 d_{j+1,j}, \forall j = 1, \cdots, J-1, \tag{22}$$

$$q_{j,j+1} \geq 0.1 d_{j,j+1}, \forall j = 1, \cdots, J-1, \tag{23}$$

$$\sum_{r=1}^{R} y_{rj+1} \geq d_{j,j+1}, \forall j = 1, \cdots, J-1, \tag{24}$$

$$\sum_{r=1}^{R} y_{rj+1} \geq d_{j+1,j}, \forall j = 1, \cdots, J-1, \tag{25}$$

$$q_{1,0}, q_{0,1}, q_{J,J+1}, q_{J+1,J} = 0, \tag{26}$$

$$q_{j,s} \geq 0, \forall j = 1, \cdots, J, \forall s = 1, \cdots, J, \tag{27}$$

$$g_{rj} \geq 0, \forall r = 1, \cdots, R, \forall j = 1, \cdots, J, \tag{28}$$

Equation (1) shows the objective function, which represented by the primary (m) and secondary objective (TEC). Constraint (2) calculates the number of workstations that are actually installed. Constraint (3) calculates the total energy consumption by summing the total operation energy consumption (Constraint (4)) and standby energy consumption (Constraint (5)) on the assembly line. Constraints (6)– (7) calculate the idle time for each station, and if the robot r is allocated to the station j ($y_{rj} = 1$), the constraints are active. Constraint (8) refers to the requirement that, at most, one robot can be allocated to the station. Constraint (9) ensures that tasks are assigned only to stations that are installed. Constraint (10) ensures that the installed station is not empty. Constraint (11) ensures the stations are installed continuously in turn. Constraint (12) ensures a task can only be assigned to one station. Constraint (13) refers to the priority constraint. Constraints (14)–(17) demonstrate the logical relationship between tasks, robots, and workstations. Constraint (18) limits the upper bound of time borrowing from one station to another. Constraints (19)–(23) ensure that two stations cannot borrow each other's time simultaneously. Constraints (24)–(25) ensure that the action of lending and borrowing cannot occur for the empty station. Constraints (26)–(28) are the domain constraints.

4. A Simulated Annealing Algorithm

In this section, a simulated annealing algorithm (SA) is designed to tackle the proposed problem. SA was first proposed by Kirkpatrick et al. [34], which is an optimization algorithm that imitates the gradual cooling of metals. It is a meta-heuristic including a random optimization approach, which is to avoid a local optimum by evaluating inferior solutions. It has the advantages of simple description, flexible use, wide application, high operation efficiency, and less affected by the input parameter. Starting from a randomly chosen initial solution S, SA seeks a candidate solution S' in the area surrounding the initial solution. S and S' correspond to fitness values Obj and Obj', respectively. Then, determine which candidate solution can be approved by comparing the fitness values of the candidate with the present solutions. The amount of change in the fitness value is referred to as Δ ($\Delta = Obj - Obj'$). If $\Delta > 0$, S' is accepted; otherwise, S' is accepted with a given probability ($p = e^{-\frac{\Delta}{T}}$), where T is the temperature parameter. At the beginning, there is a high probability of accepting the inferior solution due to the higher T. In each iteration, T is decreased by a cooling schedule until a predetermined stopping requirement is satisfied.

4.1. The General Framework of SA

In this study, the optimal objective is solved by designing the iterative mechanism and developing an improvement rule based on the traditional SA. Obj represents the fitness value. The notation of SA is given below.

T	Temperature parameter
α	Cooling rate
ite	The iteration index
ite_{max}	The maximum number of iterations of temperature
dn_{max}	The maximum number of iterations per restart
Sq_{task}	A feasible assignment sequences of task
Sq_{robot}	A feasible assignment sequences of robot
rn, pr	Uniform random numbers between $[0, 1]$

Specifically, SA searches for the Sq_{task} and Sq_{robot} that result in the best fitness. Variable neighborhood search (VNS) is another characteristic of the SA. In SA, VNS chooses between two neighborhoods and systematically searches them. A random number (pr) determines whether to alter the robot sequence or the task sequence for each iteration. The iterative steps of SA are listed below.

Step 1: Input $P(i)$, T, Nn, α, OPC_r, SPC_r, t_{ir}, J, c, dn_{max}, TEC_{max}, set $ite = 1$, $dn = 0$.

Step 2: Initialize sequences Sq_{task} and Sq_{robot}.

Step 3: Generate the initial objective value Obj. Set the first three optimal objective values equal to Obj.

Step 4: Generate pr, if $pr \leq 0.5$, launch the neighborhood generation mechanism of the task sequence to obtain a new Sq'_{task}; otherwise, launch the neighborhood generation mechanism of the robot sequence, and obtain a new Sq'_{robot}.

Step 5: Launch decoding process to obtain a new objective value Obj'.

Step 6: Calculate $\Delta = Obj' - Obj$. If $\Delta < 0$, update and retain the first three optimal solutions, update $dn = 0$, go to step 8; otherwise, update $dn = dn + 1$, and go to step 7.

Step 7: Generate rn, if $rn < e^{-\frac{\Delta}{T}}$, go to step 8; otherwise, go to step 9.

Step 8: Accept and update $Sq_{task} = Sq'_{task}$, $Sq_{robot} = Sq'_{robot}$, $Obj = Obj'$.

Step 9: If $dn == dn_{max}$, activate the restart mechanism, update $dn = 0$, go to step 2; otherwise, update $T = \alpha T$, $ite = ite + 1$, and go to step 10.

Step 10: If $ite == ite_{max}$, launch the improvement mechanism, output optimal solution, done.

4.2. Initial Sequence Encoding

In the initial sequence encoding process, we employ Sq_{task} and Sq_{robot} to express the feasible assignment sequences of task and robot, respectively. In contrast to the task sequence, a feasible robot sequence can be generated randomly because the robot sequence is not constrained by precedence. The encoding details are provided below for generating Sq_{task} and Sq_{robot}.

Step 1: Generate Sq_{robot}, which is an array containing J random integers taken from 1 to R.

Step 2: Set $Sq_{task} = [\ \ ]$.

Step 3: Generate Sq'_{task}, which is an array containing random permutation of the integers from 1 to n.

Step 4: Based on the Sq'_{task} and the precedence constraints, assign the task i to $Sq_{task} = [\ \ ]$, then delete the task i in Sq'_{task}.

Step 5: If task i cannot be assigned because of violating precedence relationship, skip it and then consider the next task i according to Sq'_{task}, then go to step 4.

Step 6: Repeat steps 4–5 until $Sq'_{task} = [\ \ ]$, obtain Sq_{task}, done.

4.3. Decoding of Objective Function

In the decoding process, each robot is allocated to a station in turn according to Sq_{robot}. The robot allocation at stations s and $s + 1$, respectively, are denoted by r and r'. The initialization process (step 1) refers to loading the input data. The task assignment process is described in steps 2 to 5, where step 2 refers to assigning the task to the current station, and steps 3–5 refer to assigning the task to the current station or next adjacent station considering the cross-station design. These three points should be noted: (1) if the final task needs to borrow time of the assignment process, it is assigned to the next adjacent station directly to prevent the situation that an empty stations lend its time; (2) the current station's available time is not Ret_s but $Ret_s + \gamma$ since the station is permitted to lend or borrow time; and (3) if the task is assigned to station $s + 1$, the idle time $\max(Ret_s - \gamma, 0)$ may incur at station s, i.e., $\min(Ret_s, \gamma)$ is the amount of time that station s lends to station $s + 1$. Step 6 computes the energy consumption for station s. Eventually, the feasible solutions Obj can be obtained in step 7.

Step 1: Input t_{ir}, J, Sq_{task}, Sq_{robot}, OPC_r, SPC_r, set $s = 1$, $id = 1$, and $Ret_s = c$.

Step 2: If all tasks are assigned, go to step 7; otherwise, assign the id_{th} task in Sq_{task} to station s, update Ret_s ($Ret_s = Ret_s - t_{ir}$) and $id = id + 1$, and go to step 3.

Step 3: If $t_{ir} > Ret_s$ and $id == n$, assign the task $i = Sq_{task}(id)$ to station $s + 1$, set $q_{s,s+1} = 0$ and $q_{s+1,s} = 0$, update $s = s + 1$, and compute OEC_s and SEC_s, go to step 7; elseif $t_{ir} > Ret_s$ and $id < n$, go to step 4; otherwise, go to step 3.

Step 4: If $t_{ir} \leq \gamma + Ret_s$ and $t_{ir'} + \max(Ret_s - \gamma, 0) \geq t_{ir}$, assign the task $i = Sq_{task}(id)$ to station s, and set $q_{s,s+1} = t_{ir} - Ret_s$, $q_{s+1,s} = 0$, update Ret_{s+1}, where $Ret_{s+1} = c - q_{s,s+1}$, go to step 6; otherwise, go to step 5.

Step 5: Assign the task $i = Sq_{task}(id)$ to station $s+1$, set $q_{s,s+1} = 0$ and $q_{s+1,s} = \min(Ret_s, \gamma)$, update Ret_{s+1}, where $Ret_{s+1} = c + q_{s+1,s} - t_{ir}$, and go to step 6.

Step 6: Compute OEC_s and SEC_s, update $id = id + 1$, and $s = s + 1$, go to step 2.

Step 7: Get m, where $m = s$, compute TCF, and output Obj.

The decoding operation of objective function is then illustrated with an example.

Example 2. *Consider the precedence relationship graph given in Figure 4 and the parameters are provided in Table 2. $Sq_{task} = \{1,2,4,3,5,6,7,8\}$, $Sq_{robot} = \{3,2,2,1,3\}$. Tasks 1 and 2 are assigned to station 1 and update Ret_1, $Ret_1 = 3$. Then we should assign task 4 according to Sq_{task} since $t_{4,3} > Ret_1$ and task 4 is not the last. We should take into account the cross-task design. By executing step 4, $4 < 3 + 2$ and $3 + \max(3 - 2, 0) = 4$, task 4 should be assigned to station 1 and station 1 shall borrow 1 unit of time from station 2. Update Ret_2, $Ret_2 = 10$, tasks 3 and 5 are assigned to station 2, and update Ret_2, $Ret_2 = 1$. Task 6 is assigned to station 3 due to $4 > 1 + 2$, and station 2 lends 1 unit of time to station 3. Task 7 is assigned to station 3 and we update Ret_3, $Ret_3 = 3$. When assigning task 8, we find $t_{8,2} > Ret_3$ and task 8 is the last, and thus task 8 is assigned to station 4 directly. Finally, a feasible solution can be found, where $m = 4$, $TCF = 9.825$, and $Obj = 4.856$.*

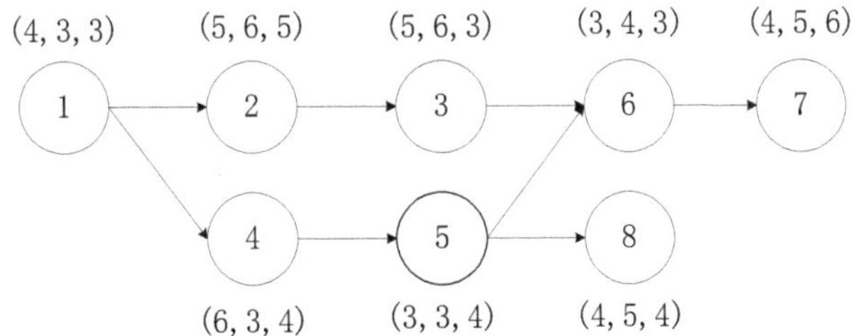

Figure 4. Priority relationship for example 2.

Table 2. Parameters for example 2.

Variables and Parameters	Value
n	8
R	3
J	5
c	11
γ	2
TEC_{max}	11.48
OEC_r	[0.3; 0.25; 0.32]
SEC_r	[0.03; 0.025; 0.032]

4.4. Neighborhood Generation and Restart Mechanism

In SA, an insert method mentioned by Khorasanian et al. [35] is used for generating a neighbor of the task sequence. Simply put, a new task sequence Sq_{task} can be generated by relocating a task to a different position. The reader can refer to the cited literature for more details on the neighborhood generation of task sequences. For neighborhood generation of robot sequences, we randomly select stations with the robot that have been installed, and then randomly select a different type of robot to replace to obtain a new robot sequence Sq_{robot}.

In SA, as we know, local optimality can be escaped by accepting inferior solutions. As the temperature decreases, the probability of accepting the inferior solution becomes smaller, and the easier it is to fall into the local optimality. To address this issue, we designed a restart mechanism with reference to Li et al. [36]. If the optima are not improved in dn_{max} consecutive iterations, the algorithm returns to the initial solution generation phase.

4.5. Improvement Mechanism

In this section, we propose a novel improvement rule, which embeds the exact algorithm into the SA algorithm to improve the quality of the solution.

As we know, the MILP model can be solved using exact algorithms or heuristic algorithms. Exact algorithms can find the optimal solution to the model, but for the complex assembly line problem, it is difficult to obtain a feasible solution when the allowable time is limited. Therefore, researchers generally use heuristic algorithms to obtain an approximate optimal solution. However, heuristic algorithms tend to fall into the local optimality, and the gap between the approximate optimal solution and the actual optimal solution cannot be measured.

In the improvement mechanism, we cut the original problem into a few small problems to find a better solution using exact algorithms. That is, most of the variables in RALBP are fixed, and only the remaining variables are relaxed. Based on the SA results, we select relaxed tasks and robots based on the three rules given in Table 3. In Example 3, the effectiveness of the improvement mechanism is shown.

Table 3. Improvement rules.

	Rules Description
Rule 1	The robots and tasks assigned to the last station are relaxed
Rule 2	Compare the first 3 approximate optimal solutions of SA, the task and robot of assigning different positions are relaxed
Rule 3	10% of the tasks and robots are randomly relaxed in the remaining sequence (upper limit rounding)

Example 3. *To test the effectiveness of the improvement mechanism, we compare the results before and after the introduction of the improvement mechanism. The sub-problems are derived from the datasets described in Section 5.1. The results of the comparison are shown in Figure 5. In Figure 5, the same color represents the same cycle time in datasets (Arcus, Heskiaoff, Scholl). Obviously, after the introduction of the improvement mechanism, although m has not changed, TEC has become smaller for datasets of different sizes. Thus, the improvement mechanism we propose is effective.*

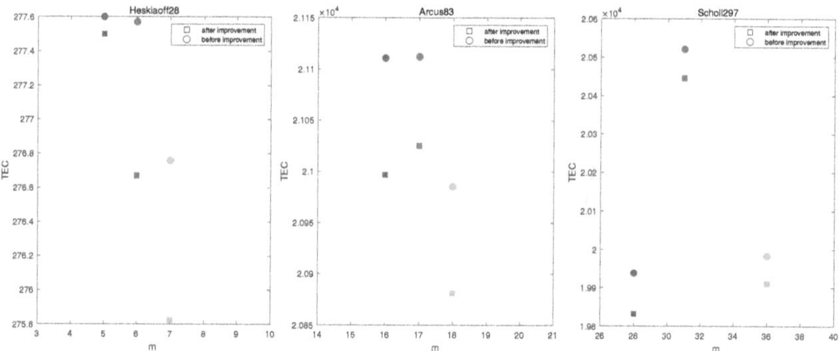

Figure 5. Example 3.

5. Computational Experiments

5.1. Design of Experiment

The basic datasets are extracted from a well-known database (https://assembly-line-balancing.de/, accessed on 30 August 2022). They are Heskiaoff (28), Kilbrid (45), Arcus (83), and Scholl (297). The numbers inside the parentheses indicate the total number of tasks of that dataset. The task time t_{ir} is randomly generated based on the original data according to the fact that the higher the energy consumption, the lower the efficiency. SPC_r and OPC_r are selected by referring to Nilakantan et al. [27], which are provided in the supplementary file. For each instance, c is fixed at six different values, and γ is set to the $0.1 \times c$. Hence, there is a total of 24 independent experiments to conduct.

The particle swarm algorithm (PSO) and the late acceptance hill-climbing algorithm (LAHC) are two traditional methods that are compared. To observe the quality of each algorithm, we used the Gurobi 9.1.2 optimizer to solve the MILP model. Due to the excessive time of the exact algorithm for solving large instances, the runtime limit is set to 3600 s and the gap value is returned. To ensure that the heuristic algorithms are comparable, constrain the algorithm runtime (rt) to $rt = 10 \times n$ seconds. These algorithms are implemented in Matlab (R2019a) and executed on an AMD Ryzen 55500U 2.10 GHz CPU.

We have seen from the preliminary experiment results that the optima are not parameter sensitive. As a result, the parameter values are taken from the literature, as in Table 4. It should be noted that the initial temperature is case-dependent. The initial temperature in SA is determined using the methods described in Li et al. [36].

Table 4. Parameter values for each algorithm.

Parameters	SA	LAHC	PSO
Cooling rate	0.9	-	-
Length of cost list	-	100	-
Learning coefficient $l_1 (l_2)$	-	-	2(2)
The number of task(robot) particles	-	-	30(30)

5.2. Results and Analysis

The computational results are displayed in Table 5. The best results each algorithm can obtain are recorded in column *Obj*. *Obj* is calculated by Formula (1), which contains information in both m and TEC. Due to TEC/TEC_{max} is in the range $[0, 1]$, before and after the decimal point are our primary objective (m) and secondly objective (TEC), respectively. When the runtime is reached but the Gurobi optimizer cannot return a result, it is denoted by $-$. The unique optimal result between the three heuristics is marked in bold.

Table 5. Computational results.

Dataset	c	Gurobi		SA	PSO	LACH
		Obj	Gap (%)	Obj	Obj	Obj
Heskiaoff	160	7.7150	0.0	**7.7153**	7.7236	7.7235
	190	6.6989	0.0	6.7057	**6.7041**	6.7085
	220	5.6799	0.0	5.6848	5.6857	**5.6832**
	250	4.7058	0.0	5.6789	**5.6782**	5.6860
	280	4.6507	0.0	**4.6518**	4.6559	4.6548
	310	4.6435	0.0	**4.6449**	4.6538	4.6590
Kilbrid	70	8.7203	0.0	**8.7528**	8.7739	9.7163
	90	7.6807	0.0	**7.6842**	7.6889	7.6863
	110	5.6852	0.0	**5.7273**	5.7293	6.6619
	130	5.6313	0.0	5.6328	**5.6334**	5.6364
	150	4.6068	0.0	**4.6105**	4.6123	4.6190
	170	4.5893	0.0	**4.5904**	4.5906	4.5935
Arcus	4200	25.7511	52.6	**18.7532**	19.7510	19.7233
	4500	–	–	17.7538	**17.7517**	18.7151
	4800	17.7018	58.6	**16.7229**	16.7377	17.7113
	5100	–	–	**15.7267**	15.7277	16.7014
	5400	–	–	**14.7218**	14.7395	15.6946
	5700	14.6807	35.4	**14.6866**	14.7082	14.6911

Table 5. *Cont.*

Dataset	c	Gurobi		SA	PSO	LACH
		Obj	Gap (%)	Obj	Obj	Obj
Scholl	2000	–	–	**36.7751**	36.8018	37.7644
	2300	–	–	31.7916	**31.7875**	32.7511
	2600	–	–	**28.7415**	28.7516	28.7494
	2900	–	–	**25.7373**	25.7472	26.7213
	3200	–	–	**22.7629**	22.7809	23.7187
	3500	–	–	**21.7004**	21.7137	21.7110

Compared with three heuristic algorithms, the Gurobi optimizer returns the optimal solution for each instance (gap = 0) of the first two datasets. Thus, we can conclude that exact algorithms can quickly get a feasible solution and is optimal for instances with a small number of tasks. Additionally, 9 out of the 12 primary objective m of the first two datasets are the same with three heuristic algorithms, indicating that the heuristic can find solutions as well as the exact algorithm for small instances.

However, when solving a problem with a large number of tasks, exact algorithms may find a solution with a large gap (Arcus) or not even find any feasible solution (Scholl), and the quality of the feasible solution may also be worse than that of the heuristic algorithms. Therefore, for large ALBP problems, heuristics are better than the exact algorithm.

Comparing the results among the three heuristic algorithms, the SA algorithm finds the best optimal solution where 18 out of 24 are the unique best, dominated by PSO in 5 instances, and is dominated by LACH in 1 instance. For the SA algorithm, the number of stations m is optimal among the three heuristics. In addition, in the first dataset, the SA algorithm finds the best optimal solutions where 3 out of 6 are the unique best. The SA algorithm finds the best optimal solutions where 5 out of 6 are the unique best in the remaining three datasets. Thus, the SA algorithm is much better for solving large instances.

6. Conclusions

A satisfactory ecological environment is an important part of people's pursuit of a better life. Pollution from energy consumption in the industrial sector is a problem that cannot be ignored. Presently, robotic assembly lines are widely applied in industrial production. Though the introduction of multi-functional robots on assembly line results in a significant improvement in production efficiency, it brings about high energy consumption. Thus, how to balance energy consumption and efficiency is the goal of our paper.

In this paper, a robotic assembly line balancing problem considering minimizing the number of workstations as the primary objective and energy consumption as the secondary objective is investigated. Our research is the first attempt to model and solve the type-1 RALBP with multi-objectives and cross-station task design. A mixed integer linear integer programming model is formulated to solve the problem. A simulated annealing algorithm which encapsulates an improvement mechanism, is designed and compared with the particle swarm algorithm and the late acceptance hill climbing algorithm. The computational study shows that SA performs better than PSO and LAHC.

This study has some limitations. Since it is the first attempt to explore this innovative research topic, a simple single product is assumed, which is less useful than a general multi-product assumption. Additionally, we only represented three classic algorithms (SA, PSO, and LAHC) to solve the straight assembly line. Algorithmic design and varieties can be further improved to solve more complex assembly line balancing problems (two-sided, U-shaped, and parallel).

Author Contributions: Conceptualization, Y.L. and Y.C.; methodology, Z.Q.; investigation, M.L. and Y.Z. All authors have read and agreed to the published version of the manuscript.

Funding: This work was supported in part by the National Natural Science Foundation of China under Grant (71901006, 72140001), and the Humanities and Social Sciences of Ministry of Education Planning Fund of China (21YJA790009).

Informed Consent Statement: Not applicable.

Data Availability Statement: The basic datasets at https://assembly-line-balancing.de/, accessed on 30 August 2022.

Conflicts of Interest: The authors declare no conflict of interest.

References

1. Li, Z.; Dey, N.; Ashour, A.S.; Tang, Q. Discrete cuckoo search algorithms for two-sided robotic assembly line balancing problem. *Neural Comput. Appl.* **2018**, *30*, 2685–2696. [CrossRef]
2. Fysikopoulos, A.; Anagnostakis, D.; Salonitis, K.; Chryssolouris, G. An empirical study of the energy consumption in automotive assembly. *Procedia CIRP* **2012**, *3*, 477–482. [CrossRef]
3. Bryton, B. *Balancing of a Continuous Production Line*; Northwestern University: Evanston, IL, USA, 1954.
4. Rubinovitz, J.; Bukchin, J. Design and balancing of robotic assembly lines. In Proceedings of the Fourth World Conference on Robotics Research, Pittsburgh, PA, USA, 17–19 September 1991.
5. Bock, S.; Boysen, N. Integrated real-time control of mixed-model assembly lines and their part feeding processes. *Comput. Oper. Res.* **2021**, *132*, 105344. [CrossRef]
6. Kilincci, O. Firing sequences backward algorithm for simple assembly line balancing problem of type 1. *Comput. Ind. Eng.* **2011**, *60*, 830–839. [CrossRef]
7. Manavizadeh, N.; Hosseini, N.S.; Rabbani, M.; Jolai, F. A Simulated Annealing algorithm for a mixed model assembly U-line balancing type-I problem considering human efficiency and Just-In-Time approach. *Comput. Ind. Eng.* **2013**, *64*, 669–685. [CrossRef]
8. Li, Z.; Tang, Q.; Zhang, L.P. Two-sided assembly line balancing problem of type I: Improvements, a simple algorithm and a comprehensive study. *Comput. Oper. Res.* **2017**, *79*, 78–93. [CrossRef]
9. Li, Y.; Wen, M.; Kang, R.; Yang, Z. Type-1 assembly line balancing considering uncertain task time. *J. Intell. Fuzzy Syst.* **2018**, *35*, 2619–2631. [CrossRef]
10. Li, Y.; Hu, X.; Tang, X.; Kucukkoc, I. Type-1 U-shaped Assembly Line Balancing under uncertain task time. *IFAC-Pap.* **2019**, *52*, 992–997. [CrossRef]
11. Zhang, H. An immune genetic algorithm for simple assembly line balancing problem of type 1. *Assem. Autom.* **2019**, *39*, 113–123. [CrossRef]
12. Baskar, A.; Xavior, M.A. Heuristics based on Slope Indices for Simple Type I Assembly Line Balancing Problems and Analyzing for a Few Performance Measures. *Mater. Today Proc.* **2020**, *22*, 3171–3180. [CrossRef]
13. Pınarbaşı, M.; Alakaş, H.M. Assembly line balancing type-1 problem with assignment restrictions: A constraint programming modeling approach. *Pamukkale Univ. J. Eng. Sci.* **2021**, *27*, 532–541. [CrossRef]
14. Huang, D.; Mao, Z.; Fang, K.; Yuan, B. Combinatorial Benders decomposition for mixed-model two-sided assembly line balancing problem. *Int. J. Prod. Res.* **2022**, *60*, 2598–2624. [CrossRef]
15. Rubinovitz, J.; Bukchin, J.; Lenz, E. RALB-A Heuristic Algorithm for Design and Balancing of Robotic Assembly Lines. *CIRP Ann.-Manuf. Technol.* **1993**, *42*, 497–500. [CrossRef]
16. Hong, D.S.; Cho, H.S. Generation of robotic assembly sequences with consideration of line balancing using simulated annealing. *Robotica* **1997**, *15*, 663–673. [CrossRef]
17. Gao, J.; Sun, L.; Wang, L.; Gen, M. An efficient approach for type II robotic assembly line balancing problems. *Comput. Ind. Eng.* **2009**, *56*, 1065–1080. [CrossRef]
18. Janardhanan, M.N.; Li, Z.; Bocewicz, G.; Banaszak, Z.; Nielsen, P. Metaheuristic algorithms for balancing robotic assembly lines with sequence-dependent robot setup times. *Appl. Math. Model.* **2019**, *65*, 256–270. [CrossRef]
19. Sun, B.Q.; Wang, L. An estimation of distribution algorithm with branch-and-bound based knowledge for robotic assembly line balancing. *Complex Intell. Syst.* **2021**, *7*, 1125–1138. [CrossRef]
20. Aslan, Ş. Mathematical model and a variable neighborhood search algorithm for mixed-model robotic two-sided assembly line balancing problems with sequence-dependent setup times. *Optim. Eng.* **2022**, 1–28. [CrossRef]
21. Michels, A.S.; Lopes, T.C.; Sikora, C.G.S.; Magatão, L. The Robotic Assembly Line Design (RALD) problem: Model and case studies with practical extensions. *Comput. Ind. Eng.* **2018**, *120*, 320–333. [CrossRef]
22. Pereira, J.; Ritt, M.; Vásquez, Ó.C. A memetic algorithm for the cost-oriented robotic assembly line balancing problem. *Comput. Oper. Res.* **2018**, *99*, 249–261. [CrossRef]
23. Rabbani, M.; Behbahan, S.Z.B.; Farrokhi-Asl, H. The Collaboration of Human-Robot in Mixed-Model Four-Sided Assembly Line Balancing Problem. *J. Intell. Robot. Syst. Theory Appl.* **2020**, *100*, 71–81. [CrossRef]
24. Koltai, T.; Dimény, I.; Gallina, V.; Gaal, A.; Sepe, C. An analysis of task assignment and cycle times when robots are added to human-operated assembly lines, using mathematical programming models. *Int. J. Prod. Econ.* **2021**, *242*. [CrossRef]

25. Lahrichi, Y.; Damand, D.; Deroussi, L.; Grangeon, N.; Norre, S. Investigating two variants of the sequence-dependent robotic assembly line balancing problem by means of a split-based approach. *Int. J. Prod. Res.* **2022**, 1–17. [CrossRef]
26. Nilakantan, J.M.; Huang, G.Q.; Ponnambalam, S.G. An investigation on minimizing cycle time and total energy consumption in robotic assembly line systems. *J. Clean. Prod.* **2015**, *90*, 311–325. [CrossRef]
27. Nilakantan, J.M.; Li, Z.; Tang, Q.; Nielsen, P. Multi-objective co-operative co-evolutionary algorithm for minimizing carbon footprint and maximizing line efficiency in robotic assembly line systems. *J. Clean. Prod.* **2017**, *156*, 124–136. [CrossRef]
28. Zhang, Z.; Tang, Q.; Zhang, L. Mathematical model and grey wolf optimization for low-carbon and low-noise U-shaped robotic assembly line balancing problem. *J. Clean. Prod.* **2019**, *215*, 744–756. [CrossRef]
29. Zhou, B.; Wu, Q. Decomposition-based bi-objective optimization for sustainable robotic assembly line balancing problems. *J. Manuf. Syst.* **2020**, *55*, 30–43. [CrossRef]
30. Zhang, B.; Xu, L.; Zhang, J. Balancing and sequencing problem of mixed-model U-shaped robotic assembly line: Mathematical model and dragonfly algorithm based approach. *Appl. Soft Comput.* **2021**, *98*, 106739. [CrossRef]
31. Belkharroubi, L.; Yahyaoui, K. Solving the energy-efficient Robotic Mixed-Model Assembly Line balancing problem using a Memory-Based Cuckoo Search Algorithm. *Eng. Appl. Artif. Intell.* **2022**, *114*, 105112. [CrossRef]
32. Grzechca, W.; Foulds, L.R. The assembly line balancing problem with task splitting: A case study. *IFAC-Pap.* **2015**, *28*, 2002–2008. [CrossRef]
33. Nanda, R.; Scher, J.M. Assembly lines with overlapping work stations. *AIIE Trans.* **1975**, *7*, 311–318. [CrossRef]
34. Kirkpatrick, S.; Gelatt, C.D.; Vecchi, M.P. Optimization by simulated annealing. *Science* **1983**, *220*, 671–680. [CrossRef]
35. Khorasanian, D.; Hejazi, S.R.; Moslehi, G. Two-sided assembly line balancing considering the relationships between tasks. *Comput. Ind. Eng.* **2013**, *66*, 1096–1105. [CrossRef]
36. Li, Y.; Kucukkoc, I.; Tang, X. Two-sided assembly line balancing that considers uncertain task time attributes and incompatible task sets. *Int. J. Prod. Res.* **2021**, *59*, 1736–1756. [CrossRef]

 *systems*

Article

An Approach for Predictive Maintenance Decisions for Components of an Industrial Multistage Machine That Fail before Their MTTF: A Case Study

Francisco Javier Álvarez García [1],* and David Rodríguez Salgado [2]

[1] Department of Mechanical, Energy and Materials Engineering, University of Extremadura, C/Sta. Teresa de Jornet 38, 06800 Mérida, Spain
[2] Department of Mechanical, Energy and Materials Engineering, University of Extremadura, Avda. Elvas s/n, 06006 Badajoz, Spain
* Correspondence: fjag@unex.es

Abstract: Making the correct maintenance strategy decision for industrial multistage machines (MSTM) is a constant challenge for industrial manufacturers. Preventive maintenance strategies are the most popular and provide interesting results but cannot prevent unexpected failures and consequences, such as time lost production (TLP). In these cases, a predictive maintenance strategy should be used to maintain the appropriate level of operation time. This research aims to present a model to identify the component that failed before its mean time to failure (MTTF) and, depending on whether the cause of the failure is known, propose the use of a predictive maintenance strategy and further decision-making to ensure the highest possible value from operating time. Also, it is necessary to check the reliable value of MTTF before taking certain decisions. For this research, a real case study of a MSTM was characterized component by component, setting the individual maintenance times. The initial maintenance strategy used for all the components is the preventive programming maintenance (PPM). If a component presents an unexpected failure, a method is proposed to decide whether the maintenance strategy should be changed, adding a predictive maintenance strategy to monitor said component. The research also provides a trust level to evaluate the reliable value of MTTF of each component. The authors consider this approach very useful for machine manufacturers and end users.

Keywords: predictive maintenance; multistage machine; sensorisation; decision-making; mean time to failure; algorithm; system

Citation: García, F.J.Á.; Salgado, D.R. An Approach for Predictive Maintenance Decisions for Components of an Industrial Multistage Machine That Fail before Their MTTF: A Case Study. *Systems* **2022**, *10*, 175. https://doi.org/10.3390/systems10050175

Academic Editor: Ed Pohl

Received: 17 July 2022
Accepted: 26 September 2022
Published: 29 September 2022

Publisher's Note: MDPI stays neutral with regard to jurisdictional claims in published maps and institutional affiliations.

1. Introduction

Multistage machines (MSTM) are quite common in the manufacturing processes industry. These machines are more complex than single-stage machines. The diversity of components and coordinated steps or successive transformations they perform entails the need to establish an adequate maintenance strategy for each component.

It is very important to bear in mind that a failure in one of the components of a multistage industrial machine can lead to a failure in the whole machine. Due to this condition, the best maintenance policy must combine the most suitable strategies for each component. Different components of the same multistage machine may well have different maintenance strategies depending on their maintenance parameters that affect their mean time to repair (MTTR). Once the component has been repaired or substituted, the machine must return to its normal work rhythm and needs time to restart the line (TTLR).

The success in the use of these machines is to meet high demands without unexpected failures that involve the loss of production in progress and a high operation recovery time. Due to this, it is very important that the components of the machine are reliable. If these components have an individual MTTF, this time must be reliable to establish maintenance policies that allow for optimizing the stop-operation time; it is necessary to have the right components, and reliable MTTF. Also it is very important that the main devices and location components used in the MSTM are correct, in order to eliminate avoidable failures. If the MTTF is reliable, it is therefore possible to program the preventive maintenance, so it is necessary to establish the adequate time to maintenance that does not affect the scheduled production.

Preventive maintenance is the most popular strategy in industrial manufacturing systems. Therefore, there must be an adequate level of stocks of components based on a mathematical model proposed in the decision-making strategy. The optimal decision process for setting time to production and time to maintenance programming is studied by A. Gharbi [1], namely, how to develop a mathematical model based on the cost for optimal decision making. A. Gharbi's [2] research also found the most appropriate production rate and preventive maintenance schedule that minimizes the total cost of maintenance and inventory/backlog in periodic preventive maintenance.

As already known, for established scheduled preventive times, is important to define what to do in these times. MTTR considers time to provisioning, time to replacement or removal of component, time to configuration or setting and time to mechanical adjustments. H. Jun-Hee [3] proposed in his research that periodic machine maintenance for single machines and flow-shop scheduling models should be based on an algorithm, minimizing the total weighted completion time. His work defines two principal maintenance actions, setup operations and removal operations, in a production system based on a sequence of single-stage machines. If a removal or setting time is required, a lateness time must be considered. As an in-line process needs to be functioning with stage coordination, is very important to measure the operation times and make the maintenance decisions. If the functioning of the MSTM requires setting maintenance operations during the times of normal operations, this can affect the cycle operation time of the whole machine, and some lateness times must be studied before the maintenance of the machine begins. Other studies have proposed how to realize the appropriate preventive maintenance with imperfect actions, while continuing the normal operation condition, as J. Zuhua [4] showed how to create function blocs in a Programming Logic Controller (PLC) with a previous data acquisition system.

But the important question for preventive maintenance is how to accomplish it, and what might be the appropriate procedure, depending on the system's definition and its complexity. Hernández, D.R. [5] modeled a discrete-time infinite horizon Markov Decision Problem, and F. Chiacchio [6] a stochastic-hybrid reliability model. Other studies, such as M. Fujishima [7], have calculated the optimal time to start preventive maintenance before an unexpected failure. A recent study by A. Irfan [8] modelled a series-parallel system, proposing a reliability model using a Lagrangian optimization method to guarantee MTTF values and avoid unexpected failures. Also, when an unexpected failure occurs, some essential information should be known. For example, the cause of the failure is important, whether the cause stems from a poor design of the machine or an incorrect location of the component, or whether the cause can be eliminated altogether to restore the machine's functioning with a higher level of availability and security. Also, it is important to determine whether the cause is a normal or an occasional (infrequent) situation.

All the components of a machine are always subject, at least, to the laws of degradation. Therefore, even working in its ideal operating conditions, the component will end up failing. In this sense, it would be appropriate to be able to calculate the reliability, as in D.M. Frangopol [9], of the component in the whole machine. However, this study is very complex and normally the manufacturer of the components only defines normal working conditions and sometimes the operating time. Therefore, it is necessary to study models that evaluate whether the component is suitable for the machine and if it is, whether it is so in the normal operation of the machine. G. Silva [10] proposes a model to decide on the most suitable maintenance strategy for the obsolescence of electronic components by creating a decision-making tool, and analyzing the risks, the obsolescence of the components and the consequences of a failure. Recent research by Garcia, F.J.Á. and Salgado, D.R. [11] has proposed a matrix to decide the optimal preventive maintenance strategy based on the individual maintenance times of all the components, their approximate location (global operation condition (GOC)) in the machine, and two key performance indicators (KPIs). The results of both papers describe situations where the component may have different maintenance strategies. If a component fails multiple times, a failure mode and effect analysis (FMEA) can be the solution for finding a design error in the machine, in the component, or an inadequate component selection for the normal operation condition required by the machine. In this way, T. Yuk-Ming [12] has highlighted the importance of product design in the future reliability of the components that must work within certain operating conditions. Product design and functional performance have been shown to be the main research foci in this area.

Predictive maintenance strategies have been shown to be able to avoid unexpected failures by monitoring the operation of the machine using sensors (P. Ponce [13]) and machine learning algorithms to know the normal behavior of the components or of the machine. Dolatabadi, S.H. [14] has provided an overview of past articles highlighting the major expectations, requirements, and challenges for small and medium-sized enterprises (SMEs) regarding the implementation of predictive maintenance (PdM). Normally, the PdM based on algorithms have several steps: data acquisition, data manipulation, configuration, aggregation, and prediction model (the condition monitoring sub-model); and maintenance decision-making, scheduling, and status (the maintenance sub-model). Sometimes, the main algorithm or calculating process is embedded in a PLC, as discussed in Cavalieri, S. [15] and Bouabdallaoui, Y.S. [16].

The study by Garcia, F.J.Á. and Salgado, D.R. [17] described a way to present the available strategies for multistage industrial machines. Their paper describes preventive strategies (with or without stock) and developed predictive strategies like digital behavior twin (DBT), composed of an algorithm with no need to learn normal behavior. S. Givnan [18] studied the normal behavior of the components of an industrial machine for early failure detection by using a machine learning model based on feed-forward neuronal networks trained to identify normal and abnormal behavior. One of the best advantages of the algorithms and the machine learning models is the time necessary to train the model to identify the normal behavior of the machine. Industries need, to the degree possible, simple, fast and reliable systems to take decisions about the availability of their machines in order to avoid unexpected failures. M.M.L. Pfaff [19] developed and tested an adaptive algorithm in a real environment. This algorithm created a dynamic limit value using an adaptive characteristic value segmentation. The paper also studied the location of the sensors for predictive maintenance and confirmed that location can significantly affect the measurement result and, thus, has a direct impact on the outcome of the data analysis. One of the advantages of this research is that there is no need to train the algorithm; the application does not require in-depth process knowledge.

As the technical decisions to take maintenance actions can be provided by the analysis of technical data, normal behavior trained, or not trained, by the adopted predictive algorithm, some authors have mixed the machine learning study with the cost of the maintenance to take global predictive maintenance decisions, as in E. Florian [20] and S.Arena [21], by using, in this case, the Decision Tree technique (DTs) process of implementing predictive maintenance (PdM) and also detecting potential failures (identified through FMEA analysis) and evaluating direct and indirect maintenance costs. It is very important to evaluate a FMEA analysis where a possible failure design of the machine can be the reason of repeated failures.

The digital twin (DT) concept, based on cyber physical systems (CBS) (C. Stary [22]), is a good way to study predictive maintenance and the behavior of the machine if it works under different operating conditions. J. O'Sullivan [23] studied the adoption of digital twins by the maintenance engineering industry to aid in predicting problems before they occur. The algorithm used provided three alarm levels to identify action before a failure.

But not all MSTM can be modelled with a digital twin, due to the fact that these machines normally are highly customized and adapted to the needs of each end user, and therefore since they are not mass-produced, they would require the development of their specific digital twin.

New models embedded in industry 4.0 and created to control corrective maintenance actions are based on a system built on the augmented reality (AR) or computer vision (CV). These systems are used when the machine must be maintained with non-expert operators, and the support of the system can drive the maintenance action with the most success, and in the optimal time. As it can understand the machine, this supporting system minimizes the MTTR and lets the availability of the machine remain in the highest degree possible. Similarly, the work of Konstantinidis, F.K. [24] and Z. Haihua [25] also attempts to solve unexpected failures that are not stored in the maintenance-experience database.

Little of the literature focuses on the simultaneous study of different preventive and predictive maintenance strategies at the same time in the same system. In the case of the MSTM, such studies are non-existent. An interesting paper of H. Wang [26] focuses on a DT-enabled integrated optimization problem of flexible job shop scheduling and flexible preventive maintenance (PM), considering both machine and worker resources. This approach is interesting, particularly if it is possible to open a flexible window to preventive maintenance actions and let the system constantly work with the monitoring of predictive maintenance policy. The architecture of a DT-enhanced job shop is developed, and then the end user has a method to take decisions for the maintenance actions.

This research aims to present a model to identify the component that has failed before its MTTF and, depending on whether or not the cause of the failure is known and the time to restart the normal functioning of the machine, propose the use of a predictive maintenance strategy and further decision-making to ensure the highest possible value from the machine's operating time. For this research, a real case study has been characterized component-by-component, studying the individual maintenance times to obtain the time lost production (TLP) for each component. Figure 1 shows the features of a multistage machine and the conditions on which the proposed maintenance strategies are based.

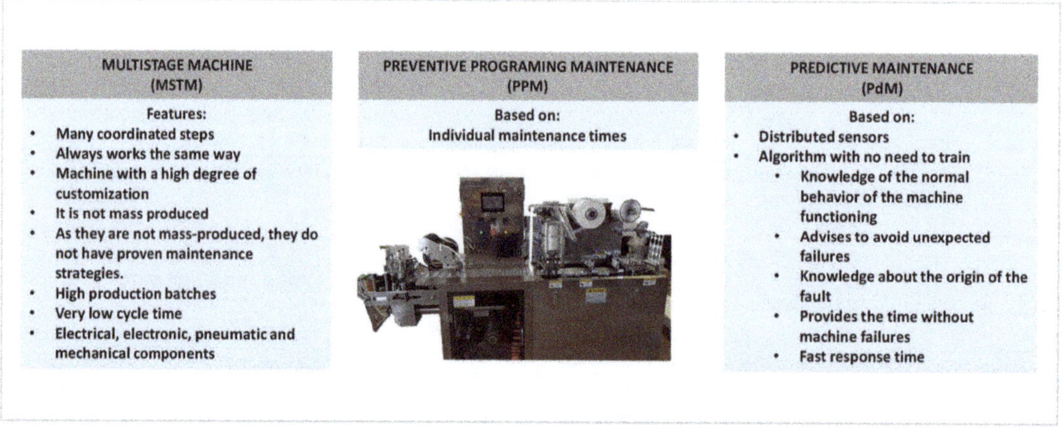

Figure 1. Features of a MSTM and main conditions of maintenance policy.

This approach determines the focus of the maintenance strategies, which are always aimed at rapid response, and calculated to avoid unexpected failures, and minimize TLP.

2. Materials and Methods

The machine worked for a year with a preventive maintenance system based on the previously characterized components. An algorithm for predictive maintenance was adopted in the beginning, but only to advise if a component had failed before its MTTF. The authors used a digital behavior twin algorithm [17] for predictive maintenance in this case. A comparison of the components that presented failures before their MTTF is given below. The results allow future users to add predictive maintenance for the components needing supervision to avoid unexpected failures and probable industrial costs for lost production time and quality production.

Below is the methodology used in this research, ordered by steps:

- Step One: The multistage thermoforming machine was selected as the case study. This machine was characterized, and all the components were identified and classified by type. See Section 2.1.
- Step Two: Reliable maintenance times were defined for each component. Importantly, an adequate MTTF was established for each component. See Section 2.2.
- Step Three: Possible preventive maintenance strategies were defined, and predictive maintenance strategies adopted. See Section 2.3.
- Step Four: The components that presented a failure before their MTTF after a year of working were studied. See Section 2.4.
- Step Five: For all of the components, the advice shown by the DBT predictive algorithm was presented to ascertain which failures could be identified before occurring unexpectedly. The advice does not entail a change of maintenance strategy. The only purpose of these dates was for use in data logging. See Section 2.5.
- Step Six: The authors proposed a procedure to make decisions for possible maintenance strategy changes in the components studied by looking for the cause of the failure and then by evaluating two key performance indicators (KPIs). See Section 2.6.

The results, discussion, conclusions, and future research are shown in Sections 3–5.

2.1. The Case Study: A Multistage Thermoforming Machine

Thermoforming and tub filling machines are one of many cases, and this study covers this type of machine. Figure 2 shows the MSTM and the placement of the components. The seven steps are identified, together with the main operation in each of them.

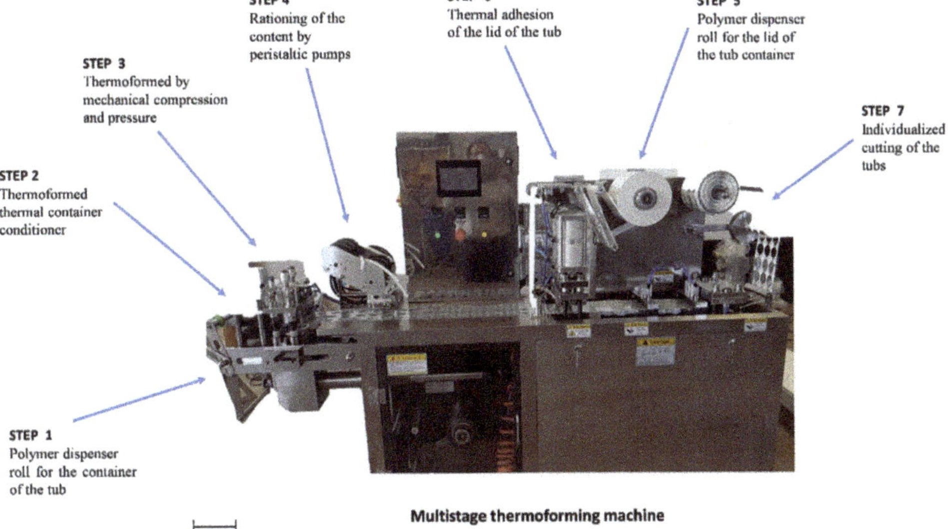

Figure 2. A multistage thermoforming machine of six terrines per cycle and its type of components.

This machine has a cycle time of 4 s, and the thermoforming mold allows the manufacture of 6 terrines for each cycle time. So, for each cycle time, seven steps constantly work in coordination.

The proper sequence of steps depends on the programmable logic controller (PLC) inside the electrical panel. The PLC receives all the information provided by sensors and takes decisions for all the actuators at the correct moment.

All the steps may have electrical, electronic, mechanical, and pneumatic components distributed for the whole industrial multistage machine. The adequate state of all the components allows for the correct functioning of the machine and avoids unexpected failures. It is easy to understand that the accumulated work time may affect the state of the components. Due to this and other considerations such as ambient conditions, power supplier events, normal degradation of mechanical components, compressed air system failure or jams in the peristaltic pumping system (step 4), unwanted mechanical shocks can be the origin of unexpected failures in the components and consequently in the industrial multistage machine.

The components of this machine and their type can be seen in Table 1. Many components may have a number greater than one. Also, the figure indicates the possible failure source and the consequences of the failure event.

Table 1. List of components in the industrial multistage machine.

Type of Component	Component	Cause of Failure	Failure Event
Electrical	Master power switch	Ambient condition, Power supplier event	Stop
	Plug-in relay	Ambient condition, Power supplier event, Unexpected hit	Malfunction Stop
	Command and signalling	Ambient condition, Power supplier event	Stop
	Safety limit switch	Ambient condition, Power supplier event, Unexpected hit	Stop
Electronic	PLC	Ambient condition, Power supplier event	Stop
	HMI	Ambient condition, Power supplier event	Stop
	Chromatic sensor	Ambient condition, Power supplier event	Stop
	Safety relay	Ambient condition, Power supplier event	Stop
	Temperature controller	Ambient condition, Power supplier event, Unexpected hit	Stop
	Solid state relay	Ambient condition, Power supplier event	Stop
	Belt drive	Ambient condition, Power supplier event	Malfunction
	Pressure sensor	Pressure failure, Global fatigue	Malfunction
	Servo drive peristaltic pump	Ambient condition, Power supplier event	Stop
	Absolute encoder	Global fatigue, Mechanical hit	Malfunction
Mechanical	Safety button	Ambient condition, Power supplier event	Stop
	Thermal resistance	Ambient condition, Power supplier event	Malfunction
	Thermocouple sensor	Global fatigue	Malfunction
	Belt motor	Global fatigue	Stop
	Bronze cap	Global fatigue	Malfunction
	Linear axis	Global fatigue	Malfunction
	Linear bearing	Global fatigue	Malfunction
	Peristaltic pump	Ambient condition, Power supplier event, compressed air system failure	Stop
	Terrine cutter	Global fatigue	Malfunction
Pneumatic	Pneumatic valve	Global fatigue	Malfunction
	Pneumatic cylinder	Pressure failure, Failure valve	Malfunction

2.2. Maintenance Times for Each Component

Once the industrial thermoforming machine has been characterized, the individual maintenance times required for each component must be studied to adopt the most appropriate preventive maintenance strategy policy accordingly. For this purpose, many individual times and equations are used and presented in this study, which has been provided by J. Jiři [27] and G. Liberopoulos [28].

- TTRP   Time to replace a component
- TTC   Time to configure
- TTMA   Time to mechanical adjustment
- TTPR   Time to provisioning
- MTTR   Mean time to repair
- MTTF   Mean time to failure

- MTBF   Mean time between failure
- TTLR   Line restart time, defined by expert knowledge
- TLP   Time lost production

MTTR (1), TLP (2), and MTBF (3) can be calculated with these equations. Efficiency and availability are used as indicators of success in preventive maintenance.

$$MTTR = TTRP + TTC + TTMA + TTPR \tag{1}$$

$$TLP = MTTR + TTLR \tag{2}$$

$$MTBF = MTTR + MTTF \tag{3}$$

After defining the times and expressions, Table 2 presents the individual maintenance times in seconds for each component in this research. For this machine, the end users and original equipment manufacturer (OEM) have suggested, with their knowledge based on the experience of use, manufacture and maintenance, the fixing of individual maintenance as its shown in Table 2 and global TTLR at 14,400 s.

Table 2. Individual maintenance times in s for all the components in the industrial multistage machine.

Component	MTTR	TTPR	MTTF	TLP
Master power switch	14,400	10,800	9,999,999	28,800
PLC	435,600	345,600	9,999,999	450,000
HMI	435,600	345,600	9,999,999	450,000
Chromatic sensor	176,520	172,800	5,000,000	190,920
Plug-in relay	14,400	10,800	5,000,000	28,800
Command and signalling	14,400	10,800	5,000,000	28,800
Safety limit switch	14,400	10,800	9,999,999	28,800
Safety relay	14,400	10,800	9,999,999	28,800
Safety button	14,400	10,800	9,999,999	28,800
Temperature controller	435,600	345,600	9,999,999	450,000
Solid state relay	176,400	172,800	5,000,000	190,800
Thermal resistance	25,500	10,800	3,700,800	39,900
Thermocouple sensor	14,700	10,800	3,700,800	29,100
Belt drive	435,600	345,600	9,999,999	450,000
Belt motor	187,200	172,800	5,000,000	201,600
Bronze cap	288,000	172,800	7,750,000	302,400
Linear axis	288,000	172,800	7,625,000	302,400
Linear bearing	288,000	172,800	7,500,000	302,400
Pneumatic valve	176,400	172,800	9,999,999	190,800
Pneumatic cylinder	176,400	172,800	9,999,999	190,800
Pressure sensor	176,700	172,800	5,000,000	191,100
Servo drive peristaltic pump	435,600	345,600	9,999,999	450,000
Peristaltic pump	547,200	518,400	5,000,000	561,600
Terrine cutter	288,000	172,800	9,999,999	302,400
Absolute encoder	360,000	172,800	5,000,000	374,400

For this type of machine, both components used at the beginning, as well as those that have presented failures, are completely new units, not ones restored by the technical service of each component manufacturer. For necessary components replacements, only in the case of the pneumatic cylinder is it possible to repair the unit by substituting internal components for new components. All other components are replaced by new units.

2.3. Maintenance Strategies

In this section, two preventive maintenance strategies are presented, and one predictive maintenance strategy is used:

- Preventive maintenance, based on the MTTF of each component, to avoid unexpected failures during the work process.

- Improved preventive maintenance, based on the above but minimizing the TTPR of each component.
- Digital behavior twin (DBT) for predictive maintenance.

2.3.1. Preventive Programming Maintenance (PPM)

This strategy is based on the MTTF of each component and proposes inspecting and replacing the component once the worked time reaches the MTTF. This is the maintenance strategy adopted for all the components at the beginning of this study.

Once the decision to replace a component is taken, a decision based on its MTTF, lost production time is necessary for the corresponding maintenance operation. As shown by Equation (1), if the MTTR is higher than the value of TTPR, a new maintenance policy can be used to minimize the MTTR. This policy entails an increase in security stocks. Figure 3 shows the ratio TTPR/MTTR in this machine. The values are provided by the machine manufacturers and shown in [17].

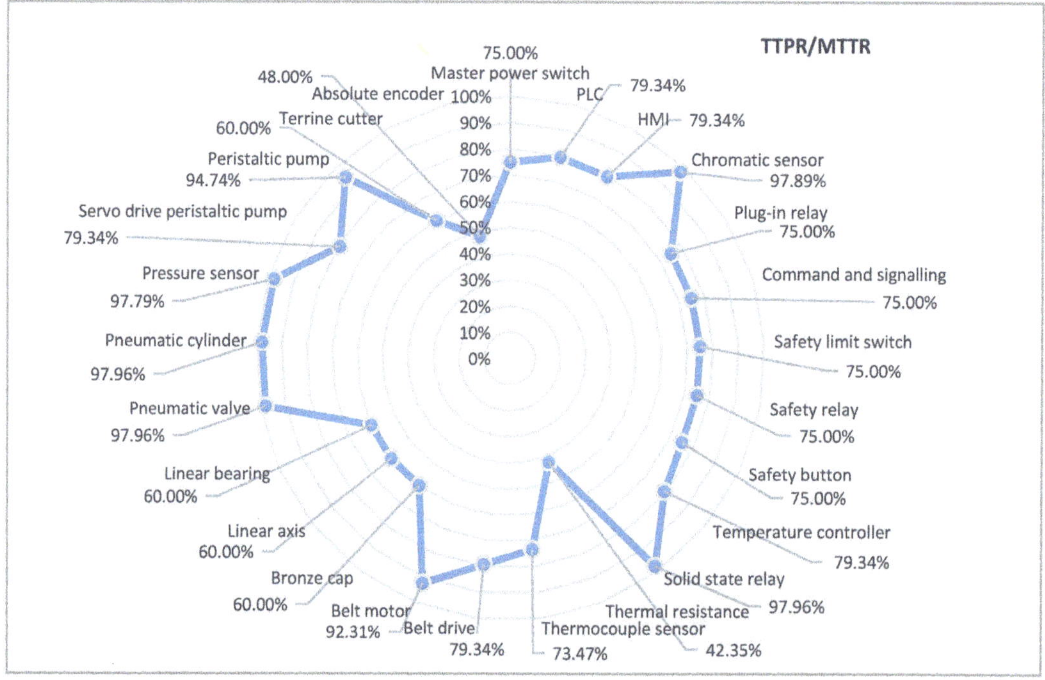

Figure 3. Comparison between TTPR and MTTR for the components of the case study.

The significant influence of TTPR value in MTTR is notable. The authors consider this ratio interesting. In Section 2.6, KPI_1 and KPI_2 will be defined by using TTPR value to propose a change in preventive maintenance strategy.

2.3.2. Improved Preventive Programming Maintenance (IPPM)

This strategy is based on the PPM strategy. When a component has a higher value of TTPR, this strategy can be used to minimize the TLP of the industrial multistage machine. In this case, the TTPR is replaced by a residual time fixed in 300s, which is the time it takes the end user of the machine to collect it from its replacement stock. Garcia, F.J.Á. and Salgado, D.R. [11] proposed a matrix to decide on the most appropriate preventive maintenance strategy but not on the component that needs a predictive maintenance strategy.

2.3.3. Digital Behaviour Twin (DBT)

This strategy is based on the sensors placed on the machine. These sensors give their values to a PLC, and the PLC uses an algorithm that triggers maintenance recommendations to avoid unexpected failures. A human interface machine (HMI) is used to show these recommendations.

This algorithm uses the signals received from the sensors and refers them to the position of a central axis by means of an absolute encoder. Since the normal operating condition is known, the algorithm detects normal operation without the need for learning, provides warnings of possible faults, and can provide the number of work cycles performed without faults.

Figure 4 shows the conceptualization of this strategy in the cited industrial multistage machine.

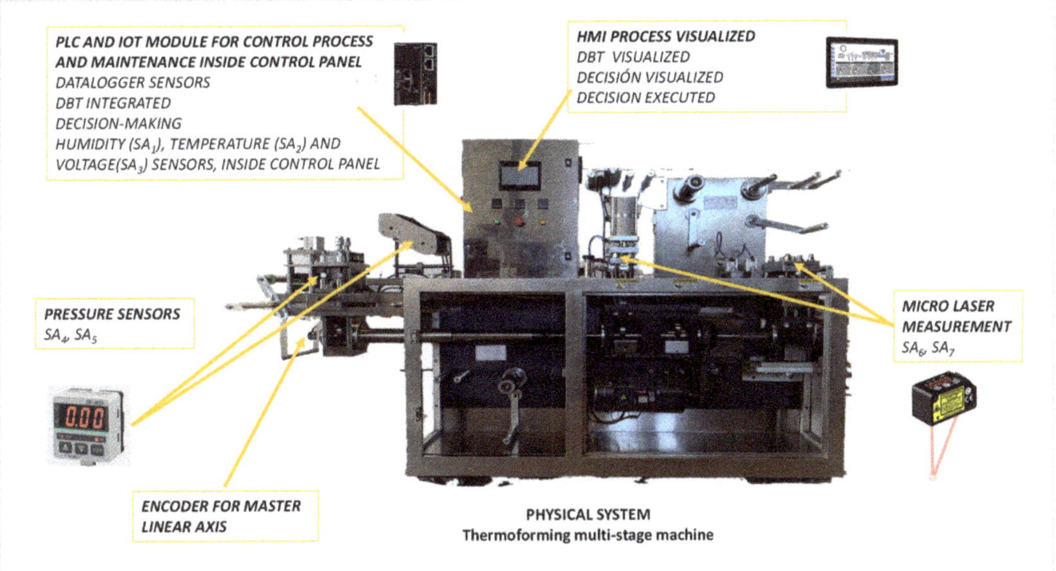

Figure 4. Conceptualization of the DBT predictive maintenance strategy.

Garcia, F.J.Á. and Salgado, D.R. [17] described this predictive maintenance strategy in detail. The objective of their research was not to define the predictive algorithm but to use it to propose a method to decide on a change of maintenance strategy from preventive to predictive.

This strategy has already been tested in this industrial multistage machine and allows the detection of potential failures within each cycle of operation of the whole machine.

2.4. Recovered Data after a Year of the Machine Working

The industrial multistage machine worked without stopping, 8 h per day, Monday to Friday, for one year, with the PPM in place and the DBT functioning only for data logging advice. Table 3 shows the list of components with the corrected MTTF if the component had failed before its MTTF and whether the cause of the failure was known or unknown.

Table 3. Individual component failures occurring within one year of the machine's working. Times in s.

Component	Fails before MTTF	Cause of Failure	Corrected MTTF
Chromatic sensor	1	Known	3,998,750
Plug-in relay	1	Known	4,056,010
Temperature controller	1	Known	7,934,710
Solid state relay	1	Known	4,678,034
Thermal resistance	1	Unknown	3,067,090
Thermocouple sensor	1	Unknown	2,890,760
Bronze cap	1	Unknown	6,500,453
Linear bearing	1	Unknown	6,375,010
Pressure sensor	1	Unknown	4,575,102
Peristaltic pump	1	Known	4,434,090
Terrine cutter	1	Unknown	8,750,778
Absolute encoder	1	Known	4,756,002

Table 3 shows that many components have presented failures before their MTTF. Figure 5 shows the results by type of component. The pneumatic components have not presented failures before their MTTF, unlike the rest of the components.

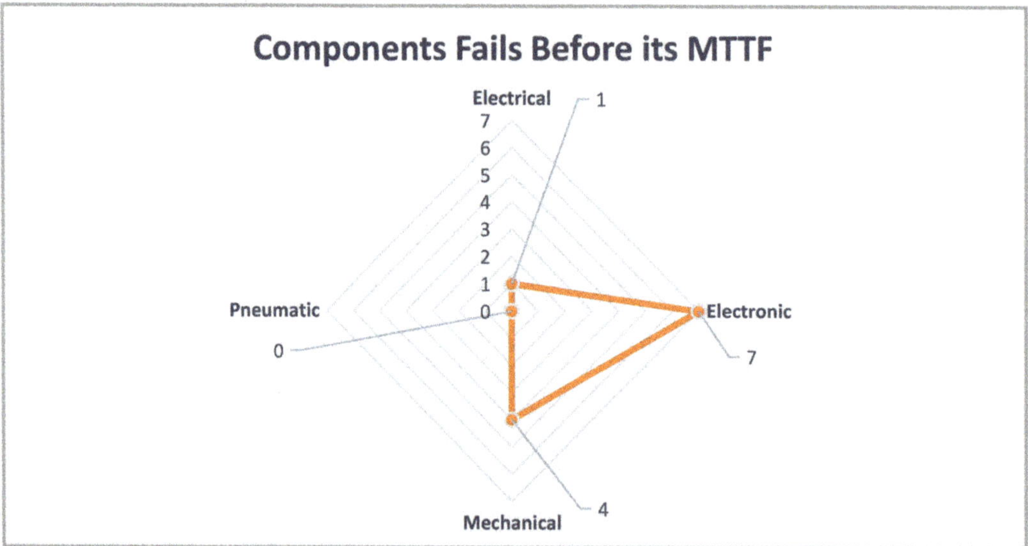

Figure 5. Component failures before their MTTF.

Table 4 shows the description of the cause of the failure of the components that presented a failure before their MTTF and also indicates, for these components, whether the cause is due to an occasional (infrequent) situation or a normal situation.

In the case of the plug-in relay, the authors believe that this component was not completely new at the beginning of the experiment. To verify this, a quality test was carried out. The rest of the components shown in Table 4, presented a failure-for-occasional-situation. In Section 2.6 a procedure to avoid the same situation is proposed.

These failures were registered. The DBT algorithm used only for data logging the advice for an unexpected failure will be compared below.

Table 4. Registered known causes of failure in components that presented a failure before their MTTF.

Component	Situation	Description of the Known Cause of Failure
Chromatic sensor	Occasional situation	The supplier of the film for the terrine lid changed the color without prior notice and made it darker and more reflective. This caused the sensor to stop seeing the mark correctly.
Plug-in relay	Normal situation	The number of commutations exceeded the mechanical endurance.
Temperature controller	Occasional situation	Mixed events of voltage RMS and high level of humidity.
Solid state relay	Occasional situation	The higher level of humidity and air dust caused a short circuit.
Peristaltic pump	Occasional situation	A higher density of fluid dosed in the terrine caused a jam.
Absolute encoder	Occasional situation	Accidental mechanical shock.

2.5. DBT Predictive Algorithm Warnings of Failure Recovered

The DBT algorithm matched the real failures that occurred when the machine was working. Table 5 shows the warning of failures obtained for the DBT predictive maintenance strategy. This table only shows components that presented failures.

Table 5. DBT warnings of failures by components within the operation time studied.

Component	DBT Warning of Failures
Chromatic sensor	1
Plug-in relay	1
Temperature controller	1
Solid state relay	1
Thermal resistance	1
Thermocouple sensor	1
Bronze cap	1
Linear bearing	1
Pressure sensor	1
Peristaltic pump	1
Terrine cutter	1
Absolute encoder	1

The coincidence of warning of failures provided by DBT and shown in Table 5 and components that failed before their MTTF, as shown in Table 3, suggests using the DBT algorithm if a component requires predictive maintenance. Due to this coincidence, a scorecard must be designed to make decisions about changes in the component maintenance strategy.

2.6. Method Proposal to Take Decisions for Maintenance Strategy Decisions

As Section 2.3.1 set forth, the PPM strategy had been adopted for all the components at the beginning of the study. With all the accumulated dates compared to the actual event, and the warning advice given by the DBT algorithm, this section explains how to make decisions for a possible change of maintenance strategy.

The objective is to identify the components that need predictive maintenance. For this purpose, there are two key questions:

- Has the component failed before its MTTF?
- Do we know why it failed?

Two key performance indicators are studied to ascertain whether the reason is known. The expressions for KPI_1 (4) and KPI_2 (5) are the following:

$$KPI_1 = (MTTR - TTPR)/MTTR \tag{4}$$

$$KPI_2 = TTPR /TLP \tag{5}$$

KPI_1 is used to ascertain the influence of TTPR in the MTTR for each component. If this ratio presents a small value, the TTRP will be higher, which is considered an important piece of information with reference to changing the maintenance strategy.

KPI_2 is used to assess the influence of TTPR in Time TLP because this ratio shows the availability and efficiency decrease for a higher value of TTPR.

2.6.1. Procedure to Set KPI$_1$ and KPI$_2$ Values

The procedure to set initial values of KPI1 and KPI2 is the following:

- Calculate KPI1 interval between PPM and IPPM strategies.
- Calculate KPI2 interval between PPM and IPPM strategies.
- Calculate average value of KPI1 and KPI2, assuming PPM strategy.
- Calculate average value of KPI1 and KPI2, assuming IPPM strategy.
- Calculate average value of TTPR/MTTR ratio assuming PPM strategy.

Individual times for PPM strategy are shown in Table 2. In the case of IPPM strategy, only TTPR is modified for a constant value fixed in 300s (see Section 2.3.2).

Table 6 shows all of the calculated values.

Table 6. Calculated ratios to define fixed values of KPI$_1$ and KPI$_2$.

Strategy	Ratio	Average KPI$_1$	Average KPI$_2$	Average TTPR/MTTR
PPM	Value Interval	22.92% [2.04–57.65%]	63.14% [27.07–92.31%]	77.08%
IPPM	Value Interval	96.04% [92.31–99.84%]	0.99% [0.15–1.64%]	3.96%

As can be observed, the intervals for KPI$_1$ and KPI$_2$ values in PPM and IPPM strategies have no common points. For initial, fixed KPI's points, we must be within the intervals provided by PPM strategy. Whether KPI$_2$ is considered the average ratio between TTRP and MTTR due to the TLP depends on a constant value (TTLR equal to 14,400 s) and the MTTR value (see Equation (2))

Figures 6 and 7 show the fixed KPI$_1$ and KPI$_2$ values for decision-making. A large dispersion of KPI$_1$ and KPI$_2$ values is observed in the case study. The correct functioning of the fixed values is evaluated by the minimization of TLP and stock cost in case of adopting IPPM strategy. The final value fixed in the case of KPI$_1$ is 25% and in the case of KPI$_2$ is 70% (value obtained by comparing average KPI2 with average TTPR/MTTR).

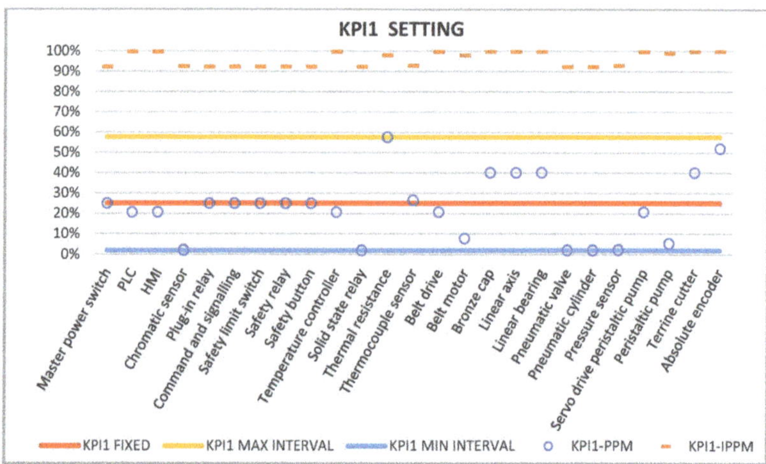

Figure 6. Comparison of KPI$_1$ values for each component in PPM and IPPM strategies, KPI$_1$ interval in PPM strategy and fixed value 25% of KPI$_1$.

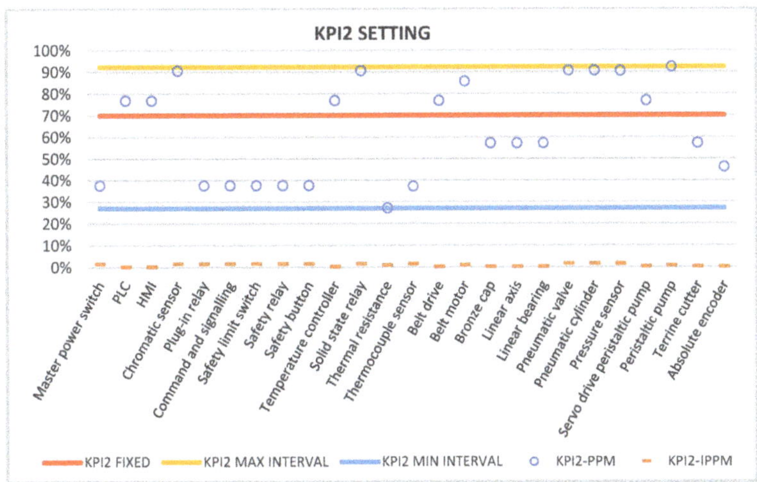

Figure 7. Comparison of KPI_2 values for each component in PPM and IPPM strategies, KPI_2 interval in PPM strategy and fixed value 70% of KPI_1.

With the fixed values, decision-making to change maintenance strategy can be adopted. So:

- If $KPI_1 < 25\%$ and $KPI_2 > 70\%$, the improved preventive maintenance strategy can be proposed, with a previous GOC evaluating the component;
- If $KPI_1 > 25\%$ and $KPI_2 < 70\%$, a preventive maintenance strategy change is unnecessary.

Of course, if a component presents a failure before its MTTF and the cause of the failure is unknown, and the value of $KPI_1 < 25\%$ and $KPI_2 > 70\%$, several changes must be made in the maintenance strategy.

2.6.2. Proposed Method for Maintenance Strategy Adoption

Figure 8 shows the method proposed by a PPM strategy adopted for all the components of the industrial multistage machine initially and the possible decisions to be taken depending on the knowledge of the fault and the KPI values.

The proposed method can be used in this machine when assessing whether to change the maintenance strategy if a component fails before its MTTF. However, feedback is useful to control the real evolution of the machine in every possible way. The authors consider that this feedback is only useful if the cause of the failure is known.

With the proposed method, all the components start operating with a PPM strategy, and if any fail before their MTTF, a change to IPPM or IPPM with DBT may be appropriate.

As mentioned in Section 2.5, if it is necessary to use a predictive maintenance strategy, DBT can be used, due to the good results offered with the advice shown in Table 5. In this way if a component fails before its MTTF, and the cause is unknown, IPPM will be adopted regardless of the value of KPIs. Later, an inspection of the location and other factors to find the cause of the failure with DBT monitoring enables a new value of MTTF to be set. If the cause of the failure is found, the method returns to starting point. If the component does not fail before its new MTTF, the maintenance strategy will be PPM. Otherwise, the way depends on the knowledge of the second failure.

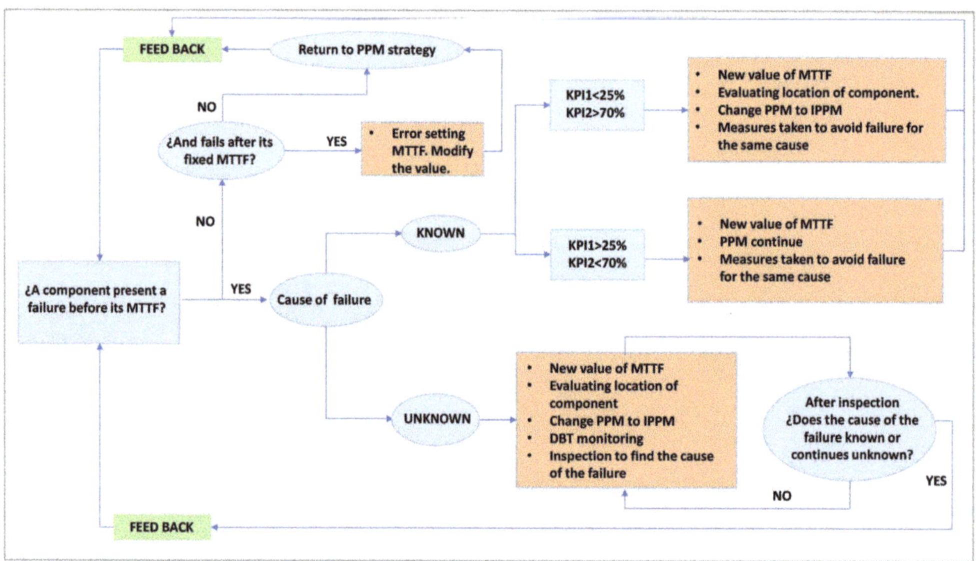

Figure 8. Proposed method to switch from a preventive maintenance strategy to a predictive strategy with a DBT algorithm.

According to the proposed method it i considered that the principal object and benefit of adopting a DBT predictive maintenance strategy is to determine the reason of a component failure before its MTTF by using distributed sensors that are part of the DBT predictive maintenance strategy. Also, when DBT monitoring is used, the behavior of the whole machine is monitored to guarantee the correct functioning of the MSTM. So, as Figure 5 shows, if the cause of the failure is found, the replaced component returns to the starting point of the proposed method and the DBT monitoring continues until the next failure of the same component. Otherwise, the method does not continue towards the starting point, and the model waits instead for a better result of the main actions proposed to find the reason of the failure.

The final object of this dynamic method is the default selection of PPM strategy for all components, and only IPPM if KPI1 < 25% and KPI2 > 70% at the same time if these components fail before its MTTF. Nevertheless, the real situations involving these types of machines, located in important factories and parts of a production process, indicates to us the need for a predictive maintenance strategy if, due to an occasional (infrequent) situation, a component fails before its initial MTTF.

3. Results

The application of the proposed method for possible maintenance strategy changes is shown in Table 7. The initial maintenance strategy was PPM for all the components. The next column shows the maintenance strategy to be adopted if the component fails before its MTTF.

The results show different changes in maintenance strategies. The PPM strategy of all the components that failed before their MTTF due to unknown causes was changed to IPPM with DBT monitoring (pressure sensor, thermal resistance, thermocouple sensor, bronze cap, linear bearing, and terrine cutter). The maintenance strategy of two components with known causes of failure remained unchanged (plug-in relay and absolute encoder), and the maintenance strategy of four components (chromatic sensor, temperature controller, solid state relay, and peristaltic pump) was changed to IPPM.

Of course, the initial PPM strategy of components that did not fail remained the same.

Table 7. Maintenance strategies for all components after a year in operation using the proposed method.

Component	KPI$_1$	KPI$_2$	Maintenance Strategy after a Year of Work
Master power switch	0.25	0.38	PPM
Plug-in relay	0.25	0.38	PPM
Command and signalling	0.25	0.38	PPM
Safety limit switch	0.25	0.38	PPM
PLC	0.21	0.77	PPM
HMI	0.21	0.77	PPM
Chromatic sensor	0.02	0.91	IPPM
Safety relay	0.25	0.38	PPM
Temperature controller	0.21	0.77	IPPM
Solid state relay	0.02	0.91	IPPM
Belt drive	0.21	0.77	PPM
Pressure sensor	0.02	0.90	IPPM + DBT monitoring
Servo drive peristaltic pump	0.21	0.77	PPM
Absolute encoder	0.52	0.46	PPM
Safety button	0.25	0.38	PPM
Thermal resistance	0.58	0.27	IPPM + DBT monitoring
Thermocouple sensor	0.27	0.37	IPPM + DBT monitoring
Belt motor	0.08	0.86	PPM
Bronze cap	0.40	0.57	IPPM + DBT monitoring
Linear axis	0.40	0.57	PPM
Linear bearing	0.40	0.57	IPPM + DBT monitoring
Peristaltic pump	0.05	0.92	IPPM
Terrine cutter	0.40	0.57	IPPM + DBT monitoring
Pneumatic valve	0.02	0.91	PPM
Pneumatic cylinder	0.91	0.91	PPM

4. Discussion

The proposed method for changing the maintenance strategy for all the components that failed does not provide a static decision criterion. For example, the same component can fail first due to an unknown cause, and again a second time due to a known cause. The method allows taking different decisions according to whether or not the cause of the failure is known.

The authors consider knowing the cause of the failure critical. An industrial multistage machine must not operate with unknown failures. Also, once the cause is known, the manufacturer must take action to avoid an unexpected failure due to the same cause. If these actions are correct and there is feedback, the industrial multistage machine can restart operating with adequate functionality guarantees.

The values of KPI$_1$ and KPI$_2$ are used to assess whether a change of preventive maintenance strategy is required. As mentioned in Section 2.6, the extreme values of both are fixed to show whether the preventive maintenance strategy should be changed from PPM to IPPM. However, if the value of time to provisioning (TTPR) of a component goes up or down, the value of its KPI$_1$ and KPI$_2$ will also change. In this scenario, if a failure occurs in this component, the method will use another way to make decisions, in a further evaluation.

A continuous application of this method for the same industrial multistage machine will allow greater failure control and higher levels of operation time without failures.

If a component supplier is changed for market reasons, the proposed method must be reassessed, and the KPIs and MTTR must be recalculated. Also, the value of MTTF must be changed accordingly before the machine resumes its operation.

The authors consider the following assessment critical, ascertaining the trust level in the component manufacturer by evaluating the ratio between the corrected and initial MTTF of all the components that failed before their initial MTTF. Table 8 shows this ratio:

Table 8. Trust levels in component manufacturers by comparing the initial and corrected MTTF in components that failed before their initial MTTF.

Type of Component	Component	Trust Level
Electronic	Chromatic sensor	79.98%
Electrical	Plug-in relay	81.12%
Electronic	Temperature controller	79.35%
Electronic	Solid state relay	93.56%
Electronic	Thermal resistance	82.88%
Electronic	Thermocouple sensor	78.11%
Mechanical	Bronze cap	83.88%
Mechanical	Linear bearing	85.00%
Electronic	Pressure sensor	91.50%
Mechanical	Peristaltic pump	88.68%
Mechanical	Terrine cutter	87.51%
Electronic	Absolute encoder	95.12%

This assessment, therefore, allows for a long operating time with an adequate selection of component manufacturers. Figure 9 shows the average trust level (ATL) by component type. The authors consider that the compared values must be similar. This indicates that the machine maintenance team is adequate for the whole machine. Obviously, the optimal value of this ATL is 100%. As the pneumatic components did not fail before their MTTF, they have not been included in Figure 9.

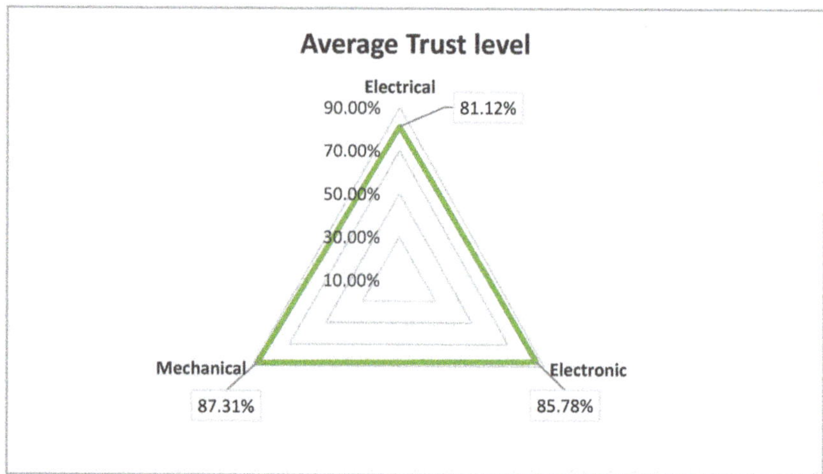

Figure 9. Average Trust level of components that failed before their initial MTTF.

The authors consider that one way to further this research would be if an ATL were to be fixed for all the components for possible decisions to change component manufacturers. Also, a new result would be obtained if the cost of the component were to be used in this proposed future research.

5. Conclusions

The proposed method for possible maintenance strategy changes for components in the same industrial multistage machine provides ways to change the maintenance strategy for PPM to IPPM or IPPM with DBT monitoring. The authors consider that this method will be useful for other industrial multistage machines.

The predictive maintenance strategy is used for constant component monitoring if an unexpected failure has occurred, so if the cause is known and the measures for avoiding a new failure for the same cause are taken, the component will probably fail at its new MTTF and will then go back to having preventive maintenance like PPM or IPPM.

If a component presents many consecutive failures before its initial MTTF and the actions proposed by the method are taken, then the ATL of the component manufacturer should be revisited to decide on whether to change the manufacturer with an objectively higher quality in this component. In this case, this could be another way to start a failure mode and effect analysis (FMEA) to redesign the function, location, and work of this stage of the machine.

All the components can be included in the study of ATL by component type. In this case, the average value will be higher than that shown in Figure 9. The authors only included the components that failed to avoid wrong results in the machine maintenance team's evaluation.

As the trust level, or the ATL, depends on the ratio between initial MTTF and real MTTF of the component or type of components, these values do not improve with a maintenance strategy change; they only improve to 1 or 100% if the actions necessary to take for avoid occasional (infrequent) situations that end with an unexpected failure work correctly. In the case of trust level or ATL improvement up to 100% the authors suggest an incorrect initial MTTF fixed at starting point of the machine's use. The proposed method suggests this way (see Figure 8)

Industrial multistage machines need a long working time without unexpected failures, so a global method for taking the appropriate decisions for maintenance strategies is needed, and adequate changes must be made to avoid such unexpected failures. The proposed method allows reaching this objective. Nevertheless, some comments for its application in the context of other multistage machines must be related:

- The case study is a multistage thermoforming machine. This machine has an absolute encoder. Its position is constantly sent to the PLC for synchronization and management of all the coordinated steps in the correct order. This encoder allows the use of a digital behavior twin algorithm for predictive maintenance strategy. Not all of the multistage machines have an encoder for this special function, so the normal behavior of the machine must be referred to with a more precise physical analogue.
- Due to the fact that the cycle time is only 4 s, the algorithm for predictive maintenance must be speedy and certain. Other machines with longer cycle times could use predictive maintenance based on the time;
- As Figure 1 indicates, the preventive maintenance strategy depends upon the individual maintenance times. It would be interesting to evaluate the sensibility of the method for an incipient change of TTPR in some components due to global market conditions.

The main contributions highlighted in this article are:

- Providing a method for deciding when to use predictive maintenance strategy and when to stop it in different components of a MSTM.
- Providing a dynamic global method to establish the maintenance strategy of any component of an MSTM.
- Providing a confidence level of a component or type of components in an MSTM that indicates whether the MTTF of said component operating in said machine is reliable.
- Because of the above, obtaining information on the reliability of the components of a MSTM to avoid unexpected failures during its operating time.

Table 9 shows the results of the comparison between the introduction citations and the proposed method. Due to the singularity of this type of multistage machine, the cited references are not alternative methods that can be used to provide other maintenance strategies for the same machine in the same working conditions, with the same components and the same evaluation time (1 year). Due to this, the comparison offered in the following table focuses on the most significant aspects found in each citation that are related to the methodology developed in this study. This comparison is, therefore, in qualitative terms, and not able to offer numerical comparisons. The first column indicates the item or relevant aspect to be compared. The second column indicates the highlighted references compared. The third column is a qualitative comparison between the cited item (column 1) in the reference (column 2) and the method proposed in this article.

Table 9. Qualitative comments highlighted between the proposed method and the state of art. (*) Improve options.

Item	References	Qualitative Comments after Comparison
Minimizing security stocks	[1,2]	Correct selection of fixed KPIs allows the optimization of stock and provides the adequate preventive maintenance policy
Stops to settings, removal actions. Imperfect maintenance	[3,4]	Settings only at the start time of the machine functioning by the temperature controller, thermal resistance, and thermocouple sensor. The maintenance actions must perform the machine functioning. The system can evaluate if the actions in each component or each type of component are imperfect by trust level or ATL.
Mathematical model for Preventive maintenance	[5–8]	Complex, very theoretical and many variables to manage. Simple, sensitive to variations of individual maintenance times.
MTTF reliable Reliability and law degradation	[9–11]	Initial MTTF fixed for all components; reliability functions not used. Possibility to change MTTF value if real MTTF lower or upper than initial MTTF fixed.
Product design and operation conditions	[12]	If a component exhibits repeated failures, an immediate FMEA analysis procedure is initiated to find design errors or component selection errors.
Mathematical model for Predictive maintenance	[13–18]	Uses PLC with embedded DBT algorithm. No need training and learning time. Quick response Very useful for a machine with fast cycle time.
Location components	[19]	(*) Possible improvement. Can be evaluated for this application
Mixed cost and technical analysis	[20,21]	(*) Possible improvement. Also is cited in future research. Coincidence in the use of FMEAS analysis
Digital Twin	[22,23]	The behavior of the machine always is the same and does not need a real digital twin since the characterization is special for each MSTM and operation conditions are always are the same. Coincidence in the event failure advises, no training and utilization of FMEAS analysis.
Augmented Reality and Computer Vision	[24,25]	(*) Possible improvement. Not used. ATL is used for evaluating the maintenance operator actions required for maintenance policy. But it is used after a maintenance action.
Preventive actions in flexible windows time. Predictive maintenance always running Method for decision-making	[26]	(*) Possible improvement to use flexible windows time for preventive maintenance actions. Predictive maintenance only works if a component fails before its MTTF, and the cause of the failure is unknow. Coincidence in the contribution of a method for decision-making
Individual preventive maintenance Times	[27,28]	Used in the article and performed by developing KPIS for preventive maintenance decisions

The method proposed is appropriate for the MSTM but can improve with respect to some items.

Future research:

- Study the influence of a fixed ATL and cost assessment for possible component manufacturer changes;

- Utilization of DBT monitoring for combined supervision in parallel of the same machine system to use Predictive Maintenance and use the advice for one machine to start DBT monitoring in other machines of the system working in the same operating conditions;
- Global cost analysis of the components, DBT monitoring system, and their influence on possible maintenance strategies for all the components in an industrial multistage machine;
- Mixed method for maintenance strategies using technical parameters and cost terms.

Author Contributions: Conceptualization, F.J.Á.G. and D.R.S.; methodology, F.J.Á.G.; software, F.J.Á.G.; validation, D.R.S. and F.J.Á.G.; formal analysis, F.J.Á.G.; investigation, F.J.Á.G. and D.R.S.; resources, F.J.Á.G.; data curation, F.J.Á.G.; writing—original draft preparation, D.R.S. and F.J.Á.G.; writing—review and editing, F.J.Á.G. and D.R.S.; visualization, F.J.Á.G.; supervision, D.R.S.; project administration, D.R.S.; funding acquisition, D.R.S. and F.J.Á.G. All authors have read and agreed to the published version of the manuscript.

Funding: This study has been carried out through the Research Project GR-21098 linked to the VI Regional Research and Innovation Plan of the Regional Government of Extremadura.

Data Availability Statement: Not applicable.

Acknowledgments: The authors wish to thank the European Regional Development Fund "Una manera de hacer Europa" for their support of this research. This study has been carried out through the Research Project GR-21098 linked to the VI Regional Research and Innovation Plan of the Regional Government of Extremadura.

Conflicts of Interest: The authors declare no conflict of interest.

References

1. Gharbi, A.; Kenne, J.P.; Beit, M. Optimal safety stocks and preventive maintenance periods in unreliable manufacturing systems. *Int. J. Prod. Econ.* **2007**, *107*, 422–434. [CrossRef]
2. Gharbi, A.; Kenne, J.P.; Boulet, J.F.; Berthaut, F. Improved joint preventive maintenance and hedging point policy. *Int. J. Prod. Econ.* **2010**, *127*, 60–72. [CrossRef]
3. Jun-Hee, H.; Tae-Sun, Y. Scheduling proportionate flow shops with preventive machine maintenance. *Int. J. Prod. Econ.* **2021**, *231*, 107874. [CrossRef]
4. Zuhua, J.; Jiawen, H.; Haitao, L. Preventive maintenance of a single machine system working under piecewise constant operating condition. *Reliab. Eng. Syst. Saf.* **2017**, *168*, 105–115. [CrossRef]
5. Ruiz-Hernández, D.; Pinar-Pérez, J.M.; Delgado-Gómez, D. Multi-machine preventive maintenance scheduling with imperfect interventions: A restless bandit approach. *Comput. Oper. Res.* **2020**, *119*, 104927. [CrossRef]
6. Chiacchio, F.; D'Urso, D.; Sinatra, A.; Compagno, L. Assesment of the optimal preventive maintenance period using stochastic hybrid modelling. *Procedia Comput. Sci.* **2022**, *200*, 1664–1673. [CrossRef]
7. Fujishima, M.; Mori, M.; Nishimura, K.; Takayama, M.; Kato, Y. Development of sensing interface for preventive maintenance of machine tools. *Procedia CIRP* **2017**, *61*, 796–799. [CrossRef]
8. Irfan, A.; Umar Muhammad, M.; Omer, A.; Mohd, A. Optimization and estimation in system reliability allocation problem. *Reliab. Eng. Syst. Saf.* **2021**, *212*, 107620. [CrossRef]
9. Yang, D.Y.; Frangopol, D.M.; Han, X. Error analysis for approximate structural life-cycle reliability and risk using machine learning methods. *Struct. Saf.* **2021**, *89*, 102033. [CrossRef]
10. Silva, G.; Ferreira, S.; Casais, R.B.; Pereira, M.T.; Ferreira, L.P. KPI development and obsolescence management in industrial maintenance. *Procedia Manuf.* **2019**, *38*, 1427–1435. [CrossRef]
11. Álvarez García, F.J.; Rodríguez Salgado, D. Analysis of the Influence of Component Type and Operating Condition on the Selection of Preventive Maintenance Strategy in Multistage Industrial Machines: A Case Study. *Machines* **2022**, *10*, 0385. [CrossRef]
12. Yuk-Ming, T.; Kai-Leung, Y.; Wai-Hung, I.; Wei-Ting, K.A. Systematic Review of Product Design for Space Instrument Innovation, Reliability, and Manufacturing. *Machines* **2021**, *9*, 244. [CrossRef]
13. Ponce, P.; Meier, A.; Miranda, J.; Molina, A.; Peffer, T. The Next Generation of Social Products Based on Sensing, Smart and Sustainable (S3) Features: A Smart Thermostat as Case Study. Science Direct. *IFAC Pap. Line* **2019**, *52*, 2390–2395. [CrossRef]
14. Hassankhani Dolatabadi, S.; Budinska, I. Systematic Literature Review Predictive Maintenance Solutions for SMEs from the Last Decade. *Machines* **2021**, *9*, 191. [CrossRef]
15. Cavalieri, S.; Salafia, M.G. A Model for Predictive Maintenance Based on Asset Administration Shell. *Sensors* **2020**, *20*, 6028. [CrossRef]

16. Bouabdallaoui, Y.; Lafhaj, Z.; Yim, P.; Ducoulombier, L.; Bennadji, B. Predictive Maintenance in Building Facilities: A Machine Learning-Based Approach. *Sensors* **2021**, *21*, 1044. [CrossRef]
17. Álvarez García, F.J.; Rodríguez Salgado, D. Maintenance Strategies for Industrial Multi-Stage Machines: The Study of a Thermo-forming Machine. *Sensors* **2021**, *21*, 6809. [CrossRef]
18. Givnan, S.; Chalmers, C.; Fergus, P.; Ortega-Martorell, S.; Whalley, T. Anomaly Detection Using Autoencoder Reconstruction upon Industrial Motors. *Sensors* **2022**, *22*, 3166. [CrossRef]
19. Pfaff, M.M.L.; Dörrer, F.; Friess, U.; Preedicow, M.; Putz, M. Adaptive Predictive Machine Condition assessment for resilient digital solutions. *Procedia CIRP* **2021**, *104*, 821–826. [CrossRef]
20. Florian, E.; Sgarbossa, F.; Zennaro, I. Machine learning-based predictive maintenance: A cost-oriented model for implementation. *Int. J. Prod. Econ.* **2021**, *236*, 108114. [CrossRef]
21. Arena, S.; Florian, E.; Zennaro, I.; Orrù, P.F.; Sgarboss, F. A novel decision support system for managing predictive maintenance strategies based on machine learning approaches. *Saf. Sci.* **2022**, *146*, 105529. [CrossRef]
22. Stary, C. Digital Twin Generation: Re-Conceptualizing Agent Systems for Behavior-Centered Cyber-Physical System Development. *Sensors* **2021**, *21*, 1096. [CrossRef] [PubMed]
23. O'Sullivan, J.; O'Sullivan, D.; Bruton, K. A case-study in the introduction of a digital twin in a large-scale smart manufacturing facility. *Procedia Manuf.* **2020**, *51*, 1523–1530. [CrossRef]
24. Konstantinidis, F.K.; Kansizoglou, I.; Santavas, N.; Mouroutsos, S.G.; Gasteratos, A. MARMA: A Mobile Augmented Reality Maintenance Assistant for Fast-Track Repair Procedures in the Context of Industry 4.0. *Machines* **2020**, *8*, 88. [CrossRef]
25. Haihua, Z.; Changchun, L.; Tang, D.; Nie, Q.; Zhou, T.; Wang, L.; Song, Y. Probing an intelligent predictive maintenance approach with deep learning and augmented reality for machine tools in IoT-enabled manufacturing. *Robot. Comput.-Integr. Manuf.* **2022**, *77*, 102357. [CrossRef]
26. Hongfeng, W.; Qi, Y.; Fang, W. Digital twin-enabled dynamic scheduling with preventive maintenance using a double-layer Q-learning algorithm. *Comput. Oper. Res.* **2022**, *144*, 105823. [CrossRef]
27. Jiři, D.; Tuhý, T.; Jančíková, Z.K. Method for optimizing maintenance location within the industrial plant. *Int. Sci. J. Logist.* **2019**, *6*, 55–62. [CrossRef]
28. Liberopoulos, G.; Tsarouhas, P. Reliability analysis of an automated pizza production line. *J. Food Eng.* **2005**, *69*, 79–96. [CrossRef]

 *systems*

MDPI

Article

The Regulatory Architecture of Digital Platforms: A Perspective of Life Cycle and Risk Management

Cong Xu [1,*] and Yu-Min Wang [2]

[1]   Intellectual Property Academy, School of Law, Shanghai University, Shanghai 200444, China
[2]   School of Law, The University of Edinburgh, Edinburgh EH8 9YL, UK
*   Correspondence: xu_cong@shu.edu.cn

Abstract: The rise of internet platforms meets people's needs for a better life. However, the platforms also pose the risk of ecological monopolies. Using the methodology of economic analysis of law, with the help of ANT theory, the laws governing the operation of the platform ecosystem are discovered, and the paper analyzes the life cycle of digital platform development and figures out that the regulatory strategy for platforms should be adjusted to follow its life cycle and adopt more intuitive evaluation criteria for assessing market power. Meanwhile, the regulatory strategy for plat-forms could fully guarantee the active participation of multiple subjects, such as operators and consumers, in the platform's governance. With the continuous advancement of data and algorithm technology, new content service providers will continue to emerge, and a new industry is developing. Besides the dynamic track analysis of platforms' life cycles, another static research outcome is also given in this research. To ensure that the algorithmic technologies developed by the platform truly contribute to economic and social development and the well-being of people, the right to interpret algorithms and the establishment of scenario-based regulation of algorithms should be established.

Keywords: digital platforms; regulation; antitrust; game theory; algorithm

check for **updates**

Citation: Xu, C.; Wang, Y.-M. The Regulatory Architecture of Digital Platforms: A Perspective of Life Cycle and Risk Management. *Systems* **2022**, *10*, 145. https://doi.org/ 10.3390/systems10050145

Academic Editor: William T. Scherer

Received: 17 July 2022
Accepted: 3 September 2022
Published: 8 September 2022

Publisher's Note: MDPI stays neutral with regard to jurisdictional claims in published maps and institutional affil-iations.

1. Introduction

One of the key features of the current digitalisation and informatisation process in China is the increasing rise of digital platforms. Although there is no consensus on the social impact of digital platforms, there is a convergence in the conceptual meaning: A digital platform is a programmable digital infrastructure that facilitates interaction between users through the aggregation of information and shapes the "platform ecology" in the information society [1]. The operation of a platform ecosystem relies on the coordination of multiple systems, including computing power, data, programming algorithms, payment systems, etc. Although the development of digital platforms has permeated every corner of society through mobile applications, their technical properties and social impacts are still debated.

Nonetheless, our current understanding of platform technology remains too narrow. Firstly, the infrastructural attributes of the 'platform ecology' have been neglected in previous studies, and the main emphasis has been on its digital technological attributes [2]. Although a platform consists of a collection of information in its presentation, it is not necessarily entirely virtual [3]. In China, there are physical elements in the rise of digital platforms, including the internet's infrastructure and the support of societal aspects such as urbanisation, the enterprise system, and the labour force. This physical reality has enabled the digital expansion of the platforms and laid the material foundation for their development. Secondly, the exploration of platform technology is still dominated by macro- and meso-level grand narratives, ignoring the micro-level movements of platform development and operation [4,5]. Even though macro-level narratives enable structural and trend-based inquiries into platform development at the levels of corporate orientation, technological governance, government policies, etc., participants often become static and singular in their analytical perspectives. Hence, technology is detached from reality and incapable of explaining social contexts and their changing patterns.

This paper reconceptualizes technology as a "dispositif" with both physical and virtual attributes from the perspective of technology production [6], and explores the production, development, and evolution of algorithms through a microscopic and dynamic lens. Therefore, algorithms, as a media technology body in the information society, are both physical and discursive, with abundant extension and expressive power in micro-social production.

The research is grounded within Actor Network Theory (ANT)—the integration of humans and "non-humans" within analysis—to explore the "algorithmic production network" of digital platforms. It breaks down the virtual and the material, the intangible and the tangible, and the dichotomy between the infrastructure and the culture of consciousness and establishes a new perspective on the analysis of digital platforms. Taking the popular take-away food delivery platform as an example, this paper further reflects on the production logic of platform algorithms and the explanatory power and limitations of ANT theory in relation to digital platforms. The main research issues include how digital platforms are produced, shaped, and defined by the actor-networks from various social forces, how the algorithmic production networks of platforms have changed and developed during this period, and to what extent the spatial-temporal production and technological governance of digital platforms have enriched and developed the actor-network theory. While previous empirical studies on ANT have tended to focus on the production of social space [7,8], few studies have dealt with the manufacture and production of virtual technologies, perhaps due to ANT being subject to many technical challenges [9] (pp. 160–162). ANT has been misunderstood as a technical network and incorrectly compared to telephony, social media, the internet, etc. However, as Latour argues, ANT is not necessarily linked to the network society or internet technology, but is more of an ontological elaboration than a technical discourse. Nonetheless, the production and development of internet technology can still be included in its analytical framework [10]. In fact, this paper maintains that, with the increasing prominence of the social attribute of technology, ANT can offer new paths for interpreting the multifaceted social meaning of technology.

2. Digital Platforms and Algorithmic Production as Actor-Networks

ANT perceives digital platforms as information aggregation infrastructures coupled with technology, society, and economics. Using smartphones, mobile payments, and information distribution and delivery systems, digital platforms provide a virtual space for information sharing, the fulfilment of material needs, and communication. Working both online and offline, the spatial-temporal construction of digital platforms highlights the interplay of infrastructure and social relations. Mainly proposed by Bruno Latour, Michel Callon, and John Law, ANT offers a theoretical and methodological lens and advocates for the adoption of an association perspective insofar as that society is an association of heterogeneous things [11]. The participants in the network of relations can be both human and non-humans [9] (p. 120). ANT relies on three core concepts in the actor-network theory: Agency, mediator, and network. Actors are wide-ranging, including human and non-human heterogeneous actors, such as biology, technologies, information, ideas, etc., and have agency. Latour considers any actor a mediator that has agency, "anything that changes the state of affairs by making a difference can be called an 'actor'" [12]. A mediator is an actor involved in the process of changing and translating the production of meaning. The term 'mediator' is opposed to the term 'intermediary', which refers to a participant with agency, while the latter is a black box of passive and undifferentiated transporting meaning. Translation consists of four stages: Problematisation, interestment, enrolment, and mobilisation [13]. During problematisation, the core actor creates an obligatory passage point (OPP) to establish relationships with potential participants by integrating them into the network system, through interestment and enrolment, and ensuring, through mobilization, that actors can represent their collective and exercise their rights. Both Calonne and

Latour take a neutral and developmental perspective on translation within ANT, conceiving translation as a process of redistribution and re-transformation of power, which may result in either functioning or breakdown of the order. ANT, as an ontological cognition, can be regarded as a combination of a string of actions. Networks in ANT emphasize the participation of actors, becoming multi-dimensional nodes that form temporary, fluid, and uncertain associations [14] (pp. 8–10). In these networks, human and non-human mediators are involved and form traceable associations.

3. Game Theory Analysis for the Development and Regulation of Internet Content Platforms

3.1. Players and Game Theory Models

Internet platforms X, whose number is assumed to be n_x, have two alternative strategies: Compliance operation and non-compliance operation. Compliance operation can obtain normal revenue R, while necessitating the payment of certain compliance costs C. Non-compliance operation does not require the payment of compliance costs but may be detected by the regulatory authorities. If so, platforms would not only lose all revenue but also face a fine F. Regulatory authorities G mainly refer to entities that retain jurisdiction over internet platforms, including the national internet information institutions. Although different regulatory authorities may have different directions, scopes of power, and levels of information, the overall objective remains the same. The establishment of a single regulatory authority will not substantially impact the outcome of games relying on equilibrium. In addition, users who are not players in the game but are stakeholders are included, assuming that the gains to users are U and the number of users is n_u. The regulator also has two strategies: Weak regulation and strong regulation.

When weak regulation is implemented, the regulator does not face regulatory costs, but is also unable to detect violations promptly. This failure generates reputational damage H. When a strong regulation strategy is implemented, the regulator must pay certain regulatory costs C_g. However, if the internet platform has infringements and violations, the regulator can impose fines F. It should be noted that this paper does not distinguish between strong and weak regulatory strategies by the scale of penalties, but rather by whether it is possible to devote sufficient regulatory resources and thus more easily detect violations. In essence, in a mature regulatory framework, the institutional requirements should be sufficiently clear that strong regulation is reflected in an increased probability of detecting violations, rather than in the number of penalties imposed when violations are detected.

3.2. Content Service Payoff Function and Strategy Collection

3.2.1. Internet Platform

Regarding the compliance operation situation, when the regulator implements a weak regulatory strategy, the internet platform can obtain normal revenue R ($R > 0$), but at the same time, it must pay certain compliance costs C ($C > 0$). The content or service payoff function is $X_{ac} = R - C$. When the regulator implements strong regulation, the internet platform can still obtain normal revenue R, but the cost of compliance increases to λ (markup coefficient) times that of the weakly regulated model, so the cost of compliance for the internet platform is λC, where $\lambda > 1$. The content or service payoff function is $X_{ad} = R - \lambda C$.

Secondly, concerning the case of non-compliance operation, when the regulator implements a weak regulatory strategy, the internet platform's compliance costs are zero and the content or service payoff function is $X_{bc} = R$. When the regulator implements a strong regulatory strategy, the internet platform compliance costs are zero, but violations will be detected and lead to a fine F. The content or service payoff function is $X_{bd} = R - F$.

3.2.2. The Regulator

It is assumed that regulator G aims to maximise the overall social gain, which, alongside fine collection, consists of two major components: Users' gains and the profit of internet platforms. When internet platforms operate in compliance with the regulations, users can obtain normal gains U. When internet platforms operate in violation of the law, it will bring certain losses θ to users, and users' gains are $U - \theta$. Serious violations of regulations may lead to $U - \theta < 0$, i.e., users suffer absolute losses.

For the weak regulatory situation, when the internet platform operates in compliance, the content/service payoff function is $G_{ca} = n_x (R - C) + n_u U$. When the internet platform chooses to operate in violation, the interests of users will be damaged, and the regulator will suffer a reputational loss H. The content/service payoff function is: $G_{cb} = n_x R + nu (U - \theta) - H$.

For the strong regulation strategy, when the internet platform chooses to operate in compliance, the regulator has to pay a certain regulatory cost C_g, and the internet platform compliance costs will rise. The content/service payoff function is $G_{da} = n_x (R - \lambda C) + n_u U - C_g$. When the internet platform chooses to operate in violation, it will cause certain losses to users, and the regulator will find violations and impose penalties. The content/service payoff function is $G_{db} = n_x R + n_u (U - \theta) - C_g$. In summary, the internet platform and the regulator game strategies can be defined as shown in Table 1.

Table 1. Internet platform and regulator gaming strategies.

	Strong Regulation Strategy	**Weak Regulation Strategy**
Compliance Operation	$X_{ac} = R - C$ $G_{ca} = n_x (R - C) + n_u U$	$X_{ad} = R - \lambda C$ $G_{da} = n_x (R - \lambda C) + n_u U - C_g$
Non-compliance Operation	$X_{bc} = R$ $G_{cb} = n_x R + n_u (U - \theta) - H$	$X_{bd} = R - F$ $G_{db} = n_x R + n_u (U - \theta) - C_g$

3.2.3. Game Strategy Equilibrium Analysis

Pure Strategy Equilibrium Analysis

For internet platforms, when the regulator chooses weak regulation, non-compliance operation is a dominant strategy for internet platforms because $X_{bc} - X_{ac} = R - (R - C) = C > 0$. When the regulator chooses strong regulation, $X_{bd} - X_{ad} = \lambda C - F$, if $\lambda C - F > 0$, the internet platform chooses to operate in non-compliance; if $\lambda C - F < 0$, the internet platform decides to comply with the regulations. Therefore, under a strong regulatory strategy, the non-compliant operation will dominate if the compliance costs λC that the internet platform needs to pay are higher than the penalty it may be subject to. For the regulator, when the internet platform operates in compliance, $G_{da} - G_{ca} = n_x (1 - \lambda)C - C_g < 0$ there is a dominant strategy of weak regulation. In contrast, when the internet platform operates in breach of the law, $G_{db} - G_{cb} = H - C_g$ and when $H - C_g > 0$, the regulator's rational choice is strong regulation. When $H - C_g < 0$, the regulator's rational choice is weak regulation. Therefore, the equilibrium point—"non-compliance operation, weak regulation"—exists between the internet platform and the regulator only when $\lambda C - F > 0$ (condition 1) and $H - C_g < 0$ (condition 2). If the regulatory authorities punish the internet platform with a too low standard, it may strengthen the incentive for the internet platform to operate non-compliance and weaken the incentive for the regulatory authorities to regulate, while the users and society as a whole lose out. Therefore, this equilibrium point is not conducive to the healthy and sustainable development of internet platforms.

Mixed Strategy Equilibrium Analysis

When $\lambda C - F < 0$ or $H - Cg > 0$, there is no pure strategy Nash equilibrium between the two sides of the game, and a mixed strategy analysis is needed based on the probability of specific behavioural strategies of the regulator and the internet platform. We assume that the probability of an internet platform choosing a compliance strategy is p and the probability of choosing a non-compliance strategy is $1 - p$. The probability of a regulator choosing a weak regulatory strategy is q and the probability of choosing a strong regulatory strategy is $1 - q$. The strategy matrix of the mixed game can be obtained as shown in Table 2.

Table 2. Strategy matrix of the mixed game.

	Weak Regulation Strategy q	**Strong Regulation Strategy** $1 - q$
Compliance Operation	$X_{ac} = R - C$ $G_{ca} = n_x\,(R\text{ -}C) + n_u U$	$X_{ad} = R - \lambda C$ $G_{da} = n_x\,(R - \lambda C) + n_u U - C_g$
Non-compliance Operation	$X_{bc} = R$ $G_{cb} = n_x R + n_u\,(U - \theta) - H$	$X_{bd} = R - F$ $G_{db} = n_x R + n_u\,(U - \theta) - C_g$

By derivation, the mixed strategy equilibrium of the internet platform and the regulator can be derived as $\left(p* = \dfrac{H - C_g}{H + nx\,(\lambda - 1)\,C},\, q* = 1 - \dfrac{C}{F - \lambda C + C}\right)$. In terms of the factors affecting the optimal probability, on the internet platform side, $\dfrac{\partial p}{\partial nx} = \dfrac{(H - C_g)\,(1 - \alpha)}{[H + nx\,(\lambda - 1)C\,]^2} < 0$; the greater the number of market service providers, the smaller the probability of the internet platform operating in compliance. $\dfrac{\partial p}{\partial \lambda} = \dfrac{(C_g - H)nxC}{[H + nx\,(\lambda - 1)C\,]^2} < 0$, meaning the larger markup of compliance costs of internet platforms under a strong regulatory strategy, the lower the probability of internet platforms operating in compliance. $\dfrac{\partial p}{\partial C} = \dfrac{(C_g - H)(\lambda - 1)\,nx}{[H + nx\,(\lambda - 1)\,C\,]^2} < 0$, i.e., the higher the compliance costs of internet platforms, the lower the probability of compliance operation by the internet platforms. $\dfrac{\partial p}{\partial H} = \dfrac{C_g + nx\,(\lambda - 1)C}{[H + nx\,(\lambda - 1)C\,]^2} > 0$, i.e., the greater the loss of the regulator's reputation, the greater the probability that the internet platform will operate in compliance. $\dfrac{\partial p}{\partial C_g} = -\dfrac{1}{H + nx\,(\lambda - 1)\,C} < 0$, i.e., the higher the costs of regulatory implementation by the regulator, the lower the probability that the internet platform will operate in compliance. For regulators, $\dfrac{\partial q}{\partial \lambda} = \dfrac{-C^2}{(F - \lambda C + C)^2} < 0$, i.e., the higher markup coefficient of the compliance costs of the internet platform when the regulator imposes strong regulation, the smaller the optimal probability of the regulator imposing weak regulation. $\dfrac{\partial q}{\partial C} = \dfrac{-F}{(F - \lambda c + c)^2} < 0$, i.e., the higher the compliance costs of the internet platform, the lower the optimal probability of the regulator imposing weak regulation. $\dfrac{\partial q}{\partial F} = \dfrac{C}{(F - \lambda C + C)^2} > 0$, i.e., the greater the value of fines imposed by the regulator on the internet platform for non-compliance, the greater the optimal probability of the regulator imposing weak regulation.

- The budding period

In 2004, Alipay separated from Taobao and became independent, gradually evolving into the largest internet service platform in China. In terms of internet platform development, the number of platforms n_x in the nascent period was relatively small, the products and businesses were not mature, and the overall market size was small. Meanwhile, the regulatory policies were more relaxed, the compliance costs C for internet platforms were relatively small, and the markup coefficient λ was not particularly high. With fewer regulation objectives, the costs C_g for regulators to implement strong regulation would also be lower. The loss of reputation H of the regulator would not have been a significant consideration in the strategy of the internet platform. Based on the conclusions of the previous analysis, $\dfrac{\partial q}{\partial \lambda} < 0$, $\dfrac{\partial q}{\partial C} < 0$, the regulator $q*$ will be larger when both λ and C are relatively small. From the perspective of the regulator, the compliance costs C for internet platforms are also smaller due to the relatively small markup coefficient λ. Moreover, the likelihood of the regulator imposing penalties during the nascent period would be relatively low, even if the penalties mostly manifest in business rectification rather than fines. Thus, the value of fines F would not have a substantial impact on the regulator's choice of strategy. Therefore, we can create the formula, $\dfrac{\partial q}{\partial \lambda} < 0$, $\dfrac{\partial q}{\partial C} < 0$, where the regulator $q*$ will be larger if λ and C are both relatively small.

- Growth period

Since the development of internet platforms, the overall number n_x has grown significantly, the complexity of business has increased, innovations have emerged, and the regulatory costs Cg to the regulators have increased. At the same time, the costs of compliance C for internet platforms have also grown due to the gradual increase in regulatory requirements while the markup coefficient λ has further increased. As the impact of internet platforms on social welfare is generally positive, the potential loss of reputation H suffered by the regulator is relatively small. Since $\frac{\partial p}{\partial n_x} < 0$, $\frac{\partial p}{\partial C} < 0$, $\frac{\partial p}{\partial \lambda} < 0$, $\frac{\partial p}{\partial Cg} < 0$, $\frac{\partial p}{\partial H} > 0$, the optimal probability p^* of an internet platform would decrease. From the perspective of the regulators, λ began to increase and the costs of compliance C for internet platforms continued to grow. Furthermore, as internet platform violations began to increase, the regulators began to impose fines in the form of fines, but the value of F is still relatively small and far from being a deterrent relative to the revenue gained from the violations. Due to $\frac{\partial q}{\partial \lambda} < 0$, $\frac{\partial q}{\partial C} < 0$, $\frac{\partial q}{\partial F} > 0$, the regulator's optimal probability q^* will be lower compared to the budding period.

- Maturity period

Once China's internet platform market enters a mature period of regulated development, new entities will continue to enter the market. Meanwhile, poorly run institutions would exit the market and n_x will remain relatively stable. However, regulatory costs Cg further increase with the scale and complexity of businesses, and the compliance costs C for internet platforms also further increase alongside, a growing markup coefficient λ.

At the same time, however, internet platforms in the maturity period already have a huge social impact, and the public demands on the regulator are so high that the potential loss of reputation H suffered by the regulator will rise rapidly. Since $\frac{\partial p}{\partial C} < 0$, $\frac{\partial p}{\partial \lambda} < 0$, $\frac{\partial p}{\partial Cg} < 0$, $\frac{\partial p}{\partial H} > 0$, the trend of the optimal probability p^* of internet platforms will be uncertain, i.e., the strategic choice of internet platforms may be differentiated. The possibility of some small and medium-sized internet platforms with weak operational capacity and insufficient compliance capability choosing to operate in violation of the law further increase while the possibility of some institutions with larger market shares, stronger operational capacity, and higher compliance capability choosing to operate in compliance will be greater.

From the perspective of the regulator, the costs of compliance C for internet platforms will further increase and the markup coefficient λ will continue to grow; according to $\frac{\partial q}{\partial \lambda} < 0$, $\frac{\partial q}{\partial C} < 0$, the regulator's optimal probability q^* will decrease because of this. As the number of non-compliance internet platforms and businesses starts to increase and the damage caused to the public increases, the value of fines F for the regulator will also rise further, according to $\frac{\partial q}{\partial F} > 0$, and the regulator's optimal probability q^* will rise again as a result of this. Therefore, in the maturity period, the regulator's optimal strategy will be influenced by both positive and negative factors, and the direction of change of its final optimal probability q^* will be fairly uncertain.

- Decline period

In 2021, the "Internet Platform Regulations (Draft for Public Comments)" was officially published, strengthening the regulatory regime for China's internet content platforms. In the future, the infrastructure role of internet platforms will become more prominent, and the decline period of China's internet industry is not forthcoming. However, from a life-cycle perspective, the profitability of the internet platform industry will further decline during the recession, and the number of internet platforms n_x is expected to decline further as mergers, restructuring and the exiting of poorly run corporations become more common. As internet platforms have formed more stable expectations of regulatory policies, the compliance costs C, markup coefficient λ, and regulatory costs Cg of internet platforms will also decline. However, at the same time, the public, having experienced the painful lessons brought about by the brutal growth of internet platforms, will demand more from the regulators, and the loss of reputation H of the regulators will further rise. Due to $\frac{\partial p}{\partial nx} < 0$, $\frac{\partial p}{\partial C} < 0$, $\frac{\partial p}{\partial \lambda} < 0$, $\frac{\partial p}{\partial Cg} < 0$, $\frac{\partial p}{\partial H} > 0$, the optimal probability p^* of an internet platform will keep rising. From the regulator's point of view, although the number of institutions choosing to operate in compliance is rising, the social impact of non-compliance will be so great that heavy fines will be necessary to maintain deterrence as the remaining internet platforms are large in terms of both numbers of users and market share. As a result, the number of fines imposed by the regulator, F, will rise further and, together with the decrease in compliance costs, C, and the markup coefficient, λ, according to $\frac{\partial q}{\partial \lambda} < 0$, $\frac{\partial q}{\partial C} < 0$, $\frac{\partial q}{\partial F} > 0$, the regulator's optimal probability q^* will increase.

Finally, to aid reader understanding, the relationship between the life cycle of platforms and regulatory strategies is depicted in Figure 1.

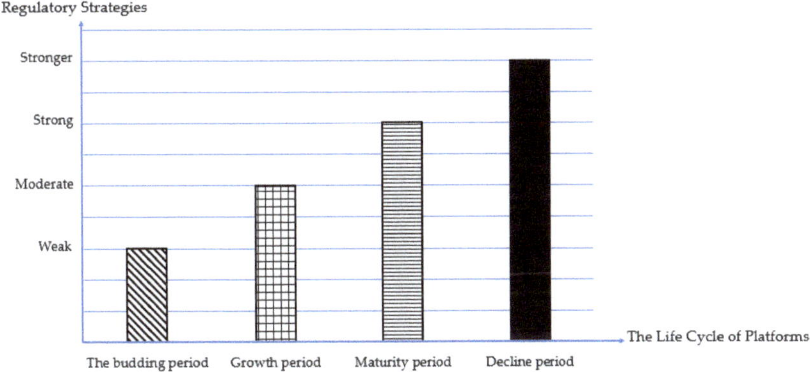

Figure 1. The relationship between the life cycle of platforms and regulatory strategies.

4. Risks Associated with Platform Ecosystem-Based Monopoly

4.1. Risk of Market Concentration Caused by Ecosystem Expansion

During development, the platform ecosystem absorbs goods and services from other fields through acquisitions and mergers, thus expanding the ecosystem. This process is not only the process of concentration of operators but also the process of data and technology aggregation. Since the platform ecosystem not only gains control over other operators but also aggregates a large amount of user data and advanced technology, it achieves a large aggregation of production factors, leading to the following two risks.

Firstly, the market becomes overly saturated with a centralized and structured pattern. "Ecological monopoly" is a process of transferring monopoly power to other fields through data and algorithms. Once formed, an ecological monopoly will threaten the monopoly in multiple markets. Moreover, these markets become controlled by a dominant platform. Compared with traditional monopolies, "ecological monopolies" have a greater impact on the market structure as well as a broader scope of influence.

Secondly, the "ecological monopoly" brings all factors under the dominant platform, creating excessive market concentration and high barriers to entry [15]. Although new enterprises' hardware and software costs are not high in the platform economy, the dominant platform has a great deal of data, algorithms, and capital, enabling new companies to enter the market effectively. Especially in the case of a lack of initial users, new enterprises cannot obtain enough data and capital support so they can easily be eliminated from the market. As it is difficult for new platform companies to obtain data and algorithms to compete with the platform ecosystem, new entrants fail to enter the market effectively and eventually close. The market structure gradually solidifies after going through centralization, which negatively impacts the order of market competition. The lack of effective competition constraints causes the market to become less competitive. The ecosystem centred around the head platform may lose the incentive to further improve the quality of goods or lower prices, which will lead to the market losing its "competitive dividend".

4.2. Anti-Competitive Risks Arising from the Dual Identity of Platforms

In the platform ecosystem, the dominant platform often has a dual identity: A "manager" and an "operator". The dual identity can enable the dominant platform to better control the operation of the ecosystem, but at the same time, it may also damage the openness of the ecosystem. As an "operator", the platform's fundamental goal is to maximize profits, so it may use its power as a "manager" to improperly interfere with competitors.

The dominant platform's status as an "operator" makes it difficult to remain neutral. As the dominant platform crosses borders into other areas with its affiliated businesses, it may restrict its competitors by engaging in self-interested, anti-competitive behaviour. Since the platform ecosystem has many users and massive data resources, it may block or exclude competitors' access to similar data to hinder development [16]. Moreover, the dominant platform's status as a "manager" makes it a rule-setter in the ecosystem. Other operators within the platform ecosystem are subject to the rules they set, and these regulations are often based on the interests of the dominant platform, possibly at the expense of other operators.

4.3. The Aggregation of Production Factors and the Lack of Competitive Constraints Trigger the Risk of Inhibiting Innovation

The formation of the platform ecological monopoly will likely inhibit the innovation of operators inside and outside the ecosystem. Firstly, the platform ecological monopoly will inhibit innovation outside the ecosystem. Under the platform ecological monopoly, it is difficult for other competitors to carry out effective innovation due to two external reasons, market structure and interference in competitive behaviour. Data, capital, and other factors will continue to gather in the platform ecosystem. Thus, other operators will be restricted from obtaining the necessary resources for innovation. They also have difficulties obtaining sufficient data and capital for innovative R&D. Moreover, the platform ecosystem performs acquisitions and mergers in an expansion process, eliminating the threat posed by innovative platforms. Ioannis Lianos points out that current "killer acquisitions" are acquisitions of startups by platforms to acquire more advanced algorithms and different kinds of data, which kill potential "disruptive innovation" in the cradle [17].

Secondly, platform ecosystem monopolies can also inhibit innovation within the ecosystem. When a platform-ecological monopoly is formed, the platform ecosystem can make large profits without maintaining a competitive advantage through innovation due to the market's lack of effective competitive constraints. At the same time, other cooperative enterprises in the ecosystem are often restricted in their innovation activities because they are controlled by the dominant platform as the "manager". Once the dominant platform finds the innovation activities of a cooperative company, it may threaten the platform ecosystem, and the platform will take action to punish the enterprise. Even if an enterprise successfully conducts innovation research and development, its achievements will likely be annexed or stolen by the dominant platform.

4.4. Risk of Algorithmic Discrimination Due to Misuse of Algorithmic Technology

The problem of algorithmic discrimination arising from the misuse of algorithmic technology has been a cause for concern. When an "ecological monopoly" is formed, the harm caused by algorithmic discrimination will be more serious. Discriminatory algorithmic technologies may be widely applied to various ecosystem components, leading to multi-dimensional, multi-group, and multi-scenario algorithmic discrimination, which, if not taken seriously, may cause irreversible damage to consumer welfare.

Firstly, under the ecological monopoly, the platform's discriminatory algorithms will increase the "exploitation" of consumers and users. Platforms have accumulated massive amounts of data and high-quality algorithms, which can grasp the actual needs of consumers, even down to the needs of individuals for specific products. In order to maximize profits, platforms may use price discrimination and other means to develop personalized pricing strategies based on individual consumer characteristics, leading to a reduction in consumer surplus. Once the "ecological monopoly" is formed, consumers and users will be affected by the lock-in effect of switching costs and user stickiness in a feedback loop with increasing effects. Even if users are dissatisfied with the "exploitation" of discriminatory algorithms, they are often forced to continue using the original platform due to the lock-in effect or the fact that there are no choices of equivalent quality in the market.

Secondly, algorithmic discrimination harms other operators. Dominant platforms can impose discriminatory restrictions on platforms that affect their interests. This discrimination may affect both operators and outside of the ecosystem. For operators inside the platform, when the dominant platform intends to develop services in a new area, it may, through its algorithm, impose certain restrictions on other operators inside the platform system involved in that area. For example, when the algorithm allocates resources, it may locate the target of discrimination and impose restrictions on the allocation of resources to the target or raise the threshold. In the case of an "ecological monopoly", the platform ecosystem is an important channel for insiders to gain access to trading opportunities. It means internal operators often accept discrimination in silence. For operators outside the platform, the dominant platform will set certain conditions or operational mechanisms for the algorithm based on its interests, blocking or shielding companies that may pose a competitive threat to its platform ecosystem.

5. Strategies for Regulating Platform Ecological Monopolies

Driven by the two wheels of "data plus algorithm", the market power of the platform ecosystem will continue to be amplified by the market mechanism and network effect. It may eventually turn the platform ecosystem into a monopoly in a single market and transfer it to adjacent markets to form second or even multiple rounds of an ecological monopoly. Once the platform ecosystem monopoly is formed, the risk to the market competition and entities within the market is multifaceted and may produce irreversible and substantial damage. Therefore, based on the systemic concept of grasping the platform ecosystem's operation rules and analysing its ecological monopoly's formation mechanism, it is necessary to update the current competition law governance system; that is, to carry out multi-dimensional governance of the data and algorithm elements and their two-wheel-drive model to avoid the platform ecosystem from tending to monopolistic and disorderly expansion. At the same time, multiple subjects in the ecosystem, including consumers and operators, should actively cooperate to achieve multi-governance and maintain the healthy and orderly development of the platform ecosystem.

5.1. Improve the Legal Identification and Recognition Method of Platform Ecological Monopoly

The platform ecosystem is based on "data + algorithms". It achieves the linkage operation and development of various non-competitive goods and services, which subverts the approach to identifying and determining the relevant markets in traditional competition law. According to the current Anti-monopoly Law, the relevant market is the market consisting of goods or services with substitutability. When analysing whether the ecosystem has a dominant position, it is only possible to define the market power in the area where the dominant platform or the associated business of the ecosystem is located, which tends to underestimate the power of the overall platform. Therefore, the existing competition law identification method needs to be optimized based on the platform ecosystem's characteristics and the development trend.

Currently, China is improving its scientific and reasonable regulatory system that responds to the new development of the platform economy. On 7 February 2021, the Anti-monopoly Committee of the State Council issued the "Platform Economy Antimonopoly Guidelines" (Guidelines). Starting from the characteristics and operation rules of the platform economy, the Guidelines further build on Article 11 of the Interim Provisions on the Prohibition of Abuse of Market Dominance issued by the State Administration of Market Supervision and Administration on 26 June 2019. To determine the dominant market position of operators in new economic sectors such as the internet stipulated in Article 11, Guidelines focus more on the impact of the dynamic process of data behaviour on assessing the market power of platforms. For example, subsections 1 and 2 of Article 11 of the Guidelines "Determination of market dominance" provide that "the number of active users", "clicks", "using hours", and "ability to control the market" can be considered to determine whether a platform has a dominant market position. However, the existing legal provisions are still weak in operability and lack clarity. It not only prevents the regulatory and enforcement authorities from effectively regulating the platform ecological monopoly but also may "misplace" the normal functioning of the platform ecosystem and frustrate its development momentum. Therefore, the current ecosystem formed by aggregating multiple non-competitive goods and services on the platform can be considered a cluster market and uses tools that directly assess market powers. It has been noted that platform ecosystems form markets not because of high cross-elasticity of demand among various goods and services but because of significant customer convenience and preferences in aggregation or joint provisioning economic models and because aggregation is difficult to replicate [18]. Hence, simply defining the relevant market and assessing market share does not quantify the market power that platform ecosystems have. Taking differentiated goods into account would expand the relevant market scope and underestimate the platform ecosystem's market power. Conversely, not taking these commodities into account may overestimate the market power of the platform. Therefore, a tool that more directly assesses market power could be used for this assessment.

More intuitive assessment criteria can be used to directly assess market power. The specific assessment criteria can be combined with identifying gatekeepers in the EU Digital Markets Act [19], i.e., when analysing and assessing the market power, the multiple factors of the areas covered and dominated by the dominant platform are assessed. Especially based on the operation rule of the platform ecosystem, the dominant platform's ability to control the market can be assessed by accounting for the components of its platform ecosystem and their relationship. Among others, user stickiness can also be measured by analysing the number of multi-attributed users of the ecosystem and the length of their usage [20]. For example, Taobao's control over the upstream and downstream markets was considered in the administrative penalty decision of the State Administration of Market Supervision and Administration against Alibaba for "two-for-one". It is in line with the vertical integration of Alibaba's ecosystem and the two-wheeled operation of "data + algorithm". In summary, the optimization of the market dominance identification criteria should be based on the operation mode and the operation rules of the ecosystem. It will further refine and supplement the existing competition law and enable the competition law to effectively identify the market power of the platform ecosystem and intervene through a legal approach before the "platform ecosystem monopoly" takes shape.

In practice, there is an abundance of behaviours that use algorithmic technology to achieve improper purposes. As stated in Article 28 of the Data Security Law, data processing activities and research and development of new data technologies should promote economic and social development, enhance people's welfare, and conform to social morality and ethics. The development and application of algorithm technology in the platform ecosystem should also be based on the optimal allocation of the ecosystem and the promotion of consumer welfare, rather than excluding competitors or exploiting consumer surplus through algorithms. In this regard, certain regulatory measures are needed to ensure that algorithmic technologies in the ecosystem are on the right path to promote economic and social development and enhance people's welfare.

The biggest obstacle to regulating algorithms is the secrecy of the algorithm itself, the "algorithmic black box". Meanwhile, the algorithm's function may change according to specific scenarios, thus making it more difficult to regulate algorithms. Therefore, we can start from the following two aspects.

5.1.1. Establish the Right to Explanation

The metaphor of the "algorithmic black box" expresses the concern that humans risk losing control over the decision-making process due to the opacity of algorithms. Humans devolving decisions that affect their rights and obligations to a black box that they cannot understand implies that algorithmic transparency is difficult, if not impossible [21] (p. 165). However, algorithmic transparency does not require platform companies to disclose algorithmic code to be "transparent". The algorithm itself is highly technical and professional. As the algorithm relates to the company's core competitiveness, requiring full disclosure of the underlying algorithm code or mechanism would be difficult for consumers and regulators to understand. It may also result in competitors' algorithm parsing, causing the company to lack competitiveness. The implementation of algorithm transparency should be realized through the right to explanation. 'The right to explanation' was clarified in the EU General Data Protection Regulation, which came into force on 25 May 2018, and became an actualized right. This right has been widely recognized within the EU and implemented in each member state through policy standards. Because of the inscrutability and non-intuitive nature of algorithms [21] (p. 169), the platform needs to articulate the basic rules and criteria by which algorithms make decisions. For example, when the takeaway platform is asked to explain a product's price differentiation, the platform can publish the factors that lead to the algorithm's pricing differentiation. For example, the price of the takeaway product is determined by the delivery distance and delivery time without providing the actual algorithm's calculation process. Based on the explanation of the algorithm, the regulator should also verify the consistency of the algorithm explanation with the actual results. Because the possibility of platforms providing false explanations cannot be ruled out, at this time, the algorithm sandbox technology can be used to conduct the actual algorithm by controlling the variables and comparing whether there are deviations between the theoretical results and the actual results. At the same time, it is also possible to improve the situation of individual consumers and operators against the abuse of ecosystem algorithms. This occurs through rights allocation, establishing the right to explanation, and the right to be forgotten while giving data subjects the option to oppose automated decision-making and establishing and improving algorithms' attribution and punishment mechanism.

5.1.2. Establishing the Scenario-Based Regulation of Algorithms

Even for the same algorithm, very different results may arise due to the differences in input variables and data in different scenarios. Therefore, it is important to establish the scenario-based regulation of algorithms. It is necessary to classify and grade the algorithm technology and the specific scenarios of algorithm technology application. For example, when the algorithm is used to optimize resource allocation within the platform ecosystem, the risk of algorithm abuse is small if it only involves allocating resources within the platform and does not involve other operators not under the platform. However, involving both operators within and not the platform is a high-risk scenario and requires focused regulation. In terms of specific regulatory schemes, the 2020 EU White Paper on AI proposes that a five-level risk-based regulatory system should be established in five dimensions: Application scenarios, deployment purposes, security protection, consumer interests, and fundamental rights; thus, clearly distinguishing the assessment criteria for various AI applications and implementing differentiated regulation [22] (p. 105). Platform algorithms cannot be regulated generally but should distinguish specific scenarios to avoid the rhetoric of filtering undesirable information to exclude competition, especially the situation of algorithms in key scenarios, such as shopping and information filtering scenarios of social platforms. Adopting the ideas of scenario governance and precise governance, the evaluation index system based on the impact dimension should be established based on the impact subject, the scope and the degree of impact. In setting the impact dimension indicators, every country's legislation also embedded the core social values to be maintained by the algorithmic governance practice. Around the core values of transparency, legitimacy, fairness, and justice, an assessment indicator system based on the algorithmic accountability dimension should be established. In addition to drawing on the above system, the regulatory thinking for algorithms can be further optimized. In the platform ecosystem, algorithms are constantly and dynamically changing due to the real-time and large amount of data "feeding" them. Therefore, there is an urgent need to develop and experiment with new ideas, set reasonable regulatory thresholds, and find a good balance between technological innovation and risk control. We should set up an intelligent supervision system to regulate, from ex-ante to ex-post, in a full cycle. At the same time, we should strengthen the accountability mechanism of algorithms, clarify the blacklist of algorithms, and require enterprises to avoid discrimination when setting up algorithms and set up certain redress and punishment mechanisms. In addition, in the regulatory strategy and technology, human supervision finds it difficult to cope with the high dynamics of algorithms. Thus, algorithms can be supervised by algorithms through algorithmic technology to collect the decision-making of the platform ecosystem in high-risk scenarios to identify whether there are violations.

5.2. Build a Multifaceted Synergy between the Inner and Outer Circles of the Platform Ecosystem

5.2.1. Dominant Platform

The dominant platform in the platform ecosystem has both corporate and market duality. When it manages platform affairs as a manager, it may harm the interests of users on the platform based on its interests. Therefore, platform autonomy is crucial, and only when the platform is autonomous can it guarantee competition compliance. Since the dominant platform controls the data and algorithms of the platform ecosystem, to avoid data security risks, the dominant platform needs to strengthen the compliance review of relevant laws, including the Data Security Law, and build a classification and grading management system for data in the platform ecosystem. It also needs to strengthen the technical protection measures in safe data transmission and flow, regularly conduct a risk assessment, and establish a data security emergency plan. For ecosystem algorithm technology, the dominant platform should regularly examine algorithms within the platform and assess the risk points of the non-compliance of algorithms in a scenario-specific manner to prevent the algorithm's elements in the self-learning process from violating relevant laws. At the same time, the dominant platform, as the rule setter of the platform ecosystem, should ensure rules do not distort market competition within the platform. In the absence of reasonable objective efficiency reasons, it cannot use its rule-making power to influence and distort the competition among operators within the platform and exploit the legitimate interests of operators or third-party partners [23] (p. 46).

5.2.2. Operators

Operators may be subject to undue interference by the dominant platform or suffer "data hegemony" detrimental to their interests. In this regard, there is an urgent need to further clarify the responsibilities and obligations of the "managers" of the ecosystem so that the cooperative operators in the platform ecosystem can have a justifiable basis and rules to appeal to the relevant authorities when they are unduly restricted.

5.2.3. Consumers

For consumers, educational activities such as consumer data security should be actively organized to raise consumers' awareness of protecting themselves and their rights. Since consumers' strengths are weak, it is necessary to streamline the consumer complaint and reporting mechanism. Furthermore, it is necessary to improve the consumer public interest litigation system and provide certain protections for those who report and complain, concerning the provisions of Article 12 of the Data Security Law, so that consumers can mitigate concerns, and form effective supervision and deterrence against illegal acts on the platform ecosystem. The external circle of the platform ecosystem contains governance subjects such as enforcement departments, judicial departments, and social organisations. For states, the focus of their regulation is to prevent the platform ecosystem from threatening market competitors, operators, consumers, and other subjects. However, the current enforcement and justice of anti-monopoly in the platform economy face problems of high professionalism and large workload, and there are unbalanced and insufficient problems in the allocation of staff and enforcement resources, so there are considerable difficulties in the process of enforcement and justice.

For this reason, it is necessary to further strengthen the training of relevant law enforcement and judicial staff and hire more personnel with professional technical knowledge to provide technical guidance for enforcement and judicial work. In addition, relevant organizations are also the external governance subjects of the platform ecosystem. As stipulated in Article 10 of the Data Security Law, relevant industry organizations, following their charters, formulate data security codes of conduct and group standards under the law and strengthen industry self-regulation. Therefore, there is an urgent need to enhance social organizations' participation in the platform's governance. Without a clear legal basis, social organizations such as industry associations can be encouraged to play a positive self-regulatory management function. They can maintain a fair, competitive order in the market by formulating industry norms, especially for multiple links, such as collecting and using data in the platform ecosystem. They can also effectively protect the legitimate rights and interests of consumers by formulating a code of conduct for the application of algorithm technology.

The emergence of the platform ecosystem is not an accident but a result of the choice of platform economy and market operation mechanism. The original platform development models based on capital gathering and initial traffic attraction are unable to break through the development bottleneck. However, the platform ecosystem based on the two-wheel drive of "data + algorithm" can realize the collection of diversified data, the diversified needs of consumers and users, and the deep excavation of consumer data and attention. It can also build a sustainable platform ecosystem around consumers' needs and realize the economy of scope to create a sustainable platform ecosystem around consumer needs and achieve economies of scope. However, with the feedback loop formed by the network effect, user lock-in effect, and transmission effect, the platform ecosystem can obtain competitive advantages through data and algorithm advantages. The relative advantages of its dual identity of "manager" and "producer" enable it to gain market dominance and economies of scope. It can gain dominant power over the market and transmit it to multiple markets, eventually forming an "ecological monopoly". To effectively regulate the ecological monopoly of the platform, a two-pronged approach based on the laws governing the operation of the platform ecosystem as a whole must be taken concerning the data and algorithms that drive the development of the platform ecosystem and that catalyse its evolution towards an "ecological monopoly". Existing competition laws should also be updated to strengthen the identification of dominant forces in the ecosystem. In addition, based on the shape of the platform ecosystem operation, it is possible to achieve all-around regulation of the platform ecosystem both inside and outside the dual circle of the platform ecosystem and the multi-subject governance model. However, in strengthening the prevention and regulation of platform ecosystem monopoly, it is necessary to grasp the intensity of intervention and avoid overkill because, combined with the operation mechanism of the platform ecosystem, the platform ecosystem, driven by data and algorithms, aggregates diverse goods and services to form an ecosystem. It not only improves the operational efficiency of the system and each component but also provides many trading opportunities for other operators, satisfies the diverse needs of consumers, and enhances their service experience.

Given this, in the regulation process, it is necessary to correct the "big is bad" mindset and correctly understand the consistency and synergy between the healthy development of the platform ecosystem and the promotion of the platform economy. The purpose of effective regulation and scientific law enforcement is not to restrict the innovative development of the platform ecosystem but to prevent it from falling into the ecological monopoly, the "development trap". After falling into this trap, the platform's innovation momentum will be suppressed. The seemingly stable growth of platform benefits over time is at the expense of the long-term growth of innovation benefits for the industry. It will eventually force the dominant platform companies to reduce their innovation benefits as they become unsustainable. However, to guarantee the effective functioning of the platform ecosystem's positive relevance, enforcement departments should be cautious of using structural means, such as splitting the company to affect the structural advantages of the platform ecosystem. Moreover, they should not overly raise the threshold of user data collection without emphasizing data security in any scenario. This is because it may inhibit the optimization of the platform ecosystem based on the effective supply of data, which will objectively lead to the continued healthy operation of the platform ecosystem. Article 7 of the Data Security Law clearly states that the state protects the rights and interests of individuals and organizations related to data. Additionally, it encourages the reasonable and effective use of data following the law, guarantees the orderly and free flow of data under the law, and promotes the development of the digital economy with data as a key element. Finally, it should be pointed out that "scientifically regulating the realistic risks and potential hazards of platform ecological monopoly, guaranteeing fair competition, and safeguarding the legitimate rights and interests of operators and consumers" does not contradict "optimizing the operation of the platform ecosystem". Promoting the legal use of data and algorithms can improve market competition efficiency and stimulate the high level of innovation and development of platforms. It is ultimately conducive to the long-term welfare of consumers and is a win–win situation for multiple subjects in the platform economy.

6. Main Conclusions and Insights

6.1. Main Conclusions

During the budding period, as the regulators, the public, and others are not too familiar with the products and services of internet platforms, the regulatory policy requirements are relatively relaxed, and the probability of internet platforms operating in compliance and weak regulation by the regulators is higher. As the payment market continues to develop, more and more internet platforms participate in the market competition, the profit level begins to decline, the rate of internet platforms that choose to operate in violation of the law begins to rise, and the number, frequency, and strength of the regulators begin to rise. In the maturity period, the operating ability of small and medium-sized internet platforms is constantly tested. The phenomenon of mergers and acquisitions increases, the overall number of internet platforms tends to stabilize, and the new platform content market enters the stock competition. The strategies of different internet platforms will be differentiated, and the optimal strategy level of the regulators will be affected by both positive and negative aspects. The change in their strategy choice tends to slow down. Entering the recession period, after experiencing the big wave and the elimination of the best, most of the remaining internet platforms will be large enterprises with strength and scale. The probability of choosing to operate in compliance will rise greatly. The probability of regulators choosing weak regulation will also increase. However, it should be noted that the "rules" and "discipline" in "compliance" will be improved with the development of the market, and the standard of weak supervision by the regulatory department will be constantly improved with the improvement of the regulatory system requirements.

6.2. Main Inspiration

Firstly, we should take a rational view of the development of the internet platform and the existence of risks. As a new type of content service provider, the birth and development of the internet platform met the actual needs of social and economic development in the era of e-commerce, which is inevitable and of great significance to meet people's needs for a better life. However, the risks brought about in the process of its development are inevitable and cannot be ignored. The internet platforms and the regulatory authorities should maintain an objective and rational understanding of this. Secondly, the regulatory system should adapt to the industry development life cycle to avoid regulatory mismatch. The social benefits and risks associated with an industry vary greatly depending on its life cycle. Thus, accurately grasping the specific life cycle of the industry's development stage and taking the regulatory policy adapted to it can maximize the overall social and economic benefits based on the healthy development of the internet platform and meet the objective needs. Thirdly, forward-looking theoretical research will reserve the regulatory toolbox for new content service providers such as digital industry operators. With the continuous advancement of data and algorithm technology, new content service providers will continue to emerge, and a new industry will develop. Therefore, relevant experts and scholars should carry out forward-looking research, clarify the rights and responsibilities of all relevant stakeholders, and provide institutional safeguards for the healthy and sustainable development of the industry.

Author Contributions: Conceptualization, C.X.; methodology, C.X.; software, C.X. and Y.-M.W.; validation, C.X.; formal analysis, C.X.; investigation, C.X.; resources, C.X.; data curation, C.X.; writing—original draft preparation, C.X.; writing—review and editing, C.X. and Y.-M.W.; supervision, C.X.; project administration, C.X. All authors have read and agreed to the published version of the manuscript.

Funding: This research is supported by "Strategy on China's participation in the global governance of intellectual property rights" (21&ZD165), the Major Projects of the National Social Science Fund in China.

Data Availability Statement: Not applicable.

Acknowledgments: The authors are grateful to the editor and anonymous referees for their constructive comments and suggestions, which sufficiently help the authors to improve the presentation of this manuscript.

Conflicts of Interest: The authors declare no conflict of interest.

References

1. Van Dijck, J.; Poell, T.; De Waal, M. *The Platform Society: Public Values in a Connective World*; Oxford University Press: Oxford, UK, 2018; p. 9.
2. Kittler, F.A.; Griffin, M. The city is a medium. *New Lit. Hist.* **1996**, *27*, 717–729.
3. Sun, P. How to Understand the Materiality of Algorithm—An exploration of platform economy and digital. *Sci. Soc.* **2019**, *3*, 50–66. [CrossRef]
4. Liu, Z.M.; Zou, W. Digital entrepreneurship ecosystem: Theoretical framework and policy thinking. *Soc. Sci. Guangdong* **2020**, *4*, 5–14.
5. Ren, T.H.; Cao, X.J. From Technology to Architecture: Driving Effects of Network Media Evolution on Social Platformization. *J. Xi'an Jiaotong Univ. (Soc. Sci. Ed.)* **2020**, *5*, 144–152. [CrossRef]
6. Foucault, M. From Torture to Cellblock. In *Foucault Live: Collected Interviews (1961–1984)*; Lotringer, S., Ed.; Hochroth, L.; Johnston, J., Translators; Semiotext(e): New York, NY, USA, 1996; pp. 146–149.
7. Shan, L. Making the Ark of the Sea: The Production of Memorial Spaces for Jewish Refugees in Shanghai since the 1990s—A Perspective Based on a Network of Actors. *Chin. J. J. Commun.* **2019**, *9*, 105–126. [CrossRef]
8. Chen, Y.S.; Wu, S.T. An exploration of actor-network theory and social affordance for the development of a tourist attraction: A case study of a Jimmy-related theme park, Taiwan. *Tour. Manag.* **2021**, *82*, 104206. [CrossRef]
9. Latour, B. *Reassembling the Social: An Introduction to Actor Network Theory*; Oxford University Press: Oxford, UK, 2007; pp. 120–162.
10. Latour, B. Factures/fractures: From the concept of network to the concept of attachment. *Res. Anthropol. Aesthet.* **1999**, *36*, 20–31. [CrossRef]
11. Latour, B. On Actor—Network Theory: A Few Clarifications Plus More Than a Few Complications. *Philos. Lit. J. Logos* **1996**, *25*, 47–64.
12. Borvil, A.D.; Natalie Kishchuk, N.; Potvin, L. Typology of actors' influence strategies in intersectoral governance process in Montreal, Canada. *Health Promot. Int.* **2022**, *37*, 4. [CrossRef] [PubMed]
13. Callon, M. Some Elements of a Sociology of Translation: Domestication of the Scallops and the Fishermen of St Brieuc Bay. *Sociol. Rev.* **1984**, *32*, 196–233. [CrossRef]
14. Latour, B. *Science in Action: How to Follow Scientists and Engineers in Society*; Liu, W.X.; Zheng, K., Translators; Oriental Press: Beijing, China, 2005; pp. 8–10.
15. Evans, D.S. The antitrust economics of multi-sided platform markets. *Yale J. Regul.* **2003**, *20*, 325–382.

16. Grunes, A.P.; Stucke, M.E. No mistake about it: The important role of antitrust in the era of big data. *The Antitrust Source* **2015**, *14*, 1–14.
17. Lianos, I. The Future of Competition Policy in Europe–Some Reflections on the Interaction between Industrial Policy and Competition Law. *CLES Policy Pap. Ser.* **2019**, *1*, 5–26. [CrossRef]
18. Hovenkamp, H. Digital Cluster Markets. *Columbia Bus. Law Rev.* **2022**, *2022*, 1–34. [CrossRef]
19. Zhang, X.B. On Imposing Obligations on the "Gatekeepers" of the Ecology of the Internet in the Protection of Personal Information. *J. Comp. Law* **2021**, *3*, 11–24.
20. The Website of the State Administration of Market Supervision. Available online: http://www.samr.gov.cn/xw/zj/202104/P020 210410285606356273.docx (accessed on 16 July 2022).
21. Wang, Q.H. The Multiple Dimensions of Algorithmic Transparency and Algorithmic Accountability. *J. Comp. Law* **2020**, *6*, 163–173.
22. Zhang, X. The Construction Mechanism of the Algorithmic Impact Assessment System and the Chinese Solution. *Stud. Law Bus.* **2021**, *2*, 102–115. [CrossRef]
23. Tang, Y.J. Research on Economic Nature and Regulatory Policy System of Digital Platform. *Econ. Rev. J.* **2021**, *4*, 43–51. [CrossRef]

systems

Article

An Integrated Entropy-COPRAS Framework for Ningbo-Zhoushan Port Logistics Development from the Perspective of Dual Circulation

Shouzhen Zeng [1,2,*,†] , Zitong Fang [1,†], Yuhang He [1,†] and Lina Huang [1]

1 School of Business, Ningbo University, Ningbo 315211, China
2 School of Statistics and Mathematics, Zhejiang Gongshang University, Hangzhou 310018, China
* Correspondence: zengshouzhen@nbu.edu.cn
† These authors contributed equally to this work.

Abstract: To promote the construction of new development patterns of dual circulation and to accelerate the smooth flow of logistics channels, port logistics has become a new growth point for the logistics industry that accelerates the connection between domestic and foreign dual circulation. Ningbo-Zhoushan Port, as one of the main hub ports in China, is facing the key issue of how to clarify its current development status and future development direction. To scientifically measure and evaluate the status quo of the logistics development of Ningbo-Zhoushan Port, clarify the advantages and disadvantages of the development and construction of the Ningbo-Zhoushan Port logistics industry, based on the situation of new standards and new requirements for the logistics industry in the dual circulation pattern, this study firstly constructs a scientific and reasonable evaluation index system of port logistics from seven aspects, including port infrastructure, international logistics capacity, and smart logistics capacity. An integrated comprehensive evaluation method based on entropy and Complex Proportional Assessment (COPRAS) is then proposed, and a comprehensive evaluation and longitudinal comparative analysis of the logistics level of Ningbo-Zhoushan Port are carried out. The results show that the development of Ningbo-Zhoushan Port in recent years is in line with that of many other ports due to the benefit of green logistics capacity, but it is seriously limited by smart logistics capabilities, and in the future, it should choose to continue to exert efforts in international logistics capabilities, green logistics capabilities, and total logistics capabilities.

Keywords: dual circulation; port logistics; Ningbo-Zhoushan Port; entropy weight; COPRAS

Citation: Zeng, S.; Fang, Z.; He, Y.; Huang, L. An Integrated Entropy-COPRAS Framework for Ningbo-Zhoushan Port Logistics Development from the Perspective of Dual Circulation. *Systems* **2022**, *10*, 131. https://doi.org/10.3390/systems10050131

Academic Editors: Zaoli Yang, Yuchen Li and Ibrahim Kucukkoc

Received: 20 July 2022
Accepted: 22 August 2022
Published: 25 August 2022

Publisher's Note: MDPI stays neutral with regard to jurisdictional claims in published maps and institutional affiliations.

1. Research Background

Faced with the dilemma of the surge of international unfavorable factors and weak domestic consumer demand, the meeting of the Standing Committee of the Political Bureau of the Central Committee on 14 May 2020, proposed to deepen the supply-side structural reform, give full play to China's super-large-scale market advantages and domestic demand potential, and build a development pattern of domestic and international "dual circulation" mutual promotion. As a basic, strategic, leading industry, the logistics industry is an important support for the operation of the national economy. Among them, port logistics is the core hub for allocating urban resources, which provides an important driving force for urban economic development and is also an important part of national economic development. In this process, some coastal ports that can lead the domestic big circulation and connect the domestic and international "dual circulation" logistics hub nodes shoulder great responsibilities [1]. Although China's port logistics construction has gradually matured, the existing domestic research on the evaluation index of port logistics still stays on the basic index measurement, lacking a more advanced index system, and the evaluation results and development suggestions obtained are no longer applicable to today's "double-cycle" development trend.

Ningbo-Zhoushan Port is a rare deep-water port in the world [2]. As a key part of the Shanghai International Shipping Center, Ningbo-Zhoushan Port is not only an important transit base for iron ore and crude oil and a storage and transportation base for liquid chemicals, coal, and grain in China, but also a core carrier for serving the Yangtze River Economic Belt, building Zhoushan River-Sea Intermodal Service Center and the development of Zhejiang's marine economy. In the development and construction of the integration of "the belt and road initiative" and the Yangtze River Delta, it is considered as the "hardcore" backbone. Promoting the development and construction of Ningbo-Zhoushan Port is conducive to promoting the development of port-adjacent industries, promoting economic development, attracting capital inflows, and enhancing the competitiveness of the city. However, at present, the port logistics development of Ningbo-Zhoushan Port is not perfect, the service level of port logistics is low, the service cost is high, the collection and transportation system and logistics information service system are not ideal, and the innovation ability is relatively backward. Under the "double-circulation" pattern, all ports formulate corresponding strategic policies and actively promote the innovative development of port logistics. However, Ningbo-Zhoushan Port lacks a systematic port logistics development evaluation system, which leads to the failure of timely feedback on the port's real situation, the difficulty of effective supervision and evaluation of rules and regulations, and the blindness of its development.

To solve the aforementioned problems, this study re-selects the evaluation indicators that are in line with the progress of the times, scientifically measures and evaluates the lo-gistics development status of Ningbo-Zhoushan Port from seven important aspects, and on the basis of which, an integrated Entropy-COPRAS framework is proposed to promote the efficient and green development of the port, and improve the formation of "dual circulation" pattern.

2. Literature Review

As the core hub for allocating urban resources, the port is an important node in the transportation network. With the integration of the world economy, its functions are constantly improved and its connotation is constantly broadened, and it is playing an increasingly important role in modern logistics. As a new term, "port logistics" has only been put forward frequently in recent years, and it has been paid more and more attention to the deepening of people's research on modern logistics. From the beginning of the 21st century, domestic and foreign scholars began to propose the quantitative analysis of port logistics, and now the theoretical research and practice of port logistics have gradually deepened and matured. Scholars have made diversified theoretical research on port logistics, which has not only achieved fruitful results in the supply chain [1] and green port efficiency [3], but also actively carried out research on port logistics analysis and forecast [4], regional port logistics [5], and port logistics optimization [6]. At the same time, the measurement of port logistics efficiency [7,8], the evaluation of the coordination degree and coordinated development of port logistics and the hinterland economy [9–11], the evaluation of green logistics development level in port [12], are gaining more and more attention.

During the research of port logistics, Feng et al. [13] took Shandong coastal ports (Qingdao, Rizhao, Yantai, Weihai, Weifang, and Jining ports) as research objects, established an evaluation index system and evaluation model of logistics transportation efficiency on the basis of the analytic hierarchy process (AHP), evaluated their logistics transportation efficiency, and put forward suggestions for improvement. Gao et al. [14] used fuzzy-AHP and ELECTRE III to evaluate the port competitiveness according to the total weights obtained based on six key criteria (port size, port location, hinterland economy, port costs, operations management, and growth potential), thus obtaining reasonable and effective evaluation results, Zhong et al. [15] selected five evaluation indexes and used the grey target model to measure the port's economic efficiency. This method is simple in the calculation, and can effectively overcome the problem which is that the previous research cannot accurately evaluate the port efficiency with small samples and multiple indexes. In addition, some scholars also used the Technique for Order of Preference by Similarity to Ideal Solution (TOPSIS) method, comprehensive weighting method, principal component analysis (PCA), and the data envelopment analysis (DEA) model for theoretical research [16].

In today's mature development of port logistics construction, most of the existing domestic and foreign studies on port logistics evaluation indexes still focus on infrastructure construction and basic throughput, and the methods used are basically the traditional evaluation methods. It should be noted that China's port construction has entered a high-quality development stage, and the "double-circulation" situation is becoming more and more fierce. Under such a domestic environment, although there are some researches on intelligent logistics and international logistics in the industry, these more advanced indicators have never been used to measure the port logistics capability, and the new comprehensive evaluation system is lacking. The Complex Proportional Assessment (COPRAS) method proposed by Zavadskas et al. [17], as an effective multi-attribute decision-making method, does not need to convert and unify attribute categories during use, can comprehensively consider the importance and validity of attributes, and is easy to operate and calculate. It has been widely used in various project selection and performance evaluations at present. For example, to overcome the shortage of water resources and the increasing water demand by using inter-basin water transfer schemes, eight inter-basin water transfer schemes from the Grand Karon Basin to the central plateau of Iran are prioritized by the COPRAS method [18], which is beneficial for decision makers to evaluate inter-basin water transfer projects under uncertain conditions; Alkan et al. [19] used the COPRAS method to evaluate and rank renewable energy in 26 regions of Turkey, which provided a theoretical reference for development institutions and investors. Hezer et al. [20] studied the security level of 100 regions around the world under the impact of the COVID-19 virus, and ranked them by using COPRAS and VIKOR methods, respectively.

The main aim of this study is to construct a set of scientific and reasonable index systems of port logistics development that is more in line with the requirements of the "double-cycle" era, take the logistics situation of Ningbo-Zhoushan Port from 2014 to 2020 as an individual plan, and make a longitudinal analysis and evaluation of its logistics situation by using the new proposed Entropy-COPRAS evaluation framework. Compared with TOPSIS, DEA, and other methods, COPRAS is suitable for the research of larger samples in the case that individual data may be missing, which is more in line with the research idea and purpose of this paper. In the evaluation process, the utility of indicators can be comprehensively considered, which makes the evaluation results more reasonable. The entropy weight method mainly determines the weight through the difference of index information, which can avoid the interference of human factors in the process of solving the weight, and has strong objectivity. Thus, the proposed Entropy-COPRAS method can make the evaluation results more realistic and facilitate the objective and reasonable analysis of the Ningbo-Zhoushan port logistics level. This will help us to better define the self-positioning of regional port logistics in the new era, and intuitively reflect the problems in the development of regional port logistics, to get targeted countermeasures and suggestions.

3. Evaluation Index System of Port Logistics Based on "Dual Circulation" Pattern

3.1. Principles of Selecting Index

Formulating a set of operable and effective comprehensive evaluation index systems is the basis for evaluating port logistics from the perspective of "dual circulation". According to existing researches of port logistics evaluation [21–26], this paper will be as close to the following principles as possible during the selection of evaluation indexes to construct a comprehensive evaluation system:

(1) Scientific principle. The scientific principle means that the selection of evaluation indexes should conform to the requirements of "dual circulation" for port logistics and

the socio-economic law of port logistics development when constructing a comprehensive evaluation system. The requirement is that the classification of indexes is accurate and comprehensive, the definition of indexes is concise and clear, and each stage of data processing must have a realistic basis of science and strong supporting theory.

(2) Principle of comprehensiveness. The selected port logistics indexes should include not only basic indexes such as port infrastructure, logistics capacity, and the economy of the city where the port is located, but also innovative indexes such as production logistics capacity, international logistics capacity, and intelligent logistics capacity that can reflect the future development direction of port logistics based on the "dual circulation" pattern. The system should comprehensively reflect the development of port logistics from the "dual circulation" perspective.

(3) Principle of operability. When designing the index system, it is necessary to consider whether the index selection is comprehensive or not, as well as the operability of the index, such as whether the index data can be collected, what method to collect, etc. We shall try to select the index with strong operability and representativeness.

3.2. Analysis of the Composition of the Index System

Combined with the new requirements of the development pattern of "constructing a large domestic circulation as the main body, and the mutual promotion of international and domestic 'dual circulation'" to port logistics, this paper takes port logistics infrastructure, logistics capacity, the economy of the city where the port is located, production logistics capacity, international logistics capacity, green logistics capacity, and intelligent logistics capacity as the seven subsystems of the index system to construct the evaluation index system of port logistics from the perspective of "dual circulation".

(1) Port logistics infrastructure

Port infrastructure refers to the facilities that must be provided to complete the most basic functions of the port, generally including port channel, breakwater, anchorage, wharf, berth, port traffic and supporting facilities, etc. Considering the related logistics, representativeness and operability of the indexes, the total number of wharf berths and berths above 10,000 tons are selected to measure the infrastructure of the port.

(2) Logistics capability

Port logistics refers to the central port city making use of its own port advantages, relying on the advanced software and hardware environment, strengthening its radiation ability to the logistics activities around the port, highlighting the advantages of port collection, inventory, and distribution, and developing a comprehensive port service system with the characteristics of covering all links of the logistics industry chain, based on the port-adjacent industry and supported by information technology, with the goal of optimizing the integration of port resources. According to the existing comprehensive evaluation system of port logistics, this paper selects the cargo throughput (100 million tons) and container throughput (10,000 TEU) to represent the logistics capacity of the port.

(3) Economy of port city

The economic situation of the port city affects all aspects of the port, and the development of the urban logistics industry is also closely related to port logistics. Therefore, the economic development of the port city is an important index to evaluate the port logistics. Based on the summary of the existing evaluation index system, this paper selects fiscal revenue (CNY 100 million), the added value of primary industry (CNY 100 million), added value of secondary industry (CNY 100 million), per capita GDP (yuan), freight volume (10,000 tons), and railway freight volume (10,000 tons) to measure the economy of the city where the port is located.

(4) Production logistics capacity

Production logistics refers to the logistics activities in the production process. Generally, it means that after raw materials and purchased parts are put into production, they are delivered to each processing point and storage point through the distributed materials. In the form of work-in-process, they flow from one production unit (warehouse) to another and are processed and stored according to the specified process. With the help of certain transportation devices, they circulate at a certain point and then flow out from a certain point, which always reflects the flow process of the physical form of materials. The productive berth refers to the place where the freighter carrying the means of production stops. In this paper, productive berths (10,000 tons or more) are selected to measure the port's production and logistics capacity.

(5) International logistics capacity

International logistics, also known as global logistics, refers to an international commodity transaction or exchange activity in which the production and consumption are carried out independently in two or more countries, in order to overcome the space and time distance between production and consumption and to physically move the materials, so as to achieve the ultimate goal of international commodity transactions. In this paper, the throughput of foreign trade goods (100 million tons) and the volume of international transit containers (10,000 TEU) are selected to measure the international logistics capability of ports.

(6) Green logistics capacity

Green logistics is the process of reducing the impact of logistics on the environment by making full use of logistics resources, adopting advanced logistics technology, and rationally planning and implementing logistics activities such as transportation, storage, loading and unloading, handling, packaging, distribution, processing, distribution, and information processing. According to some existing research and news reports, this paper selects coal throughput (10,000 tons), ore throughput (10,000 tons), and sea-rail intermodal container volume (10,000 TUE) to measure the port's green logistics capability. It should be noted that coal throughput (ten thousand tons) and ore throughput (ten thousand tons) may seem to have nothing to do with green logistics, but in fact, when analyzing the entire domestic logistics chain, it is very important to change "bulk transportation" to "container transportation" in key places, and ports are such places. The main advantage of "bulk transportation" to "container transportation" is that the use of fully enclosed transportation results in less pollution to the environment, less affected by transportation loss, and effective use of empty containers for return journey, etc. It is undeniable that the efficiency improvement brought by "bulk transportation" to "container transportation" is related to the green construction of a complete domestic logistics chain, and the overall efficiency improvement will promote the green construction of the port part. The annual export volume of coal and ore in our country has reached a certain level, which will not increase or drop sharply in a short time. Therefore, as a key point where "bulk transportation" is changed from "container transportation" to "container transportation" in logistics chain, the port green logistics capacity can be measured by these two indicators to a certain extent.

(7) Intelligent logistics capacity

Intelligent logistics refers to improving the ability of analysis and decision-making and intelligent execution of logistics systems through intelligent technologies and means such as intelligent hardware, the Internet of Things, big data, and improving the intelligence and automation level of the whole logistics system. On 7 June 2017, the Ministry of Transport issued the "Notice on Publishing the List of Smart Port Demonstration Projects and Related Matters", which indicates that all work on China's smart port demonstration projects has officially entered the implementation stage. As important content and the most direct embodiment of smart port, smart logistics is the strategic direction and key area of China's port development during the 13th Five-Year Plan period and in the future. However, at present, when measuring the port's intelligent logistics capability, there is still a problem that most indexes are not representative enough, and there is also a lack of complete evaluation system research in the field. However, considering the scientific nature of the data, this paper still chooses two operable indicators: R&D expenditure (10,000 yuan) and the number of winning projects (items) for the "China Port Association Science and Technology Award" to measure the intelligent logistics capability of ports.

The detailed evaluation index of port logistics is listed in Table 1.

Table 1. Evaluation Index System of Port Logistics.

First-Level Index	Second-Level Index	Unit	References
Port infrastructure construction (X_1)	Total number of berths (C_{11}) Berths above 10,000 tons (C_{12})	Individual Individual	Mo et al. [21] Feng et al. [13]
Logistics capability (X_2)	Cargo throughput (C_{21}) Container throughput (C_{22})	Tons Wanteu	Mo et al. [21] Qing et al. [27]
Economy of port city (X_3)	State revenue (C_{31}) Added value of primary industry (C_{32}) Added value of secondary industry (C_{33}) Per capita GDP (C_{34}) Volume of goods transported (C_{35}) Railway freight volume (C_{36})	One hundred million yuan One hundred million yuan One hundred million yuan Yuan Dynasty (1206–1368) Ten thousand tons Ten thousand tons	Mo et al. [21] Wang et al. [22] Yang et al. [28]
Production logistics capacity (X_4)	Productive berth (C_{41}) Productive berths with a tonnage of over 10,000 tons (C_{42})	individual individual	Mo et al. [21] Jiang et al. [23]
International logistics capability (X_5)	Foreign trade throughput (C_{51}) International transit volume (C_{52})	Tons Wanteu	Mo et al. [21] Jiang et al. [23]
Green logistics capability (X_6)	Coal throughput (C_{61}) Ore throughput (C_{62}) Container volume of sea-rail combined transport (C_{63})	Ten thousand tons Ten thousand tons Wanteu	Hu et al. [24] Hua et al. [12]
Smart logistics capability (X_7)	R&D expenditure (C_{71}) Number of winning projects of the Science and Technology Award of China Port Association (C_{72})	Ten thousand yuan Item	Hua et al. [12] Meng et al. [26] Yang et al. [28]

4. An Integrated Entropy-COPRAS Framework for the Port Logistics Development

4.1. The Weights of Index System

In this paper, from the perspective of vertical development, the weights of nineteen indexes in seven subsystems of port logistics are determined by the entropy method. The evaluation information of the entropy weight method mostly comes from the original data of the evaluated object, which can reflect the development course of the port in the last seven years to the greatest extent and is comparable in time, so it is an objective comprehensive evaluation method. The main steps are described as follows.

Step 1. It is assumed that various indexes in the year are to be evaluated. The collection of different annual monomer is $A = \{A_1, A_2, \ldots, A_m\}$, and the index $C = \{C_1, C_2, \ldots, C_n\}$ with the quantity n is used to evaluate the port logistics status every year. As the measurement units, economic significance and manifestation of each index are different in the measurement index system, these indexes cannot be compared directly, and they must be standardized first. The collection of positive indexes is recorded as J_1, and the collection of negative indexes is recorded as J_2. The standardized formula is given as:

$$x_{ij} = I_{ij} / \sum_{i=1}^{m} I_{ij}, \ i = 1, 2, \ldots, m; \ j = 1, 2, \ldots, n. \tag{1}$$

where I_{ij} is the evaluation value of index C_j in year i, x_{ij} is the standardized result. Therefore,

the standardized decision matrix $\overline{X} = \begin{pmatrix} x_{11} & \cdots & x_{1n} \\ x_{21} & \ddots & x_{2n} \\ x_{n1} & \cdots & x_{nn} \end{pmatrix}$ is obtained.

Step 2. Calculate the entropy e_j of index C_j in year i:

$$e_j = -k \sum_{i=1}^{n} x_{ij} \ln x_{ij} \tag{2}$$

Step 3. Calculate the difference coefficient g_j of index C_j:

$$g_j = 1 - e_j \tag{3}$$

Step 4. Normalize g_j and calculate the weight w_j of index C_j in year i:

$$w_j = g_j / \sum_{j=1}^{n} g_j \tag{4}$$

Obviously, $0 \leq w_j \leq 1$ and $\sum_{j=1}^{n} w_j = 1$.

4.2. Evaluation Process of COPRAS Method

After obtaining the index weight, we use the COPRAS method to transform the evaluation of each index into the overall characteristics of the reaction index. The main steps are as follows:

Step 1. Calculate the weighted standardized decision matrix $\hat{X}$ based on the standardized decision matrix.

$$\hat{X} = \begin{pmatrix} \hat{x}_{11} & \cdots & \hat{x}_{1n} \\ \vdots & \ddots & \vdots \\ \hat{x}_{m1} & \cdots & \hat{x}_{mn} \end{pmatrix} \tag{5}$$

where the weighted standardized assessment is:

$$\hat{x}_{ij} = w_j \times x_{ij} \tag{6}$$

Step 2. Calculate the comprehensive benefit value of the set A_i composed of each annual monomer by using the following formula for the positive index of each year:

$$p_i^a = \sum_{j=1}^{n} \hat{x}_{+ij}, \ i = 1, 2, \ldots, m; \ j = 1, 2, \ldots, n. \tag{7}$$

where, $\hat{x}_{+ij} = \begin{cases} \hat{x}_{ij}, \ if \ C_j \in J_1 \\ 0, \ if \ C_j \in J_2 \end{cases}$. Obviously, p_i^a should be as large as possible.

For the reverse index of each year, the following formula is used to calculate its comprehensive cost value:

$$p_i^b = \sum_{j=1}^{n} \hat{x}_{-ij}, \ i = 1, 2, \ldots, m; \ j = 1, 2, \ldots, n. \tag{8}$$

where, $\hat{x}_{-ij} = \begin{cases} 0, if \ C_j \in J_1 \\ \hat{x}_{ij}, if \ C_j \in J_2 \end{cases}$. Obviously, p_i^b should be as small as possible.

Step 3. Calculate the comprehensive evaluation value of each annual monomer set A_i, by using the following formula:

$$Q_i = p_i^a + \frac{p_{min}^b \times \sum_{i=1}^{m} p_i^b}{p_i^b \times \sum_{i=1}^{m} \frac{p_{min}^b}{p_i^b}} = p_i^a + \frac{\sum_{i=1}^{m} p_i^b}{p_i^b \times \sum_{i=1}^{m} \frac{1}{p_i^b}}, \ i = 1, 2, \ldots, m \tag{9}$$

where $p_{min}^b = \min\left\{p_i^b\right\}$. Obviously, the larger the value of Q_i, the better the annual monomer set A_i.

Step 4. Define the utility degree of each annual monomer set A_i as the ratio between the comprehensive evaluation value of the annual monomer itself and the maximum comprehensive evaluation value:

$$U_i = \frac{Q_i}{Q_{max}} \times 100\%, \ i = 1, 2, \ldots, m \tag{10}$$

where $Q_{max} = \max\{Q_i\}$.

5. Evaluation and Analysis of Ningbo-Zhoushan Port Logistics Development Level

5.1. Evaluation Results of Ningbo-Zhoushan Port Logistics Level

In this subsection, the proposed integrated Entropy-COPRAS method is used to measure the logistics development level of Ningbo-Zhoushan port from 2014 to 2020, so as to clarify the development trend and path of the Ningbo-Zhoushan port logistics industry, and thus provide targeted countermeasures and suggestions for its new development. According to the designed evaluation index system of port logistics, most of the index data are derived from China Port Yearbook, Ningbo Statistical Yearbook, and Zhoushan Statistical Yearbook. In addition, this paper uses the isolated method to interpolate and extrapolate the existing data to fill in a small number of missing data. Considering that the original data of each index are different in measurement units and economic significance, the original data of each index of Ningbo-Zhoushan Port shall be standardized.

Firstly, the original data of indexes are standardized into a same type according to Equation (1), and then the standardized decision matrix X is constructed.

Secondly, according to Equations (2)–(4), the entropy and difference coefficient of each index of Ningbo-Zhoushan port logistics can be calculated, and the weights of each index can be obtained. The results are shown in Table 2.

Table 2. Index weight of port logistics in Ningbo-Zhoushan Port.

w_1	0.001	w_8	0.026	w_{15}	0.024
w_2	0.010	w_9	0.053	w_{16}	0.072
w_3	0.014	w_{10}	0.009	w_{17}	0.480
w_4	0.022	w_{11}	0.001	w_{18}	0.055
w_5	0.028	w_{12}	0.010	w_{19}	0.135
w_6	0.005	w_{13}	0.010		
w_7	0.029	w_{14}	0.019		

Then, based on the standardized decision matrix, the COPRAS method is used to measure the standardized data according to Equation (5), and the utility degree of each subsystem evaluation index of Ningbo-Zhoushan port logistics is calculated according to Equations (7)–(10), respectively. The evaluation result of each subsystem evaluation index of Ningbo-Zhoushan port logistics is shown in Table 3.

Table 3. Evaluation results of each subsystem of port logistics in Ningbo-Zhoushan Port.

	2014	2015	2016	2017	2018	2019	2020
X_1	0.769	0.804	0.834	0.875	0.910	0.970	1
X_2	0.703	0.733	0.764	0.858	0.920	0.957	1
X_3	0.634	0.669	0.717	0.818	0.906	0.971	1
X_4	0.769	0.804	0.834	0.874	0.910	0.970	1
X_5	0.732	0.766	0.793	0.883	0.939	0.977	1
X_6	0.208	0.216	0.362	0.456	0.634	0.818	1
X_7	1	0.591	0.836	0.612	0.549	0.726	0.663

Similarly, on the basis of Equations (7)–(10), the comprehensive benefit value, comprehensive cost value, comprehensive evaluation value, and utility degree corresponding to each index value of port logistics of Ningbo-Zhoushan Port in each year are calculated, and the comprehensive evaluation of different years is sorted according to the obtained utility degree. The results are shown in Table 4.

5.2. Analysis of Evaluation Results of Each Subsystem Index

According to the utility obtained in Table 3, we make the development trend chart of Ningbo-Zhoushan Port Logistics from 2014 to 2020, shown in Figure 1.

Table 4. Ningbo-Zhoushan Port Logistics Evaluation Results.

	Comprehensive Benefit Value	Comprehensive Cost Value	Comprehensive Evaluation Value	Utility Degree	Sort
2014	0.084	0.014	0.097	44.01%	6
2015	0.075	0.019	0.084	38.13%	7
2016	0.098	0.009	0.118	53.39%	5
2017	0.114	0.013	0.128	57.80%	4
2018	0.143	0.014	0.156	70.64%	3
2019	0.181	0.014	0.194	87.71%	2
2020	0.209	0.014	0.221	100.00%	1

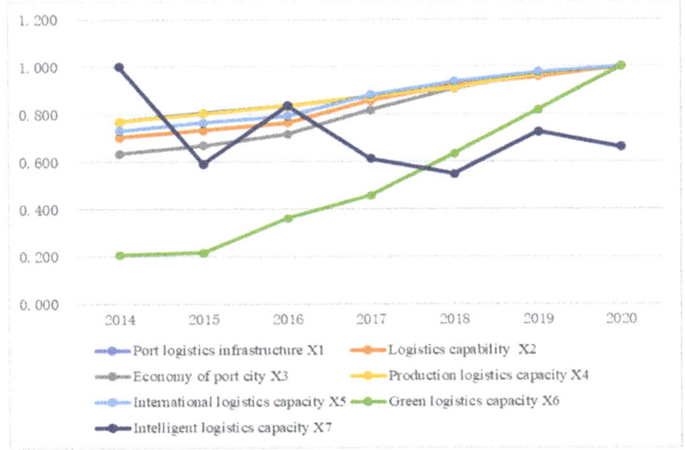

Figure 1. Utility degree of port logistics of Ningbo-Zhoushan Port from 2014 to 2020.

It can be seen from the changes in the utility degree of Ningbo-Zhoushan Port's logistics evaluation indexes from 2014 to 2020, the development level of the five indexes of Ningbo-Zhoushan Port's infrastructure, logistics capacity, city economy where the port is located, production logistics capacity and international logistics capacity are equivalent, with the average utility degree reaching more than 80%. Correspondingly, the levels of green logistics capability and intelligent logistics capability vary greatly, but the average utility is not high. This shows that from 2014 to 2020, Ningbo-Zhoushan Port's port logistics industry has achieved great success in port infrastructure construction, logistics capacity, the economy of the city where the port is located, production logistics capacity, and international logistics capacity, and it is still progressing with a weak growth trend. The construction of port logistics infrastructure and production logistics capacity is the material basis for the vigorous development of the port logistics industry. Ningbo-Zhoushan Port's emphasis on logistics infrastructure construction, logistics capacity, and production logistics capacity has provided a strong guarantee for the rapid development of the port logistics industry in recent years. The international logistics capability reflects the competitiveness of the port logistics industry in overseas markets, and at the same time, it responds to the new development pattern of "dual circulation" and promotes the international and domestic dual circulation policy. Secondly, the vigorous development of Ningbo-Zhoushan Port's port logistics industry is inextricably linked with the improvement of Ningbo's economic level. However, the low level of green logistics capability and intelligent logistics capability indicates that the process of vigorously promoting low-carbon development, reinforcing the construction of ecological civilization, promoting the greening and recycling of the logistics industry, and combining high-tech with the logistics industry to raise the intelligence and automation level of the whole logistics system is slow, which is more difficult than other aspects.

From the growth rate, the five indexes of port infrastructure, logistics capacity, the economy of the city where the port is located, production logistics capacity, and interna-tional logistics capacity are all showing an increasing trend year by year, and the development paths are consistent. It can be seen that the logistics industry of Ningbo-Zhoushan Port has a good development prospect and shows a steady growth trend. However, due to some achievements in these aspects, there is not much room for improvement. Among them, after 2016, the economic growth rate of the city where the port is located has increased compared with previous years, because the merger of Ningbo Port and Zhoushan Port was completed in 2015, which further promoted the integration of Ningbo-Zhoushan Port, and the port economy has been rapidly improved, thus promoting the rapid economic development of Ningbo. At the same time, compared with previous years, Ningbo-Zhoushan Port's international logistics capability has also improved significantly. The reason is that in recent years, Ningbo-Zhoushan Port has paid attention to expanding the international market of the port logistics industry, striving to build a world-class strong port and an international shipping logistics hub. Especially after the new development pattern of "dual circulation" was put forward, in the process of promoting dual circulation at home and abroad, expanding the international market became an indispensable condition for the further development of the port logistics industry under the new pattern of "dual circulation". However, the development level of intelligent logistics capability fluctuated greatly from 2014 to 2020, except for 2015–2016 and 2018–2019, which showed a downward trend. According to the collected data analysis, the R&D expenditure of Ningbo-Zhoushan Port's logistics industry has increased year by year, but the number of "China Port Association Science and Technology Award" winning projects has not increased year by year, and even dropped from 11 in 2014 to 5 in 2020, thus affecting the development of intelligent logistics capabilities. Therefore, to promote the combination of the high-tech and port logistics industry, and push forward the digital and intelligent development of the port logistics industry requires not only the investment of technical research and development, but also the progress of scientific and technological capability and technology. Moreover, with the country's vigorous implementation of the "innovation for all" policy, the competitive pressure among ports increases, and the progress of technological innovation tends to fall into the bottleneck period. Therefore, the development level of intelligent logistics capability will show such a fluctuation range. The development level of green logistics capability in 2014–2020 is also increasing year by year, and the growth rate is faster than other indexes. This shows that in recent years, Ningbo-Zhoushan Port has responded to the national policy, vigorously promoted the greening and recycling of the logistics industry, and achieved remarkable results.

5.3. Analysis of Comprehensive Evaluation Results of Logistics Level

According to the utility obtained in Table 4, we make the overall development trend chart of Ningbo-Zhoushan Port Logistics from 2014 to 2020, as shown in Figure 2.

From the vertical perspective, the development level of the port logistics industry in Ningbo-Zhoushan Port decreased from 2014 to 2015. From the perspective of subsystem indicators, the reason is that the development of intelligent logistics capability slowed down, which indicated that the decline of it would affect the overall development speed of the port logistics industry. Except for 2014–2015, the port logistics level of Ningbo-Zhoushan Port has improved at a certain speed, especially after 2017, which shows that the port logistics industry of Ningbo-Zhoushan Port has a good development prospect, a large development space, and a more intuitive and remarkable development results. Among them, the growth rate of Ningbo-Zhoushan Port's port logistics level is larger than that of five subsystems such as port infrastructure, which is equivalent to the growing trend of green logistics capability. It can be seen that the improvement of green logistics capability has a strong promoting effect on Ningbo-Zhoushan Port. To sum up, under the "dual circulation" development pattern, maintaining the development of port infrastructure, logistics capacity, the economy of the city where the port is located, production logistics capacity and international logistics capacity, and increasing investment in smart logistics and green logistics can promote the high-quality development of Ningbo-Zhoushan port logistics industry.

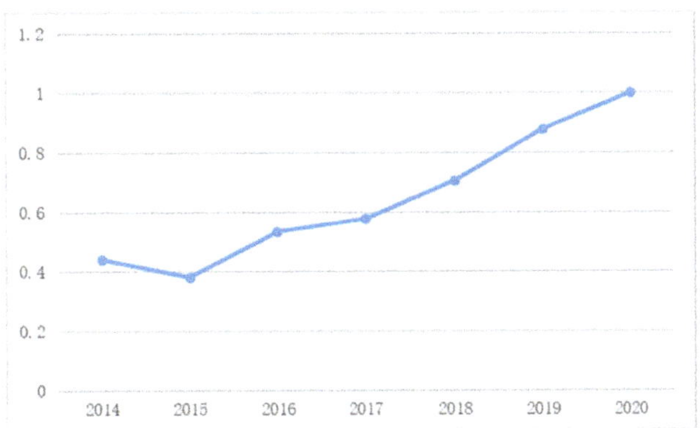

Figure 2. Development Trend of Ningbo-Zhoushan Port Logistics from 2014 to 2020.

5.4. Comparative Analysis with Major Domestic Ports

In this paper, the development level of the logistics industry of the four major ports in China, i.e., Shenzhen Port (Y_1), Shanghai Port (Y_2), Dalian Port (Y_3), and Ningbo-Zhoushan Port (Y_4), is analyzed and compared over the past seven years. On this basis, the horizontal comparison among ports is made, the logistics development level of major domestic ports is analyzed, and the development status of Ningbo-Zhoushan Port is explored.

The collected raw data of various indexes of Shenzhen Port, Shanghai Port, Dalian Port, and Ningbo-Zhoushan Port are processed, and the comprehensive benefit value, comprehensive cost value, comprehensive evaluation value, and utility degree corresponding to each port's logistics index value are obtained every year. Then the index values of each port are sorted according to the obtained utility degree, and the results are shown in Table 5. The utility degree of port logistics evaluation index in different years of each port is shown in Table 6 and Figure 3.

Table 5. Comprehensive evaluation results of port logistics in each port.

Port	Comprehensive Benefit Value	Comprehensive Cost Value	Comprehensive Evaluation Value	Utility Degree	Sort
Y_1	0.097	0.009	0.186	63.52%	4
Y_2	0.278	0.049	0.294	100.00%	1
Y_3	0.214	0.012	0.276	94.00%	2
Y_4	0.237	0.105	0.244	83.09%	3

Table 6. Utility degree of port logistics evaluation index in different years of each port.

Year	Y_1	Y_2	Y_3	Y_4
2014	1.000	0.865	0.913	0.816
2015	1.000	0.774	0.798	0.607
2016	1.000	0.956	0.890	0.916
2017	0.708	0.958	1.000	0.883
2018	0.606	1.000	0.920	0.839
2019	0.537	1.000	0.848	0.854
2020	0.467	1.000	0.768	0.790

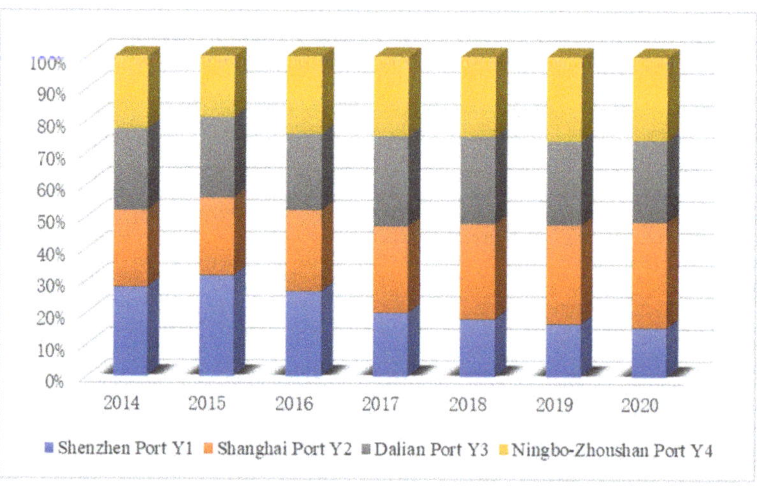

Figure 3. Utility degree of port logistics evaluation index in different years of each port.

As can be seen from Table 5, compared with Shenzhen Port, Shanghai Port, and Dalian Port, Ningbo-Zhoushan Port's port logistics industry has not achieved remarkable development results, and the utility of comprehensive evaluation is 83.09%, which is only higher than Shenzhen Port. However, it can be seen from Figure 3, among the four ports, only Ningbo-Zhoushan Port and Shanghai Port's comprehensive evaluation utility ratio shows an increasing trend year by year. That is to say, compared with other ports, Ningbo-Zhoushan Port and Shanghai Port's port logistics industry can maintain a better development trend. On the whole, although the development level of Ningbo-Zhoushan Port's port logistics industry is not the best, its future development trend is good, and it has more room for development. Therefore, taking Ningbo-Zhoushan Port as the research object of this paper has stronger practical significance. At the same time, combined with the collected port logistics data and the evaluation results in Table 5, it can be seen that the comprehensive benefit value of Ningbo-Zhoushan Port's port logistics index is second only to Shanghai Port, and the benefit performance is good, but the comprehensive cost value ranks first among the four ports, which indicates that the treatment of Ningbo-Zhoushan Port's port logistics in the reverse index is not in place. For example, the coal throughput of Ningbo-Zhoushan Port is much higher than that of other ports. Similarly, for some positive indicators, Ningbo-Zhoushan Port is still weak and needs to be improved. For example, the added value, fiscal revenue, and R&D expenditure of the secondary industry are not as good as those of Shanghai Port and Shenzhen Port. On the whole, compared with other ports, Ningbo-Zhoushan Port's development achievements are not significant, and its green logistics capability and intelligent logistics capability have not been greatly improved. At the same time, the economic development of Ningbo, the hinterland of the port, is also relatively slow.

6. Conclusions

6.1. Main Contributions

Port logistics is an important part of modern logistics. Due to the vigorous promotion of governments at all levels, the active exploration of port enterprises and the entry of foreign multinational logistics enterprises, the development of domestic port logistics is full of vitality. However, development is difficult and there are still many problems. In this case, it is of great theoretical and practical significance to comprehensively and systematically study and evaluate the development state of Ningbo-Zhoushan port logistics from a new perspective of "dual circulation" development pattern. This paper makes an in-depth study on the development of Ningbo-Zhoushan port logistics under the new development pattern of "dual circulation" from both theoretical and empirical aspects. The main research contributions are as follows:

(1) Construct the analytical framework of port logistics development under the "dual circulation" development pattern. On the basis of clearly defining the concept of port logistics and the background of "double-circulation" development, this paper combined the development of port logistics with the pattern of "double-circulation" and discussed the development trend and direction of port logistics under the development pattern of "double-circulation", which provided a broader perspective for theoretical analysis and practical exploration of port logistics evaluation in the new era and new environment.

(2) Under the guidance of fully embodying the principles of economy, technology, timeliness, and sustainable development of port logistics system, this paper tentatively discussed the evaluation index factors of port logistics under the "double-circulation" development pattern, and then subdivided the major factors into twenty indexes. Following the principle of establishing port logistics evaluation index, the comprehensive evaluation index system of port logistics under the background of "dual circulation" was established, and the applicability of the evaluation index system of port logistics was analyzed and explained. It lays a certain foundation for further evaluation of port logistics under the background of "dual circulation".

(3) A mathematical Entropy-COPRAS evaluation framework was tentatively constructed by using qualitative analysis and quantitative analysis, wherein the weights of multiple indicators were determined by the entropy method, and the comprehensive evaluation of port logistics was achieved by the COPRAS approach. Then, the logistics level of Ningbo-Zhoushan Port and Ningbo-Zhoushan Port and other ports was empirically studied and analyzed. The evaluation results are in good agreement with the actual development of these ports. It provides some guidance and theoretical support for better solving the practice of port logistics evaluation under the background of "dual circulation", and further enriches the existing theoretical system of port logistics evaluation.

6.2. Main Research Conclusions

According to the qualitative analysis and quantitative analysis, and the logistics level of Ningbo-Zhoushan Port and Ningbo-Zhoushan Port and other ports, the evaluation results are in good agreement with the actual development of these ports. It provides some guidance and theoretical support for better solving the practice of port logistics evaluation under the background of "dual circulation", and further enriches the existing theoretical system of port logistics evaluation.

Specifically, the empirical research results show that the development of Ningbo-Zhoushan Port in recent years meets the requirements of "dual circulation", but it still needs to keep forging ahead and seeking new ideas, and actively responding to the new call for development. It should focus on the construction of international logistics, green logistics, intelligent logistics, production logistics, etc., and make continuous efforts in international logistics capacity, green logistics capacity, total logistics capacity, etc., to realize the construction of a new port logistics system and adapt the port development to the new logistics demand of "dual circulation".

Based on the analysis of the development status of Ningbo-Zhoushan Port, the following suggestions are given.

(1) In terms of international logistics, it is necessary to further optimize the supply system, improve the construction of port management division of labor system, enhance the comprehensive service awareness of ports, vigorously develop multimodal transport based on ports, and provide various modes of transportation such as sea, road, railway, and inland river, promote the effective connection of different modes, reinforce the bonded function of parks, and set up special channels for customs supervision between terminals to improve the international transportation capacity of ports.

(2) In terms of smart logistics, major ports need to enhance their independent innovation capability, grasp the essence of smart logistics, accurately understand the connotation

of smart logistics, accelerate the informatization construction of port logistics, and make infrastructure scientific and technological. It also shall improve the construction of port information networks, integrate the existing transportation, warehousing, and other logistics infrastructure, revitalize existing assets, expand functions, improve services, and better meet the needs of port logistics development. On this basis, strengthen the integrated services of the logistics value chain, do a good job in the efficient collaborative operation of the logistics service chain, realize intelligent operation based on the data drive, and finally build an open, shared, and interconnected port ecosystem.

(3) Increase investment in green logistics, apply the new concept of low-carbon development and green economy to port construction, reduce pollution and strengthen governance through technological transformation and technological innovation, strengthen resource recycling and ecological protection, improve energy efficiency, and vigorously build a resource-saving, environment-friendly, and quality-efficient port.

Author Contributions: Writing—original draft preparation, S.Z. and Z.F.; methodology, Y.H. and L.H. All authors have read and agreed to the published version of the manuscript.

Funding: This work is supported by the Social Sciences Planning Projects of Zhejiang (21QNYC11ZD), Statistical Scientific Key Research Project of China (2021LZ33), Major Humanities and Social Sciences Research Projects in Zhejiang Universities (2018QN058), the First Class Discipline of Zhejiang-A (Zhejiang Gongshang University-Statistics) and Fundamental Research Funds for the Provincial Universities of Zhejiang (SJWZ2020002).

Institutional Review Board Statement: Not applicable.

Informed Consent Statement: Not applicable.

Data Availability Statement: Not applicable.

Conflicts of Interest: The authors declare no conflict of interest.

References

1. Jiang, M.; Lu, J.; Qu, Z.; Yang, Z. Port vulnerability assessment from a supply chain perspective. *Ocean. Coast. Manag.* **2021**, *213*, 105851. [CrossRef]
2. Feng, H.X.; Grifoll, M.; Zheng, P.J. From a feeder port to a hub port: The evolution pathways, dynamics and perspectives of Ningbo-Zhoushan port (China). *Transp. Policy* **2019**, *76*, 21–35. [CrossRef]
3. Wang, L.; Zhou, Z.; Yang, Y.; Wu, J. Green efficiency evaluation and improvement of Chinese ports: A cross-efficiency model. *Transp. Res. Part D-Transp. Environ.* **2020**, *88*, 102590. [CrossRef]
4. Tian, X.; Xu, L.; Liu, L.; Wang, S. Analysis and forecasting of port, logistics using TEI@ I methodology. *Transp. Plan. Technol.* **2013**, *36*, 685–702. [CrossRef]
5. Fouda, R.A.N.; Romeo, N.D.; Azizi, M.; Fernandez, S.R. Port logistics in west and central Africa: A strategic development under globalization. *Open J. Appl. Sci.* **2014**, *4*, 76–84. [CrossRef]
6. Ji, M.J.; Chu, Y.L. Optimization for hub-and-spoke port logistics network of dynamic hinterland. *Phys. Procedia* **2012**, *33*, 827–832.
7. Chen, J.; Wan, Z.; Zhang, F.; Park, N.-K.; He, X.; Yin, W. Operational efficiency evaluation of iron ore logistics at the ports of Bohai Bay in China: Based on the PCA-DEA model. *Math. Probl. Eng.* **2016**, *2016*, 9604819. [CrossRef]
8. Ha, M.H.; Yang, Z.L.; Lam, J.S.L. Port performance in container transport logistics: A multi-stakeholder perspective. *Transp. Policy* **2019**, *73*, 25–40. [CrossRef]
9. Raimbault, N. From regional planning to port regionalization and urban logistics. The inland port and the governance of logistics development in the Paris region. *J. Transp. Geogr.* **2019**, *78*, 205–213. [CrossRef]
10. Shi, X.; Li, H. Developing the port hinterland: Different perspectives and their application to Shenzhen Port, China. *Res. Transp. Bus. Manag.* **2016**, *19*, 42–50. [CrossRef]
11. Zhou, L.J. Research on coordinated development of Nanjing Port logistics and hinterland economy based on gray correlation analysis. *J. Coast. Res.* **2020**, *105*, 215–218. [CrossRef]
12. Hua, C.; Chen, J.; Wan, Z.; Xu, L.; Bai, Y.; Zheng, T.; Fei, Y. Evaluation and governance of green development practice of port: A sea port case of China. *J. Clean. Prod.* **2020**, *249*, 119434. [CrossRef]
13. Feng, H.; Ye, C.; Gao, R.W. AHP-based evaluation of port logistics transportation efficiency. *J. Coast. Res.* **2020**, *106*, 477–480. [CrossRef]
14. Gao, T.L.; Na, S.Y.; Dang, X.H.; Zhang, Y. Study of the competitiveness of Quanzhou Port on the belt and road in China based on a fuzzy-AHP and ELECTRE III model. *Sustainability* **2018**, *10*, 1253. [CrossRef]
15. Ming, Z.; Yuan, Y.L. Port efficiency evaluation based on grey target model. *J. Dalian Marit. Univ.* **2007**, *6*, 130–132.

16. Davarzani, H.; Fahimnia, B.; Bell, M.; Sarkis, J. Greening ports and maritime logistics: A review. *Transp. Res. Part D Transp. Environ.* **2016**, *43*, 473–487. [CrossRef]
17. Zavadskas, E.K.; Kaklauskas, A.; Sarka, V. The new method of multicriteria complex proportional assessment of projects. *Technol. Econ. Dev. Econ.* **1994**, *1*, 131–139.
18. Roozbahani, A.; Ghased, H.; Shahedany, M.H. Inter-basin water transfer planning with grey COP-RAS and fuzzy COPRAS techniques: A case study in Iranian Central Plateau. *Sci. Total Environ.* **2020**, *726*, 138499. [CrossRef]
19. Alkan, O.; Albayrak, O.K. Ranking of renewable energy sources for regions in Turkey by fuzzy entropy based fuzzy COPRAS and fuzzy MULTIMOORA. *Renew. Energy* **2020**, *162*, 712–726. [CrossRef]
20. Hezer, S.; Gelmez, E.; Ozceylan, E. Comparative analysis of TOPSIS, VIKOR and COPRAS methods for the COVID-19 Regional Safety Assessment. *J. Infect. Public Health* **2021**, *14*, 775–786. [CrossRef]
21. Mo, Y.; Zhao, Y.; Li, S.; Liu, X. Evaluation of port logistics competitiveness based on DEA. *IOP Conf. Ser. Earth Environ. Sci.* **2018**, *189*, 062041. [CrossRef]
22. Wang, C.S.; Yang, Q.; Wu, S.H. Coordinated development relationship between port cluster and its hinterland economic system based on improved coupling coordination degree model: Empirical study from China's port integration. *Sustainability* **2022**, *14*, 4963. [CrossRef]
23. Jiang, Z.R.; Pi, C.F.; Zhu, H.Y.; Ye, S. Temporal and spatial evolution and influencing factors of the port system in Yangtze River Delta Region from the perspective of dual circulation: Comparing port domestic trade throughput with port foreign trade throughput. *Transp. Policy* **2022**, *118*, 79–90.
24. Hu, Z.B.; Lan, F.; Xu, H. Green TFP heterogeneity in the ports of China's pilot free trade zone under environmental constraints. *Int. J. Environ. Res. Public Health* **2021**, *18*, 12910. [CrossRef]
25. Zhang, N.; Su, W.H.; Zhang, C.H.; Zeng, S.Z. Evaluation and selection model of community group purchase platform based on WEPLPA-CPT-EDAS method. *Comput. Ind. Eng.* **2022**, *172*, 108573. [CrossRef]
26. Meng, B.; Kuang, H.; Niu, E.; Li, J.; Li, Z. Research on the transformation path of the green intelligent port: Outlining the perspective of the evolutionary game "government–port–third-party organization". *Sustainability* **2020**, *12*, 8072. [CrossRef]
27. Qing, P.P.; Cong, L.Y. The realistic logic, implementation path and value of the times of building a new development pattern of "dual circulation". *J. Chongqing Univ.* **2020**, *26*, 24–34.
28. Yang, S.; Pan, Y.; Zeng, S.Z. Decision making framework based Fermatean fuzzy integrated weighted distance and TOPSIS for green low-carbon port evaluation. *Eng. Appl. Artif. Intell.* **2022**, *114*, 105048. [CrossRef]

 *systems*

Article

Modeling Emergency Logistics Location-Allocation Problem with Uncertain Parameters

Hui Li [1], Bo Zhang [2,*] and Xiangyu Ge [2]

1   School of Mathematics and Economics, Hubei University of Education, Wuhan 430205, China; linna_li@hue.edu.cn
2   School of Statistics and Mathematics, Zhongnan University of Economics and Law, Wuhan 430073, China; xiangyu_ge@zuel.edu.cn
*   Correspondence: bozhang@zuel.edu.cn

Abstract: In order to model the emergency facility location-allocation problem with uncertain parameters, an uncertain multi-objective model is developed within the framework of uncertainty theory. The proposed model minimizes time penalty cost, distribution cost and carbon dioxide emissions. The equivalents of the model are discussed via operational laws of uncertainty distribution. By employing the goal attainment technique, a series of Pareto-optimal solutions are generated that can be used for decision-making. Finally, several numerical experiments are presented to verify the validity of the proposed model and to illustrate decision-making strategy.

Keywords: uncertainty modeling; emergency logistics; location-allocation; sustainability; uncertainty theory

Citation: Li, H.; Zhang, B.; Ge, X. Modeling Emergency Logistics Location-Allocation Problem with Uncertain Parameters. *Systems* **2022**, *10*, 51. https://doi.org/10.3390/systems10020051

Academic Editor: Edward Pohl

Received: 16 March 2022
Accepted: 14 April 2022
Published: 17 April 2022

Publisher's Note: MDPI stays neutral with regard to jurisdictional claims in published maps and institutional affiliations.

1. Introduction

Emergency logistics is one of the most active fields in Operations Research and Management Science, which involve planning, managing, and controlling the flow of resources to provide relief to the people affected (Sheu [1]). Providing an adequate service level is a struggle for many emergency response systems, but it is of special concern in the field of emergency management. Despite the recent progress in hardware equipment, information access, human resources, some pressing issues are worth to be further concerned: the reduction of the urgency of the emergency response, the funding shortfalls, the carbon emission issues, to mention a few. As a critical part of emergency logistics, the location-allocation problem concerns to open a set of candidate emergency distribution centers such that the transportation cost from centers to demand points is minimized.

As its general practical application backgrounds, location-allocation problem has received considerable attention since Cooper [2] studied it for the first time. For reviews on optimization models of emergency location-allocation problem, we may consult Beaumont [3] and de Camargo and Miranda [4], where various cases were discussed. In Murtagh and Niwattisyawong [5], a capacitated location-allocation problem, in which the capacities of facilities are limited, was proposed and a new model was introduced. Then Badri [6] and Fang and Li [7] studied multi-objective location-allocation problems using goal programming. Recently, Fan et al. [8] formulated a reliable location-allocation model for hazardous materials by considering the depot disruption. Chu and Chen [9] proposed a multi-objective location-allocation model for a three-level logistics network. There is considerable research that has been carried out for emergency facility location-allocation problems; the interested readers may refer to Wang et al. [10].

A limitation of most existing studies on the location-allocation problem is that most work is done for the deterministic case. In practice, many parameters, such as the timing, location and magnitude of a natural disaster are unpredictable rather than deterministic.

The highly unpredictable of natural disasters causes the real decisions are usually surrounded in the state of indeterminacy. Some scholars thought that parameters in this case can be described as random variables. Accordingly, probability theory (Kolmogorov [11]) is introduced into location-allocation problems and a wide range of stochastic programming models and a series of heuristic algorithms have been developed. For instance, Zhou and Liu [12] formulated three kinds of stochastic programming models according to different criteria and presented a hybrid intelligent algorithm on the basis of stochastic simulation, network simplex algorithm and genetic algorithm. Alizadeh et al. [13] studied a location-allocation problem in the heterogeneous environment and employed the normal approximation method to assess the probability distribution of the demand. Cheng and Wang [14] developed a stochastic chance-constrained programming location model for capacitated alternative-fuel stations by considering road condition. Hu et al. [15] proposed a stochastic programming model for a capacitated single allocation hub location problem.

Although the stochastic programming models have been widely used, and tally with the facts in widespread cases, it is also not suitable in a great many situations. As a matter of fact, the probability distributions for indeterminate quantities are not easy to obtain due to the lack of sufficient data. Instead, experts' belief degrees are usually employed to describe the quantities with indeterministic information. Considering this, how can we deal with belief degrees? Fuzziologists believe belief degrees can be interpreted as fuzziness. This calls for the incorporation of fuzzy set theory (Zadeh [16]) into location-allocation problem. A fuzzy set A in X is characterized by its membership function which assigns to each element x a real number $\mu_A(x) \in [0, 1]$. The value of $\mu_A(x)$ is interpreted as the degree of membership of x in fuzzy set A for each $x \in X$ (Zadeh [16]). Within the framework of fuzzy set theory, location-allocation problem has been widely studied and a variety of fuzzy programming models have been carried out. For example, Zhou and Liu [17] investigated a capacitated location-allocation problem with fuzzy demands and proposed three types of fuzzy programming models. Wen and Iwamura [18] formulated an α-cost model for fuzzy location-allocation problem under the Hurwicz criterion. Liu et al. [19] proposed a two-stage fuzzy 0–1 mixed integer programming model for a three-level location-allocation transfer center.

However, is it reasonable to use fuzzy set theory to deal with belief degrees? Through a lot of investigation and analysis, Kahneman and Tversky [20] concluded that human beings usually overweight some unlikely events. With the further research, Liu [21] reveals that paradoxes may appear when we use fuzzy set theory to handle belief degrees. That is to say, it is not suitable to employ fuzzy set theory to handle belief degrees. As a breakthrough to cope with indeterminate information, uncertainty theory, which was created by Liu [22] based on normality, duality, subadditivity and product axioms, provides a powerful alternative to address belief degrees. In turn, uncertain programming was proposed by Liu [23] as a spectrum of mathematical programming to handle optimization problems involving uncertain variables. Subsequently, uncertain programming has been gradually developed and successfully applied to a series of optimization problems, such as network optimization (Gao [24]), supply chain design (Ma and Li [25]), and optimal control problems (Sheng et al. [26]), and so forth.

In terms of emergency logistics network design, Gao [27] first proposed two types of uncertain models for a single facility location problem. Then Wen et al. [28] formulated an uncertain location-allocation model via chance-constraints, and designed a hybrid intelligent algorithm to solve the model. Wang and Yang [29] studied a hierarchical facility location problem from different decision criteria and proposed two types of uncertain models. Zhang et al. [30] developed three types of covering location models for emergency facilities and discussed the analytical solutions of the models based on operational laws of uncertainty distribution. Soltanpour et al. [31] investigated an inverse 1-median location problem with uncertain parameters, and discussed the necessary and sufficient condition for the α-1-median on uncertain trees. Wang and Qin [32] studied a hub maximal covering location problem in the presence of partial coverage with uncertain travel times, proposed two types of uncertain models and presented a greedy variable neighborhood search heuristic. Gradually, more researchers investigated emergency facility location problems, see Wen et al. [33], Gao and Qin [34], Zhang et al. [35] and Zhang et al. [36], for example.

The above discussion makes it clear that a whole range of research has been carried out to address location-allocation problem within the framework of uncertainty theory. However, most of studies did not take into account the urgency of emergency response. Additionally, sustainability is currently a societal concern for development, especially in the aspect of the environment. However, large-scale logistics activities will inevitably aggravate the environment. This fact provides our motivation to design a sustainable emergency logistics network, which can not only timely rescue the affected people but also effectively reduce the pollution to the environment.

Inspired by the above discussion, this article mainly concerns the following two issues: First, how can we quantitatively characterize urgency and environmental pollution? Second, how can we provide a good trade-off the relationship among urgency, cost and environmental pollution? To answer the questions, the time penalty cost function and carbon dioxide (CO_2) emissions are used to describe urgency and environmental pollution, respectively. Then, we formulate a multi-objective uncertain programming model that includes (i) minimization of the time penalty cost, (ii) minimization of the total distribution cost, and (iii) minimization of the CO_2 emissions. To solve the model, we apply goal attainment technique to obtain Pareto-optimal solutions. As a consequence, a set of Pareto-optimal solutions is provided for decision-makers to seek the best strategy in accordance with their preferences.

In comparison to the existing works, this paper mainly focuses on optimizing facility location-allocation strategies with uncertain information. The main contributions of this paper can be summarized as follows. First, compared to the use of other approaches, such as probability theory and fuzzy set theory, for location-allocation with uncertain information, this paper provides an opportunity to advance uncertainty theory to address location-allocation problem. Second, compared to the existing works related to the uncertain facility location-allocation problem, this paper develops a novel multi-objective location-allocation model. This paper thus complements existing literature, and it will certainly further improve the efficiency of emergency response.

The rest of the paper is organized as follows. Section 2 briefly reviews some relevant basic knowledge about uncertainty theory for readers to better understand this paper. Section 3 introduces the location-allocation problem and presents a novel multi-objective model for the deterministic case. In Section 4, an uncertain multi-objective location-allocation model is developed. After that, some crisp equivalents of the model are discussed and a solution methodology is introduced. Section 5 discusses the main contributions of the paper by comparing it with the existing works. Section 6 presents the computational experiments and illustrates the results of the analysis in detail. Finally, conclusion remarks and future research directions are provided in Section 7.

2. Preliminaries

Since its introduction in 2007 by Liu [22], uncertainty theory has been well studied and applied in a wide variety of fields. In the following, we briefly review the basic concepts, such as uncertainty space, uncertain variable, and uncertainty distribution, of uncertainty theory.

Let Γ be a nonempty set, $\mathcal{L}$ a σ-algebra over Γ, and $\mathcal{M}\{\Lambda\} \in [0,1]$ an uncertain measure. Then the triplet $(\Gamma, \mathcal{L}, \mathcal{M})$ is said to be an uncertainty space. The uncertain measure $\mathcal{M}$ satisfies the following three axioms (Liu [22]) : (i) $\mathcal{M}\{\Gamma\} = 1$ for the universal set Γ; (ii) $\mathcal{M}\{\Lambda\} + \mathcal{M}\{\Lambda^c\} = 1$ for any event Λ; (iii) For every countable sequence of events $\Lambda_1, \Lambda_2, \cdots$, we have $\mathcal{M}\{\bigcup_{i=1}^{\infty} \Lambda_i\} \leq \sum_{i=1}^{\infty} \mathcal{M}\{\Lambda_i\}$.

To obtain an uncertain measure of compound event, Liu [37] defined a product uncertain measure on multiple uncertainty spaces: Let $(\Gamma_k, \mathcal{L}_k, \mathcal{M}_k)$ be uncertainty spaces for $k = 1, 2, \cdots$. The product uncertain measure $\mathcal{M}$ is an uncertain measure satisfying

$$\mathcal{M}\left\{\prod_{k=1}^{\infty}\Lambda_k\right\} = \bigwedge_{k=1}^{\infty}\mathcal{M}_k\{\Lambda_k\}$$

where Λ_k are arbitrarily chosen events from $\mathcal{L}_k$ for $k = 1, 2, \cdots$, respectively.

An uncertain variable (Liu [22]) is defined as a measurable function ξ from an uncertainty space to the set of real numbers such that $\{\xi \in B\}$ is an event for any Borel set B of real numbers. The uncertainty distribution Φ (Liu [22]) of an uncertain variable ξ is defined as

$$\Phi(x) = \mathcal{M}\{\xi \leq x\}, \qquad \text{for any} \quad x \in \Re.$$

If an uncertain variable ξ has a linear uncertainty distribution

$$\Phi(x) = \begin{cases} 0, & \text{if } x \leq a \\ \dfrac{x-a}{b-a}, & \text{if } a < x \leq b \\ 1, & \text{if } b < x \end{cases}$$

where a and b are both real numbers with $a < b$, we name it linear and denote this by $\xi \sim L(a, b)$ (Liu [22]). It is easy to verify that the inverse uncertainty distribution of $\xi \sim L(a, b)$ is

$$\Phi^{-1}(\alpha) = (1 - \alpha)a + \alpha b, \quad \alpha \in (0, 1).$$

Theorem 1 (Liu [38]). *Let $\xi_1, \xi_2, \cdots, \xi_n$ be independent uncertain variables with regular uncertainty distributions $\Phi_1, \Phi_2, \cdots, \Phi_n$, respectively. If $f(\xi_1, \xi_2, \cdots, \xi_n)$ is a continuous and strictly increasing function, then $\xi = f(\xi_1, \xi_2, \cdots, \xi_n)$ is an uncertain variable with inverse uncertainty distribution*

$$\Psi^{-1}(\alpha) = f(\Phi_1^{-1}(\alpha), \Phi_2^{-1}(\alpha), \cdots, \Phi_n^{-1}(\alpha)).$$

Definition 1 (Liu [22]). *Let ξ be an uncertain variable. Then, the expected value of ξ is defined by*

$$E[\xi] = \int_0^{+\infty} \mathcal{M}\{\xi \geq x\}dx - \int_{-\infty}^0 \mathcal{M}\{\xi \leq x\}dx$$

provided that at least one of the two integrals is finite.

It can be verified that the expected value of the linear uncertain variable $\xi \sim L(a, b)$ is $(a + b)/2$, i.e., $E[\xi] = \frac{a+b}{2}$. For detailed expositions, the interested reader may refer to Liu [21,22,38].

3. Problem Description and Formulation

In this paper, the location-allocation problem contains two levels (emergency distribution centers and demand points) and aims to open some centers to supply resources to demand points. In consideration of urgency, relief cost and CO_2 emissions, the goals of the problem are (i) to determine the subset of candidate emergency distribution centers to open and (ii) to draw up the resource allocation plan from the centers to the demand points.

3.1. Notations

To clearly describe the problem by a mathematical model, Table 1 lists some parameters and variables which are used to formulate the problem.

Table 1. List of notations.

Parameters	
I	The set of candidate emergency distribution centers with index $i = 1, 2, \ldots, n$
J	The set of demand points with index $j = 1, 2, \ldots, m$
C_i	The ith candidate emergency distribution center, $i \in I$
D_j	The jth demand point, $j \in J$
p_{ij}	The unit material time penalty cost from C_i to D_j, $i \in I, j \in J$
c_{ij}	The unit material distribution cost from C_i to D_j, $i \in I, j \in J$
d_{ij}	The distance from C_i to D_j, $i \in I, j \in J$
e_{ij}	The unit material CO_2 emissions per kilometer from C_i to D_j, $i \in I, j \in J$
s_j	The minimal demand of materials in D_j, $j \in J$
q_i	The capacity of C_i, $i \in I$
p	The maximum number of emergency distribution centers selected
f_i	The operation cost of opening C_i, $i \in I$
W	The budget for opening the emergency distribution centers
Variables	
y_i	1 if C_i is selected to open, 0 otherwise, $i \in I$
x_{ij}	The material distribution volume from C_i to D_j, $i \in I, j \in J$
Y	The set of y_i, $i \in I$
X	The set of x_{ij}, $i \in I, j \in J$

3.2. Multi-Objective Location-Allocation Model

In our model, three objectives are considered: (i) urgency, i.e., the urgency of emergency response; (ii) cost, i.e., the distribution costs for supplying the demand points with relief goods; and (iii) sustainability, i.e., the carbon dioxide emissions from transportation.

3.2.1. Objective 1: Minimization of the Time Penalty Cost

The urgency of demand points often has a significant effect on emergency response decision-making. Now, the critical issue is how to measure urgency. It follows from Wan et al. [39] that the time penalty cost approach can be used for the quantization of urgency. Denote p_{ij} as the unit material time penalty cost from the candidate emergency distribution center C_i to the demand point D_j, and use x_{ij} to denote the material distribution volume from C_i to D_j. Therefore, the time penalty cost from C_i to D_j is $p_{ij}x_{ij}$. Then, the objective 1 can be formulated as:

$$\min \quad F_1(X, Y) = \max_{j \in J} \sum_{i \in I} p_{ij}x_{ij}, \tag{1}$$

which tries to minimize the maximal time penalty cost.

3.2.2. Objective 2: Minimization of the Total Distribution Cost

Our model tries to simultaneously determine the locations of emergency distribution centers and the material distribution volume such that the total material distribution cost is minimized. Denote c_{ij} as the unit material distribution cost from C_i to D_j. Objective 2 is as follows:

$$\min \quad F_2(X, Y) = \max_{j \in J} \sum_{i \in I} c_{ij}x_{ij}. \tag{2}$$

3.2.3. Objective 3: Minimization of the CO_2 Emissions

In the past few years, a growing number of authors have paid attention to address the issue of carbon emissions in emergency response. Road transportation has become a major factor causing the rapid growing of carbon emissions. Denote e_{ij} and d_{ij} as the unit material CO_2 emissions per kilometer and distance from C_i to D_j, respectively. Thus, the formulation for objective 3 is given by

$$\min \quad F_3(X, Y) = \max_{j \in J} \sum_{i \in I} d_{ij}e_{ij}x_{ij}. \tag{3}$$

3.3. Constraints

By considering some realistic constraints, the constraints shown in (4)–(8) are established. Constraint (4) requires that the total goods transported from n emergency distribution centers should satisfy the demand of D_j. Constrain (5) guarantees that the amount of the relief distributed from C_i does not exceed its capacity; it also prevents materials from being distributed from an emergency distribution center that is not open. Constraint (6) ensures that all expenditures on the operations of the distribution centers cannot exceed the budget. Constraint (7) limits the maximum number of emergency distribution centers opened at p. Constraint (8) provides the range of the decision variables x_{ij} and y_i.

$$\sum_{i \in I} x_{ij} \geq s_j, \quad \forall j \in J, \tag{4}$$

$$\sum_{j \in J} x_{ij} \leq q_i y_i, \quad \forall i \in I, \tag{5}$$

$$\sum_{i \in I} f_i y_i \leq W, \tag{6}$$

$$\sum_{i \in I} y_i \leq p, \tag{7}$$

$$x_{ij} \geq 0, y_i \in \{0,1\}, \quad \forall i \in I, j \in J. \tag{8}$$

4. Uncertain Location-Allocation Model

Notice that the parameters in the above model are all assumed to be constants. However, in the real world, especially in the emergency response system, indeterminate factors may exist, and it is difficult for us to describe them as random variables due to the lack of history data. In this case, we usually invite some domain experts to evaluate the values of the parameters. As a result, we may get states such as "about 10 km", "about 5 tons" and "between 3 and 5 h". Uncertainty theory is a mathematical subject specially used to address this kind of expert's experimental data. To formulate the location-allocation problem with uncertain information, we assume that p_{ij}, c_{ij}, d_{ij}, s_j, q_i, and f_i are all uncertain variables, and rewrite them as ξ_{ij}, η_{ij}, ζ_{ij}, ψ_j, φ_i, and ω_i, respectively. That is,

ξ_{ij}: Uncertain unit material time penalty cost from C_i to $D_j, i \in I, j \in J$
η_{ij}: Uncertain unit material distribution cost from C_i to $D_j, i \in I, j \in J$
ζ_{ij}: Uncertain distance from C_i to $D_j, i \in I, j \in J$
ψ_j: Uncertain minimal demand of materials in $D_j, j \in J$
φ_i: Uncertain capacity of $C_i, i \in I$
ω_i: Uncertain operation cost of opening $C_i, i \in I$.

Accordingly, the model formulated above becomes an uncertain model. Employing expected value and chance-constraint, the uncertain location-allocation model can be developed as follows:

$$\min \quad \widetilde{F}_1(X, Y) = \max_{j \in J} E\left[\sum_{i \in I} \zeta_{ij} x_{ij}\right] \tag{9}$$

$$\min \quad \widetilde{F}_2(X, Y) = \max_{j \in J} E\left[\sum_{i \in I} \eta_{ij} x_{ij}\right] \tag{10}$$

$$\min \quad \widetilde{F}_3(X, Y) = \max_{j \in J} E\left[\sum_{i \in I} \zeta_{ij} e_{ij} x_{ij}\right] \tag{11}$$

$$\text{s.t.} \quad \mathcal{M}\left\{\sum_{i \in I} x_{ij} \geq \psi_j\right\} \geq \alpha_j, \quad \forall j \in J, \tag{12}$$

$$\mathcal{M}\left\{\sum_{j \in J} x_{ij} \leq \varphi_i y_i\right\} \geq \beta_i, \quad \forall i \in I, \tag{13}$$

$$\mathcal{M}\left\{\sum_{i \in I} \omega_i y_i \leq W\right\} \geq \gamma, \tag{14}$$

$$\sum_{i \in I} y_i \leq p, \tag{15}$$

$$x_{ij} \geq 0, y_i \in \{0,1\}, \quad \forall i \in I, j \in J, \tag{16}$$

where α_j, β_i and γ are predetermined confidence levels.

4.1. Deterministic Transformations

To solve the proposed uncertain model, it is necessary for us to discuss the equivalence of the model. Using θ, δ and μ as auxiliary variables, which are defined as follows:

$$\theta = \max_{j \in J} E\left[\sum_{i \in I} \xi_{ij} x_{ij}\right],$$

$$\delta = \max_{j \in J} E\left[\sum_{i \in I} \eta_{ij} x_{ij}\right],$$

$$\mu = \max_{j \in J} E\left[\sum_{i \in I} \zeta_{ij} e_{ij} x_{ij}\right].$$

Then, Models (9)–(16) is reformulated as

$$\min \quad \theta \tag{17}$$

$$\min \quad \delta \tag{18}$$

$$\min \quad \mu \tag{19}$$

$$\text{s.t.} \quad E\left[\sum_{i \in I} \xi_{ij} x_{ij}\right] \leq \theta, \quad \forall j \in J, \tag{20}$$

$$E\left[\sum_{i \in I} \eta_{ij} x_{ij}\right] \leq \delta, \quad \forall j \in J, \tag{21}$$

$$E\left[\sum_{i \in I} \zeta_{ij} e_{ij} x_{ij}\right] \leq \mu, \quad \forall j \in J, \tag{22}$$

$$\text{s.t.} \quad \mathcal{M}\left\{\sum_{i \in I} x_{ij} \geq \psi_j\right\} \geq \alpha_j, \quad \forall j \in J, \tag{23}$$

$$\mathcal{M}\left\{\sum_{j \in J} x_{ij} \leq \varphi_i y_i\right\} \geq \beta_i, \quad \forall i \in I,$$

$$\mathcal{M}\left\{\sum_{i \in I} \omega_i y_i \leq W\right\} \geq \gamma,$$

$$\sum_{i \in I} y_i \leq p,$$

$$x_{ij} \geq 0, y_i \in \{0,1\}, \quad \forall i \in I, j \in J,$$

Theorem 2. *Let ξ_{ij}, η_{ij}, and ζ_{ij} be independent uncertain variables with finite expected values. If ψ_j, φ_i, and ω_i are independent uncertain variables with regular uncertainty distributions Φ_j, Ψ_i, and Ω_i, respectively. Then, Models (17)–(23) can be converted into the following form:*

$$\min \quad \theta \tag{24}$$

$$\min \quad \delta \tag{25}$$

$$\min \quad \mu \tag{26}$$

$$\text{s.t.} \quad \sum_{i\in I} E[\xi_{ij}]x_{ij} \leq \theta, \quad \forall j \in J, \tag{27}$$

$$\sum_{i\in I} E[\eta_{ij}]x_{ij} \leq \delta, \quad \forall j \in J, \tag{28}$$

$$\sum_{i\in I} E[\zeta_{ij}]e_{ij}x_{ij} \leq \mu, \quad \forall j \in J, \tag{29}$$

$$\sum_{i\in I} x_{ij} \geq \Phi_j^{-1}(\alpha_j), \quad \forall j \in J, \tag{30}$$

$$\sum_{j\in J} x_{ij} \leq y_i \Psi_i^{-1}(1-\beta_i), \quad \forall i \in I, \tag{31}$$

$$\sum_{i\in I} y_i \Omega_i^{-1}(\gamma) \leq W, \tag{32}$$

$$\sum_{i\in I} y_i \leq p, \tag{33}$$

$$x_{ij} \geq 0, y_i \in \{0,1\}, \quad \forall i \in I, j \in J,$$

Proof. It follows from the linearity of expected value operator of uncertain variable that

$$E\left[\sum_{i\in I} \xi_{ij}x_{ij}\right] = \sum_{i\in I} E[\xi_{ij}]x_{ij},$$

which shows that constraint (27) holds for any $j \in J$. Similarly, we may verify that constraints (21) and (22) can be equivalently transformed into constraints (28) and (29), respectively.

Next, we prove that constraint (13) can be transformed into constraint (31). Denote Y_i as the uncertainty distribution of $\varphi_i y_i$. That is,

$$Y_i(x) = \mathcal{M}\{\varphi_i y_i \leq x\}$$

for any real number x. It follows from Theorem 1 that Y_i is regular since φ_i has a regular uncertainty distribution Ψ_i. In other words, for any $\beta_i \in (0,1)$, we have

$$Y_i^{-1}(\beta_i) = y_i \Psi_i^{-1}(\beta_i).$$

In addition,

$$\mathcal{M}\{\varphi_i y_i \leq y_i \Psi_i^{-1}(1-\beta_i)\} = \mathcal{M}\{\varphi_i y_i \leq Y_i^{-1}(1-\beta_i)\} = 1-\beta_i, \tag{34}$$

since Y_i is regular. According to constraint (13), for each i, we know

$$\mathcal{M}\left\{\varphi_i y_i \leq \sum_{j\in J} x_{ij}\right\} \leq 1-\beta_i. \tag{35}$$

As a result, it follows from (34) and (35) that constraint (31) holds. Similarly, we may also verify that constraints (12) and (14) can be converted into constraints (30) and (32), respectively. Thus, the proof is completed. $\square$

4.2. Solution Methodology

As mentioned above, to solve the proposed uncertain Models (9)–(16), the most important but most difficult step is to solve the Models (24)–(33), which is essentially a multi-objective programming model. Typically, no single optimal solution can be found to optimize all of the objectives at the same time. Instead, the efficient solution/Pareto optimal is commonly used for multi-objective optimization. To date, an assortment of methods, such as heuristic algorithms (Zhang and Xiong [40], Majumder et al. [41]), goal programming methods (Liu and Chen [42]) and multi-objective evolutionary algorithms

(Alcaraz et al. [43]), have been developed to solve multi-objective problems. Goal attainment technique (Azaron et al. [44]) is one of the multi-objective techniques, and has been successfully applied to production systems (Azaron et al. [45]) and supply chain design (Azaron et al. [44]). In the following, we will use the goal attainment technique to solve Models (24)–(33) and to generate Pareto-optimal solutions. Different from interactive multi-objective technology, the goal attainment technique is a one-stage method, so it will be computationally faster.

By using the goal attainment technique, Models (24)–(33) can be reformulated as follows:

$$\min \quad \pi \tag{36}$$

$$\text{s.t.} \quad \theta - w_1 \pi \leq g_1, \tag{37}$$

$$\delta - w_2 \pi \leq g_2, \tag{38}$$

$$\mu - w_3 \pi \leq g_3, \tag{39}$$

$$\text{s.t.} \quad \sum_{i \in I} E[\xi_{ij}] x_{ij} \leq \theta, \quad \forall j \in J, \tag{40}$$

$$\sum_{i \in I} E[\eta_{ij}] x_{ij} \leq \delta, \quad \forall j \in J,$$

$$\sum_{i \in I} E[\zeta_{ij}] e_{ij} x_{ij} \leq \mu, \quad \forall j \in J,$$

$$\sum_{i \in I} x_{ij} \geq \Phi_j^{-1}(\alpha_j), \quad \forall j \in J,$$

$$\sum_{j \in J} x_{ij} \leq y_i \Psi_i^{-1}(1 - \beta_i), \quad \forall i \in I,$$

$$\sum_{i \in I} y_i \Omega_i^{-1}(\gamma) \leq W,$$

$$\sum_{i \in I} y_i \leq p,$$

$$x_{ij} \geq 0, y_i \in \{0,1\}, \quad \forall i \in I, j \in J,$$

In Models (36)–(40), g_i and w_i, $i = 1,2,3$, are goals and weights for the three objective functions. The values of w_i are generally normalized such that $\sum_{i=1}^{3} w_i = 1$ and $w_i \geq 0$ for $i = 1,2,3$. The weights w_i relate the relative under-attainment of the goals g_i. In other words, the more important the goal g_i, the smaller the weight w_i.

Notice that if (X^*, Y^*) is Pareto-optimal, then there exists a pair of $(g = \{g_i\}, w = \{w_i\})$ such that (X^*, Y^*) is an optimal solution to the optimization Problems (36)–(40) (Azaron et al. [44]). Obviously, the optimal solution to model (36)–(40) is sensitive to (g, w). In the numerical experiments, we will generate different Pareto-optimal solutions by changing the values of g and w.

5. Innovations and Comparisons

In order to highlight the contributions of the proposed work, we will compare the article with existing works in two main aspects. First, the modeling method of this paper is discussed by comparing it with stochastic optimization models. Second, the innovations of the paper are further illustrated by comparing it with the existing works within the framework of uncertainty theory.

As previously discussed, we are usually in the state of indeterminacy since the highly unpredictable nature of emergencies. To date, a wide range of innovative studies on facility location and resource allocation problem have been conducted within the framework of probability theory. Although stochastic models have been widely accepted and applied to practical problems, it is not suitable to regard every indeterminate factor as a random factor. It is universally acknowledged that a premise of applying probability theory is that we can obtain sufficient historical data to estimate probability distribution. However, we often have no access to get the required observational data due to economical or technical reasons. In this case, we have no choice but to rely on the subjective-intuitive opinions of experts, which is named "belief degree". Obviously, it is not suitable to deal with belief degree by probability theory. In order to address belief degree rationally, uncertainty theory was founded by Liu [22] based on normality, duality, subadditivity and product axioms. Liu [22] also pointed out that belief degree follows the laws of uncertainty theory. Subsequently, uncertainty theory has been successfully applied to the fields of information science, management science and engineering. In this paper, uncertainty theory is employed to address location-allocation problem with uncertain parameters, which are described as uncertain variables. An uncertain multi-objective model is proposed by using of uncertainty theory. In more detail, the main difference between the proposed model and stochastic location-allocation models are demonstrated in Table 2.

Table 2. Comparison between the proposed model and stochastic models.

	Stochastic Location-Allocation Models	The Proposed Model
Sample size	Large enough	Too small (even no-sample)
Type of indeterminacy	Stochastic factors	Subjective-intuitive opinions of experts
Uncertain parameter	Random variable	Uncertain variable
Theoretical basis	Probability theory	Uncertainty theory

As a powerful tool to address belief degrees, uncertainty theory has been applied to the field of location problem and resource allocation problem. Compared with existing articles within the framework of uncertainty theory, the main innovation of the proposed paper is to introduce an approach to quantitatively characterizing urgency, and further develop a novel multi-objective model. The main difference among them are illustrated in Table 3 for better readability.

Table 3. Comparison between the proposed model and some related uncertain models.

References	Location Type	Uncertain Parameters	Modeling Approach	Critical Factors
Gao [27]	Single facility	Demand	Single objective	Satisfaction degree
Wen et al. [28]	Capacitated facility	Demand	Single objective; α-optimistic criterion	Transportation cost
Wang and Yang [29]	Hierarchical facility	Cost; the amount of waste	Single objective	Logistics cost
Zhang et al. [30]	Covering location	Demand; time	Single objective	Number of utilized facilities; the covered demand
Soltanpour et al. [31]	Inverse 1-median	Vertex weight; modification cost	Single objective	Total cost; tail value at risk
Wang and Qin [32]	Hub maximal covering	Travel time	Multi-objective	The total flows covered; total cost
Wen et al. [33]	Capacitated facility	Demand	Single objective; expected value	Transportation cost
Gao and Qin [34]	p-hub center	Travel time	Minimax single objective	The maximal travel time
Zhang et al. [35]	Emergency facility	Demand; travel time; opening cost	Multi-objective	Travel time; relief cost; carbon dioxide emission
Zhang et al. [36]	Capacitated p-center	Demand; distance	Minimax single objective	The maximal travel distance; capacity
The proposed model	Emergency distribution center	Demand; distance; time penalty cost; distribution cost; capacity; opening cost	Minimax multi-objective	Time penalty cost; distribution cost; carbon dioxide emission

6. Numerical Experiments

To verify the validity of the proposed model and test the efficiency of the solution method, numerical experiments are conducted in this section. To further illustrate the sensitivity of the optimal solution with respect to the values of goals g_i and weights w_i, more experiments with different values of g_i and w_i are provided.

We consider an instance with 12 demand points and 6 candidate emergency distribution centers, i.e., $m = 12$ and $n = 6$. Without loss of generally, the unit material time penalty cost ξ_{ij}, unit material distribution cost η_{ij} and distance ζ_{ij} from candidate emergency distribution center C_i to demand point D_j are assumed to be independent linear uncertain variables, which are shown in Table 4, Table 5 and Table 6, respectively. The minimal demands ψ_j are given in Table 7. The capacities φ_i and the operation costs ω_i are also assumed to be independent linear uncertain variables, which are shown in Table 8. In addition, set $p = 5$, $W = 120$, $e_{ij} = 6$ for any $i \in I, j \in J$.

Table 4. Unit material time penalty cost ξ_{ij}.

C_i \ D_j	D_1	D_2	D_3	D_4	D_5	D_6
C_1	$\mathcal{L}(1,5)$	$\mathcal{L}(2,6)$	$\mathcal{L}(1,4)$	$\mathcal{L}(3,7)$	$\mathcal{L}(1,7)$	$\mathcal{L}(3,8)$
C_2	$\mathcal{L}(2,5)$	$\mathcal{L}(1,6)$	$\mathcal{L}(1,5)$	$\mathcal{L}(2,7)$	$\mathcal{L}(4,7)$	$\mathcal{L}(2,8)$
C_3	$\mathcal{L}(3,6)$	$\mathcal{L}(1,4)$	$\mathcal{L}(2,4)$	$\mathcal{L}(3,5)$	$\mathcal{L}(2,5)$	$\mathcal{L}(1,4)$
C_4	$\mathcal{L}(2,4)$	$\mathcal{L}(3,6)$	$\mathcal{L}(3,6)$	$\mathcal{L}(1,5)$	$\mathcal{L}(2,7)$	$\mathcal{L}(1,5)$
C_5	$\mathcal{L}(4,8)$	$\mathcal{L}(3,6)$	$\mathcal{L}(2,5)$	$\mathcal{L}(2,6)$	$\mathcal{L}(3,8)$	$\mathcal{L}(1,3)$
C_6	$\mathcal{L}(3,7)$	$\mathcal{L}(2,4)$	$\mathcal{L}(2,6)$	$\mathcal{L}(1,5)$	$\mathcal{L}(2,6)$	$\mathcal{L}(2,5)$

C_i \ D_j	D_7	D_8	D_9	D_{10}	D_{11}	D_{12}
C_1	$\mathcal{L}(2,7)$	$\mathcal{L}(3,9)$	$\mathcal{L}(1,4)$	$\mathcal{L}(2,5)$	$\mathcal{L}(3,6)$	$\mathcal{L}(2,6)$
C_2	$\mathcal{L}(2,4)$	$\mathcal{L}(1,5)$	$\mathcal{L}(2,6)$	$\mathcal{L}(1,5)$	$\mathcal{L}(1,4)$	$\mathcal{L}(1,6)$
C_3	$\mathcal{L}(2,6)$	$\mathcal{L}(2,6)$	$\mathcal{L}(2,4)$	$\mathcal{L}(4,8)$	$\mathcal{L}(3,6)$	$\mathcal{L}(3,8)$
C_4	$\mathcal{L}(2,5)$	$\mathcal{L}(2,4)$	$\mathcal{L}(3,5)$	$\mathcal{L}(3,8)$	$\mathcal{L}(1,4)$	$\mathcal{L}(1,6)$
C_5	$\mathcal{L}(1,5)$	$\mathcal{L}(2,5)$	$\mathcal{L}(3,7)$	$\mathcal{L}(2,5)$	$\mathcal{L}(2,8)$	$\mathcal{L}(4,6)$
C_6	$\mathcal{L}(4,8)$	$\mathcal{L}(2,7)$	$\mathcal{L}(5,8)$	$\mathcal{L}(1,5)$	$\mathcal{L}(2,6)$	$\mathcal{L}(1,3)$

Table 5. Unit material distribution cost η_{ij}.

C_i \ D_j	D_1	D_2	D_3	D_4	D_5	D_6
C_1	$\mathcal{L}(13,16)$	$\mathcal{L}(10,16)$	$\mathcal{L}(12,14)$	$\mathcal{L}(13,15)$	$\mathcal{L}(11,13)$	$\mathcal{L}(14,17)$
C_2	$\mathcal{L}(12,16)$	$\mathcal{L}(11,17)$	$\mathcal{L}(12,14)$	$\mathcal{L}(10,16)$	$\mathcal{L}(13,18)$	$\mathcal{L}(13,16)$
C_3	$\mathcal{L}(10,15)$	$\mathcal{L}(12,16)$	$\mathcal{L}(15,18)$	$\mathcal{L}(10,14)$	$\mathcal{L}(12,16)$	$\mathcal{L}(13,15)$
C_4	$\mathcal{L}(15,18)$	$\mathcal{L}(12,16)$	$\mathcal{L}(14,16)$	$\mathcal{L}(10,15)$	$\mathcal{L}(10,13)$	$\mathcal{L}(12,15)$
C_5	$\mathcal{L}(13,16)$	$\mathcal{L}(11,14)$	$\mathcal{L}(11,16)$	$\mathcal{L}(12,17)$	$\mathcal{L}(10,15)$	$\mathcal{L}(11,13)$
C_6	$\mathcal{L}(10,15)$	$\mathcal{L}(12,14)$	$\mathcal{L}(12,16)$	$\mathcal{L}(11,16)$	$\mathcal{L}(10,16)$	$\mathcal{L}(14,16)$

C_i \ D_j	D_7	D_8	D_9	D_{10}	D_{11}	D_{12}
C_1	$\mathcal{L}(10,15)$	$\mathcal{L}(12,16)$	$\mathcal{L}(11,13)$	$\mathcal{L}(13,15)$	$\mathcal{L}(12,14)$	$\mathcal{L}(13,16)$
C_2	$\mathcal{L}(15,18)$	$\mathcal{L}(10,14)$	$\mathcal{L}(13,17)$	$\mathcal{L}(17,19)$	$\mathcal{L}(11,16)$	$\mathcal{L}(11,15)$
C_3	$\mathcal{L}(11,16)$	$\mathcal{L}(14,19)$	$\mathcal{L}(12,14)$	$\mathcal{L}(12,16)$	$\mathcal{L}(10,13)$	$\mathcal{L}(11,17)$
C_4	$\mathcal{L}(15,17)$	$\mathcal{L}(12,14)$	$\mathcal{L}(11,14)$	$\mathcal{L}(15,18)$	$\mathcal{L}(11,14)$	$\mathcal{L}(15,18)$
C_5	$\mathcal{L}(13,16)$	$\mathcal{L}(13,15)$	$\mathcal{L}(10,13)$	$\mathcal{L}(13,17)$	$\mathcal{L}(12,15)$	$\mathcal{L}(12,14)$
C_6	$\mathcal{L}(14,18)$	$\mathcal{L}(12,14)$	$\mathcal{L}(13,16)$	$\mathcal{L}(12,16)$	$\mathcal{L}(12,14)$	$\mathcal{L}(11,16)$

Table 6. Distance ζ_{ij}.

C_i \ D_j	D_1	D_2	D_3	D_4	D_5	D_6
C_1	$\mathcal{L}(5,10)$	$\mathcal{L}(4,8)$	$\mathcal{L}(10,15)$	$\mathcal{L}(3,9)$	$\mathcal{L}(10,15)$	$\mathcal{L}(8,12)$
C_2	$\mathcal{L}(10,14)$	$\mathcal{L}(8,10)$	$\mathcal{L}(6,9)$	$\mathcal{L}(8,12)$	$\mathcal{L}(12,15)$	$\mathcal{L}(10,14)$
C_3	$\mathcal{L}(13,16)$	$\mathcal{L}(8,13)$	$\mathcal{L}(6,10)$	$\mathcal{L}(10,14)$	$\mathcal{L}(12,14)$	$\mathcal{L}(5,8)$
C_4	$\mathcal{L}(11,15)$	$\mathcal{L}(7,10)$	$\mathcal{L}(6,10)$	$\mathcal{L}(11,16)$	$\mathcal{L}(7,12)$	$\mathcal{L}(13,16)$
C_5	$\mathcal{L}(12,15)$	$\mathcal{L}(9,14)$	$\mathcal{L}(10,13)$	$\mathcal{L}(10,15)$	$\mathcal{L}(8,10)$	$\mathcal{L}(12,14)$
C_6	$\mathcal{L}(6,11)$	$\mathcal{L}(14,19)$	$\mathcal{L}(13,17)$	$\mathcal{L}(9,13)$	$\mathcal{L}(13,18)$	$\mathcal{L}(13,17)$

C_i \ D_j	D_7	D_8	D_9	D_{10}	D_{11}	D_{12}
C_1	$\mathcal{L}(12,16)$	$\mathcal{L}(6,10)$	$\mathcal{L}(4,10)$	$\mathcal{L}(7,13)$	$\mathcal{L}(14,17)$	$\mathcal{L}(8,14)$
C_2	$\mathcal{L}(8,13)$	$\mathcal{L}(9,12)$	$\mathcal{L}(8,13)$	$\mathcal{L}(7,9)$	$\mathcal{L}(13,16)$	$\mathcal{L}(14,17)$
C_3	$\mathcal{L}(5,7)$	$\mathcal{L}(9,13)$	$\mathcal{L}(8,13)$	$\mathcal{L}(8,16)$	$\mathcal{L}(11,14)$	$\mathcal{L}(7,11)$
C_4	$\mathcal{L}(12,18)$	$\mathcal{L}(14,17)$	$\mathcal{L}(9,15)$	$\mathcal{L}(15,18)$	$\mathcal{L}(7,16)$	$\mathcal{L}(13,17)$
C_5	$\mathcal{L}(10,17)$	$\mathcal{L}(7,9)$	$\mathcal{L}(13,17)$	$\mathcal{L}(10,19)$	$\mathcal{L}(14,17)$	$\mathcal{L}(10,14)$
C_6	$\mathcal{L}(6,12)$	$\mathcal{L}(7,12)$	$\mathcal{L}(13,16)$	$\mathcal{L}(7,13)$	$\mathcal{L}(7,12)$	$\mathcal{L}(10,13)$

Table 7. The minimal demand ψ_j.

D_j	D_1	D_2	D_3	D_4	D_5	D_6
ψ_j	$\mathcal{L}(25,50)$	$\mathcal{L}(30,40)$	$\mathcal{L}(25,60)$	$\mathcal{L}(35,65)$	$\mathcal{L}(20,50)$	$\mathcal{L}(30,55)$

D_j	D_7	D_8	D_9	D_{10}	D_{11}	D_{12}
ψ_j	$\mathcal{L}(35,70)$	$\mathcal{L}(25,55)$	$\mathcal{L}(30,40)$	$\mathcal{L}(35,50)$	$\mathcal{L}(20,55)$	$\mathcal{L}(20,35)$

Table 8. The capacity φ_i and the operation cost ω_i.

C_i	C_1	C_2	C_3	C_4	C_5	C_6
φ_i	$\mathcal{L}(184,194)$	$\mathcal{L}(190,210)$	$\mathcal{L}(165,185)$	$\mathcal{L}(220,240)$	$\mathcal{L}(200,210)$	$\mathcal{L}(175,190)$
ω_i	$\mathcal{L}(15,30)$	$\mathcal{L}(20,30)$	$\mathcal{L}(10,30)$	$\mathcal{L}(25,35)$	$\mathcal{L}(20,40)$	$\mathcal{L}(20,25)$

The problem attempts to minimize the time penalty cost, the total material distribution cost and the CO_2 emissions in the sense of expected value while making the following determinations: (i) which of the candidate emergency distribution centers to open; and (ii) for each demand point, which distribution centers are assigned to it? We use the multi-objective uncertain programming approach to this multi-objective optimization problem and employ the goal attainment technique to solve the multi-objective model. Then, a mathematical model can be formulated as model (36)–(40), which contains 82 variables and 132 constraints, is essentially a deterministic linear programming model.

To generate the Pareto-optimal solutions, the values of g_i and w_i ($i = 1, 2, 3$) are varied manually. In order to illustrate the sensitivity of parameters, when one of the parameters is varied, the others are fixed. According to the obtained absolute minimum values for the maximal time penalty cost (i.e., "urgency"), the maximum material distribution cost and the maximum CO_2 emissions, by solving the corresponding single objective models, g_1 is varied from 200 (close to the absolute minimum value for the maximal time penalty cost) to 400, g_2 is varied from 850 (close to the absolute minimum value for the maximal material distribution cost) to 1700, g_3 is varied from 3000 (close to the absolute minimum value for the maximal CO_2 emissions) to 6000. Similarly, w_i are varied from 0.001 to 0.998 for $i = 1, 2, 3$. Let $\alpha_j = \beta_i = \gamma = 0.9$, by using of LINGO 10, 30 Pareto-optimal strategies and the corresponding objective values for time penalty cost, material distribution cost and the CO_2 emissions are generated and are revealed in Table 9. Obviously, each strategy generates a location-allocation strategy. For example, according to strategy 1, the candidate emergency distribution centers 1, 3, 5 and 6 should be selected. In more detail, the resource allocation strategy is illustrated in Figure 1.

For each pair of (g, w), the Pareto-optimal solution is obtained. If we are not satisfied with any Pareto-optimal solution, then the value of w should be modified. As mentioned above, the values of w_i relate the relative under-attainment of the objective goals g_i. The more important the goal g_i, the smaller the value of w_i. For example, if the obtained value of objective 1 is much greater than 200, w_1 should be decreased. Accordingly, both w_2 and w_3 should also be modified such that the summation of w_i equals to 1. Obviously, this process can be repeated with different pairs of (g, w) until the decision-maker obtains a satisfactory Pareto-optimal solution.

For example, strategy 1 in Table 9 implies that one unit deviation of the time penalty cost from 200 is about 1000 times as important as one unit deviation of the CO_2 emissions from 3000 and the same important as one unit deviation of the material distribution cost from 850. In this strategy, the weights for time penalty cost and material distribution cost are relatively low, which result in the strategy has low time penalty cost and material distribution cost. As shown in strategies 1–6, as the value of w_2 increases, the value of material distribution cost also increases gradually. For better readability, the results are shown in Figure 2.

Table 9. Pareto-optimal solutions.

Strategy	g_1	g_2	g_3	w_1	w_2	w_3	Selected Centers	Obj. 1	Obj. 2	Obj. 3
1	200	850	3000	0.001	0.001	0.998	$C_1 C_3 C_5 C_6$	248.68	898.68	5484.86
2	200	850	3000	0.001	0.005	0.994	$C_1 C_3 C_5 C_6$	217.93	939.67	5423.37
3	200	850	3000	0.001	0.01	0.989	$C_1 C_3 C_5 C_6$	210.02	950.22	5407.54
4	200	850	3000	0.001	0.05	0.949	$C_2 C_3 C_5 C_6$	202.33	966.58	5212.65
5	200	850	3000	0.001	0.1	0.899	$C_2 C_3 C_5 C_6$	201.83	1032.83	4643.60
6	200	850	3000	0.001	0.5	0.499	$C_1 C_2 C_3 C_6$	202.24	1089.04	4115.64
7	200	850	3000	0.001	0.998	0.001	$C_1 C_3 C_5 C_6$	255.13	995.30	3055.13
8	200	850	3000	0.01	0.989	0.001	$C_1 C_3 C_5 C_6$	258.41	932.25	3005.84
9	200	850	3000	0.1	0.899	0.001	$C_1 C_3 C_5 C_6$	258.76	927.74	3000.59
10	200	850	3000	0.998	0.001	0.001	$C_1 C_3 C_4 C_6$	310.00	884.41	3034.41
11	200	850	3000	0.989	0.01	0.001	$C_1 C_3 C_4 C_6$	297.96	885.05	3003.51
12	200	850	3000	0.899	0.1	0.001	$C_1 C_3 C_4 C_6$	297.96	885.12	3000.35
13	400	850	3000	0.001	0.001	0.998	$C_1 C_2 C_3 C_6$	289.07	851.61	4608.60
14	400	850	3000	0.001	0.005	0.994	$C_1 C_2 C_3 C_6$	288.75	856.83	4358.09
15	400	850	3000	0.001	0.01	0.989	$C_1 C_2 C_3 C_6$	288.75	861.48	4135.09
16	400	850	3000	0.001	0.05	0.949	$C_1 C_3 C_5 C_6$	288.75	875.17	3477.76
17	400	850	3000	0.001	0.1	0.899	$C_1 C_3 C_5 C_6$	295.83	879.58	3265.96
18	400	850	3000	0.001	0.5	0.499	$C_1 C_3 C_4 C_6$	272.67	884.41	3034.34
19	200	1700	3000	0.001	0.998	0.001	$C_1 C_3 C_5 C_6$	255.13	932.25	3055.13
20	200	1700	3000	0.01	0.989	0.001	$C_1 C_3 C_5 C_6$	258.41	914.76	3005.84
21	200	1700	3000	0.1	0.899	0.001	$C_1 C_3 C_5 C_6$	258.76	914.35	3000.59
22	200	1700	3000	0.998	0.001	0.001	$C_1 C_3 C_5 C_6$	258.80	902.95	3000.06
23	200	1700	3000	0.989	0.01	0.001	$C_1 C_3 C_5 C_6$	258.80	911.22	3000.06
24	200	1700	3000	0.899	0.1	0.001	$C_1 C_3 C_5 C_6$	258.80	911.22	3000.07
25	200	850	6000	0.001	0.998	0.001	$C_1 C_2 C_5 C_6$	200.11	963.43	5387.73
26	200	850	6000	0.01	0.989	0.001	$C_1 C_2 C_5 C_6$	201.13	962.07	5389.77
27	200	850	6000	0.1	0.899	0.001	$C_1 C_3 C_5 C_6$	211.00	948.91	5409.51
28	200	850	6000	0.998	0.001	0.001	$C_1 C_3 C_5 C_6$	285.12	850.09	5557.75
29	200	850	6000	0.989	0.01	0.001	$C_1 C_3 C_5 C_6$	284.55	850.85	5556.60
30	200	850	6000	0.899	0.1	0.001	$C_1 C_3 C_5 C_6$	278.63	858.75	5544.76

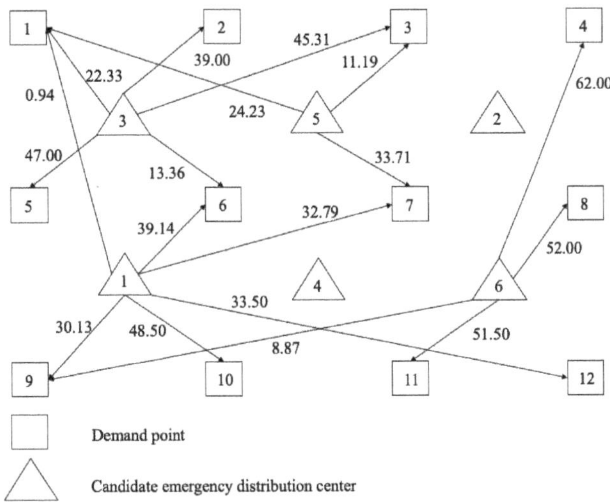

Figure 1. Location-allocation for strategy 1 in Table 9.

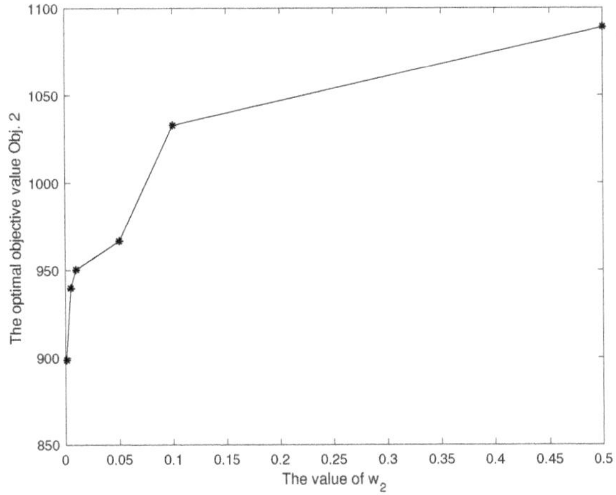

Figure 2. The value of material distribution cost with different w_2.

It follows from Table 9 that the lowest time penalty cost comes from strategy 25, with a value of 200.11. The strategy has relatively low material distribution cost, but high CO_2 emissions. The lowest CO_2 emissions is obtained in strategy 22 and strategy 23, with a time penalty cost of 258.80. The optimal material distribution cost is obtained in strategy 28, with a time penalty cost of 285.12. So, the time penalty cost ranges from 200.11 to 285.12. To find an appropriate solution, the decision-maker needs insight into this range of outcomes, to be able to trade-off different criteria in terms of the results. According to the numerical results, it is concluded that the relationship among time penalty cost, material distribution cost and CO_2 emissions is not clear and it is not easy to define a function to handle the relation among them. That is why we formulate a multi-objective model to solve this location-allocation problem.

It is also can be seen from Table 9 that increasing goal for the time penalty cost (i.e., g_1) and keeping other parameters unchanged cause the material distribution cost and CO_2 emissions to be decreased (see strategies 1–6 and strategies 13–18). In addition, the results are shown in Figures 3 and 4. Similarly, increasing goal for the material distribution cost causes both the time penalty cost and the CO_2 emissions to be decreased (see strategies 7–12 and strategies 19–24). In addition, the increasing goal for the CO_2 emissions may cause time penalty cost or material distribution cost to be decreased (see strategies 7–12 and strategies 25–30). In other words, it seems that by increasing the goal of any one of the objectives, we can provide more space for other objectives to be improved.

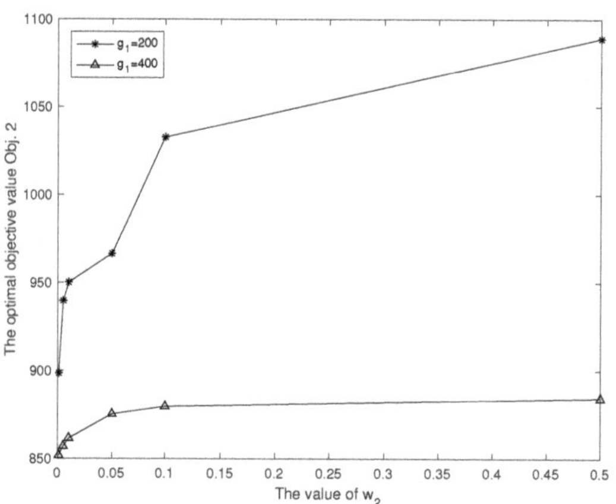

Figure 3. The material distribution cost with different g_1.

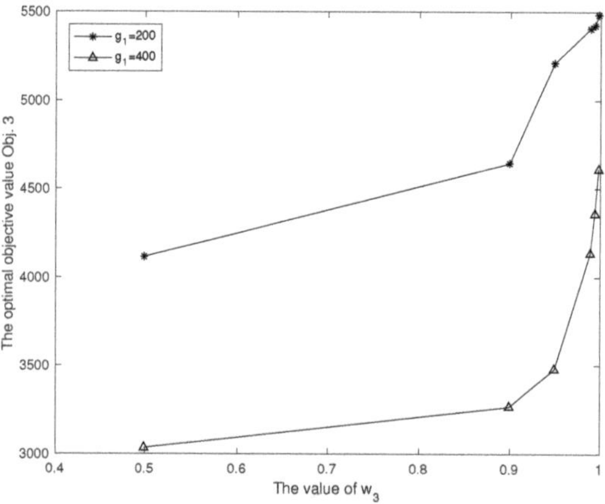

Figure 4. The CO_2 emissions with different g_1.

7. Conclusions and Future Research

This paper studies the location-allocation problem within the framework of uncertainty theory. Its main contribution, the time penalty cost function, is introduced to characterize the urgency of emergency response and the formulation of a multi-objective model to consider urgency, relief costs and CO_2 emissions simultaneously. In the solving procedures, we transform the model from an uncertain programming problem to a deterministic programming problem.

In the numerical experiments, a series of Pareto-optimal solutions are generated by goal attainment technique. The results reveal that the goal attainment technique can provide an effective approach to trade-off urgency, relief costs and CO_2 emissions. It also shows that the goal attainment technique will help the decision-makers to find the best strategy based on their preferences by varying parameter (g, w). However, for large-scale problems, this technique may no longer be effective, in terms of computational time. In this case, some heuristic algorithms, such as genetic algorithms or simulated annealing, would be suitable.

Several directions for further studies are pointed out here. Future research could take the choices by beneficiaries into account, which will improve the allocation of relief goods and effectively enhance the use of scarce resources. The beneficiaries remain free in their choice of the distribution centers, so the preferences of beneficiaries rather than cost should be considered for decision-makers. Apart from that, for large-scale emergencies, the relief goods should be delivered to distant demand points. Therefore, it is necessary to develop a series of meta-heuristic approaches such as genetic algorithm or tabu search algorithm to solve large-scale problems. In addition, we could consider the problem under a more complex environment. For example, in an uncertain random environment, some indeterminate quantities can be interpreted as uncertain variables, while others may be interpreted as random variables.

Author Contributions: Methodology, H.L. and B.Z.; validation, B.Z. and X.G.; investigation, H.L. and B.Z.; writing—original draft preparation, H.L.; writing—review and editing, B.Z. and X.G. All authors have read and agreed to the published version of the manuscript.

Funding: This work was supported by the National Natural Science Foundation of China (Grant Nos. 61703438 and 71974204), and the Scientific Research Project of Education Department of Hubei Provincial of China (Grant No. Q20203001).

Institutional Review Board Statement: Not applicable.

Informed Consent Statement: Not applicable.

Data Availability Statement: Not applicable.

Conflicts of Interest: The authors declare no conflict of interest.

References

1. Sheu, J.B. Challenges of emergency logistics management. *Transport. Res. E Log.* **2007**, *43*, 655–659. [CrossRef]
2. Cooper, L. Location-allocation problems. *Oper. Res.* **1963**, *11*, 331–344. [CrossRef]
3. Beaumont, J.R. Location-allocation problems in a plane a review of some models. *Socio-Econ. Plan. Sci.* **1981**, *15*, 217–229. [CrossRef]
4. de Camargo, R.S.; Miranda, G. Single allocation hub location problem under congestion: Network owner and user perspectives. *Expert. Syst. Appl.* **2012**, *39*, 3385–3391. [CrossRef]
5. Murtagh, B.A.; Niwattisyawong, S.R. An efficient method for the multi-depot location-allocation problem. *J. Oper. Res. Soc.* **1982**, *33*, 629–634.
6. Badri, M.A. Combining the analytic hierarchy process and goal programming for global facility location-allocation problem. *Int. J. Prod. Econ.* **1999**, *62*, 237–248. [CrossRef]
7. Fang, L.; Li, H. Multi-criteria decision analysis for efficient location-allocation problem combining DEA and goal programming. *RAIRO Oper. Res.* **2015**, *49*, 753–772. [CrossRef]
8. Fan, J.; Yu, L.; Li, X.; Shang, C.; Ha, M. Reliable location allocation for hazardous materials. *Inf. Sci.* **2019**, *501*, 688–707. [CrossRef]
9. Chu, H.; Chen, Y. A novel hybrid algorithm for multiobjective location-allocation problem in emergency logistics. *Comput. Intell. Neurosci.* **2021**, *2021*, 1951161. [CrossRef]
10. Wang, W.; Wu, S.; Wang, S.; Zhen, L.; Qu, X. Emergency facility location problems in logistics: Status and perspectives. *Transport. Res. E Log.* **2021**, *154*, 102465. [CrossRef]
11. Kolmogorov, A.N. *Grundbegriffe der Wahrscheinlichkeitsrechnung*; Springer: Berlin, Germany, 1933.
12. Zhou, J.; Liu, B. New stochastic models for capacitated location-allocation problem. *Comput. Ind. Eng.* **2003**, *45*, 111–125. [CrossRef]
13. Alizadeh, M.; Ma, J.; Mahdavi-Amiri, N.; Marufuzzaman, M.; Jaradat, R. A stochastic programming model for a capacitated location-allocation problem with heterogeneous demands. *Comput. Ind. Eng.* **2019**, *137*, 106055. [CrossRef]
14. Cheng, Y.; Wang, L. A location model for capacitated alternative-fuel stations with uncertain traffic flows. *Comput. Ind. Eng.* **2020**, *145*, 106486. [CrossRef]
15. Hu, Q.; Hu, S.; Wang, J.; Li, X. Stochastic single allocation hub location problems with balanced utilization of hub capacities. *Transport. Res. B Meth.* **2021**, *153*, 204–227. [CrossRef]
16. Zadeh, L.A. Fuzzy sets. *Inform. Control* **1965**, *8*, 338–353. [CrossRef]
17. Zhou, J.; Liu, B. Modeling capacitated location-allocation problem with fuzzy demands. *Comput. Ind. Eng.* **2007**, *53*, 454–468. [CrossRef]

18. Wen, M.; Iwamura, K. Fuzzy facility location-allocation problem under the Hurwicz criterion. *Eur. J. Oper. Res.* **2008**, *184*, 627–635. [CrossRef]
19. Liu, Z.; Qu, S.; Wu, Z.; Ji, Y. Two-stage fuzzy mixed integer optimization model for three-level location allocation problems under uncertain environment. *J. Intell. Fuzzy Syst.* **2020**, *39*, 1–16. [CrossRef]
20. Kahneman, D.; Tversky, A. Prospect theory: An analysis of decision under risk. *Econometrica* **1979**, *47*, 263–292. [CrossRef]
21. Liu, B. *Uncertainty Theory*, 4th ed.; Springer: Berlin, Germany, 2015.
22. Liu, B. *Uncertainty Theory*, 2nd ed.; Springer: Berlin, Germany, 2007.
23. Liu, B. *Theory and Practice of Uncertain Programming*, 2nd ed.; Springer: Berlin, Germany, 2009.
24. Gao, Y. Shortest path problem with uncertain arc lengths. *Comput. Math. Appl.* **2011**, *62*, 2591–2600. [CrossRef]
25. Ma, H.; Li, X. Closed-loop supply chain network design for hazardous products with uncertain demands and returns. *Appl. Soft Comput.* **2018**, *68*, 889–899. [CrossRef]
26. Sheng, L.; Zhu, Y.; Wang, K. Analysis of a class of dynamic programming models for multi-stage uncertain systems. *Appl. Math. Model.* **2020**, *86*, 446–459. [CrossRef]
27. Gao, Y. Uncertain models for single facility location problems on networks. *Appl. Math. Model.* **2012**, *36*, 2592–2599. [CrossRef]
28. Wen, M.; Qin, Z.; Kang, R. The α-cost minimization model for capacitated facility location-allocation problem with uncertain demands. *Fuzzy Optim. Decis. Mak.* **2014**, *13*, 345–356. [CrossRef]
29. Wang, K.; Yang, Q. Hierarchical facility location for the reverse logistics network design under uncertainty. *J. Uncertain Syst.* **2014**, *8*, 255–270.
30. Zhang, B.; Peng, J.; Li, S. Covering location problem of emergency service facilities in an uncertain environment. *Appl. Math. Model.* **2017**, *51*, 429–447. [CrossRef]
31. Soltanpour, A.; Baroughi, F.; Alizadeh, B. The inverse 1-median location problem on uncertain tree networks with tail value at risk criterion. *Inf. Sci.* **2020**, *506*, 383–394. [CrossRef]
32. Wang, J.; Qin, Z. Modelling and analysis of uncertain hub maximal covering location problem in the presence of partial coverage. *J. Intell. Fuzzy Syst.* **2021**, *40*, 1–16. [CrossRef]
33. Wen, M.; Qin, Z.; Kang, R.; Yang, Y. The capacitated facility location-allocation problem under uncertain environment. *J. Intell. Fuzzy Syst.* **2015**, *29*, 2217–2226. [CrossRef]
34. Gao, Y.; Qin, Z. A chance constrained programming approach for uncertain p-hub center location problem. *Comput. Ind. Eng.* **2016**, *102*, 10–20. [CrossRef]
35. Zhang, B.; Li, H.; Li, S.; Peng, J. Sustainable multi-depot emergency facilities location-routing problem with uncertain information. *Appl. Math. Comput.* **2018**, *333*, 506–520. [CrossRef]
36. Zhang, B.; Peng, J.; Li, S. Minimax models for capacitated p-center problem in uncertain environment. *Fuzzy Optim. Decis. Mak.* **2021**, *20*, 273–292. [CrossRef]
37. Liu, B. Some research problems in uncertainty theory. *J. Uncertain Syst.* **2009**, *3*, 3–10.
38. Liu, B. *Uncertainty Theory: A Branch of Mathematics for Modeling Human Uncertainty*; Springer: Berlin, Germany, 2010.
39. Wan, S.; Chen, Z.; Dong, J. Bi-objective trapezoidal fuzzy mixed integer linear program-based distribution center location decision for large-scale emergencies. *Appl. Soft Comput.* **2021**, *110*, 107757. [CrossRef]
40. Zhang, Q.; Xiong, S. Routing optimization of emergency grain distribution vehicles using the immune ant colony optimization algorithm. *Appl. Soft Comput.* **2018**, *71*, 917–925. [CrossRef]
41. Majumder, S.; Kar, M.B.; Kar, S.; Pal, T. Uncertain programming models for multi-objective shortest path problem with uncertain parameters. *Soft Comput.* **2020**, *24*, 8975–8996. [CrossRef]
42. Liu, B.; Chen, X. Uncertain multiobjective programming and uncertain goal programming. *J. Uncertain. Anal. Appl.* **2015**, *3*, 10. [CrossRef]
43. Alcaraz, J.; Landete, M.; Monge, J.F.; Sainz-Pardo, J.L. Multi-objective evolutionary algorithms for a reliability location problem. *Eur. J. Oper. Res.* **2020**, *283*, 83–93. [CrossRef]
44. Azaron, A.; Brown, K.N.; Tarim, S.A.; Modarres, M. A multi-objective stochastic programming approach for supply chain design considering risk. *Int. J. Prod. Econ.* **2008**, *116*, 129–138. [CrossRef]
45. Azaron, A.; Katagiri, H.; Kato, K.; Sakawa, M. Modelling complex assemblies as a queueing network for lead time control. *Eur. J. Oper. Res.* **2006**, *174*, 150–168. [CrossRef]

Article

Optimal Asynchronous Dynamic Policies in Energy-Efficient Data Centers

Jing-Yu Ma [1,*] , **Quan-Lin Li** [2] **and Li Xia** [3]

1   Business School, Xuzhou University of Technology, Xuzhou 221018, China
2   School of Economics and Management, Beijing University of Technology, Beijing 100124, China; liquanlin@tsinghua.edu.cn
3   School of Business, Sun Yat-sen University, Guangzhou 510275, China; xiali5@sysu.edu.cn
*   Correspondence: mjy0501@126.com

Abstract: In this paper, we apply a Markov decision process to find the optimal asynchronous dynamic policy of an energy-efficient data center with two server groups. Servers in Group 1 always work, while servers in Group 2 may either work or sleep, and a fast setup process occurs when the server's states are changed from sleep to work. The servers in Group 1 are faster and cheaper than those of Group 2 so that Group 1 has a higher service priority. Putting each server in Group 2 to sleep can reduce system costs and energy consumption, but it must bear setup costs and transfer costs. For such a data center, an asynchronous dynamic policy is designed as two sub-policies: The setup policy and the sleep policy, both of which determine the switch rule between the work and sleep states for each server in Group 2. To find the optimal asynchronous dynamic policy, we apply the sensitivity-based optimization to establish a block-structured policy-based Markov process and use a block-structured policy-based Poisson equation to compute the unique solution of the performance potential by means of the RG-factorization. Based on this, we can characterize the monotonicity and optimality of the long-run average profit of the data center with respect to the asynchronous dynamic policy under different service prices. Furthermore, we prove that a bang–bang control is always optimal for this optimization problem. We hope that the methodology and results developed in this paper can shed light on the study of more general energy-efficient data centers.

Keywords: asynchronous dynamic policy; energy-efficient data center; Markov decision process; RG-factorization; sensitivity-based optimization

check for **updates**

Citation: Ma, J.-Y.; Li, Q.-L.; Xia, L. Optimal Asynchronous Dynamic Policies in Energy-Efficient Data Centers. *Systems* **2022**, *10*, 27. https://doi.org/10.3390/systems10020027

Academic Editor: Khac Duc Do

Received: 24 January 2022
Accepted: 28 February 2022
Published: 2 March 2022

Publisher's Note: MDPI stays neutral with regard to jurisdictional claims in published maps and institutional affiliations.

1. Introduction

Over the last two decades considerable attention has been given to studying energy- efficient data centers. On the one hand, as the number and size of data centers increase rapidly, energy consumption becomes one main part of the operating costs of data centers. On the other hand, data centers have become a fundamental part of the IT infrastructure in today's Internet services, in which a huge number of servers are deployed in each data center such that the data centers can provide cloud computing environments. Therefore, finding optimal energy-efficient policies and designing optimal energy-efficient mechanisms are always interesting, difficult, and challenging in the energy-efficient management of data centers. Readers may refer to recent excellent survey papers, such as Masanet et al. [1], Zhang et al. [2], Nadjahi et al. [3], Koot and Wijnhoven [4], Shirmarz and Ghaffari [5], Li et al. [6], and Harchol-Balter [7].

Barroso and Hölzle [8] demonstrated that many data centers were designed to handle peak loads effectively, but this directly caused a significant number of servers (approximately 20%) in the data centers to be idle because no work was done in the off-peak period. Although the idle servers do not provide any services, they still continue to consume a notable amount of energy, which is approximately 70% of the servers working in the on-peak period. Therefore, it is necessary and useful to design an energy-efficient mechanism for specifically dealing with idle servers. In this case, a key technique, the energy-efficient states of "Sleep" or "Off" were introduced such that the idle servers can take the sleep or off state, which always consumes less energy, in which the sleep state consumes very little energy, while the energy consumption of the off state is zero. See Kuehn and Mashaly [9] for more interpretation. To analyze the energy-efficient states, thus far, some queueing systems either with server energy-efficient states (e.g., work, idle, sleep, and off) or with server control policies (e.g., vacation, setup, and N-policy) have been developed in the study of energy-efficient data centers. Important examples include the survey papers by Gandhi [10] and Li et al. [6], and the research papers by Gandhi et al. [11–13], Maccio and Down [14,15], Phung-Duc [16] and Phung-Duc and Kawanishi [17].

It is always necessary to design an optimal energy-efficient mechanism for data centers. To do this, several static optimization methods have been developed using two basic approaches. The first is to construct a suitable utility function for a performance-energy trade-off or a performance cost with respect to the synchronous optimization of different performance measures, for example, reducing energy consumption, reducing system response time, and improving quality of service. The second is to minimize the performance cost by means of some optimal methods, including linear programming, nonlinear programming, and integer programming. Gandhi et al. [12] provided two classes of the performance-energy trade-offs: **(a) ERWS**, the weighted sum $\beta_1 \mathbf{E}[R] + \beta_2 \mathbf{E}[E]$ of the mean response time $\mathbf{E}[R]$ and the mean power cost $\mathbf{E}[E]$, where $\beta_1, \beta_2 \geq 0$ are weighted coefficients; **(b) ERP**, the product $\mathbf{E}[R]\mathbf{E}[E]$ of the mean response time and the mean power cost. See Gandhi et al. [12] and Gandhi [10] for systematical research. Maccio and Down [14] generalized the **ERP** to a more general performance cost function that considers the expected cycle rate. Gebrehiwot et al. [18] made some interesting generalizations of the **ERP** and **ERWS** by introducing multiple intermediate sleep states. Additionally, Gebrehiwot et al. [19,20] generalized the FCFS queue of the data center with multiple intermediate sleep states to the processor-sharing discipline and to the shortest remaining processing time (SRPT) discipline, respectively. In addition, Mitrani [21,22] provided another interesting method to discuss the data center of N identical servers that contain m reserves, while the idle or work of the servers is controlled by an up threshold U and a down threshold D. He established a new performance cost $C = c_1 \mathbf{E}[L] + c_2 \mathbf{E}[S]$ and provided expressions for the average numbers $\mathbf{E}[L]$ and $\mathbf{E}[S]$, so that the performance cost C can be optimized with respect to the three key parameters m, U, and D.

To date, little work has been done on applications of Markov decision processes (MDPs) to find the optimal dynamic control policies of energy-efficient data centers. Readers may refer to recent publications for details, among which Kamitsos et al. [23] constructed a discrete-time MDP and proved that the optimal sleep energy-efficient policy is simply hysteretic, so that it has a double threshold structure. Note that such an optimal hysteretic policy follows Hipp and Holzbaur [24] and Lu and Serfozo [25]. For policy optimization and dynamic power management for electronic systems or equipment, the MDPs and stochastic network optimization were developed from five different perspectives: (a) the discrete-time MDPs by Yang et al. [26]; (b) the continuous-time MDPs by Ding et al. [27]; (c) stochastic network optimization by Liang et al. [28]; (d) the sensitivity-based optimization by Xia and Chen [29], Xia et al. [30], and Ma et al. [31]; and (e) the deep reinforcement learning by Chi et al. [32].

The purpose of this paper is to apply continuous-time MDPs to set up optimal dynamic energy-efficient policies for data centers, in which the sensitivity-based optimization is developed to find the optimal solution. Note that the sensitivity-based optimization is greatly refined from the MDPs by dealing with the Poisson equation by means of some novel tools, for instance, performance potential and performance difference. To date, the sensitivity-based optimization has also been sufficiently related to the Markov reward processes (see Li [33]), thus, it is an effective method for the performance optimization of many practical systems (see an excellent book by Cao [34] for more details). The sensitivity-based optimization theory has been applied to performance optimization in many practical areas. For example, in energy-efficient data centers by Xia et al. [35]; inventory rationing by Li et al. [36]; the blockchain selfish mining by Ma and Li [37]; and in finance by Xia [38].

The main contributions of this paper are threefold. The first contribution is to apply the sensitivity-based optimization (and the MDPs) to study a more general energy-efficient data center with key practical factors, for example, a finite buffer, a fast setup process, and transferring some incomplete service jobs to the idle servers in Group 1 or to the finite buffer, if any. Although practical factors will not increase any difficulty in performance evaluation (e.g., modeling by means of queueing systems or Markov processes, also see Gandhi [10] for more details), they can largely cause substantial difficulties and challenges in finding optimal dynamic energy-efficient policies and, furthermore, in determining threshold-type policies by using the sensitivity-based optimization. For instance, the finite buffer makes the policy-based Markov process appear as the two-dimensional block-structured Markov process from the one-dimensional birth–death process given in Ma et al. [31] and Xia et al. [35].

Note that this paper has two related works: Ma et al. [31] and Xia et al. [35], and it might be necessary to set up some useful relations between this paper and each of the two papers. Compared with Ma et al. [31], this paper considers more practical factors in the energy-efficient data centers such that the policy-based Markov process is block-structured, which makes solving the block-structured Poisson equation more complicated. Compared with Xia et al. [35], this paper introduces a more detailed cost and reward structure, which makes an analysis of the monotonicity and optimality of dynamic energy-efficient policies more difficult and challenging. Therefore, this paper is a necessary and valuable generalization of Ma et al. [31] and Xia et al. [35] through extensively establishing the block-structured policy-based Markov processes, which in fact are the core part of the sensitivity-based optimization theory and its applications in various practical systems.

The second contribution of this paper is that it is the first to find an optimal asynchronous dynamic policy in the study of energy-efficient data centers. Note that the two groups of servers in the data center have "the setup actions from the sleep state to the work state" and "the close actions from the work state to the sleep state", thus, we follow the two action steps to form an asynchronous dynamic policy, which is decomposed into two sub-policies: the setup policy (or the setup action) and the sleep policy (or the close action). Crucially, one of the successes of this paper is to find the optimal asynchronous dynamic policy from many asynchronous dynamic policies by means of the sensitivity-based optimization. To date, it has still been very difficult and challenging in the MDPs.

The third contribution of this paper is to provide a unified framework for applying the sensitivity-based optimization to study the optimal asynchronous dynamic policy of the energy-efficient data center. For such a more complicated energy-efficient data center, we first establish a policy-based block-structured Markov process as well as a more detailed cost and reward structure, and provide an expression for the unique solution to the block-structured Poisson equation by means of the RG-factorization. Then, we show the monotonicity of the long-run average profit with respect to the setup and sleep policies and the asynchronous policy, respectively. Based on this, we find the optimal asynchronous policy when the service price is higher (or lower) than a key threshold. Finally, we indicate that the optimal control is a bang–bang control. Such a structure of the optimal asynchronous energy-efficient policy reduces the search space, which is a significant reduction of the optimization complexity and effectively alleviates the curse of the dimensionality of MDPs. Therefore, the optimal asynchronous dynamic policy is the threshold-type in the energy-efficient data center. Note that the optimality of the threshold-type policy can realize a large reduction for the search space, thus, the optimal threshold-type policy is of great significance to solve the mechanism design problem of energy-efficient data centers. Therefore, the methodology and results developed in this paper provide new highlights for understanding dynamic energy-efficient policy optimization and mechanism design in the study of more general data centers.

The organization of this paper is as follows. In Section 2, we give a problem description for an energy-efficient data center with more practical factors. In Section 3, we establish a policy-based continuous-time block-structured Markov process and define a suitable reward function with respect to both states and policies of the Markov process. In Section 4, we set up a block-structured Poisson equation and provide an expression for its unique solution by means of the RG-factorization. In Section 5, we study a perturbation realization factor of the policy-based continuous-time block-structured Markov process for the asynchronous dynamic policy, and analyze how the service price impacts on the perturbation realization factor. In Section 6, we discuss the monotonicity and optimality of the long-run average profit of the energy-efficient data center with respect to the asynchronous policy. Based on this, we can give the optimal asynchronous dynamic policy of the energy-efficient data center. In Section 7, if the optimal asynchronous dynamic policy is the threshold-type, then we can compute the maximal long-run average profit of the energy-efficient data center. In Section 8, we give some concluding remarks. Finally, three appendices are given, both for the state-transition relation figure of the policy-based block-structured continuous-time Markov process and for the block entries of its infinitesimal generator.

2. Model Description

In this section, we provide a problem description for setting up and optimizing an asynchronous dynamic policy in an energy-efficient data center with two groups of different servers, a finite buffer, and a fast setup process. Additionally, we provide the system structure, operational mode, and mathematical notations in the energy-efficient data center.

Server groups: The data center contains two server groups: Groups 1 and 2, each of which is also one interactive subsystem of the data center. Groups 1 and 2 have m_1 and m_2 servers, respectively. Servers in the same group are homogeneous, while those in different groups are heterogeneous. Note that Group 1 is viewed as a base-line group whose servers are always at the work state even if some of them are idle, the purpose of which is to guarantee a necessary service capacity in the data center. Hence, each server in Group 1 always works regardless of whether it has a job or not, so that it must consume an amount of energy at any time. In contrast, Group 2 is regarded as a reserved group whose servers may either work or sleep so that each of the m_2 servers can switch its state between work and sleep. If one server in Group 2 is at the sleep state, then it consumes a smaller amount of energy than the work state, as maintaining only the sleep state requires very little energy.

A finite buffer: The data center has a finite buffer of size m_3. Jobs must first enter the buffer, and then they are assigned to the groups (Group 1 is prior to Group 2) and subsequently to the servers. To guarantee that the float service capacity of Group 2 can be fully utilized when some jobs are taken from the buffer to Group 2, we assume that $m_3 \geq m_2$, i.e., the capacity of the buffer must be no less than the server number of Group 2. Otherwise, if there are more jobs waiting in the buffer, the jobs transferred from Group 2 to the buffer will be lost.

Arrival processes: The arrivals of jobs at the data center are a Poisson process with arrival rate λ. If the buffer is full, then any arriving job has to be lost immediately. This leads to an opportunity cost C_5 per unit of time for each lost job due to the full buffer.

Service processes: The service times provided by each server in Groups 1 and 2 are i.i.d. and exponential with service rates μ_1 and μ_2, respectively. We assume that $\mu_1 \geq \mu_2$, which makes the prior use of servers in Group 1. The service discipline of each server in the data center is First Come First Serve (FCFS). If a job finishes its service at a server, then it immediately leaves the system. At the same time, the data center can obtain a fixed service reward (or service price) R from the served job.

Once a job enters the data center for its service, it has to pay holding costs per unit of time $C_2^{(1)}, C_2^{(2)}$, and $C_2^{(3)}$ in Group 1, Group 2, and the buffer, respectively. We assume that $C_2^{(1)} \leq C_2^{(2)}$ also guarantees the prior use of servers in Group 1. Therefore, to support the service priority, each server in Group 1 is not only faster but also cheaper than that in Group 2.

Switching between work and sleep: To save energy, the servers in Group 2 can switch between the work and sleep states. On the one hand, if there are more jobs waiting in the buffer, then Group 2 sets up and turns on some sleeping servers. This process usually involves a setup cost $C_3^{(1)}$. However, the setup time is very short as it directly begins from the sleep state and it can be ignored. On the other hand, if the number of jobs in Group 2 is smaller, then the working servers are switched to the sleep state, while the incomplete-service jobs are transferred to the buffer and served as the arriving ones.

Transfer rules: (1) To Group 1. Based on the prior use of servers in Group 1, if a server in Group 1 becomes idle and there is no job in the buffer, then an incomplete-service job (if it exists) in Group 2 must immediately be transferred to the idle server in Group 1. Additionally, the data center needs to pay a transferred cost C_4 to the transferred job.

(2) To the buffer. If some servers in Group 2 are closed to the sleep state, then those jobs in the servers closed at the sleep state are transferred to the buffer, and a transferred cost $C_3^{(2)}$ is paid by the data center.

To keep the transferred jobs that can enter the buffer, we need to control the new jobs arriving at the buffer. If the sum of the job number in the buffer and the job number in Group 2 is equal to m_3, then the newly arriving jobs must be lost immediately.

Power Consumption: The power consumption rates $P_{1,W}$ and $P_{2,W}$ are for the work states of servers in Groups 1 and 2, respectively, while $P_{2,S}$ is only for the sleep state of a server in Group 2. Note that each server in Group 1 does not have the sleep state and it is clear that $P_{1,S} = 0$. We assume that $0 < P_{2,S} < P_{2,W}$. There is no power consumption for keeping the jobs in the buffer. C_1 is the power consumption price per unit of the power consumption rate and per unit of time.

Independence: We assume that all the random variables in the data center defined above are independent.

Finally, to aid reader understanding, the data center, together with its operational mode and mathematical notations, is depicted in Figure 1. Table 1 summarizes some notations involved in the model. This will be helpful in our later study.

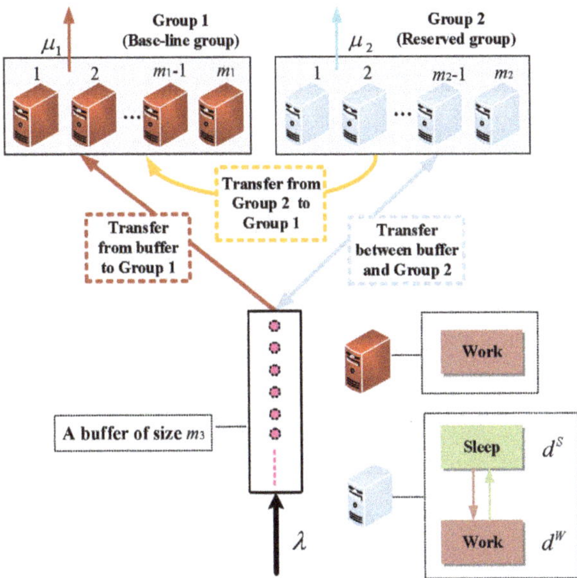

Figure 1. Energy-efficient management of a data center.

Table 1. Cost notation in the data center.

Cost	Necessary Interpretation
C_1	The power consumption price
$C_2^{(1)}$	The holding cost for a job in Group 1 per unit of sojourn time
$C_2^{(2)}$	The holding cost for a job in Group 2 per unit of sojourn time
$C_2^{(3)}$	The holding cost for a job in the buffer per unit of sojourn time
$C_3^{(1)}$	The setup cost for a server switching from the sleep state to the work state
$C_3^{(2)}$	The transferred cost for a incomplete-service job returning to the buffer
C_4	The transferred cost for a job in Group 2 is transferred to Group 1
C_5	The opportunity cost for each lost job
R	The service price from the served job

In the remainder of this section, it might be useful to provide some comparison of the above model assumptions with those in Ma et al. [31] and in Xia et al. [35].

Remark 1. *(1) Compared with our previous paper [31], this paper considers several new practical factors, such as a finite buffer, a fast setup process, and a job transfer rule. The new factors make our MDP modeling more practical and useful in the study of energy-efficient data centers. Although the new factors do not increase any difficulty in performance evaluation through modeling by means of queueing systems or Markov processes, they can cause substantially more difficulties and challenges in finding optimal dynamic energy-efficient policies and, furthermore, in determining threshold-type policies by using the sensitivity-based optimization. Note that the difficulties mainly grow out of establishing the policy-based block-structured Markov process and solving the block-structured Poisson equation. On this occasion, we have simplified the above model descriptions: for example, the setup is immediate, the jobs can be transferred without delay either between the slow and fast servers or between the slow servers and the buffer, the jobs must be transferred as soon as the fast server becomes free, the finite buffer space is reserved for jobs in progress, and so on.*

(2) For the energy-efficient data center operating with some buffer, it is seen from Figure A3 in Appendix B that the main challenge of our work is to focus on how to describe the policy-based block-structured Markov process. Obviously, (a) if there are more than two groups of servers, then it is easy to check that the policy-based Markov process will become multi-dimensional so that its analysis is very difficult; (b) if the buffer is infinite, then we have to deal with the policy-based block-structured Markov process with infinitely many levels, for which the discussion and computation are very complicated.

Remark 2. *Compared with Xia et al. [35], this paper introduces a more detailed cost and reward structure, which makes analysis for the monotonicity and optimality of the dynamic energy-efficient policies more difficult and challenging. Therefore, many cost and reward factors make the MDP analysis and the sensitivity-based optimization more complicated.*

3. Optimization Model Formulation

In this section, for the energy-efficient data center, we first establish a policy-based continuous-time Markov process with a finite block structure. Then, we define a suitable reward function with respect to both states and policies of the Markov process. Note that this will be helpful and useful for setting up a MDP to find the optimal asynchronous dynamic policy in the energy-efficient data center.

3.1. A Policy-Based Block-Structured Continuous-Time Markov Process

The data center in Figure 1 shows Group 1 of m_1 servers, Group 2 of m_2 servers, and a buffer of size m_3. We need to introduce both "states" and "policies" to express the stochastic dynamics of this data center. Let $N_1(t)$, $N_2(t)$, and $N_3(t)$ be the numbers of jobs in Group 1, Group 2, and the buffer, respectively. Therefore, $(N_1(t), N_2(t), N_3(t))$ is regarded as a state of the data center at time t. Let all the cases of such a state $(N_1(t), N_2(t), N_3(t))$ form a set as follows:

$$\Omega = \Omega_0 \cup \Omega_1 \cup \Omega_2 \cup \cdots \cup \Omega_{m_2} \cup \Omega_{m_2+1} \cup \Omega_{m_2+2},$$

where

$$\Omega_0 = \{(0,0,0), (1,0,0), \ldots, (m_1,0,0)\},$$
$$\Omega_1 = \{(m_1,0,1), (m_1,0,2), \ldots, (m_1,0,m_3)\},$$
$$\Omega_2 = \{(m_1,1,0), (m_1,1,1), \ldots, (m_1,1,m_3-1)\},$$

$$\vdots \qquad \qquad \vdots$$

$$\Omega_{m_2} = \{(m_1,m_2-1,0), (m_1,m_2-1,1), \ldots, (m_1,m_2-1,m_3-m_2+1)\},$$
$$\Omega_{m_2+1} = \{(m_1,m_2,0), (m_1,m_2,1), \ldots, (m_1,m_2,m_3-m_2)\},$$
$$\Omega_{m_2+2} = \{(m_1,m_2,m_3-m_2+1), (m_1,m_2,m_3-m_2+2), \ldots, (m_1,m_2,m_3)\}.$$

For a state (n_1, n_2, n_3), it is seen from the model description that there are four different cases: (a) By using the transfer rules, if $n_1 = 0, 1, \ldots, m_1 - 1$, then $n_2 = n_3 = 0$; if either $n_2 \neq 0$ or $n_3 \neq 0$, then $n_1 = m_1$. (b) If $n_1 = m_1$ and $n_2 = 0$, then the jobs in the buffer can increase until the waiting room is full, i.e., $n_3 = 1, 2, \ldots, m_3$. (c) If $n_1 = m_1$ and $n_2 = 1, 2, \ldots, m_2 - 1$, then the total numbers of jobs in Group 2 and the buffer are no more than the buffer size, i.e., $n_2 + n_3 \leq m_3$. (d) If $n_1 = m_1$ and $n_2 = m_2$, then the jobs in the buffer can also increase until the waiting room is full, i.e., $n_3 = 1, 2, \ldots, m_3$.

Now, for Group 2, we introduce an asynchronous dynamic policy, which is related to two dynamic actions (or sub-policies): from sleep to work (setup) and from work to sleep (close). Let $d^W_{n_1,n_2,n_3}$ and $d^S_{n_1,n_2,n_3}$ be the numbers of working servers and of sleeping servers in Group 2 at State (n_1, n_2, n_3), respectively. By observing the state set Ω, we call d^W and d^S the setup policy (i.e., from sleep to work) and the sleep policy (i.e., from work to sleep), respectively.

Note that the servers in Group 2 can only be set up when all of them are idle, while we cannot simultaneously have the setup policy (d^W) because the servers in Group 2 are always affected by the sleep policy (d^S) if they still work for some jobs. This is what we call asynchronous dynamic policies. Here, we consider the control optimization of the total system. For such sub-policies, we provide an interpretation of four different cases as follows:

(1) In Ω_0, if $n_1 = 0, 1, \ldots, m_1 - 1$, then $n_2 = n_3 = 0$ due to the transfer rule. Thus, there are no jobs in Group 2 or in the buffer, so that no policy in Group 2 is used;

(2) In Ω_1, the states will affect *how to use the setup policy*. If $n_1 = m_1, n_2 = 0, n_3 = 1, 2, \ldots, m_3$, then $d^W_{m_1,0,n_3}$ is the number of working servers in Group 2 at State $(m_1, 0, n_3)$. Note that some of the slow servers need to first start, so that some jobs in the buffer can enter the activated slow servers, thus, $d^W_{m_1,0,n_3} \in \{0, 1, \ldots, m_2\}$, each of which can possibly take place under an optimal dynamic policy;

(3) From Ω_2 to Ω_{m_2+1}, the states will affect *how to use the sleep policy*. If $n_1 = m_1$, $n_2 = 1, 2, \ldots, m_2, n_3 = 0, 1, \ldots, m_3 - n_2$, then $d^S_{m_1,n_2,n_3}$ is the number of sleeping servers in Group 2 at State (m_1, n_2, n_3). We assume that the number of sleeping servers is no less than $m_2 - n_2$. Note that the sleep policy is independent of the work policy. Once the sleep policy is set up, the servers without jobs must enter the sleep state. At the same time, some working servers with jobs are also closed to the sleep state, and the jobs in those working servers are transferred to the buffer. It is easy to see that $d^S_{m_1,n_2,n_3} \in \{m_2 - n_2, m_2 - n_2 + 1, \ldots, m_2\}$;

(4) In Ω_{m_2+2}, if $n_1 = m_1$ and $n_2 = m_2$, then n_3 may be any element in the set $\{m_3 - m_2 + 1, m_3 - m_2 + 2, \ldots, m_3\}$, it is clear that $n_2 + n_3 > m_3$.

Our aim is to determine when or under what conditions an optimal number of servers in Group 2 switch between the sleep state and the work state such that the long-run average profit of the data center is maximal. From the state space Ω, we define an asynchronous dynamic energy-efficient policy d as

$$d = d^W \boxtimes d^S, \tag{1}$$

where d^W and d^S are the setup and sleep policies, respectively; '$\boxtimes$' denotes that the policies d^W and d^S occur asynchronously; and

$$
\begin{aligned}
d^W &= \left(d^W_{m_1,0,1}, d^W_{m_1,0,2}, \ldots, d^W_{m_1,0,m_3} \right), \\
d^S &= \left(d^S_{m_1,1,0}, d^S_{m_1,1,1}, \ldots, d^S_{m_1,1,m_3-1}; d^S_{m_1,2,0}, d^S_{m_1,2,1}, \ldots, d^S_{m_1,2,m_3-2}; \cdots; \right. \\
&\qquad \left. d^S_{m_1,m_2,0}, d^S_{m_1,m_2,1}, \ldots, d^S_{m_1,m_2,m_3-m_2} \right).
\end{aligned}
$$

Note that d^W is related to the fact that if there is no job in Group 2 at the initial time, then all the servers in Group 2 are at the sleep state. Once there are jobs in the buffer, we quickly set up some servers in Group 2 such that they enter the work state to serve the jobs. Similarly, we can understand the sleep policy d^S. In the state subset $\bigcup_{i=2}^{m_2+2}\Omega_i$, it is seen that the setup policy d^W will not be needed because some servers are kept at the work state.

For all the possible policies d given in (1), we compose a policy space as follows:

$$
\begin{aligned}
\mathcal{D} := \Big\{ d = d^W \boxtimes d^S : \; & d^W_{m_1,0,n_3} \in \{0,1,\ldots,m_2\} \text{ for } 1 \le n_3 \le m_3; \\
& d^S_{m_1,n_2,n_3} \in \{m_2 - n_2, m_2 - n_2 + 1, \ldots, m_2\} \text{ for } (m_1, n_2, n_3) \in \Omega_2 \cup \Omega_3 \cup \cdots \cup \Omega_{m_2+1} \Big\}.
\end{aligned}
$$

Let $\mathbf{X}^{(d)}(t) = (N_1(t), N_2(t), N_3(t))^{(d)}$ for any given policy $d \in \mathcal{D}$. Then $\{\mathbf{X}^{(d)}(t) : t \ge 0\}$ is a policy-based block-structured continuous-time Markov process on the state space Ω whose state transition relations are given in Figure A3 in Appendix B (we provide two simple special cases to understand such a policy-based block-structured continuous-time Markov process in Appendix A). Based on this, the infinitesimal generator of the Markov process $\{\mathbf{X}^{(d)}(t) : t \ge 0\}$ is given by

$$
\mathbf{Q}^{(d)} =
\begin{pmatrix}
Q_{0,0} & Q_{0,1} & & & & & & \\
Q_{1,0} & Q_{1,1} & Q_{1,2} & Q_{1,3} & Q_{1,4} & \cdots & Q_{1,m_2+1} & \\
Q_{2,0} & Q_{2,1} & Q_{2,2} & & & & & \\
& Q_{3,1} & Q_{3,2} & Q_{3,3} & & & & \\
& Q_{4,1} & Q_{4,2} & Q_{4,3} & Q_{4,4} & & & \\
& \vdots & \vdots & \vdots & \vdots & \ddots & & \\
& Q_{m_2+1,1} & Q_{m_2+1,2} & Q_{m_2+1,3} & Q_{m_2+1,4} & \cdots & Q_{m_2+1,m_2+1} & Q_{m_2+1,m_2+2} \\
& & & & & & Q_{m_2+2,m_2+1} & Q_{m_2+2,m_2+2}
\end{pmatrix}, \tag{2}
$$

where every block element $Q_{i,j}$ depends on the policy d (for simplification of description, here we omit "d") and it is expressed in Appendix C.

It is easy to see that the infinitesimal generator $\mathbf{Q}^{(d)}$ has finite states, and it is irreducible with $\mathbf{Q}^{(d)} e = 0$, thus, the Markov process $\mathbf{Q}^{(d)}$ is a positive recurrent. In this case, we write the stationary probability vector of the Markov process $\{\mathbf{X}^{(d)}(t) : t \ge 0\}$ as

$$\pi^{(d)} = \left(\pi_0^{(d)}, \pi_1^{(d)}, \ldots, \pi_{m_2+2}^{(d)} \right), \quad d \in \mathcal{D}, \tag{3}$$

where

$$\boldsymbol{\pi}_0^{(d)} = \left(\pi^{(d)}(0,0,0), \pi^{(d)}(1,0,0), \ldots, \pi^{(d)}(m_1,0,0) \right),$$

$$\boldsymbol{\pi}_1^{(d)} = \left(\pi^{(d)}(m_1,0,1), \pi^{(d)}(m_1,0,2), \ldots, \pi^{(d)}(m_1,0,m_3) \right),$$

$$\boldsymbol{\pi}_2^{(d)} = \left(\pi^{(d)}(m_1,1,0), \pi^{(d)}(m_1,1,1), \ldots, \pi^{(d)}(m_1,1,m_3-1) \right),$$

$$\vdots \qquad\qquad \vdots$$

$$\boldsymbol{\pi}_{m_2}^{(d)} = \left(\pi^{(d)}(m_1,m_2-1,0), \pi^{(d)}(m_1,m_2-1,1), \ldots, \pi^{(d)}(m_1,m_2-1,m_3-m_2+1) \right),$$

$$\boldsymbol{\pi}_{m_2+1}^{(d)} = \left(\pi^{(d)}(m_1,m_2,0), \pi^{(d)}(m_1,m_2,1), \ldots, \pi^{(d)}(m_1,m_2,m_3-m_2) \right),$$

$$\boldsymbol{\pi}_{m_2+2}^{(d)} = \left(\pi^{(d)}(m_1,m_2,m_3-m_2+1), \pi^{(d)}(m_1,m_2,m_3-m_2+2), \ldots, \pi^{(d)}(m_1,m_2,m_3) \right).$$

Note that the stationary probability vector $\boldsymbol{\pi}^{(d)}$ can be obtained by means of solving the system of linear equations $\boldsymbol{\pi}^{(d)}\mathbf{Q}^{(d)} = 0$ and $\boldsymbol{\pi}^{(d)}e = 1$, where e is a column vector of the ones with a suitable size. To this end, the RG-factorizations play an important role in our later computation. Note that some computational details are given in Chapter 2 in Li [33].

Now, we use UL-type RG-factorization to compute the stationary probability vector $\boldsymbol{\pi}^{(d)}$ as follows. For $0 \leq i, j \leq k$ and $0 \leq k \leq m_2 + 2$, we write

$$Q_{i,j}^{[\leq k]} = Q_{i,j} + \sum_{n=k+1}^{m_2+2} Q_{i,n}^{[\leq n]} \left\{ -Q_{n,n}^{[\leq n]} \right\}^{-1} Q_{n,j}^{[\leq n]}.$$

Clearly, $Q_{i,j}^{[\leq m_2+2]} = Q_{i,j}$ and $Q_{i,j}^{[\leq 0]} = Q_{i,j}^{[0]}$. Let

$$U_n^{(d)} = Q_{n,n}^{[\leq n]}, \qquad\qquad 0 \leq n \leq m_2 + 2,$$

$$R_{i,j}^{(d)} = Q_{i,j}^{[\leq j]} \left(-U_j^{(d)} \right)^{-1}, \quad 0 \leq i < j \leq m_2 + 2,$$

and

$$G_{i,j}^{(d)} = \left(-U_i^{(d)} \right)^{-1} Q_{i,j}^{[\leq i]}, \quad 0 \leq j < i \leq m_2 + 2.$$

Then the UL-type RG-factorization is given by

$$\mathbf{Q}^{(d)} = \left(I - \mathbf{R}_U^{(d)} \right) \left(I - \mathbf{U}_D^{(d)} \right) \left(I - \mathbf{G}_L^{(d)} \right),$$

where

$$\mathbf{R}_U^{(d)} = \begin{pmatrix} 0 & R_{0,1}^{(d)} & R_{0,2}^{(d)} & \cdots & R_{0,m_2+1}^{(d)} & R_{0,m_2+2}^{(d)} \\ & 0 & R_{1,2}^{(d)} & \cdots & R_{1,m_2+1}^{(d)} & R_{1,m_2+2}^{(d)} \\ & & 0 & \ddots & \vdots & \vdots \\ & & & 0 & R_{m_2,m_2+1}^{(d)} & R_{m_2,m_2+2}^{(d)} \\ & & & & 0 & R_{m_2+1,m_2+2}^{(d)} \\ & & & & & 0 \end{pmatrix},$$

$$\mathbf{U}_D^{(d)} = \mathrm{diag}\left(U_0^{(d)}, U_1^{(d)}, \ldots, U_{m_2+1}^{(d)}, U_{m_2+2}^{(d)} \right)$$

and

$$\mathbf{G}_L^{(d)} = \begin{pmatrix} 0 & & & & & \\ G_{1,0}^{(d)} & 0 & & & & \\ G_{2,0}^{(d)} & G_{2,1}^{(d)} & 0 & & & \\ G_{3,0}^{(d)} & G_{3,1}^{(d)} & G_{3,2}^{(d)} & 0 & & \\ \vdots & \vdots & \vdots & \ddots & \ddots & \\ G_{m_2+2,0}^{(d)} & G_{m_2+2,1}^{(d)} & G_{m_2+2,2}^{(d)} & \cdots & G_{m_2+2,m_2+1}^{(d)} & 0 \end{pmatrix}.$$

By using Theorem 2.9 of Chapter 2 in Li [33], the stationary probability vector of the Markov process $\mathbf{Q}^{(d)}$ is given by

$$\begin{cases} \pi_0^{(d)} = \tau^{(d)} x_0^{(d)}, \\ \pi_k^{(d)} = \sum_{i=0}^{k-1} \pi_i^{(d)} R_{i,k}^{(d)}, \quad 1 \le k \le m_2 + 2, \end{cases}$$

where $x_0^{(d)}$ is the stationary probability vector of the censored Markov chain $U_0^{(d)}$ to level 0, and the positive scalar $\tau^{(d)}$ is uniquely determined by $\sum_{k=0}^{m_2+2} \pi_k^{(d)} e = 1$.

Remark 3. *The RG-factorizations provide a unified, constructive and algorithmic framework for the numerical computation of many practical stochastic systems. It can be applied to provide effective solutions for the block-structured Markov processes, and are shown to be also useful for the optimal design and dynamical decision-making of many practical systems. See more details in Li [33].*

The following theorem provides some useful observations on some special policies $d^W \boxtimes d^S \in \mathcal{D}$, in which the special policies will have no effect on the infinitesimal generator $\mathbf{Q}^{(d^W \boxtimes d^S)}$ or the stationary probability vector $\pi^{(d^W \boxtimes d^S)}$.

Theorem 1. *Suppose that two asynchronous energy-efficient policies $d^{W_1} \boxtimes d^S, d^{W_2} \boxtimes d^S \in \mathcal{D}$ satisfy one of the following two conditions: (a) for each $n_3 = 1, 2, \ldots, m_2$, if $d_{m_1,0,n_3}^{W_1} \in \{n_3, n_3+1, \ldots, m_2\}$, then we take $d_{m_1,0,n_3}^{W_2}$ as any element of the set $\{1, 2, \ldots, m_2\}$; (b) for each $n_3 = m_2+1, m_2+2, \ldots, m_3$, if $d_{m_1,0,n_3}^{W_1} \in \{1, 2, \ldots, m_2\}$, then we take $d_{m_1,0,n_3}^{W_2} = d_{m_1,0,n_3}^{W_1}$. Under both such conditions, we have*

$$\mathbf{Q}^{(d^{W_1} \boxtimes d^S)} = \mathbf{Q}^{(d^{W_2} \boxtimes d^S)}, \quad \pi^{(d^{W_1} \boxtimes d^S)} = \pi^{(d^{W_2} \boxtimes d^S)}.$$

Proof of Theorem 1. It is easy to see from (2) that all the levels of the matrix $\mathbf{Q}^{(d^{W_1} \boxtimes d^S)}$ are the same as those of the matrix $\mathbf{Q}^{(d^{W_2} \boxtimes d^S)}$, except level 1. Thus, we only need to compare level 1 of the matrix $\mathbf{Q}^{(d^{W_1} \boxtimes d^S)}$ with that of the matrix $\mathbf{Q}^{(d^{W_2} \boxtimes d^S)}$.

For the two asynchronous energy-efficient policies $d^{W_1} \boxtimes d^S, d^{W_2} \boxtimes d^S \in \mathcal{D}$ satisfying the conditions (a) and (b), by using $1_{\left\{d_{m_1,0,n_3}^{W} \ge n_3\right\}}$ in (2), it is clear that for $n_3 = 1, 2, \ldots, m_2$, if $d_{m_1,0,n_3}^{W_1}, d_{m_1,0,n_3}^{W_2} \in \{n_3, n_3+1, \ldots, m_2\}$, then

$$1_{\left\{d_{m_1,0,n_3}^{W_1} \ge n_3\right\}} = 1_{\left\{d_{m_1,0,n_3}^{W_2} \ge n_3\right\}}.$$

Thus, it follows from (2) that $Q_{1,k}^{(d^{W_1} \boxtimes d^S)} = Q_{1,k}^{(d^{W_2} \boxtimes d^S)}, k = 1, 2, \ldots, m_2 + 1$. This also gives that $\mathbf{Q}^{(d^{W_1} \boxtimes d^S)} = \mathbf{Q}^{(d^{W_2} \boxtimes d^S)}$, and thus $\pi^{(d^{W_1} \boxtimes d^S)} = \pi^{(d^{W_2} \boxtimes d^S)}$. This completes the proof. $\square$

Note that Theorem 1 will be necessary and useful for analyzing the policy monotonicity and optimality in our later study. Furthermore, see the proof of Theorem 4.

Remark 4. *This paper is the first to introduce and consider the asynchronous dynamic policy in the study of energy-efficient data centers. We highlight the impact of the two asynchronous sub-policies: the setup and sleep policies on the long-run average profit of the energy-efficient data center.*

3.2. The Reward Function

For the Markov process $\mathbf{Q}^{(d)}$, now we define a suitable reward function for the energy-efficient data center.

Based on the above costs and price notations in Table 1, a reward function with respect to both states and policies is defined as a profit rate (i.e., the total revenues minus the total costs per unit of time). Therefore, according to the impact of the asynchronous dynamic policy on the profit rate, the reward function at State (n_1, n_2, n_3) under policy d is divided into four cases as follows:

Case (a): For $n_1 = 0, 1, \ldots, m_1$ and $n_2 = n_3 = 0$, the profit rate is not affected by any policy, and we have

$$f(n_1, 0, 0) = R n_1 \mu_1 - (m_1 P_{1,W} + m_2 P_{2,S}) C_1 - n_1 C_2^{(1)}. \tag{4}$$

Note that in Case (a), there is no job in Group 2 or in the buffer. Thus, it is clear that each server in Group 2 is at the sleep state.

However, in the following two cases (b) and (c), since there are some jobs either in Group 2 or in the buffer, the policy d will play a key role in opening (i.e., setup) or closing (i.e., sleep) some servers of Group 2 to save energy efficiently.

Case (b): For $n_1 = m_1$, $n_2 = 0$ and $n_3 = 1, 2, \ldots, m_3$, the profit rate is affected by the setup policy d^W, we have

$$f^{(d^W)}(m_1, 0, n_3) = R m_1 \mu_1 - \left[m_1 P_{1,W} + d^W_{m_1,0,n_3} P_{2,W} + \left(m_2 - d^W_{m_1,0,n_3} \right) P_{2,S} \right] C_1$$
$$- \left[m_1 C_2^{(1)} + n_3 C_2^{(3)} \right] - d^W_{m_1,0,n_3} C_3^{(1)} - \lambda 1_{\{n_1 = m_1, n_2 = 0, n_3 = m_3\}} C_5, \tag{5}$$

where $1_{\{\cdot\}}$ is an indicator function whose value is 1 when the event is in $\{\cdot\}$, otherwise its value is 0.

Case (c): For $n_1 = m_1$, $n_2 = 1, 2, \ldots, m_2$ and $n_3 = 0, 1, \ldots, m_3 - n_2$, the profit rate is affected by the sleep policy d^S, we have

$$f^{(d^S)}(m_1, n_2, n_3) = R \left[m_1 \mu_1 + \left(m_2 - d^S_{m_1,n_2,n_3} \right) \mu_2 \right]$$
$$- \left[m_1 P_{1,W} + d^S_{m_1,n_2,n_3} P_{2,S} + \left(m_2 - d^S_{m_1,n_2,n_3} \right) P_{2,W} \right] C_1$$
$$- \left[m_1 C_2^{(1)} + n_2 C_2^{(2)} + n_3 C_2^{(3)} \right] - \left[n_2 - \left(m_2 - d^S_{m_1,n_2,n_3} \right) \right] C_3^{(2)}$$
$$- m_1 \mu_1 1_{\{n_2 > 0, n_3 = 0\}} C_4 - \lambda 1_{\{n_1 = m_1, n_2 + n_3 = m_3\}} C_5. \tag{6}$$

Note that the job transfer rate from Group 2 to Group 1 is given by $n_1 \mu_1 1_{\{n_2 > 0, n_3 = 0\}}$. If $0 \leq n_1 \leq m_1 - 1$, then $n_2 = 0$ and $n_1 \mu_1 1_{\{n_2 > 0, n_3 = 0\}} C_4 = 0$. If $n_1 = m_1$ and $n_2 = 0$, then $n_1 \mu_1 1_{\{n_2 > 0, n_3 = 0\}} C_4 = 0$. If $n_1 = m_1, 1 \leq n_2 \leq m_2$ and $n_3 = 0$, then $n_1 \mu_1 1_{\{n_2 > 0, n_3 = 0\}} C_4 = m_1 \mu_1 C_4$.

Case (d): For $n_1 = m_1$, $n_2 = m_2$ and $n_3 = m_3 - m_2 + 1, m_3 - m_2 + 2, \ldots, m_3$, the profit rate is not affected by any policy, we have

$$f(m_1, m_2, n_3) = R(m_1 \mu_1 + m_2 \mu_2) - (m_1 P_{1,W} + m_2 P_{2,W}) C_1$$
$$- \left[m_1 C_2^{(1)} + m_2 C_2^{(2)} + n_3 C_2^{(3)} \right] - \lambda 1_{\{n_1 = m_1, n_2 = m_2, n_3 = m_3\}} C_5. \tag{7}$$

We define a column vector composed of the elements $f(n_1, n_2, n_3)$, $f^{(d^W)}(n_1, n_2, n_3)$ and $f^{(d^S)}(n_1, n_2, n_3)$ as

$$f^{(d)} = \left((f_0)^T, \left(f_1^{(d^W)}\right)^T, \left(f_2^{(d^S)}\right)^T, \ldots, \left(f_{m_2+1}^{(d^S)}\right)^T, (f_{m_2+2})^T \right)^T, \tag{8}$$

where

$$f_0 = (f(0,0,0), f(1,0,0), \ldots, f(m_1,0,0))^T,$$

$$f_1^{(d^W)} = \left(f^{(d^W)}(m_1,0,1), f^{(d^W)}(m_1,0,2), \ldots, f^{(d^W)}(m_1,0,m_3) \right)^T,$$

$$f_2^{(d^S)} = \left(f^{(d^S)}(m_1,1,0), f^{(d^S)}(m_1,1,1), \ldots, f^{(d^S)}(m_1,1,m_3-1) \right)^T,$$

$$\vdots \qquad\qquad \vdots$$

$$f_{m_2+1}^{(d^S)} = \left(f^{(d^S)}(m_1,m_2,0), f^{(d^S)}(m_1,m_2,1), \ldots, f^{(d^S)}(m_1,m_2,m_3-m_2) \right)^T,$$

$$f_{m_2+2} = (f(m_1,m_2,m_3-m_2+1), f(m_1,m_2,m_3-m_2+2), \ldots, f(m_1,m_2,m_3))^T.$$

In the remainder of this section, the long-run average profit of the data center (or the policy-based continuous-time Markov process $\left\{ \mathbf{X}^{(d)}(t) : t \geq 0 \right\}$) under an asynchronous dynamic policy d is defined as

$$\eta^d = \lim_{T \to +\infty} E\left\{ \frac{1}{T} \int_0^T f^{(d)}\left(\mathbf{X}^{(d)}(t)\right) dt \right\} = \pi^{(d)} f^{(d)}, \tag{9}$$

where $\pi^{(d)}$ and $f^{(d)}$ are given by (3) and (8), respectively.

We observe that as the number of working servers in Group 2 decreases, the total revenues and the total costs in the data center will decrease synchronously, and vice versa. On the other hand, as the number of sleeping servers in Group 2 increases, the total revenues and the total costs in the data center will decrease synchronously, and vice versa. Thus, there is a tradeoff between the total revenues and the total costs for a suitable number of working and/or sleeping servers in Group 2 by using the setup and sleep policies, respectively. This motivates us to study an optimal dynamic control mechanism for the energy-efficient data center. Thus, our objective is to find an optimal asynchronous dynamic policy d^* such that the long-run average profit η^d is maximized, that is,

$$d^* = \arg\max_{d \in \mathcal{D}} \left\{ \eta^d \right\}. \tag{10}$$

Since the setup and sleep policies d^W and d^S occur asynchronously, they cannot interact with each other at any time. Therefore, it is seen that the optimal policy can be decomposed into

$$d^* = d^{W*} \boxtimes d^{S*} = \arg\max_{d^W \in \mathcal{D}} \left\{ \eta^{d^W} \right\} \boxtimes \arg\max_{d^S \in \mathcal{D}} \left\{ \eta^{d^S} \right\}.$$

In fact, it is difficult and challenging to analyze the properties of the optimal asynchronous dynamic policy $d^* = d^{W*} \boxtimes d^{S*}$, and to provide an effective algorithm for computing the optimal policy d^*. To do this, in the next sections we will introduce the sensitivity-based optimization theory to study this energy-efficient optimization problem.

4. The Block-Structured Poisson Equation

In this section, for the energy-efficient data center, we set up a block-structured Poisson equation which provides a useful relation between the sensitivity-based optimization and the MDP. Additionally, we use the RG-factorization, given in Li [33], to solve the block-structured Poisson equation and provide an expression for its unique solution.

For $d \in \mathcal{D}$, it follows from Chapter 2 in Cao [34] that for the policy-based continuous-time Markov process $\left\{ \mathbf{X}^{(d)}(t) : t \geq 0 \right\}$, we define the performance potential as

$$g^{(d)}(n_1, n_2, n_3) = E\left\{ \int_0^{+\infty} \left[f^{(d)}\left(\mathbf{X}^{(d)}(t) \right) - \eta^d \right] dt \,\middle|\, \mathbf{X}^{(d)}(0) = (n_1, n_2, n_3) \right\}, \tag{11}$$

where η^d is defined in (9). It is seen from Cao [34] that for any policy $d \in \mathcal{D}$, $g^{(d)}(n_1, n_2, n_3)$ quantifies the contribution of the initial state (n_1, n_2, n_3) to the long-run average profit of the data center. Here, $g^{(d)}(n_1, n_2, n_3)$ is also called the relative value function or the bias in the traditional Markov decision process theory, see, e.g., Puterman [39] for more details. We further define a column vector $g^{(d)}$ with elements $g^{(d)}(n_1, n_2, n_3)$ for $(n_1, n_2, n_3) \in \Omega$

$$g^{(d)} = \left(g_0^{(d)}, g_1^{(d)}, g_2^{(d)}, \ldots, g_{m_2+1}^{(d)}, g_{m_2+2}^{(d)} \right)^T, \tag{12}$$

where

$$g_0^{(d)} = \left(g^{(d)}(0,0,0), g^{(d)}(1,0,0), \ldots, g^{(d)}(m_1,0,0) \right)^T,$$

$$g_1^{(d)} = \left(g^{(d)}(m_1,0,1), g^{(d)}(m_1,0,2), \ldots, g^{(d)}(m_1,0,m_3) \right)^T,$$

$$g_2^{(d)} = \left(g^{(d)}(m_1,1,0), g^{(d)}(m_1,1,1), \ldots, g^{(d)}(m_1,1,m_3-1) \right)^T,$$

$$\vdots \qquad \qquad \vdots$$

$$g_{m_2+1}^{(d)} = \left(g^{(d)}(m_1,m_2,0), g^{(d)}(m_1,m_2,1), \ldots, g^{(d)}(m_1,m_2,m_3-m_2) \right)^T,$$

$$g_{m_2+2}^{(d)} = \left(g^{(d)}(m_1,m_2,m_3-m_2+1), g^{(d)}(m_1,m_2,m_3-m_2+2), \ldots, g^{(d)}(m_1,m_2,m_3) \right)^T.$$

A similar computation to that in Ma et al. [31], the block-structured Poisson equation is given by

$$\mathbf{Q}^{(d)} g^{(d)} = \eta^d e - f^{(d)}, \tag{13}$$

where η^d is defined in (9), $f^{(d)}$ is given in (8), and $\mathbf{Q}^{(d)}$ is given in (2).

To solve the system of linear equations (13), we note that $\mathrm{rank}\left(\mathbf{Q}^{(d)} \right) = m_1 + 3m_2/2 - m_2^2/2 + m_2 m_3 + m_3$ and $\det\left(\mathbf{Q}^{(d)} \right) = 0$ because the size of the matrix $\mathbf{Q}^{(d)}$ is $m_1 + 3m_2/2 - m_2^2/2 + m_2 m_3 + m_3 + 1$. Hence, this system (13) of linear equations exists with infinitely many solutions with a free constant of an additive term. Let $\widetilde{\mathbf{Q}}$ be a matrix obtained through omitting the first row and the first column vectors of the matrix $\mathbf{Q}^{(d)}$. Then,

$$\widetilde{\mathbf{Q}}^{(d)} = \begin{pmatrix} \widetilde{Q}_{0,0} & \widetilde{Q}_{0,1} & & & & & \\ \widetilde{Q}_{1,0} & Q_{1,1} & Q_{1,2} & Q_{1,3} & Q_{1,4} & \cdots & Q_{1,m_2+1} & \\ \widetilde{Q}_{2,0} & Q_{2,1} & Q_{2,2} & & & & & \\ & Q_{3,1} & Q_{3,2} & Q_{3,3} & & & & \\ & Q_{4,1} & Q_{4,2} & Q_{4,3} & Q_{4,4} & & & \\ & \vdots & \vdots & \vdots & \vdots & \ddots & & \\ & Q_{m_2+1,1} & Q_{m_2+1,2} & Q_{m_2+1,3} & Q_{m_2+1,4} & \cdots & Q_{m_2+1,m_2+1} & Q_{m_2+1,m_2+2} \\ & & & & & & Q_{m_2+2,m_2+1} & Q_{m_2+2,m_2+2} \end{pmatrix}, \tag{14}$$

where

$$
\widetilde{Q}_{0,0} = \begin{pmatrix}
-(\lambda + \mu_1) & \lambda & & & \\
2\mu_1 & -(\lambda + 2\mu_1) & \lambda & & \\
& \ddots & \ddots & & \ddots \\
& (m_1 - 1)\mu_1 & -[\lambda + (m_1 - 1)\mu_1] & \lambda \\
& & m_1\mu_1 & -(\lambda + m_1\mu_1)
\end{pmatrix},
$$

$\widetilde{Q}_{0,1}$ is obtained by means of omitting the first row vector of $Q_{0,1}$, $\widetilde{Q}_{1,0}$ and $\widetilde{Q}_{2,0}$ are obtained from omitting the first column vectors of $Q_{1,0}$ and $Q_{2,0}$, respectively. The other block entries in $\widetilde{\mathbf{Q}}^{(d)}$ are the same as the corresponding block entries in the matrix $\mathbf{Q}^{(d)}$.

Note that, $\mathrm{rank}\left(\widetilde{\mathbf{Q}}^{(d)}\right) = m_1 + 3m_2/2 - m_2^2/2 + m_2 m_3 + m_3$ and the size of the matrix $\widetilde{\mathbf{Q}}^{(d)}$ is $m_1 + 3m_2/2 - m_2^2/2 + m_2 m_3 + m_3$. Hence, the matrix $\widetilde{\mathbf{Q}}^{(d)}$ is invertible.

Let $h^{(d)}$ and $\varphi^{(d)}$ be two column vectors of size $m_1 + 3m_2/2 - m_2^2/2 + m_2 m_3 + m_3$ obtained through omitting the first element of the two column vectors $f^{(d)} - \eta^d e$ and $g^{(d)}$ of size $m_1 + 3m_2/2 - m_2^2/2 + m_2 m_3 + m_3 + 1$, respectively, and

$$
\begin{aligned}
l_0 &= m_1, \\
l_1 &= m_1 + m_3, \\
l_2 &= m_1 + 2m_3, \\
&\;\;\vdots \\
l_{m_2+1} &= m_1 + \frac{m_2}{2} - \frac{m_2^2}{2} + m_2 m_3 + m_3, \\
L &= m_1 + \frac{3m_2}{2} - \frac{m_2^2}{2} + m_2 m_3 + m_3.
\end{aligned}
$$

Then,

$$
h^{(d)} = \begin{pmatrix}
\widetilde{f}_0 - \eta^d \\
f_1^{(d_W)} - \eta^d \\
f_2^{(d_S)} - \eta^d \\
\vdots \\
f_{m_2+1}^{(d_S)} - \eta^d \\
f_{m_2+2} - \eta^d
\end{pmatrix}
\stackrel{\text{def}}{=}
\begin{pmatrix}
h_0^{(d)} \\
h_1^{(d)} \\
h_2^{(d)} \\
\vdots \\
h_{m_2+1}^{(d)} \\
h_{m_2+2}^{(d)}
\end{pmatrix},
\quad
\varphi^{(d)} \stackrel{\text{def}}{=}
\begin{pmatrix}
\widetilde{g}_0^{(d)} \\
g_1^{(d)} \\
g_2^{(d)} \\
\vdots \\
g_{m_2+1}^{(d)} \\
g_{m_2+2}^{(d)}
\end{pmatrix},
\tag{15}
$$

where $\widetilde{f}_0$ and $\widetilde{g}_0^{(d)}$ are the two column vectors, which are obtained through omitting the scale entries $f(0,0,0)$ and $g^{(d)}(0,0,0)$ of f_0 and $g_0^{(d)}$, respectively, and

$$
\begin{aligned}
h_0^{(d)} &= (h_1, h_2, \ldots, h_{l_0})^T, \\
h_1^{(d)} &= (h_{l_0+1}, h_{l_1+2}, \ldots, h_{l_1})^T, \\
h_2^{(d)} &= (h_{l_1+1}, h_{l_2+2}, \ldots, h_{l_2})^T, \\
&\;\;\vdots \\
h_{m_2+1}^{(d)} &= \left(h_{l_{m_2}+1}, h_{l_{m_2}+1+2}, \ldots, h_{l_{m_2+1}}\right)^T, \\
h_{m_2+2}^{(d)} &= \left(h_{l_{m_2+1}+1}, h_{l_{m_2}+2+2}, \ldots, h_L\right)^T.
\end{aligned}
$$

Therefore, it follows from (13) that

$$
-\widetilde{\mathbf{Q}}^{(d)}\varphi^{(d)} = h^{(d)} + \mu_1 e_1 g^{(d)}(0,0,0),
\tag{16}
$$

where e_1 is a column vector with the first element being one and all the others being zero. Note that the matrix $-\tilde{\mathbf{Q}}^{(d)}$ is invertible and $\left(-\tilde{\mathbf{Q}}^{(d)}\right)^{-1} > 0$, thus the system (16) of linear equations always exists with one unique solution:

$$\varphi^{(d)} = \left(-\tilde{\mathbf{Q}}^{(d)}\right)^{-1} h^{(d)} + \mu_1 \left(-\tilde{\mathbf{Q}}^{(d)}\right)^{-1} e_1 \cdot \Im, \tag{17}$$

where $g^{(d)}(0,0,0) = \Im$ is any given positive constant. For the convenience of computation, we take $g^{(d)}(0,0,0) = \Im = 1$. In this case, we have

$$\varphi^{(d)} = \left(-\tilde{\mathbf{Q}}^{(d)}\right)^{-1} h^{(d)} + \mu_1 \left(-\tilde{\mathbf{Q}}^{(d)}\right)^{-1} e_1. \tag{18}$$

Note that the expression of the invertible matrix $\left(-\tilde{\mathbf{Q}}^{(d)}\right)^{-1}$ can be obtained by means of the RG-factorization, which is given in Li [33] for general Markov processes.

For convenience of computation, we write

$$\left(-\tilde{\mathbf{Q}}^{(d)}\right)^{-1} = \left(\mathbb{Q}_0^T, \mathbb{Q}_1^T, \mathbb{Q}_2^T, \ldots, \mathbb{Q}_{m_2+1}^T, \mathbb{Q}_{m_2+2}^T\right)^T,$$

and every element of the matrix $\mathbb{Q}_r$ is written by a scalar $q_{n,l}^{(r)}$, we denote by n a system state under the certain block, and l the index of element, where $r = 0, 1, \ldots, m_2 + 2, l = 1, 2, \ldots, L$, for $L = m_1 + 3m_2/2 - m_2^2/2 + m_2 m_3 + m_3$, and

$$n = \begin{cases} 1, 2, \ldots, m_1, & \text{for } r = 0, \\ 1, 2, \ldots, m_3, & \text{for } r = 1, \\ 0, 1, \ldots, m_3 - r + 1, & \text{for } 2 \leq r \leq m_2 + 1, \\ m_3 - m_2 + 1, m_3 - m_2 + 2, \ldots, m_3, & \text{for } r = m_2 + 2. \end{cases}$$

It is easy to check that

$$\mathbb{Q}_0 = \begin{pmatrix} q_{1,1}^{(0)} & q_{1,2}^{(0)} & \cdots & q_{1,L}^{(0)} \\ q_{2,1}^{(0)} & q_{2,2}^{(0)} & \cdots & q_{2,L}^{(0)} \\ \vdots & \vdots & & \vdots \\ q_{m_1,1}^{(0)} & q_{m_1,2}^{(0)} & \cdots & q_{m_1,L}^{(0)} \end{pmatrix}_{L \times m_1},$$

$$\mathbb{Q}_1 = \begin{pmatrix} q_{1,1}^{(1)} & q_{1,2}^{(1)} & \cdots & q_{1,L}^{(1)} \\ q_{2,1}^{(1)} & q_{2,2}^{(1)} & \cdots & q_{2,L}^{(1)} \\ \vdots & \vdots & & \vdots \\ q_{m_3,1}^{(1)} & q_{m_3,2}^{(1)} & \cdots & q_{m_3,L}^{(1)} \end{pmatrix}_{L \times m_3},$$

for $2 \leq r \leq m_2 + 1$,

$$\mathbb{Q}_r = \begin{pmatrix} q_{0,1}^{(r)} & q_{0,2}^{(r)} & \cdots & q_{0,L}^{(r)} \\ q_{1,1}^{(r)} & q_{1,2}^{(r)} & \cdots & q_{1,L}^{(r)} \\ \vdots & \vdots & & \vdots \\ q_{m_3-r+1,1}^{(r)} & q_{m_3-r+1,2}^{(r)} & \cdots & q_{m_3-r+1,L}^{(r)} \end{pmatrix}_{L \times (m_3-r+2)},$$

and

$$\mathbb{Q}_{m_2+2} = \begin{pmatrix} q^{(m_2+2)}_{m_3-m_2+1,1} & q^{(m_2+2)}_{m_3-m_2+1,2} & \cdots & q^{(m_2+2)}_{m_3-m_2+1,L} \\ q^{(m_2+2)}_{m_3-m_2+2,1} & q^{(m_2+2)}_{m_3-m_2+2,2} & \cdots & q^{(m_2+2)}_{m_3-m_2+2,L} \\ \vdots & \vdots & & \vdots \\ q^{(m_2+2)}_{m_3,1} & q^{(m_2+2)}_{m_3,2} & \cdots & q^{(m_2+2)}_{m_3,L} \end{pmatrix}_{L\times m_2}.$$

The following theorem provides an expression for the vector $\boldsymbol{\varphi}^{(d)}$ under a constraint condition $g^{(d)}(0,0,0) = \Im = 1$. Note that this expression is very useful for applications of the sensitivity-based optimization theory to the study of Markov decision processes in our later study.

Theorem 2. *If $g^{(d)}(0,0,0) = 1$, then for $n_1 = 1,2,\ldots,m_1$,*

$$g^{(d)}(n_1,0,0) = \sum_{l=1}^{L} q^{(0)}_{n_1,l}h_l + \mu_1 q^{(0)}_{n_1,1};$$

for $n_3 = 1,2,\ldots,m_3$,

$$g^{(d)}(m_1,0,n_3) = \sum_{l=1}^{L} q^{(1)}_{n_3,l}h_l + \mu_1 q^{(1)}_{n_3,1};$$

for $n_2 = r-1, n_3 = 0,1,\ldots,m_3 - n_2$, and $2 \leq r \leq m_2+1$,

$$g^{(d)}(m_1,n_2,n_3) = \sum_{l=1}^{L} q^{(r)}_{n_3,l}h_l + \mu_1 q^{(r)}_{n_3,1};$$

for $n_3 = m_3 - m_2 + 1, m_3 - m_2 + 2, \ldots, m_3$,

$$g^{(d)}(m_1,m_2,n_3) = \sum_{l=1}^{L} q^{(m_2+2)}_{n_3,l}h_l + \mu_1 q^{(m_2+2)}_{n_3,1}.$$

Proof of Theorem 2. It is seen from (18) that we need to compute two parts: $\left(-\tilde{\mathbf{Q}}^{(d)}\right)^{-1}h^{(d)}$ and $\mu_1\left(-\tilde{\mathbf{Q}}^{(d)}\right)^{-1}e_1$. Note that

$$\left(-\tilde{\mathbf{Q}}\right)^{-1}h^{(d)} = \begin{pmatrix} \sum_{l=1}^{L} q^{(0)}_{1,l}h_l \\ \vdots \\ \sum_{l=1}^{L} q^{(0)}_{m_1,l}h_l \\ \vdots \\ \sum_{l=1}^{L} q^{(r)}_{0,l}h_l \\ \vdots \\ \sum_{l=1}^{L} q^{(r)}_{m_3-r+1,l}h_l \\ \vdots \\ \sum_{l=1}^{L} q^{(m_2+2)}_{m_3-m_2+1,l}h_l \\ \vdots \\ \sum_{l=1}^{L} q^{(m_2+2)}_{m_3,l}h_l \end{pmatrix}_{L\times 1} \quad\text{and}\quad \mu_1\left(-\tilde{\mathbf{Q}}\right)^{-1}e_1 = \mu_1 \begin{pmatrix} q^{(0)}_{1,1} \\ \vdots \\ q^{(0)}_{m_1,1} \\ \vdots \\ q^{(r)}_{0,1} \\ \vdots \\ q^{(r)}_{m_3-r+1,1} \\ \vdots \\ q^{(m_2+2)}_{m_3-m_2+1,1} \\ \vdots \\ q^{(m_2+2)}_{m_3,1} \end{pmatrix}_{L\times 1}.$$

thus a simple computation for the vector $\boldsymbol{\varphi}^{(d)} = \left(-\tilde{\mathbf{Q}}^{(d)}\right)^{-1}h^{(d)} + \mu_1\left(-\tilde{\mathbf{Q}}^{(d)}\right)^{-1}e_1$ can obtain our desired results. This completes the proof. $\square$

5. Impact of the Service Price

In this section, we study the perturbation realization factor of the policy-based continuous-time Markov process both for the setup policy and for the sleep policy (i.e., they form the asynchronous energy-efficient policy), and analyze how the service price impacts on the perturbation realization factor. To do this, our analysis includes the following two cases: the setup policy and the sleep policy. Note that the results given in this section will be useful for establishing the optimal asynchronous dynamic policy of the energy-efficient data center in later sections.

It is a key in our present analysis that the setup policy and the sleep policy are asynchronous at any time; thus, we can discuss the perturbation realization factor under the asynchronous dynamic policy from two different computational steps.

5.1. The Setup Policy

For the performance potential vector $\varphi^{(d)}$ under a constraint condition $g^{(d)}(0,0,0) = 1$, we define a perturbation realization factor as

$$G^{(d)}(n, n') \stackrel{\text{def}}{=} g^{(d)}(n') - g^{(d)}(n), \tag{19}$$

where $n = (n_1, n_2, n_3)$, $n' = (n'_1, n'_2, n'_3)$. We can see that $G^{(d)}(n, n')$ quantifies the difference between two performance potentials $g^{(d)}(n_1, n_2, n_3)$ and $g^{(d)}(n'_1, n'_2, n'_3)$. It measures the long-run effect on the average profit of the data center when the system state is changed from $n' = (n'_1, n'_2, n'_3)$ to $n = (n_1, n_2, n_3)$. For our next discussion, through observing the state space, it is necessary to define some perturbation realization factors as follows:

$$
\begin{aligned}
G_1^{(d^W)} &\stackrel{\text{def}}{=} g^{(d)}(m_1, i_1, n_3 - i_1) - g^{(d)}(m_1, i_2, n_3 - i_2), \\
G_2^{(d^W)} &\stackrel{\text{def}}{=} g^{(d)}(m_1, 0, n_3) - g^{(d)}(m_1, i_1, n_3 - i_1), \\
G_3^{(d^W)} &\stackrel{\text{def}}{=} g^{(d)}(m_1, 0, n_3) - g^{(d)}(m_1, i_2, n_3 - i_2),
\end{aligned}
\tag{20}
$$

where $0 \leq i_2 < i_1 \leq m_2$ and $n_3 = 0, 1, \ldots, m_3$.

It follows from Theorem 2 that

$$g^{(d)}(m_1, 0, n_3) = \sum_{l=1}^{L} q_{n_3,l}^{(1)} h_l + \mu_1 q_{n_3,1}^{(1)},$$

$$g^{(d)}(m_1, i_1, n_3 - i_1) = \sum_{l=1}^{L} q_{n_3-i_1,l}^{(i_1+1)} h_l + \mu_1 q_{n_3-i_1,1}^{(i_1+1)},$$

$$g^{(d)}(m_1, i_2, n_3 - i_2) = \sum_{l=1}^{L} q_{n_3-i_2,l}^{(i_2+1)} h_l + \mu_1 q_{n_3-i_2,1}^{(i_2+1)}.$$

To express the perturbation realization factor by means of the service price R, we write

$$A_0 = 0, \quad B_0 = (m_1 P_{1,W} + m_2 P_{2,S}) C_1 > 0;$$

for $1 \leq l \leq l_0$ and $n_1 = 1, 2, \ldots, m_1$, $n_2 = n_3 = 0$,

$$A_l = n_1 \mu_1 > 0, \quad B_l = (m_1 P_{1,W} + m_2 P_{2,S}) C_1 + n_1 C_2^{(1)} > 0;$$

for $l_0 + 1 \leq l \leq l_1$, and $n_1 = m_1$, $n_2 = 0$, $n_3 = 1, 2, \ldots, m_3$,

$$A_l = m_1 \mu_1 > 0,$$

$$B_l^{(d^W)} = \left[m_1 P_{1,W} + d_{m_1,0,n_3}^W P_{2,W} + \left(m_2 - d_{m_1,0,n_3}^W \right) P_{2,S} \right] C_1$$
$$+ \left[m_1 C_2^{(1)} + n_3 C_2^{(3)} \right] + d_{m_1,0,n_3}^W C_3^{(1)} + \lambda 1_{\{n_1 = m_1, n_2 = 0, n_3 = m_3\}} C_5 > 0;$$

for $l_1 + 1 \leq l \leq l_{m_2+1}$, and $n_1 = m_1$, $n_2 = 1, 2, \ldots, m_2$, $n_3 = 0, 1, \ldots, m_3 - n_2$,

$$A_l^{(d^S)} = m_1 \mu_1 + \left(m_2 - d_{m_1,n_2,n_3}^S \right) \mu_2 > 0,$$

$$B_l^{(d^S)} = \left[m_1 P_{1,W} + d_{m_1,n_2,n_3}^S P_{2,S} + \left(m_2 - d_{m_1,n_2,n_3}^S \right) P_{2,W} \right] C_1 + \left[m_1 C_2^{(1)} + n_2 C_2^{(2)} + n_3 C_2^{(3)} \right]$$
$$+ \left[n_2 - \left(m_2 - d_{m_1,n_2,n_3}^S \right) \right] C_3^{(2)} + m_1 \mu_1 1_{\{n_3 = 0\}} C_4 + \lambda 1_{\{n_2 + n_3 = m_3\}} C_5 > 0;$$

for $l_{m_2+1} + 1 \leq l \leq L$, and $n_1 = m_1$, $n_2 = m_2$, $n_3 = m_3 - m_2 + 1, m_3 - m_2 + 2, \ldots, m_3$,

$$A_l = m_1 \mu_1 + m_2 \mu_2 > 0,$$

$$B_l = (m_1 P_{1,W} + m_2 P_{2,W}) C_1 + \left[m_1 C_2^{(1)} + m_2 C_2^{(2)} + n_3 C_2^{(3)} \right] + \lambda 1_{\{n_1 = m_1, n_2 = m_2, n_3 = m_3\}} C_5 > 0.$$

Then for $1 \leq l \leq l_0$ and $n_1 = 0, 1, \ldots, m_1$, $n_2 = n_3 = 0$,

$$f(n_1, 0, 0) = RA_l - B_l;$$

for $l_0 + 1 \leq l \leq l_1$, and $n_1 = m_1$, $n_2 = 0$, $n_3 = 1, 2, \ldots, m_3$,

$$f^{(d^W)}(m_1, 0, n_3) = RA_l - B_l^{(d^W)};$$

for $l_1 + 1 \leq l \leq l_{m_2+1}$, and $n_1 = m_1$, $n_2 = 1, 2, \ldots, m_2$, $n_3 = 0, 1, \ldots, m_3 - n_2$,

$$f^{(d^S)}(m_1, n_2, n_3) = RA_l^{(d^S)} - B_l^{(d^S)};$$

for $l_{m_2+1} + 1 \leq l \leq L$, and $n_1 = m_1$, $n_2 = m_2$, $n_3 = m_3 - m_2 + 1, m_3 - m_2 + 2, \ldots, m_3$,

$$f^{(d)}(m_1, m_2, n_3) = RA_l - B_l.$$

We rewrite $\pi^{(d)}$ as

$$\pi_0^{(d)} = \left(\pi_0; \pi_1, \pi_2, \ldots, \pi_{l_0} \right),$$
$$\pi_1^{(d)} = \left(\pi_{l_0+1}, \pi_{l_0+2}, \ldots, \pi_{l_1} \right),$$
$$\pi_2^{(d)} = \left(\pi_{l_1+1}, \pi_{l_1+2}, \ldots, \pi_{l_2} \right),$$
$$\vdots$$
$$\pi_{m_2+1}^{(d)} = \left(\pi_{l_{m_2}+1}, \pi_{l_{m_2}+2}, \ldots, \pi_{l_{m_2+1}} \right),$$
$$\pi_{m_2+2}^{(d)} = \left(\pi_{l_{m_2+1}+1}, \pi_{l_{m_2+1}+2}, \ldots, \pi_L \right).$$

Then it is easy to check that

$$D^{(d)} = \pi_0 A_0 + \sum_{l=0}^{l_0} \pi_l A_l + \sum_{l=l_0+1}^{l_1} \pi_l A_l + \sum_{l=l_1+1}^{l_{m_2+1}} \pi_l A_l^{(d^S)} + \sum_{l=l_{m_2+1}+1}^{L} \pi_l A_l > 0,$$

and

$$F^{(d)} = \pi_0 B_0 + \sum_{l=0}^{l_0} \pi_l B_l + \sum_{l=l_0+1}^{l_1} \pi_l B_l^{(d^W)} + \sum_{l=l_1+1}^{l_{m_2+1}} \pi_l B_l^{(d^S)} + \sum_{l=l_{m_2+1}+1}^{L} \pi_l B_l > 0.$$

Thus, we obtain

$$\begin{aligned}
\eta^d &= \pi^{(d)} f^{(d)} \\
&= \sum_{n_1=0}^{m_1} \pi^{(d)}(n_1,0,0) f(n_1,0,0) + \sum_{n_3=1}^{m_3} \pi^{(d)}(m_1,0,n_3) f^{(d^W)}(m_1,0,n_3) \\
&\quad + \sum_{n_3=0}^{m_3-n_2} \sum_{n_2=0}^{m_2} \pi^{(d)}(m_1,n_2,n_3) f^{(d^S)}(m_1,n_2,n_3) \\
&\quad + \sum_{n_3=m_3-m_2+1}^{m_3} \pi^{(d)}(m_1,m_2,n_3) f(m_1,m_2,n_3) \\
&= RD^{(d)} - F^{(d)}.
\end{aligned}$$

It follows from (15) that for $1 \leq l \leq l_0$,

$$h_l = R\left[A_l - D^{(d)}\right] - \left[B_l - F^{(d)}\right];$$

for $l_0 + 1 \leq l \leq l_1$,

$$h_l = R\left[A_l - D^{(d)}\right] - \left[B_l^{(d^W)} - F^{(d)}\right];$$

for $l_1 + 1 \leq l \leq l_{m_2+1}$,

$$h_l = R\left[A_l^{(d^S)} - D^{(d)}\right] - \left[B_l^{(d^S)} - F^{(d)}\right];$$

for $l_{m_2+1} + 1 \leq l \leq L$,

$$h_l = R\left[A_l - D^{(d)}\right] - \left[B_l - F^{(d)}\right].$$

If a job finishes its service at a server and leaves this system immediately, then the data center can obtain a fixed revenue (i.e., the service price) R from such a served job. Now, we study the influence of the service price R on the perturbation realization factor. Note that all the numbers $q_{n,l}^{(r)}$ are positive and are independent of the service price R, while all the numbers h_l are the linear functions of R. We write

$$\begin{aligned}
W_n^{(r)} &= \sum_{l=1}^{l_0} q_{n,l}^{(r)} h_l \left[A_l - D^{(d)}\right] + \sum_{l=l_0+1}^{l_1} q_{n,l}^{(r)} h_l \left[A_l - D^{(d)}\right] \\
&\quad + \sum_{l=l_1+1}^{l_{m_2+1}} q_{n,l}^{(r)} h_l \left[A_l^{(d^S)} - D^{(d)}\right] + \sum_{l=l_{m_2+1}+1}^{L} q_{n,l}^{(r)} h_l \left[A_l - D^{(d)}\right] + \mu_1 q_{n,1}^{(r)}
\end{aligned}$$

and

$$\begin{aligned}
V_n^{(r)} &= \sum_{l=1}^{l_0} q_{n,l}^{(r)} h_l \left[B_l - F^{(d)}\right] + \sum_{l=l_0+1}^{l_1} q_{n,l}^{(r)} h_l \left[B_l^{(d^W)} - F^{(d)}\right] \\
&\quad + \sum_{l=l_1+1}^{l_{m_2+1}} q_{n,l}^{(r)} h_l \left[B_l^{(d^S)} - F^{(d)}\right] + \sum_{l=l_{m_2+1}+1}^{L} q_{n,l}^{(r)} h_l \left[B_l - F^{(d)}\right],
\end{aligned}$$

then for $i_1, i_2 = 0, 1, \ldots, m_2$, we obtain

$$
\begin{aligned}
G_1^{\left(d^W\right)} &= R\left[W_{n_3-i_1}^{(i_1+1)} - W_{n_3-i_2}^{(i_2+1)}\right] - \left[V_{n_3-i_1}^{(i_1+1)} - V_{n_3-i_2}^{(i_2+1)}\right], \\
G_2^{\left(d^W\right)} &= R\left[W_{n_3}^{(1)} - W_{n_3-i_1}^{(i_1+1)}\right] - \left[V_{n_3}^{(1)} - V_{n_3-i_1}^{(i_1+1)}\right], \\
G_3^{\left(d^W\right)} &= R\left[W_{n_3}^{(1)} - W_{n_3-i_2}^{(i_2+1)}\right] - \left[V_{n_3}^{(1)} - V_{n_3-i_2}^{(i_2+1)}\right].
\end{aligned}
\tag{21}
$$

Now, we define

$$
G^{\left(d^W\right)} = m_1\mu_1 G_1^{\left(d^W\right)} - i_1\mu_2\left(G_2^{\left(d^W\right)} + \beta_1\right) + i_2\mu_2\left(G_3^{\left(d^W\right)} + \beta_1\right),
$$

where β_1 is defined as

$$
\beta_1 = \frac{(P_{2,W} - P_{2,S})C_1 + C_3^{(1)}}{\mu_2}.
$$

From the later discussion in Section 6, we will see that $G^{\left(d^W\right)}$ plays a fundamental role in the performance optimization of data centers, and the sign of $G^{\left(d^W\right)}$ directly determines the selection of decision actions, as shown in (38) later. To this end, we analyze how the service price can impact on $G^{\left(d^W\right)}$ as follows. Substituting (21) into the linear equation $G^{\left(d^W\right)} = 0$, we obtain

$$
R = \frac{\varphi(i_1)V_{n_3-i_1}^{(i_1+1)} - \varphi(i_2)V_{n_3-i_2}^{(i_2+1)} - \psi(i_1, i_2)\left(V_{n_3}^{(1)} - \beta_1\right)}{\varphi(i_1)W_{n_3-i_1}^{(i_1+1)} - \varphi(i_2)W_{n_3-i_2}^{(i_2+1)} - \psi(i_1, i_2)W_{n_3}^{(1)}},
\tag{22}
$$

where $\varphi(i_1) = m_1\mu_1 + i_1\mu_2$, $\varphi(i_2) = m_1\mu_1 + i_2\mu_2$ and $\psi(i_1, i_2) = (i_1 - i_2)\mu_2$.

Thus, the unique solution of the price R in (22) is given by

$$
\Re^{\left(d^W\right)}(i_1, i_2) = \frac{\varphi(i_1)V_{n_3-i_1}^{(i_1+1)} - \varphi(i_2)V_{n_3-i_2}^{(i_2+1)} - \psi(i_1, i_2)\left(V_{n_3}^{(1)} - \beta_1\right)}{\varphi(i_1)W_{n_3-i_1}^{(i_1+1)} - \varphi(i_2)W_{n_3-i_2}^{(i_2+1)} - \psi(i_1, i_2)W_{n_3}^{(1)}},
\tag{23}
$$

It is easy to see from (22) that (a) if $R \geq \Re^{\left(d^W\right)}(i_1, i_2)$, then $G^{\left(d^W\right)} \geq 0$; and (b) if $R \leq \Re^{\left(d^W\right)}(i_1, i_2)$, then $G^{\left(d^W\right)} \leq 0$.

In the energy-efficient data center, we define two critical values, related to the service price, as

$$
R_H^W = \max_{d \in \mathcal{D}}\left\{0, \Re^{\left(d^W\right)}(1, 0), \Re^{\left(d^W\right)}(2, 0), \ldots, \Re^{\left(d^W\right)}(m_2, m_2 - 1)\right\}
\tag{24}
$$

and

$$
R_L^W = \min_{d \in \mathcal{D}}\left\{\Re^{\left(d^W\right)}(1, 0), \Re^{\left(d^W\right)}(2, 0), \ldots, \Re^{\left(d^W\right)}(m_2, m_2 - 1)\right\}
\tag{25}
$$

The following proposition uses the two critical values, which are related to the service price, to provide a key condition whose purpose is to establish a sensitivity-based optimization framework of the energy-efficient data center in our later study. Additionally, this proposition will be useful in the next section for studying the monotonicity of the asynchronous energy-efficient policies.

Proposition 1. *(1) If* $R \geq R_H^W$, *then for any* $d \in \mathcal{D}$ *and for each couple* (i_1, i_2) *with* $0 \leq i_2 < i_1 \leq m_2$, *we have*

$$
G^{\left(d^W\right)} \geq 0.
\tag{26}
$$

(2) *If $0 \leq R \leq R_L^W$, then for any $d \in \mathcal{D}$ and for each couple (i_1, i_2) with $0 \leq i_2 < i_1 \leq m_2$, we have*

$$G^{(d^W)} \leq 0. \tag{27}$$

Proof of Proposition 1. (1) For any $d \in \mathcal{D}$ and for each couple (i_1, i_2) with $0 \leq i_2 < i_1 \leq m_2$, since $R \geq R_H^W$ and $R_H^W = \max_{d \in \mathcal{D}} \left\{ 0, \Re^{(d^W)}(1,0), \Re^{(d^W)}(2,0), \ldots, \quad \Re^{(d^W)}(m_2, m_2 - 1) \right\}$, this gives

$$R \geq \Re^{(d^W)}(i_1, i_2).$$

Thus, for any couple (i_1, i_2) with $0 \leq i_2 < i_1 \leq m_2$ this makes that $G^{(d^W)} \geq 0$.

(2) For any $d \in \mathcal{D}$ and for each couple (i_1, i_2) with $0 \leq i_2 < i_1 \leq m_2$, if $0 \leq R \leq R_L^W$, we get

$$R \leq \Re^{(d^W)}(i_1, i_2),$$

this gives that $G^{(d^W)} \leq 0$. This completes the proof. $\square$

5.2. The Sleep Policy

The analysis for the sleep policy is similar to that of the setup policy given in the above subsection. Here, we shall provide only a simple discussion.

We define the perturbation realization factor for the sleep policy as follows:

$$
\begin{aligned}
G_1^{(d^S)} &\stackrel{\text{def}}{=} g^{(d)}(m_1, j_2, n_3 + n_2 - j_2) - g^{(d)}(m_1, j_1, n_3 + n_2 - j_1), \\
G_2^{(d^S)} &\stackrel{\text{def}}{=} g^{(d)}(m_1, n_2, n_3) - g^{(d)}(m_1, j_1, n_3 + n_2 - j_1), \\
G_3^{(d^S)} &\stackrel{\text{def}}{=} g^{(d)}(m_1, n_2, n_3) - g^{(d)}(m_1, j_2, n_3 + n_2 - j_2),
\end{aligned}
\tag{28}
$$

where $0 \leq j_2 < j_1 \leq n_2$, $n_2 = 0, 1, \ldots, m_2$ and $n_3 = 0, 1, \ldots, m_3$.

It follows from Theorem 2 that

$$g^{(d)}(m_1, n_2, n_3) = \sum_{l=1}^{L} q_{n_3,l}^{(n_2+1)} h_l + \mu_1 q_{n_3,1}^{(n_2+1)},$$

$$g^{(d)}(m_1, j_1, n_3 + n_2 - j_1) = \sum_{l=1}^{L} q_{n_3+n_2-j_1,l}^{(j_1+1)} h_l + \mu_1 q_{n_3+n_2-j_1,1}^{(j_1+1)},$$

$$g^{(d)}(m_1, j_2, n_3 + n_2 - j_2) = \sum_{l=1}^{L} q_{n_3+n_2-j_2,l}^{(j_2+1)} h_l + \mu_1 q_{n_3+n_2-j_2,1}^{(j_2+1)}.$$

Similarly, to express the perturbation realization factor by means of the service price R, we write

$$G^{(d^S)} = m_1 \mu_1 G_1^{(d^S)} + j_1 \mu_2 \left(G_2^{(d^S)} + \beta_2 \right) - j_2 \mu_2 \left(G_3^{(d^S)} + \beta_2 \right),$$

where

$$
\begin{aligned}
G_1^{(d^S)} &= R \left[W_{n_3+n_2-j_2}^{(j_2+1)} - W_{n_3+n_2-j_1}^{(j_1+1)} \right] - \left[V_{n_3+n_2-j_2}^{(j_2+1)} - V_{n_3+n_2-j_1}^{(j_1+1)} \right], \\
G_2^{(d^S)} &= R \left[W_{n_3}^{(n_2+1)} - W_{n_3+n_2-j_1}^{(j_1+1)} \right] - \left[V_{n_3}^{(n_2+1)} - V_{n_3+n_2-j_1}^{(j_1+1)} \right], \\
G_3^{(d^S)} &= R \left[W_{n_3}^{(n_2+1)} - W_{n_3+n_2-j_2}^{(j_2+1)} \right] - \left[V_{n_3}^{(n_2+1)} - V_{n_3+n_2-j_2}^{(j_2+1)} \right],
\end{aligned}
\tag{29}
$$

and

$$\beta_2 = -R + \frac{(P_{2,W} - P_{2,S})C_1 + C_3^{(2)}}{\mu_2}.$$

Now, we analyze how the service price impacts on $G^{(d^S)}$ as follows: Substituting (29) into the linear equation $G^{(d^S)} = 0$, we obtain

$$R = \frac{\varphi(j_1)V^{(j_1+1)}_{n_3+n_2-j_1} - \varphi(j_2)V^{(j_2+1)}_{n_3+n_2-j_2} - \psi(j_1,j_2)\left(V^{(n_2+1)}_{n_3} - \beta_2\right)}{\varphi(j_1)W^{(j_1+1)}_{n_3+n_2-j_1} - \varphi(j_2)W^{(j_2+1)}_{n_3+n_2-j_2} - \psi(j_1,j_2)\mu_2 W^{(n_2+1)}_{n_3}}. \tag{30}$$

Then, the unique solution of the price R in (30) is given by

$$\Re^{(d^S)}(j_1,j_2) = \frac{\varphi(j_1)V^{(j_1+1)}_{n_3+n_2-j_1} - \varphi(j_2)V^{(j_2+1)}_{n_3+n_2-j_2} - \psi(j_1,j_2)\left(V^{(n_2+1)}_{n_3} - \beta_2\right)}{\varphi(j_1)W^{(j_1+1)}_{n_3+n_2-j_1} - \varphi(j_2)W^{(j_2+1)}_{n_3+n_2-j_2} - \psi(j_1,j_2)\mu_2 W^{(n_2+1)}_{n_3}}.$$

It is easy to see from Equation (30) that (a) if $R \geq \Re^{(d^S)}(j_1,j_2)$, then $G^{(d^S)} \geq 0$; and (b) if $R \leq \Re^{(d^S)}(j_1,j_2)$, then $G^{(d^S)} \leq 0$.

In the energy-efficient data center, we relate to the service price and define two critical values as

$$R^S_H = \max_{d \in \mathcal{D}}\left\{0, \Re^{(d^S)}(1,0), \Re^{(d^S)}(2,0),\ldots, \Re^{(d^S)}(m_2, m_2 - 1)\right\} \tag{31}$$

and

$$R^S_L = \min_{d \in \mathcal{D}}\left\{\Re^{(d^S)}(1,0), \Re^{(d^S)}(2,0),\ldots, \Re^{(d^S)}(m_2, m_2 - 1)\right\} \tag{32}$$

The following proposition is similar to Proposition 1, thus its proof is omitted here.

Proposition 2. *(1) If $R \geq R^S_H$, then for any $d \in \mathcal{D}$ and for each couple (j_1, j_2) with $0 \leq j_2 < j_1 \leq n_2$, we have*

$$G^{(d^S)} \geq 0.$$

(2) If $0 \leq R \leq R^S_L$, then for any $d \in \mathcal{D}$ and for each couple (j_1, j_2) with $0 \leq j_2 < j_1 \leq n_2$, we have

$$G^{(d^S)} \leq 0.$$

From Propositions 1 and 2, we relate to the service price and define two new critical values as

$$R_H = \max\left\{R^W_H, R^S_H\right\} \text{ and } R_L = \min\left\{R^W_L, R^S_L\right\}. \tag{33}$$

The following theorem provides a simple summarization from Propositions 1 and 2, and it will be useful for studying the monotonicity and optimality of the asynchronous dynamic policy in our later sections.

Theorem 3. *(1) If $R \geq R_H$, then for any asynchronous dynamic policy $d \in \mathcal{D}$, we have*

$$G^{(d^W)} \geq 0 \text{ and } G^{(d^S)} \geq 0.$$

(2) If $0 \leq R \leq R_L$, then for any asynchronous policy $d \in \mathcal{D}$, we have

$$G^{(d^W)} \leq 0 \text{ and } G^{(d^S)} \leq 0.$$

6. Monotonicity and Optimality

In this section, we use the block-structured Poisson equation to derive a useful performance difference equation, and discuss the monotonicity and optimality of the long-run average profit of the energy-efficient data center with respect to the setup and sleep policies, respectively. Based on this, we can give the optimal asynchronous dynamic policy of the energy-efficient data center.

The standard Markov model-based formulation suffers from a number of drawbacks. First and foremost, the state space is usually too large for practical problems. That is, the number of potentials to be calculated or estimated is too large for most problems. Secondly, the generally applicable Markov model does not reflect any special structure of a particular problem. Thus, it is not clear whether and how potentials can be aggregated to save computation by exploring the special structure of the system. The sensitivity point of view and the flexible construction of the sensitivity formulas provide us a new perspective to explore alternative approaches for the performance optimization of systems with some special features.

For any given asynchronous energy-efficient policy $d \in \mathcal{D}$, the policy-based continuous-time Markov process $\{\mathbf{X}^{(d)}(t) : t \geq 0\}$ with infinitesimal generator $\mathbf{Q}^{(d)}$ given in (2) is an irreducible, aperiodic, and positive recurrent. Therefore, by using a similar analysis to Ma et al. [31], the long-run average profit of the data center is given by

$$\eta^d = \pi^{(d)} f^{(d)},$$

and the Poisson equation is written as

$$\mathbf{Q}^{(d)} g^{(d)} = \eta^d e - f^{(d)}.$$

For State (n_1, n_2, n_3), it is seen from (2) that the asynchronous energy-efficient policy d directly affects not only the elements of the infinitesimal generator $\mathbf{Q}^{(d)}$ but also the reward function $f^{(d)}$. That is, if the asynchronous policy d changes, then the infinitesimal generator $\mathbf{Q}^{(d)}$ and the reward function $f^{(d)}$ will have their corresponding changes. To express such a change mathematically, we take two different asynchronous energy-efficient policies d and d', both of which correspond to their infinitesimal generators $\mathbf{Q}^{(d)}$ and $\mathbf{Q}^{(d')}$, and to their reward functions $f^{(d)}$ and $f^{(d')}$.

The following lemma provides a useful equation for the difference $\eta^{d'} - \eta^d$ of the long-run average performances η^d and $\eta^{d'}$ for any two asynchronous policies $d, d' \in \mathcal{D}$. Here, we only restate it without proof, while readers may refer to Ma et al. [31] for more details.

Lemma 1. *For any two asynchronous energy-efficient policies $d, d' \in \mathcal{D}$, we have*

$$\eta^{d'} - \eta^d = \pi^{(d')} \left[\left(\mathbf{Q}^{(d')} - \mathbf{Q}^{(d)} \right) g^{(d)} + \left(f^{(d')} - f^{(d)} \right) \right]. \tag{34}$$

Now, we describe the first role played by the performance difference, in which we set up a partial order relation in the policy set $\mathcal{D}$ so that the optimal asynchronous energy-efficient policy in the finite set $\mathcal{D}$ can be found by means of finite comparisons. Based on the performance difference $\eta^{d'} - \eta^d$ for any two asynchronous energy-efficient policies $d, d' \in \mathcal{D}$, we can set up a partial order in the policy set $\mathcal{D}$ as follows. We write $d' \succ d$ if $\eta^{d'} > \eta^d$; $d' \approx d$ if $\eta^{d'} = \eta^d$; $d' \prec d$ if $\eta^{d'} < \eta^d$. Furthermore, we write $d' \succeq d$ if $\eta^{d'} \geq \eta^d$; $d' \preceq d$ if $\eta^{d'} \leq \eta^d$. By using this partial order, our research target is to find an optimal asynchronous policy $d^* \in \mathcal{D}$ such that $d^* \succeq d$ for any asynchronous energy-efficient policy $d \in \mathcal{D}$, or

$$d^* = \arg\max_{d \in \mathcal{D}} \left\{ \eta^d \right\}.$$

Note that the policy set $\mathcal{D}$ and the state space Ω are both finite, thus an enumeration method is feasible for finding the optimal asynchronous energy-efficient policy d^* in the policy set $\mathcal{D}$. Since

$$\mathcal{D} = \left\{ d = d^W \boxtimes d^S : d^W_{m_1,0,n_3} \in \{0, 1, \ldots, m_2\} \text{ for } 1 \leq n_3 \leq m_3; \right.$$

$$\left. d^S_{m_1,n_2,n_3} \in \{m_2 - n_2, m_2 - n_2 + 1, \ldots, m_2\} \text{ for } (m_1, n_2, n_3) \in \Omega_2 \cup \Omega_3 \cup \cdots \cup \Omega_{m_2+1} \right\}.$$

It is seen that the policy set $\mathcal{D}$ contains $(m_2+1)^{m_3} \times 2^{m_3} \times 3^{m_3-1} \times \cdots \times (m_2+1)^{m_2+1}$ elements so that the enumeration method used to find the optimal policy will require a huge enumeration workload. However, our following work will be able to greatly reduce the amount of searching for the optimal asynchronous policy d^* by means of the sensitivity-based optimization theory.

Now, we discuss the monotonicity of the long-run average profit η^d with respect to any asynchronous policy d under the different service prices. Since the setup and sleep policies d^W and d^S occur asynchronously and they will not interact with each other at any time, we can, respectively, study the impact of the policies d^W and d^S on the long-run average profit η^d. To this end, in what follows we shall discuss three different cases: $R \geq R_H$, $0 \leq R \leq R_L$, and $R_L < R < R_H$.

6.1. The Service Price $R \geq R_H$

In the case of $R \geq R_H$, we discuss the monotonicity and optimality with respect to two different policies: the setup policy and the sleep policy, respectively.

6.1.1. The Setup Policy with $R \geq R_H^W$

The following theorem analyzes the right-half part of the unimodal structure (see Figure 2) of the long-run average profit η^d with respect to the setup policy—either $d_{m_1,0,n_3}^W \in \{n_3, n_3+1, \ldots, m_2\}$ if $n_3 \leq m_2$ or $d_{m_1,0,n_3}^W = m_2$ if $n_3 > m_2$.

Theorem 4. *For any setup policy d^W with $d^W \boxtimes d^S \in \mathcal{D}$ and for each $n_3 = 1, 2, \ldots, m_3$, the long-run average profit $\eta^{d^W \boxtimes d^S}$ is linearly increasing with respect to the setup policy either $d_{m_1,0,n_3}^W \in \{n_3, n_3+1, \ldots, m_2\}$ if $n_3 \leq m_2$ or $d_{m_1,0,n_3}^W = m_2$ if $n_3 > m_2$.*

Proof of Theorem 4. For each $n_3 = 1, 2, \ldots, m_3$, we consider two interrelated policies $d^W \boxtimes d^S, d^{W'} \boxtimes d^S \in \mathcal{D}$ as follows:

$$d^W = \left(d_{m_1,0,1}^W, d_{m_1,0,2}^W, \ldots, d_{m_1,0,n_3-1}^W, \underline{d_{m_1,0,n_3}^W}, d_{m_1,0,n_3+1}^W, \ldots, d_{m_1,0,m_3}^W\right),$$

$$d^{W'} = \left(d_{m_1,0,1}^W, d_{m_1,0,2}^W, \ldots, d_{m_1,0,n_3-1}^W, \underline{n_3 \wedge m_2}, d_{m_1,0,n_3+1}^W, \ldots, d_{m_1,0,m_3}^W\right),$$

where $d_{m_1,0,n_3}^W \leq n_3 \wedge m_2$. It is seen that the two policies $d^W, d^{W'}$ have one difference only between their corresponding decision elements $d_{m_1,0,n_3}^W$ and $n_3 \wedge m_2$. In this case, it is seen from Theorem 1 that $Q^{(d^W \boxtimes d^S)} = Q^{(d^{W'} \boxtimes d^S)}$ and $\pi^{(d^W \boxtimes d^S)} = \pi^{(d^{W'} \boxtimes d^S)}$. Furthermore, it is easy to check from (4) to (7) that

$$f^{(d)} - f^{(d')} = \left(0, 0, \ldots, 0, -\left(d_{m_1,0,n_3}^W - n_3 \wedge m_2\right)\left[(P_{2,W} - P_{2,S})C_1 + C_3^{(1)}\right], 0, \ldots, 0\right)^T.$$

Thus, it follows from Lemma 1 that

$$\eta^{d^W \boxtimes d^S} - \eta^{d^{W'} \boxtimes d^S}$$

$$= \pi^{(d^W \boxtimes d^S)}\left[\left(Q^{(d^W \boxtimes d^S)} - Q^{(d^{W'} \boxtimes d^S)}\right)g^{(d^{W'} \boxtimes d^S)} + \left(f^{(d^W)} - f^{(d^{W'})}\right)\right]$$

$$= -\pi^{(d^W \boxtimes d^S)}(m_1, 0, n_3)\left(d_{m_1,0,n_3}^W - n_3 \wedge m_2\right)\left[(P_{2,W} - P_{2,S})C_1 + C_3^{(1)}\right]$$

or

$$\eta^{d^W \boxtimes d^S} = \eta^{d^{W'} \boxtimes d^S} - \pi^{(d^W \boxtimes d^S)}(m_1, 0, n_3)\left(d_{m_1,0,n_3}^W - n_3 \wedge m_2\right)\left[(P_{2,W} - P_{2,S})C_1 + C_3^{(1)}\right]. \quad (35)$$

Since $\pi^{\left(d^W \boxtimes d^S\right)} = \pi^{\left(d^{W'} \boxtimes d^S\right)}$, it is easy to see that $\pi^{\left(d^W \boxtimes d^S\right)}(m_1, 0, n_3) = \pi^{\left(d^{W'} \boxtimes d^S\right)}$ $(m_1, 0, n_3)$ can be determined by $d^{W'}_{m_1,0,n_3} = n_3 \wedge m_2$. This indicates that $\pi^{\left(d^W \boxtimes d^S\right)}(m_1, 0, n_3)$ is irrelevant to the decision element $d^W_{m_1,0,n_3}$. Furthermore, note that $\eta^{d^{W'} \boxtimes d^S}$ is irrelevant to the decision element $d^W_{m_1,0,n_3}$, and $P_{2,W} - P_{2,S}$, C_1 and $C_3^{(1)}$ are all positive constants, thus it is easy to see from (35) that the long-run average profit $\eta^{d^W \boxtimes d^S}$ is linearly decreasing with respect to each decision element $d^W_{m_1,0,n_3}$ for $d^W_{m_1,0,n_3} \in \{(n_3+1) \wedge m_2, (n_3+2) \wedge m_2, \ldots, m_2\}$. It is worth noting that if $m_2 \leq n_3 \leq m_3$, then $d^W_{m_1,0,n_3} \in \{m_2, m_2, \ldots, m_2\}$. This completes the proof. $\quad\square$

In what follows, we discuss the left-half part of the unimodal structure (see Figure 2) of the long-run average profit η^d with respect to each decision element $d^W_{m_1,0,n_3} \in \{0, 1, \ldots, n_3\}$ if $n_3 < m_2$. Compared to analysis of its right-half part, our discussion for the left-half part is a little bit complicated.

Let the optimal setup policy $d^{W^*} = \underset{d^W \boxtimes d^S \in \mathcal{D}}{\arg\max} \left\{ \eta^{d^W \boxtimes d^S} \right\}$ be

$$d^{W^*} = \left(d^{W^*}_{m_1,0,1}, d^{W^*}_{m_1,0,2}, \ldots, d^{W^*}_{m_1,0,m_3} \right).$$

Then, it is seen from Theorem 4 that

$$d^{W^*}_{m_1,0,n_3} = \begin{cases} 0, 1, \ldots, n_3, & 1 \leq n_3 \leq m_2, \\ m_2, & m_2 \leq n_3 \leq m_3. \end{cases}$$

Hence, Theorem 4 takes the area of finding the optimal setup policy d^{W^*} from a large set $\{0, 1, \ldots, m_2\}^{m_3}$ to a greatly shrunken area $\{0,1\} \times \{0,1,2\} \times \cdots \times \{0,1,\ldots,m_2-1\} \times \{0,1,\ldots,m_2\}^{m_3-m_2+1}$.

To find the optimal setup policy d^{W^*}, we consider two setup policies with an interrelated structure as follows:

$$d^W = \left(d^W_{m_1,0,1}, \ldots, d^W_{m_1,0,n_3-1}, \underline{d^W_{m_1,0,n_3}}, d^W_{m_1,0,n_3+1}, \ldots, d^W_{m_1,0,m_2}, \ldots, d^W_{m_1,0,m_3} \right),$$

$$d^{W'} = \left(d^W_{m_1,0,1}, \ldots, d^W_{m_1,0,n_3-1}, \underline{d^{W'}_{m_1,0,n_3}}, d^W_{m_1,0,n_3+1}, \ldots, d^W_{m_1,0,m_2}, \ldots, d^W_{m_1,0,m_3} \right),$$

where $d^{W'}_{m_1,0,n_3} = i_1 > d^W_{m_1,0,n_3} = i_2$, and $d^W_{m_1,0,n_3}, d^{W'}_{m_1,0,n_3} \in \{1, 2, \ldots, n_3 \wedge m_2\}$. It is easy to check from (2) that

$$Q^{\left(d^{W'} \boxtimes d^S\right)} - Q^{\left(d^W \boxtimes d^S\right)} =$$

$$\begin{pmatrix} 0 & & 0 & & 0 & & \\ & \ddots & & \ddots & & \ddots & \\ & & 0 & & 0 & & 0 \\ & & -(i_1 - i_2)\mu_2 \cdots & -(m_1\mu_1 + i_2\mu_2) \cdots & m_1\mu_1 + i_1\mu_2 & \\ & & 0 & & 0 & & 0 \\ & \ddots & & \ddots & & \ddots & \\ & & 0 & & 0 & & 0 \end{pmatrix}. \tag{36}$$

On the other hand, from the reward functions given in (8), it is seen that for $n_3 = 1, 2, \ldots, m_2$, and $d^W_{m_1,0,n_3} \in \{0, 1, \ldots, n_3\}$,

$$f^{\left(d^W\right)}(m_1, 0, n_3) = -\left[(P_{2,W} - P_{2,S})C_1 + C_3^{(1)} \right] d^W_{m_1,0,n_3} + Rm_1\mu_1$$

$$- (m_1 P_{1,W} + m_2 P_{2,S})C_1 - \left[m_1 C_2^{(1)} + n_3 C_2^{(3)} \right] - \lambda 1_{\{n_3 = m_3\}} C_5$$

and

$$f^{\left(d^{W'}\right)}(m_1, 0, n_3) = -\left[(P_{2,W} - P_{2,S})C_1 + C_3^{(1)}\right]d^{W'}_{m_1,0,n_3} + Rm_1\mu_1$$
$$- (m_1 P_{1,W} + m_2 P_{2,S})C_1 - \left[m_1 C_2^{(1)} + n_3 C_2^{(3)}\right] - \lambda 1_{\{n_3 = m_3\}}C_5.$$

Hence, we have

$$f^{\left(d^{W'}\right)} - f^{\left(d^W\right)} = (0, 0, \ldots, 0, -(i_2 - i_1)\mu_2\beta_1, 0, \ldots, 0)^T. \tag{37}$$

We write that

$$\eta^d\big|_{d^W_{m_1,0,n_3}=i} = \pi^{(d)}\big|_{d^W_{m_1,0,n_3}=i} \cdot f^{(d)}\big|_{d^W_{m_1,0,n_3}=i}.$$

The following theorem discusses the left-half part (see Figure 2) of the unimodal structure of the long-run average profit η^d with respect to each decision element $d^W_{m_1,0,n_3} \in \{0, 1, \ldots, m_2\}$.

Theorem 5. *If $R \geq R_H^W$, then for any setup policy d^W with $d^W \boxtimes d^S \in \mathcal{D}$ and for each $n_3 = 1, 2, \ldots, m_2$, the long-run average profit $\eta^{d^W \boxtimes d^S}$ is strictly monotone and increasing with respect to each decision element $d^W_{m_1,0,n_3}$ for $d^W_{m_1,0,n_3} \in \{0, 1, \ldots, n_3\}$.*

Proof of Theorem 5. For each $n_3 = 1, 2, \ldots, m_2$, we consider two setup policies with an interrelated structure as follows:

$$d^W = \left(d^W_{m_1,0,1}, \ldots, d^W_{m_1,0,n_3-1}, d^W_{m_1,0,n_3}, d^W_{m_1,0,n_3+1}, \ldots, d^W_{m_1,0,m_2}, \ldots, d^W_{m_1,0,m_3}\right),$$
$$d^{W'} = \left(d^W_{m_1,0,1}, \ldots, d^W_{m_1,0,n_3-1}, d^{W'}_{m_1,0,n_3}, d^W_{m_1,0,n_3+1}, \ldots, d^W_{m_1,0,m_2}, \ldots, d^W_{m_1,0,m_3}\right),$$

where $d^{W'}_{m_1,0,n_3} = i_1 > d^W_{m_1,0,n_3} = i_2$, and $d^W_{m_1,0,n_3}, d^{W'}_{m_1,0,n_3} \in \{0, 1, \ldots, n_3\}$. Applying Lemma 1, it follows from (36) and (37) that

$$\eta^{d^{W'} \boxtimes d^S} - \eta^{d^W \boxtimes d^S}$$
$$= \pi^{\left(d^{W'} \boxtimes d^S\right)}\left[\left(Q^{\left(d^{W'} \boxtimes d^S\right)} - Q^{\left(d^W \boxtimes d^S\right)}\right)g^{(d)} + \left(f^{\left(d^{W'}\right)} - f^{\left(d^W\right)}\right)\right]$$
$$= \pi^{\left(d^{W'} \boxtimes d^S\right)}(m_1, 0, n_3)\left[-(i_1 - i_2)\mu_2 g^{(d)}(m_1, 0, n_3)\right.$$
$$\left. -(m_1\mu_1 + i_2\mu_2)g^{(d)}(m_1, i_2, n_3 - i_2) \right. \tag{38}$$
$$\left. +(m_1\mu_1 + i_1\mu_2)g^{(d)}(m_1, i_1, n_3 - i_1) - (i_1 - i_2)\beta_1\right]$$
$$= \pi^{\left(d^{W'} \boxtimes d^S\right)}(m_1, 0, n_3)\left[m_1\mu_1 G_1^{\left(d^W\right)} - i_1\mu_2\left(G_2^{\left(d^W\right)} + \beta_1\right) + i_2\mu_2\left(G_3^{\left(d^W\right)} + \beta_1\right)\right]$$
$$= \pi^{\left(d^{W'} \boxtimes d^S\right)}(m_1, 0, n_3)G^{\left(d^W\right)}.$$

If $R \geq R_H^W$, then it is seen from Proposition 1 that $G^{\left(d^W\right)} \geq 0$. Thus, we get that for the two policies $d^W \boxtimes d^S, d^{W'} \boxtimes d^S \in \mathcal{D}$ with $d^{W'}_{m_1,0,n_3} > d^W_{m_1,0,n_3}$ and $d^W_{m_1,0,n_3}, d^{W'}_{m_1,0,n_3} \in \{0, 1, \ldots, n_3\}$,

$$\eta^{d^{W'} \boxtimes d^S} > \eta^{d^W \boxtimes d^S}.$$

This shows that

$$\eta^d\big|_{d^W_{m_1,0,n_3}=1} < \eta^d\big|_{d^W_{m_1,0,n_3}=2} < \cdots < \eta^d\big|_{d^W_{m_1,0,n_3}=n_3-1} < \eta^d\big|_{d^W_{m_1,0,n_3}=n_3}.$$

This completes the proof. $\square$

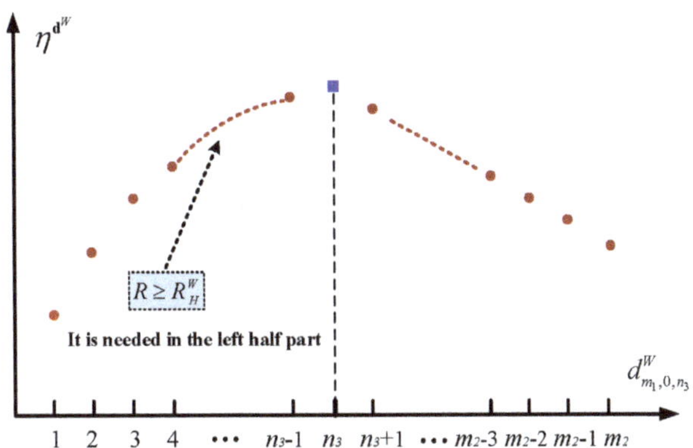

Figure 2. The unimodal structure of the long-run average profit by the setup policy.

When $R \geq R_H^W$, now we use Figure 2 to provide an intuitive summary for the main results given in Theorems 4 and 5. In the right-half part of Figure 2,

$$\eta^{d^W \boxtimes d^S} = \eta^{d^{W'} \boxtimes d^S} - \pi^{\left(d^W \boxtimes d^S\right)}(m_1, 0, n_3)\left(d_{m_1,0,n_3}^W - n_3\right)\left[(P_{2,W} - P_{2,S})C_1 + C_3^{(1)}\right]$$

shows that $\eta^{d^W \boxtimes d^S}$ is a linear function of the decision element $d_{m_1,0,n_3}^W$. By contrast, in the right-half part of Figure 2, we need to first introduce a restrictive condition: $R \geq R_H^W$, under which

$$\eta^{d^{W'} \boxtimes d^S} - \eta^{d^W \boxtimes d^S} = \pi^{\left(d^{W'} \boxtimes d^S\right)}(m_1, 0, n_3)G^{\left(d^W\right)}.$$

Since $G^{\left(d^W\right)}$ also depends on the decision element $d_{m_1,0,n_3}^W$, it is clear that $\eta^{d^W \boxtimes d^S}$ is a nonlinear function of the decision element $d_{m_1,0,n_3}^W$.

6.1.2. The Sleep Policy with $R \geq R_H^S$

It is different from the setup policy in that, for the sleep policy, each decision element is $d_{m_1,n_2,n_3}^S \in \{m_2 - n_2, m_2 - n_2 + 1, \ldots, m_2\}$. Hence, we just consider the structural properties of the long-run average profit $\eta^{d^W \boxtimes d^S}$ with respect to each decision element d_{m_1,n_2,n_3}^S.

We write the optimal sleep policy as $d^{S^*} = \arg\max_{d^W \boxtimes d^S \in \mathcal{D}}\left\{\eta^{d^W \boxtimes d^S}\right\}$, where

$$d^{S^*} = \left(d_{m_1,1,0}^{S^*}, d_{m_1,1,1}^{S^*}, \ldots, d_{m_1,1,m_3-1}^{S^*}; d_{m_1,2,0}^{S^*}, \ldots, d_{m_1,2,m_3-2}^{S^*}; \cdots; d_{m_1,m_2,0}^{S^*}, \ldots, d_{m_1,m_2,m_3-m_2}^{S^*}\right).$$

Then, it is seen that

$$d_{m_1,1,0}^{S^*} \in \{m_2 - 1, m_2\}, \cdots, d_{m_1,1,m_3-1}^{S^*} \in \{m_2 - 1, m_2\};$$

$$d_{m_1,2,0}^{S^*} \in \{m_2 - 2, m_2 - 1, m_2\}, \cdots, d_{m_1,2,m_3-2}^{S^*} \in \{m_2 - 2, m_2 - 1, m_2\};$$

$$\vdots$$

$$d_{m_1,m_2,0}^{S^*} \in \{0, 1, \ldots, m_2\}, \cdots, d_{m_1,m_2,m_3-m_2}^{S^*} \in \{0, 1, \ldots, m_2\}.$$

It is easy to see that the area of finding the optimal sleep policy d^{S^*} is $\{m_2 - 1, m_2\}^{m_3} \times \{m_2 - 2, m_2 - 1, m_2\}^{m_3-1} \times \cdots \times \{0, 1, \ldots, m_2\}^{m_3-m_2}$.

To find the optimal sleep policy d^{S^*}, we consider two sleep policies with an interrelated structure as follows:

$$d^S = \left(d^S_{m_1,1,0}, d^S_{m_1,1,1}, \ldots, d^S_{m_1,n_2,n_3-1}, \underline{d^S_{m_1,n_2,n_3}}, d^S_{m_1,n_2,n_3+1}, \ldots, d^S_{m_1,m_2,m_3-m_2}\right),$$

$$d^{S'} = \left(d^S_{m_1,1,0}, d^S_{m_1,1,1}, \ldots, d^S_{m_1,n_2,n_3-1}, \underline{d^{S'}_{m_1,n_2,n_3}}, d^S_{m_1,n_2,n_3+1}, \ldots, d^S_{m_1,m_2,m_3-m_2}\right),$$

where $d^{S'}_{m_1,n_2,n_3} = m_2 - j_2 > d^S_{m_1,n_2,n_3} = m_2 - j_1$, $0 \le j_2 < j_1 \le n_2$, it is easy to check from (2) that

$$\mathbf{Q}^{\left(d^W \boxtimes d^{S'}\right)} - \mathbf{Q}^{\left(d^W \boxtimes d^S\right)} =$$

$$\begin{pmatrix}
0 & & 0 & & 0 & \\
& \ddots & & \ddots & & \ddots & \\
& 0 & & 0 & & 0 & \\
m_1\mu_1 + j_2\mu_2 & \cdots & -(m_1\mu_1 + j_1\mu_2) & \cdots & -(j_2 - j_1)\mu_2 & \\
& 0 & & 0 & & 0 & \\
& & \ddots & & \ddots & & \ddots \\
& 0 & & 0 & & 0 & 0
\end{pmatrix}. \quad (39)$$

On the other hand, from the reward functions given in (6), $d^S_{m_1,n_2,n_3}, d^{S'}_{m_1,n_2,n_3}$ is in either $\{m_2 - n_2, m_2 - n_2 + 1, \ldots, m_2\}$ for $1 \le n_2 \le m_2$ and $0 \le n_3 \le m_3 - n_2$ or $\{0, 1, \ldots, m_2\}$ for $n_2 = m_2$ and $0 \le n_3 \le m_3 - m_2$, we have

$$f^{\left(d^S\right)}(m_1, n_2, n_3) = -\left[R\mu_2 - (P_{2,W} - P_{2,S})C_1 + C_3^{(2)}\right]d^S_{m_1,n_2,n_3} + R(m_1\mu_1 + m_2\mu_2)$$
$$- (m_1 P_{1,W} + m_2 P_{2,W})C_1 - \left[m_1 C_2^{(1)} + n_2 C_2^{(2)} + n_3 C_2^{(3)}\right]$$
$$- (n_2 - m_2)C_3^{(2)} - m_1\mu_1 \mathbf{1}_{\{n_3=0\}} C_4$$

and

$$f^{\left(d^{S'}\right)}(m_1, n_2, n_3) = -\left[R\mu_2 - (P_{2,W} - P_{2,S})C_1 + C_3^{(2)}\right]d^{S'}_{m_1,n_2,n_3} + R(m_1\mu_1 + m_2\mu_2)$$
$$- (m_1 P_{1,W} + m_2 P_{2,W})C_1 - \left[m_1 C_2^{(1)} + n_2 C_2^{(2)} + n_3 C_2^{(3)}\right]$$
$$- (n_2 - m_2)C_3^{(2)} - m_1\mu_1 \mathbf{1}_{\{n_3=0\}} C_4.$$

Hence, we have

$$f^{\left(d^{S'}\right)} - f^{\left(d^S\right)} = (0, 0, \ldots, 0, -(j_2 - j_1)\mu_2\beta_2, 0, \ldots, 0)^T. \quad (40)$$

We write

$$\eta^d \big|_{d^S_{m_1,n_2,n_3}=m_2-j} = \pi^{(d)} \big|_{d^S_{m_1,n_2,n_3}=m_2-j} \cdot f^{(d)} \big|_{d^S_{m_1,n_2,n_3}=m_2-j}.$$

The following theorem discusses the structure of the long-run average profit $\eta^{d^W \boxtimes d^S}$ with respect to each decision element $d^S_{m_1,n_2,n_3}$.

Theorem 6. *If $R \ge R^S_H$, then for any sleep policy d^S with $d^W \boxtimes d^S \in \mathcal{D}$ and for each $n_2 = 1, 2, \ldots, m_2$, $n_3 = 0, 1, \ldots, m_3 - n_2$, the long-run average profit η^d is strictly monotone decreasing with respect to each decision element $d^S_{m_1,n_2,n_3} \in \{m_2 - n_2, m_2 - n_2 + 1, \ldots, m_2\}$.*

Proof of Theorem 6. For each $n_2 = 1, 2, \ldots, m_2$ and $n_3 = 1, 2, \ldots, m_3 - n_2$, we consider two sleep policies with an interrelated structure as follows:

$$
\boldsymbol{d}^S = \left(d^S_{m_1,1,0}, \ldots, d^S_{m_1,n_2,n_3-1}, \underline{d^S_{m_1,n_2,n_3}}, d^S_{m_1,n_2,n_3+1}, \ldots, d^S_{m_1,m_2,m_3-m_2} \right),
$$
$$
\boldsymbol{d}^{S'} = \left(d^S_{m_1,1,0}, \ldots, d^S_{m_1,n_2,n_3-1}, \underline{d^{S'}_{m_1,n_2,n_3}}, d^S_{m_1,n_2,n_3+1}, \ldots, d^S_{m_1,m_2,m_3-m_2} \right),
$$

where $d^{S'}_{m_1,n_2,n_3} = m_2 - j_2 > d^S_{m_1,n_2,n_3} = m_2 - j_1$, $0 \le j_2 < j_1 \le n_2$ and $d^S_{m_1,n_2,n_3}, d^{S'}_{m_1,n_2,n_3} \in \{m_2 - n_2, m_2 - n_2 + 1, \ldots, m_2\}$. Applying Lemma 1, it follows from (39) and (40) that

$$
\eta^{\boldsymbol{d}^W \boxtimes \boldsymbol{d}^{S'}} - \eta^{\boldsymbol{d}^W \boxtimes \boldsymbol{d}^S}
$$
$$
= \pi^{\left(\boldsymbol{d}^W \boxtimes \boldsymbol{d}^{S'}\right)} \left[\left(\mathbf{Q}^{\left(\boldsymbol{d}^W \boxtimes \boldsymbol{d}^{S'}\right)} - \mathbf{Q}^{\left(\boldsymbol{d}^W \boxtimes \boldsymbol{d}^S\right)} \right) g^{(\boldsymbol{d})} + \left(f^{\left(\boldsymbol{d}^{S'}\right)} - f^{\left(\boldsymbol{d}^S\right)} \right) \right]
$$
$$
= \pi^{\left(\boldsymbol{d}^W \boxtimes \boldsymbol{d}^{S'}\right)} (m_1, n_2, n_3) \left[(m_1\mu_1 + j_2\mu_2) g^{(\boldsymbol{d})}(m_1, j_2, n_3 + n_2 - j_2) \right.
$$
$$
\left. - (m_1\mu_1 + j_1\mu_2) g^{(\boldsymbol{d})}(m_1, j_1, n_3 + n_2 - j_1) - (j_2 - j_1)\mu_2 g^{(\boldsymbol{d})}(m_1, n_2, n_3) - (j_2 - j_1)\mu_2\beta_2 \right] \quad (41)
$$
$$
= \pi^{\left(\boldsymbol{d}^W \boxtimes \boldsymbol{d}^{S'}\right)} (m_1, n_2, n_3) \left[m_1\mu_1 G_1^{\left(\boldsymbol{d}^S\right)} + j_1\mu_2 \left(G_2^{\left(\boldsymbol{d}^S\right)} + \beta_2 \right) - j_2\mu_2 \left(G_3^{\left(\boldsymbol{d}^S\right)} + \beta_2 \right) \right]
$$
$$
= \pi^{\left(\boldsymbol{d}^W \boxtimes \boldsymbol{d}^{S'}\right)} (m_1, n_2, n_3) G^{\left(\boldsymbol{d}^S\right)}.
$$

It is worthwhile to note that (41) has the same form as (38), since the perturbation of the sleep policy, j_1, and j_2 denote the number of the working servers. If $R \ge R_H^S$, then it is seen from Proposition 1 that for $j_2 < j_1$, we have $G^{\left(\boldsymbol{d}^S\right)} \ge 0$. Thus, we get that for $j_2 < j_1$ and $j_1, j_2 \in \{0, 1, \ldots, n_2\}$,

$$
\eta^{\boldsymbol{d}^W \boxtimes \boldsymbol{d}^{S'}} > \eta^{\boldsymbol{d}^W \boxtimes \boldsymbol{d}^S},
$$

this shows that $\eta^{\boldsymbol{d}^W \boxtimes \boldsymbol{d}^S}$ is strictly monotone increasing with respect to $m_2 - d^S_{m_1,n_2,n_3}$. Thus, we get that for the two policies $\boldsymbol{d}^W \boxtimes \boldsymbol{d}^S, \boldsymbol{d}^W \boxtimes \boldsymbol{d}^{S'} \in \mathcal{D}$ with $d^S_{m_1,n_2,n_3} > d^{S'}_{m_1,n_2,n_3}$ and $d^S_{m_1,n_2,n_3}, d^{S'}_{m_1,n_2,n_3} \in \{m_2 - n_2, m_2 - n_2 + 1, \ldots, m_2\}$,

$$
\eta^{\boldsymbol{d}^W \boxtimes \boldsymbol{d}^{S'}} < \eta^{\boldsymbol{d}^W \boxtimes \boldsymbol{d}^S}.
$$

It is easy to see that

$$
\eta^{\boldsymbol{d}} \big|_{d^S_{m_1,n_2,n_3} = m_2 - n_2} > \eta^{\boldsymbol{d}} \big|_{d^S_{m_1,n_2,n_3} = m_2 - n_2 + 1} > \cdots > \eta^{\boldsymbol{d}} \big|_{d^S_{m_1,n_2,n_3} = m_2}.
$$

This completes the proof. $\square$

When $R \ge R_H^S$, now we use Figure 3 to provide an intuitive summary for the main results given in Theorems 6. According to (41), $G^{\left(\boldsymbol{d}^S\right)}$ depends on $d^S_{m_1,n_2,n_3}$, and it is clear that $\eta^{\boldsymbol{d}^W \boxtimes \boldsymbol{d}^S}$ is a nonlinear function of $d^S_{m_1,n_2,n_3}$.

As a simple summarization of Theorems 5 and 6, we obtain the monotone structure of the long-run average profit $\eta^{\boldsymbol{d}}$ with respect to the asynchronous energy-efficient policy, while its proof is easy only on the condition that $R \ge R_H$ makes $R \ge R_H^W$ and $R \ge R_H^S$.

Theorem 7. *If $R \ge R_H$, then for any policy $\boldsymbol{d} \in \mathcal{D}$, the long-run average profit $\eta^{\boldsymbol{d}}$ is strictly monotone with respect to each decision element of $\boldsymbol{d}^W$ and of $\boldsymbol{d}^S$, respectively.*

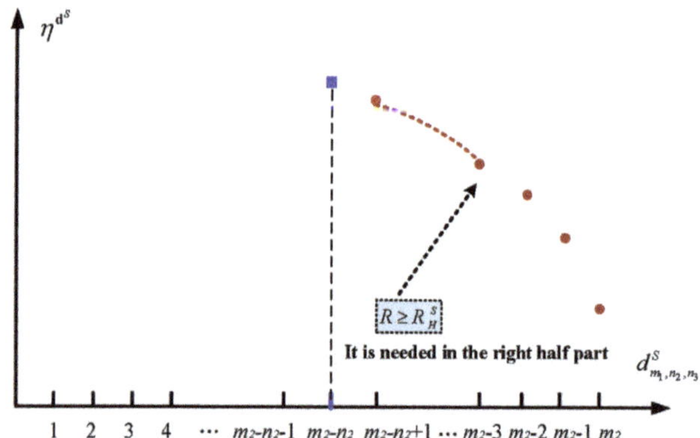

Figure 3. The monotone structure of the long-run average profit by the sleep policy.

6.2. The Service Price $0 \leq R \leq R_L$

A similar analysis to that of the case $R \geq R_H$, we simply discuss the service price $0 \leq R \leq R_L$ for the monotonicity and optimality for two different policies: the setup policy and the sleep policy.

6.2.1. The Setup Policy with $0 \leq R \leq R_L^W$

Theorem 8. *If $0 \leq R \leq R_L^W$, for any setup policy d^W with $d^W \boxtimes d^S \in \mathcal{D}$ and for each $n_3 = 1, 2, \ldots, m_2$, then the long-run average profit $\eta^{d^W \boxtimes d^S}$ is strictly monotone decreasing with respect to each decision element $d_{m_1,0,n_3}^W \in \{0, 1, \ldots, n_3\}$.*

Proof of Theorem 8. This proof is similar to that of Theorem 5. For each $n_3 = 1, 2, \ldots, m_2$, we consider two setup policies with an interrelated structure as follows:

$$d^W = \left(d_{m_1,0,1}^W, \ldots, d_{m_1,0,n_3-1}^W, \underline{d_{m_1,0,n_3}^W}, d_{m_1,0,n_3+1}^W, \ldots, d_{m_1,0,m_2}^W, \ldots, d_{m_1,0,m_3}^W \right),$$

$$d^{W'} = \left(d_{m_1,0,1}^W, \ldots, d_{m_1,0,n_3-1}^W, \underline{d_{m_1,0,n_3}^{W'}}, d_{m_1,0,n_3+1}^W, \ldots, d_{m_1,0,m_2}^W, \ldots, d_{m_1,0,m_3}^W \right),$$

where $d_{m_1,0,n_3}^{W'} = i_1 > d_{m_1,0,n_3}^W = i_2$, and $d_{m_1,0,n_3}^W, d_{m_1,0,n_3}^{W'} \in \{0, 1, \ldots, n_3\}$ for $1 \leq n_3 \leq m_2$. It is clear that

$$\eta^{d^{W'} \boxtimes d^S} - \eta^{d^W \boxtimes d^S} = \pi^{\left(d^{W'} \boxtimes d^S \right)} (m_1, 0, n_3) G^{\left(d^W \right)}.$$

If $0 \leq R \leq R_L^W$, then it is seen from Proposition 1 that $G^{\left(d^W \right)} \leq 0$. Thus, we get that for the two setup policies with $d_{m_1,0,n_3}^{W'} < d_{m_1,0,n_3}^W$ and $d_{m_1,0,n_3}^W, d_{m_1,0,n_3}^{W'} \in \{0, 1, \ldots, n_3\}$,

$$\eta^{d^{W'} \boxtimes d^S} < \eta^{d^W \boxtimes d^S}.$$

This shows that for $1 \leq n_3 \leq m_2$,

$$\eta^d \big|_{d_{m_1,0,n_3}^W = 1} > \eta^d \big|_{d_{m_1,0,n_3}^W = 2} > \cdots > \eta^d \big|_{d_{m_1,0,n_3}^W = n_3 - 1} > \eta^d \big|_{d_{m_1,0,n_3}^W = n_3}.$$

This completes the proof. $\square$

When $0 \leq R \leq R_L^W$, we also use Figure 4 to provide an intuitive summary for the main results given in Theorems 4 and 8.

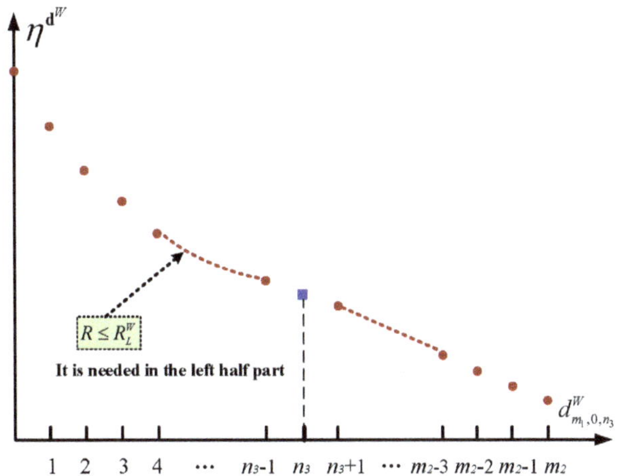

Figure 4. The decreasing structure of the long-run average profit by the setup policy.

6.2.2. The Sleep Policy with $0 \leq R \leq R_L^S$

Theorem 9. *If $0 \leq R \leq R_L^S$, then for any sleep policy $\boldsymbol{d}^S$ with $\boldsymbol{d}^W \boxtimes \boldsymbol{d}^S \in \mathcal{D}$ and for each $n_2 = 1, 2, \ldots, m_2$, $n_3 = 0, 1, \ldots, m_3 - n_2$, the long-run average profit $\eta^{\boldsymbol{d}^W \boxtimes \boldsymbol{d}^S}$ is strictly monotone increasing with respect to each decision element $d^S_{m_1, n_2, n_3}$, for $d^S_{m_1, n_2, n_3} \in \{m_2 - n_2, m_2 - n_2 + 1, \ldots, m_2\}$.*

Proof of Theorem 9. This proof is similar to that of Theorem 6. For each $n_2 = 1, 2, \ldots, m_2$ and $n_3 = 0, 1, \ldots, m_3 - n_2$, we consider two sleep policies with an interrelated structure as follows:

$$\boldsymbol{d}^S = \left(d^S_{m_1,1,0}, d^S_{m_1,1,1}, \ldots, d^S_{m_1,n_2,n_3-1}, \underline{d^S_{m_1,n_2,n_3}}, d^S_{m_1,n_2,n_3+1}, \ldots, d^S_{m_1,m_2,m_3-m_2} \right),$$

$$\boldsymbol{d}^{S'} = \left(d^S_{m_1,1,0}, d^S_{m_1,1,1}, \ldots, d^S_{m_1,n_2,n_3-1}, \underline{d^{S'}_{m_1,n_2,n_3}}, d^S_{m_1,n_2,n_3+1}, \ldots, d^S_{m_1,m_2,m_3-m_2} \right),$$

where $d^{S'}_{m_1,n_2,n_3} = m_2 - j_2 > d^S_{m_1,n_2,n_3} = m_2 - j_1, 0 \leq j_2 < j_1 \leq n_2$ and $d^S_{m_1,n_2,n_3}, d^{S'}_{m_1,n_2,n_3} \in \{m_2 - n_2, m_2 - n_2 + 1, \ldots, m_2\}$. It is clear that

$$\eta^{\boldsymbol{d}^W \boxtimes \boldsymbol{d}^{S'}} - \eta^{\boldsymbol{d}^W \boxtimes \boldsymbol{d}^S} = \pi^{\left(\boldsymbol{d}^W \boxtimes \boldsymbol{d}^{S'} \right)}(m_1, n_2, n_3) G^{\left(\boldsymbol{d}^S \right)}.$$

By a similar analysis to that in Theorem 6, if $0 \leq R \leq R_L^S$, then it is seen from Proposition 1 that for $j_2 < j_1$, we have $G^{\left(\boldsymbol{d}^S \right)} \leq 0$. Thus, we get that for $j_2 < j_1$ and $j_1, j_2 \in \{0, 1, \ldots, n_2\}$, $\eta^{\boldsymbol{d}^W \boxtimes \boldsymbol{d}^S}$ is strictly monotone decreasing with respect to $m_2 - d^S_{m_1,n_2,n_3}$, hence it is also strictly monotone increasing with respect to $d^S_{m_1,n_2,n_3}$. Thus, we get that for the two sleep policies with $d^S_{m_1,n_2,n_3} < d^{S'}_{m_1,n_2,n_3}$ and $d^S_{m_1,n_2,n_3}, d^{S'}_{m_1,n_2,n_3} \in \{m_2 - n_2, m_2 - n_2 + 1, \ldots, m_2\}$,

$$\eta^{\boldsymbol{d}^W \boxtimes \boldsymbol{d}^{S'}} < \eta^{\boldsymbol{d}^W \boxtimes \boldsymbol{d}^S}.$$

This shows that

$$\eta^{\boldsymbol{d}} \big|_{d^S_{m_1,n_2,n_3} = m_2 - n_2} < \eta^{\boldsymbol{d}} \big|_{d^S_{m_1,n_2,n_3} = m_2 - n_2 + 1} < \cdots < \eta^{\boldsymbol{d}} \big|_{d^S_{m_1,n_2,n_3} = m_2}.$$

This completes the proof. $\square$

When $0 \leq R \leq R_L^S$, we also use Figure 5 to provide an intuitive summary for the main results given in Theorem 9.

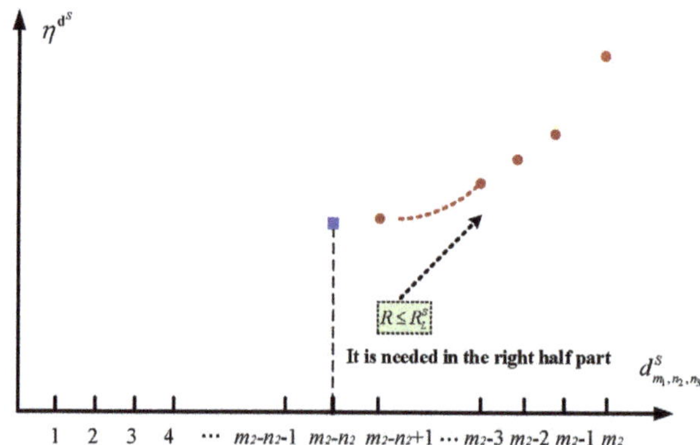

Figure 5. The increasing structure of the long-run average profit by the sleep policy.

As a simple summarization of Theorems 8 and 9, the following theorem further describes monotone structure of the long-run average profit η^d with respect to the asynchronous energy-efficient policy, while its proof is easy only through using the condition that $0 \leq R \leq R_L$ makes $0 \leq R \leq R_L^W$ and $0 \leq R \leq R_L^S$.

Theorem 10. *If $0 \leq R \leq R_L$, then for any asynchronous energy-efficient policy $\boldsymbol{d} \in \mathcal{D}$, the long-run average profit η^d is strictly monotone with respect to each decision element of $\boldsymbol{d}^W$ and of $\boldsymbol{d}^S$, respectively.*

In the remainder of this section, we discuss a more complicated case with the service price $R_L < R < R_H$. In this case, we use the bang–bang control and the asynchronous structure of $\boldsymbol{d} \in \mathcal{D}$ to prove that the optimal asynchronous energy-efficient polices $\boldsymbol{d}^{W*}$ and $\boldsymbol{d}^{S*}$ both have bang–bang control forms.

6.3. The Service Price $R_L < R < R_H$

For the price $R_L < R < R_H$, we can further derive the following theorems about the monotonicity of η^d with respect to the setup policy and the sleep policy, respectively.

6.3.1. The Setup Policy with $R_L^W < R < R_H^W$

For the service price $R_L^W < R < R_H^W$, the following theorem provides the monotonicity of η^d with respect to the decision element $d^W_{m_1,0,n_3}$.

Theorem 11. *If $R_L^W < R < R_H^W$, then the long-run average profit $\eta^{d^W \boxtimes d^S}$ is monotone (either increasing or decreasing) with respect to the decision element $d^W_{m_1,0,n_3}$, where $n_3 = 1, 2, \ldots, m_3$ and $d^W_{m_1,0,n_3} \in \{0, 1, \ldots, n_3\}$.*

Proof of Theorem 11. Similarly to the first part of the proof for Theorem 5, we consider any two setup policies with an interrelated structure as follows:

$$\boldsymbol{d}^W = \left(d^W_{m_1,0,1}, \ldots, d^W_{m_1,0,n_3-1}, \underline{d^W_{m_1,0,n_3}}, d^W_{m_1,0,n_3+1}, \ldots, d^W_{m_1,0,m_2}, \ldots, d^W_{m_1,0,m_3} \right),$$

$$\boldsymbol{d}^{W'} = \left(d^W_{m_1,0,1}, \ldots, d^W_{m_1,0,n_3-1}, \underline{d^{W'}_{m_1,0,n_3}}, d^W_{m_1,0,n_3+1}, \ldots, d^W_{m_1,0,m_2}, \ldots, d^W_{m_1,0,m_3} \right),$$

where $d^W_{m_1,0,n_3}, d^{W'}_{m_1,0,n_3} \in \{0, 1, \ldots, n_3\}$. Applying Lemma 1, we obtain

$$\eta^{d^{W'} \boxtimes d^S} - \eta^{d^W \boxtimes d^S} = \pi^{\left(d^{W'} \boxtimes d^S \right)}(m_1, 0, n_3) G^{\left(d^W \right)}. \tag{42}$$

On the other hand, we can similarly obtain the following difference equation

$$\eta^{d^W \boxtimes d^S} - \eta^{d^{W'} \boxtimes d^S} = -\pi^{\left(d^W \boxtimes d^S\right)}(m_1, 0, n_3) G^{\left(d^{W'}\right)}. \tag{43}$$

By summing (42) and (43), we have

$$\pi^{\left(d^{W'} \boxtimes d^S\right)}(m_1, 0, n_3) G^{\left(d^W\right)} - \pi^{\left(d^W \boxtimes d^S\right)}(m_1, 0, n_3) G^{\left(d^{W'}\right)} = 0.$$

Therefore, we have the *sign conservation equation*

$$\frac{G^{\left(d^W\right)}}{G^{\left(d^{W'}\right)}} = \frac{\pi^{\left(d^W \boxtimes d^S\right)}(m_1, 0, n_3)}{\pi^{\left(d^{W'} \boxtimes d^S\right)}(m_1, 0, n_3)} > 0. \tag{44}$$

The above equation means that the sign of $G^{\left(d^W\right)}$ and $G^{\left(d^{W'}\right)}$ are always identical when a particular decision element $d^W_{m_1,0,n_3}$ is changed to any $d^{W'}_{m_1,0,n_3}$. With the sign conservation Equation (44) and the performance difference Equation (43), we can directly derive that the long-run average profit $\eta^{d^W \boxtimes d^S}$ is monotone with respect to $d^W_{m_1,0,n_3}$. This completes the proof. $\square$

Based on Theorem 11, the following corollary directly derives that the optimal decision element $d^{W*}_{m_1,0,n_3}$ has a bang–bang control form (see more details in Cao [34] and Xia and Chen [35]).

Corollary 1. *For the setup policy, the optimal decision element $d^{W*}_{m_1,0,n_3}$ is either 0 or n_3, i.e., the bang–bang control is optimal.*

With Corollary 1, we should either keep all servers in sleep mode or turn on the servers such that the number of working servers equals the number of waiting jobs in the buffer. In addition, we can see that the search space of $d^W_{m_1,0,n_3}$ can be reduced from $\{0, 1, \ldots, n_3\}$ to a two-element set $\{0, n_3\}$, which is a significant reduction of search complexity.

6.3.2. The Sleep Policy with $R^S_L < R < R^S_H$

For the service price $R^S_L < R < R^S_H$, the following theorem provides the monotonicity of η^d with respect to the decision element $d^S_{m_1,n_2,n_3}$.

Theorem 12. *If $R^S_L < R < R^S_H$, then the long-run average profit $\eta^{d^W \boxtimes d^S}$ is monotone (either increasing or decreasing) with respect to the decision element $d^S_{m_1,n_2,n_3}$, where $n_2 = 1, 2, \ldots, m_2$, $n_3 = 0, 1, \ldots, m_3 - n_2$ and $d^S_{m_1,n_2,n_3} \in \{m_2 - n_2, m_2 - n_2 + 1, \ldots, m_2\}$.*

Proof of Theorem 12. Similar to the proof for Theorem 11, we consider any two sleep policies with an interrelated structure as follows:

$$d^S = \left(d^S_{m_1,1,0}, d^S_{m_1,1,1}, \ldots, d^S_{m_1,n_2,n_3-1}, \underline{d^S_{m_1,n_2,n_3}}, d^S_{m_1,n_2,n_3+1}, \ldots, d^S_{m_1,m_2,m_3-m_2}\right),$$

$$d^{S'} = \left(d^S_{m_1,1,0}, d^S_{m_1,1,1}, \ldots, d^S_{m_1,n_2,n_3-1}, \underline{d^{S'}_{m_1,n_2,n_3}}, d^S_{m_1,n_2,n_3+1}, \ldots, d^S_{m_1,m_2,m_3-m_2}\right),$$

where $d^S_{m_1,n_2,n_3}, d^{S'}_{m_1,n_2,n_3} \in \{0, 1, \ldots, m_2\}$. Applying Lemma 1, we obtain

$$\eta^{d^W \boxtimes d^{S'}} - \eta^{d^W \boxtimes d^S} = \pi^{\left(d^W \boxtimes d^{S'}\right)}(m_1, n_2, n_3) G^{\left(d^S\right)}. \tag{45}$$

On the other hand, we can also obtain the following difference equation:

$$\eta^{d^W \boxtimes d^S} - \eta^{d^W \boxtimes d^{S'}} = -\pi^{\left(d^W \boxtimes d^S\right)}(m_1, n_2, n_3) G^{\left(d^{S'}\right)}. \tag{46}$$

Thus, the *sign conservation equation* is given by

$$\frac{G^{\left(d^S\right)}}{G^{\left(d^{S'}\right)}} = \frac{\pi^{\left(d^W \boxtimes d^S\right)}(m_1, n_2, n_3)}{\pi^{\left(d^W \boxtimes d^{S'}\right)}(m_1, n_2, n_3)} > 0. \tag{47}$$

This means that the signs of $G^{\left(d^S\right)}$ and $G^{\left(d^{S'}\right)}$ are always identical when a particular decision element $d^S_{m_1,n_2,n_3}$ is changed to any $d^{S'}_{m_1,n_2,n_3}$. We can directly derive that the long-run average profit $\eta^{d^W \boxtimes d^S}$ is monotone with respect to $d^S_{m_1,n_2,n_3}$. This completes the proof. $\square$

Corollary 2. *For the sleep policy, the optimal decision element $d^{S*}_{m_1,n_2,n_3}$ is either $m_2 - n_2$ or m_2, i.e., the bang–bang control is optimal.*

With Corollary 2, we should either keep all in sleep or turn off the servers such that the number of sleeping servers equals the number of servers without jobs in Group 2. We can see that the search space of $d^S_{m_1,n_2,n_3}$ can be reduced from $\{m_2 - n_2, m_2 - n_2 + 1, \ldots, m_2\}$ to a two-element set $\{m_2 - n_2, m_2\}$, hence this is also a significant reduction of search complexity.

It is seen from Corollaries 1 and 2 that the form of the bang–bang control is very simple and easy to adopt in practice, while the optimality of the bang–bang control guarantees the performance confidence of such simple forms of control. This makes up the threshold-type of the optimal asynchronous energy-efficient policy in the data center.

7. The Maximal Long-Run Average Profit

In this section, we provide the optimal asynchronous dynamic policy d^* of the threshold-type in the energy-efficient data center and further compute the maximal long-run average profit.

We introduce some notation as follows:

$$c_0 = (P_{2,W} - P_{2,S})C_1,$$
$$c_1 = (m_1 P_{1,W} + m_2 P_{2,W})C_1,$$
$$c_2 = (m_1 P_{1,W} + m_2 P_{2,W})C_1 + m_1 C_2^{(1)}, \tag{48}$$
$$c_3 = (m_1 P_{1,W} + m_2 P_{2,W})C_1 + m_1 C_2^{(1)} + m_2 C_3^{(1)},$$
$$c_4 = (m_1 P_{1,W} + m_2 P_{2,W})C_1 + m_1 C_2^{(1)} + m_1 \mu_1 1_{\{n_2>0,n_3=0\}} C_4,$$
$$c_5 = (m_1 P_{1,W} + m_2 P_{2,W})C_1 + m_1 C_2^{(1)} + m_2 C_2^{(2)} + m_1 \mu_1 1_{\{n_2>0,n_3=0\}} C_4 + \lambda 1_{\{n_2+n_3=m_3\}} C_5.$$

Now, we express the optimal asynchronous energy-efficient policy d^* of the threshold-type and compute the maximal long-run average profit η^{d^*} under three different service prices as follows:

Case 1. *The service price $R \geq R_H$*

It follows from Theorem 7 that

$$d^{W*} = (1, 2, \ldots, n_3, \ldots, m_2, \ldots, m_2),$$
$$d^{S*} = (m_2 - 1, \ldots, m_2 - 1; \ldots; m_2 - n_2, \ldots, m_2 - n_2; \ldots; 1, \ldots, 1; 0, \ldots, 0),$$

thus we have

$$\eta^{d^*} = \sum_{n_1=0}^{m_1} \pi^{(d)}(n_1,0,0)\left[\left(R\mu_1 - C_2^{(1)}\right)n_1 - c_1\right]$$

$$+ \sum_{n_3=1}^{m_3} \pi^{(d)}(m_1,0,n_3)\left[Rm_1\mu_1 - c_2 - \left(c_0 + C_3^{(1)}\right)(n_3 \wedge m_2) - C_2^{(3)}n_3\right]$$

$$+ \sum_{n_3=0}^{m_3-n_2}\sum_{n_2=0}^{m_2} \pi^{(d)}(m_1,n_2,n_3)\left[Rm_1\mu_1 - c_4 + \left(R\mu_2 - c_0 - C_2^{(2)}\right)n_2 - C_2^{(3)}n_3\right]$$

$$+ \sum_{n_3=m_3-m_2+1}^{m_3} \pi^{(d)}(m_1,m_2,n_3)\left[R(m_1\mu_1 + m_2\mu_2) - c_5 - C_2^{(3)}n_3\right].$$

Case 2. *The service price $0 \leq R \leq R_L$*

It follows from Theorem 10 that

$$\boldsymbol{d}^{W^*} = (0,0,\ldots,0),$$

$$\boldsymbol{d}^{S^*} = (m_2,m_2,\ldots,m_2),$$

thus we have

$$\eta^{d^*} = \sum_{n_1=0}^{m_1} \pi^{(d)}(n_1,0,0)\left[\left(R\mu_1 - C_2^{(1)}\right)n_1 - c_1\right] + \sum_{n_3=1}^{m_3} \pi^{(d)}(m_1,0,n_3)\left[Rm_1\mu_1 - c_2 - C_2^{(3)}n_3\right]$$

$$+ \sum_{n_3=0}^{m_3-n_2}\sum_{n_2=0}^{m_2} \pi^{(d)}(m_1,n_2,n_3)\left[Rm_1\mu_1 - c_4 - \left(C_2^{(2)} + C_3^{(2)}\right)n_2 - C_2^{(3)}n_3\right]$$

$$+ \sum_{n_3=m_3-m_2+1}^{m_3} \pi^{(d)}(m_1,m_2,n_3)\left[R(m_1\mu_1 + m_2\mu_2) - c_5 - C_2^{(3)}n_3\right].$$

Remark 5. *The above results are intuitive because when the service price is suitably high, the number of working servers is equal to a crucial number related to waiting jobs both in Group 2 and in the buffer; when the service price is lower, each server at the work state must pay a high energy consumption cost, but they receive only a low revenue. In this case, the profit of the data center cannot increase, so that all the servers in Group 2 would like to be closed at the sleep state.*

Case 3. *The service price $R_L < R < R_H$*

In Section 6.3, we have, respectively, proven the optimality of the bang–bang control for the setup and sleep policies, regardless of the service price R. However, if $R_L < R < R_H$, we cannot exactly determine the monotone form (i.e., increasing or decreasing) of the optimal asynchronous energy-efficient policy. This makes the threshold-type of the optimal asynchronous energy-efficient policy in the data center. In fact, such a threshold-type policy also provides us with a choice to compute the optimal setup and sleep policies, they not only have a very simple form but are also widely adopted in numerical applications.

In what follows, we focus our study on the threshold-type asynchronous policy, although its optimality is not yet proven in our next analysis.

We define a couple of threshold-type control parameters as follows:

$$\{(\theta_1,\theta_2) : \theta_1,\theta_2 = 0,1,\ldots,m_2\},$$

where θ_1 and θ_2 are setup and sleep thresholds, respectively. Furthermore, we introduce two interesting subsets of the policy set $\mathcal{D}$. We write $d_{\theta_1}^W$ as a threshold-type setup policy and $d_{\theta_2}^S$ as a threshold-type sleep policy. Let

$$d_{\theta_1}^W \overset{def}{=} \left(\underbrace{0,0,\ldots,0}_{\theta_1-1 \; zeros}, \theta_1, \theta_1+1, \ldots, m_2, \ldots, m_2 \right),$$

$$d_{\theta_2}^S \overset{def}{=} \left(0,0,\ldots,0; \ldots; m_2-(\theta_2+1),\ldots,m_2-(\theta_2+1); m_2-\theta_2,\ldots,m_2-\theta_2; \underbrace{m_2,\ldots,m_2}_{\left(m_3-\frac{1}{2}\theta_2\right)(\theta_2-1)\; m_2 s} \right).$$

Then

$$\mathcal{D}^\triangle = \left\{ d_{(\theta_1,\theta_2)} : d_{(\theta_1,\theta_2)} = d_{\theta_1}^W \boxtimes d_{\theta_2}^S, \theta_1, \theta_2 = 0, 1, \ldots, m_2 \right\}.$$

It is easy to see that $\mathcal{D}^\triangle \subset \mathcal{D}$.

For an asynchronous energy-efficient policy $d_{(\theta_1,\theta_2)}$, it follows from (4) to (7) that for $n_1 = 0, 1, \ldots, m_1$ and $n_2 = n_3 = 0$,

$$f(n_1, 0, 0) = Rn_1\mu_1 - (m_1 P_{1,W} + m_2 P_{2,S})C_1 - n_1 C_2^{(1)};$$

for $n_1 = m_1, n_2 = 0$ and $n_3 = 1, 2, \ldots, \theta_1 - 1$,

$$f^{\left(d_{\theta_1}^W\right)}(m_1, 0, n_3) = Rm_1\mu_1 - (m_1 P_{1,W} + m_2 P_{2,S})C_1 - \left[m_1 C_2^{(1)} + n_3 C_2^{(3)} \right];$$

for $n_1 = m_1, n_2 = 0$ and $n_3 = \theta_1, \theta_1+1, \ldots, m_3$,

$$\begin{aligned} f^{\left(d_{\theta_1}^W\right)}(m_1, 0, n_3) = Rm_1\mu_1 &- \{ m_1 P_{1,W} + (n_3 \wedge m_2) P_{2,W} \\ &+ [m_2 - (n_3 \wedge m_2)] P_{2,S} \} C_1 \\ &- \left[m_1 C_2^{(1)} + n_3 C_2^{(3)} \right] - (n_3 \wedge m_2) C_3^{(1)}; \end{aligned}$$

for $n_1 = m_1, n_2 = 1, 2, \ldots, \theta_2 - 1$ and $n_3 = 0, 1, \ldots, m_3 - n_2$,

$$\begin{aligned} f^{\left(d_{\theta_2}^S\right)}(m_1, n_2, n_3) = Rm_1\mu_1 &- (m_1 P_{1,W} + m_2 P_{2,S})C_1 \\ &- \left[m_1 C_2^{(1)} + n_2 C_2^{(2)} + n_3 C_2^{(3)} \right] \\ &- n_2 C_3^{(2)} - m_1 \mu_1 1_{\{n_3=0\}} C_4; \end{aligned}$$

for $n_1 = m_1, n_2 = \theta_2, \theta_2+1, \ldots, m_2$ and $n_3 = 0, 1, \ldots, m_3 - n_2$,

$$\begin{aligned} f^{\left(d_{\theta_2}^S\right)}(m_1, n_2, n_3) = R(m_1\mu_1 + n_2\mu_2) &- [m_1 P_{1,W} + (m_2 - n_2) P_{2,S} + n_2 P_{2,W}] C_1 \\ &- \left[m_1 C_2^{(1)} + n_2 C_2^{(2)} + n_3 C_2^{(3)} \right] - m_1 \mu_1 1_{\{n_3=0\}} C_4; \end{aligned}$$

for $n_1 = m_1, n_2 = m_2$ and $n_3 = m_3 - m_2, m_3 - m_2 + 1, \ldots, m_3$,

$$\begin{aligned} f(m_1, m_2, n_3) = R(m_1\mu_1 + m_2\mu_2) &- (m_1 P_{1,W} + m_2 P_{2,W}) C_1 \\ &- \left[m_1 C_2^{(1)} + m_2 C_2^{(2)} + n_3 C_2^{(3)} \right] - \lambda 1_{\{n_3=m_3\}} C_5. \end{aligned}$$

Note that

$$
\eta^{d_{(\theta_1,\theta_2)}} = \sum_{n_1=0}^{m_1} \pi^{\left(d_{(\theta_1,\theta_2)}\right)}(n_1,0,0)f(n_1,0,0)
$$

$$
+ \sum_{n_3=1}^{m_3} \pi^{\left(d_{(\theta_1,\theta_2)}\right)}(m_1,0,n_3)f^{\left(d_{\theta_1}^W\right)}(m_1,0,n_3)
$$

$$
+ \sum_{n_3=0}^{m_3-n_2}\sum_{n_2=0}^{m_2} \pi^{\left(d_{(\theta_1,\theta_2)}\right)}(m_1,n_2,n_3)f^{\left(d_{\theta_2}^S\right)}(m_1,n_2,n_3)
$$

$$
+ \sum_{n_3=m_3-m_2+1}^{m_3} \pi^{\left(d_{(\theta_1,\theta_2)}\right)}(m_1,m_2,n_3)f(m_1,m_2,n_3).
$$

It follows from (48) that the long-run average profit under policy $d_{(\theta_1,\theta_2)}$ is given by

$$
\eta^{d_{(\theta_1,\theta_2)}} = \sum_{i=0}^{n} \pi^{\left(d_{(\theta_1,\theta_2)}\right)}(n_1,0,0)\left[\left(R\mu_1 - C_2^{(1)}\right)n_1 - c_1\right]
$$

$$
+ \sum_{n_3=1}^{\theta_1-1} \pi^{\left(d_{\theta_1}^W\right)}(m_1,0,n_3)\left[Rm_1\mu_1 - c_2 - C_2^{(3)}n_3\right]
$$

$$
+ \sum_{n_3=\theta_1}^{m_3} \pi^{\left(d_{\theta_1}^W\right)}(m_1,0,n_3)\left[Rm_1\mu_1 - c_2 - \left(c_0 + C_3^{(1)}\right)(n_3 \wedge m_2) - n_3 C_2^{(3)}\right]
$$

$$
+ \sum_{n_3=0}^{m_3-n_2}\sum_{n_2=1}^{\theta_2} \pi^{\left(d_{\theta_2}^S\right)}(m_1,n_2,n_3)\left[Rm_1\mu_1 - c_4 + \left(R\mu_2 - c_0 - C_2^{(2)}\right)n_2 - C_2^{(3)}n_3\right]
$$

$$
+ \sum_{n_3=0}^{m_3-n_2}\sum_{n_2=\theta_2+1}^{m_2} \pi^{\left(d_{\theta_2}^S\right)}(m_1,n_2,n_3)\left[Rm_1\mu_1 - c_4 - \left(C_2^{(2)} + C_3^{(2)}\right)n_2 - C_2^{(3)}n_3\right]
$$

$$
+ \sum_{n_3=m_3-m_2+1}^{m_3} \pi^{\left(d_{(\theta_1,\theta_2)}\right)}(m_1,m_2,n_3)\left[R(m_1\mu_1 + m_2\mu_2) - c_5 - C_2^{(3)}n_3\right].
$$

Let

$$
(\theta_1^*,\theta_2^*) = \mathop{\arg\max}\limits_{(\theta_1,\theta_2)\in\{0,1,\dots,m_2\}}\left\{\eta^{d_{(\theta_1,\theta_2)}}\right\}.
$$

Then, we call $d_{\left(\theta_1^*,\theta_2^*\right)}$ the optimal threshold-type asynchronous energy-efficient policy in the policy set $\mathcal{D}^\triangle$. Since $\mathcal{D}^\triangle \subset \mathcal{D}$, the partially ordered set $\mathcal{D}$ shows that $\mathcal{D}^\triangle$ is also a partially ordered set. Based on this, it is easy to see from the two partially ordered sets $\mathcal{D}$ and $\mathcal{D}^\triangle$ that

$$
\eta^{d_{\left(\theta_1^*,\theta_2^*\right)}} \leq \eta^{d^*}.
$$

For the energy-efficient data center, if $\eta^{d_{\left(\theta_1^*,\theta_2^*\right)}} = \eta^{d^*}$, then we call $d_{\left(\theta_1^*,\theta_2^*\right)}$ the optimal threshold-type asynchronous energy-efficient policy in the original policy set $\mathcal{D}$; if $\eta^{d_{\left(\theta_1^*,\theta_2^*\right)}} < \eta^{d^*}$, then we call $d_{\left(\theta_1^*,\theta_2^*\right)}$ the suboptimal threshold-type asynchronous energy-efficient policy in the original policy set $\mathcal{D}$.

Remark 6. *This paper is a special case of the group-server queue (see Li et al. [6]), but it provides a new theoretical framework for the performance optimization of such queueing systems. It is also more applicable to large-scale service systems, such as data centers for efficiently allocating service resources.*

Remark 7. *In this paper, we discuss the optimal asynchronous dynamic policy of the energy-efficient data center deeply, and such types of policies are widespread in practice. It would be interesting to* *extend our results to a more general situation. Although the sensitivity-based optimization theory can effectively overcome the drawbacks of MDPs, it still has some limitations. For example, it cannot discuss the optimization of two or more dynamic control policies synchronously, which is a very important research direction in dynamic optimization.*

8. Conclusions

In this paper, we highlight the optimal asynchronous dynamic policy of an energy-efficient data center by applying sensitivity-based optimization theory and RG-factorization. Such an asynchronous policy is more important and necessary in the study of energy-efficient data centers, and it largely makes an optimal analysis of energy-efficient management more interesting, difficult, and challenging. To this end, we consider a more practical model with several basic factors, for example, a finite buffer, a fast setup process from sleep to work, and the necessary cost of transferring jobs from Group 2 either to Group 1 or to the buffer. To find the optimal asynchronous dynamic policy in the energy-efficient data center, we set up a policy-based block-structured Poisson equation and provide an expression for its solution by means of the RG-factorization. Based on this, we derive the monotonicity and optimality of the long-run average profit with respect to the asynchronous dynamic policy under different service prices. We prove the optimality of the bang–bang control, which significantly reduces the action search space, and study the optimal threshold-type asynchronous dynamic policy. Therefore, the results of this paper provide new insights to the discussion of the optimal dynamic control policies of more general energy-efficient data centers.

Along such a line, there are a number of interesting directions for potential future research, for example:

- Analyzing non-Poisson inputs such as Markovian arrival processes (MAPs) and/or non-exponential service times, e.g., the PH distributions;
- Developing effective algorithms for finding the optimal dynamic policies of the policy-based block-structured Markov process (i.e., block-structured MDPs);
- Discussing the fact that the long-run performance is influenced by the concave or convex reward (or cost) function;
- Studying individual optimization for the energy-efficient management of data centers from the perspective of game theory.

Author Contributions: Conceptualization, J.-Y.M. and Q.-L.L.; methodology, L.X. All authors have read and agreed to the published version of the manuscript.

Funding: This research was funded in part by the National Natural Science Foundation of China under grant No. 71932002, by the Beijing Social Science Foundation Research Base Project under grant No. 19JDGLA004, and in part by the National Natural Science Foundation of China under grant No. 61573206 and No. 11931018.

Institutional Review Board Statement: Not applicable.

Informed Consent Statement: Not applicable.

Acknowledgments: The authors are grateful to the editor and anonymous referees for their constructive comments and suggestions, which sufficiently help the authors to improve the presentation of this manuscript.

Conflicts of Interest: The authors declare no conflict of interest.

Appendix A. Special Cases

In this appendix, we provide two simple special cases to understand the state transition relations and the infinitesimal generator of the policy-based block-structured continuous-time Markov process $\left\{ \mathbf{X}^{(d)}(t) : t \geq 0 \right\}$.

Case A1. $m_1 = m_2 = 2$ and $m_3 = 3$

To understand the block-structured continuous-time Markov process under an asynchronous energy-efficient policy, we first take the simplest case of $m_1 = m_2 = 2$ and $m_3 = 3$ as an example to illustrate the state transition relations and infinitesimal generator in detail.

Figure A1 shows the arrival and service rates without any policy for the Markov process $\left\{ \mathbf{X}^{(d)}(t) : t \geq 0 \right\}$. The transitions in which the transfer rates and the policies are simultaneously involved are a bit difficult, and such transition relations appear in a Markov chain with sync jumps of multi-events (either transition rates or policies (see Budhiraja and Friedlander [40])). For example, what begins a setup policy at State $(2,0,1)$ is the entering Poisson process to State $(2,0,1)$, whose inter-entering times are i.i.d. and exponential with the entering rate $\lambda + 2\mu_1 + \mu_2$. Hence, if there exists one or two server set ups, then the transition rate from State $(2,0,1)$ to State $(2,1,0)$ is $1_{\left\{ d_{2,0,1}^{W} \geq 1 \right\}} (\lambda + 2\mu_1 + \mu_2)$. Such an entering process is easy to see in Figure A1.

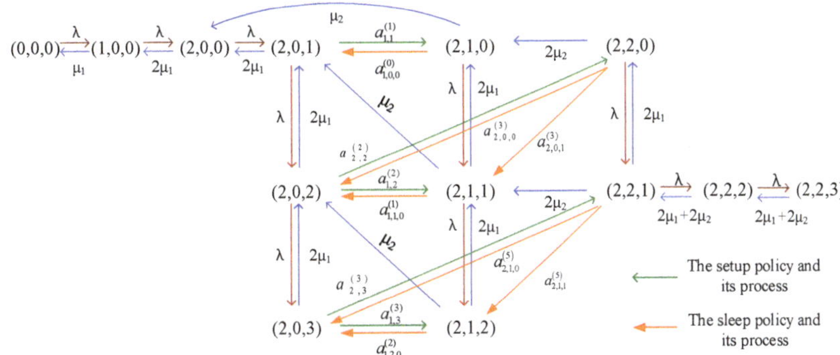

Figure A1. State transition relations for the case of $m_1 = m_2 = 2$ and $m_3 = 3$.

We denote $a_{k_1,k_2}^{(i)}$ as the transitions with the setup policy

$$a_{1,1}^{(1)} = 1_{\left\{ d_{2,0,1}^{W} \geq 1 \right\}} (\lambda + 2\mu_1 + \mu_2),$$
$$a_{1,2}^{(2)} = 1_{\left\{ d_{2,0,2}^{W} = 1 \right\}} (\lambda + 2\mu_1 + \mu_2),$$
$$a_{2,2}^{(2)} = 1_{\left\{ d_{2,0,2}^{W} = 2 \right\}} (\lambda + 2\mu_1 + \mu_2),$$
$$a_{1,3}^{(3)} = 1_{\left\{ d_{2,0,3}^{W} = 1 \right\}} \lambda, \quad a_{2,3}^{(3)} = 1_{\left\{ d_{2,0,3}^{W} = 2 \right\}} \lambda.$$

Similarly, $a_{k_3,k_4,k_5}^{(i)}$ denotes the transitions with the sleep policy from

$$a_{1,0,0}^{(0)} = 1_{\left\{ d_{2,1,0}^{S} = 2 \right\}} (2\mu_1 + 2\mu_2),$$
$$a_{1,1,0}^{(1)} = 1_{\left\{ d_{2,1,1}^{S} = 2 \right\}} (\lambda + 2\mu_1 + 2\mu_2),$$
$$a_{1,2,0}^{(2)} = 1_{\left\{ d_{2,1,2}^{S} = 2 \right\}} \lambda,$$
$$a_{2,0,0}^{(3)} = 1_{\left\{ d_{2,2,0}^{S} = 2 \right\}} 2\mu_1,$$
$$a_{2,0,1}^{(3)} = 1_{\left\{ d_{2,2,0}^{S} = 1 \right\}} 2\mu_1,$$
$$a_{2,1,0}^{(5)} = 1_{\left\{ d_{2,2,1}^{S} = 2 \right\}} (\lambda + 2\mu_1 + 2\mu_2),$$
$$a_{2,1,1}^{(5)} = 1_{\left\{ d_{2,2,1}^{S} = 1 \right\}} (\lambda + 2\mu_1 + 2\mu_2).$$

Furthermore, we write the diagonal entries

$$
\begin{aligned}
b_1^{(1)} &= \lambda + 2\mu_1 + a_{1,1}^{(1)},\\
b_2^{(2)} &= \lambda + 2\mu_1 + a_{1,2}^{(2)} + a_{2,2}^{(2)},\\
b_3^{(3)} &= 2\mu_1 + a_{1,3}^{(3)} + a_{2,3}^{(3)};\\
b_{1,0}^{(0)} &= \lambda + \mu_2 + a_{1,0,0}^{(0)},\\
b_{1,1}^{(1)} &= \lambda + 2\mu_1 + \mu_2 + a_{1,1,0}^{(1)},\\
b_{1,2}^{(2)} &= 2\mu_1 + \mu_2 + a_{1,2,0}^{(2)}.
\end{aligned}
$$

Therefore, its infinitesimal generator is given by

$$
\mathbf{Q}^{(d)} =
\begin{pmatrix}
Q_{0,0} & Q_{0,1} & & & \\
Q_{1,0} & Q_{1,1} & Q_{1,2} & Q_{1,3} & \\
Q_{2,0} & Q_{2,1} & Q_{2,2} & & \\
& Q_{3,1} & Q_{3,2} & Q_{3,3} & Q_{3,4}\\
& & & Q_{4,3} & Q_{4,4}
\end{pmatrix},
$$

where for level 0, it is easy to see that

$$
Q_{0,0} =
\begin{pmatrix}
-\lambda & \lambda & \\
\mu_1 & -(\lambda+\mu_1) & \lambda\\
& 2\mu_1 & -(\lambda+\mu_1)
\end{pmatrix}
\quad\text{and}\quad
Q_{0,1} =
\begin{pmatrix}
& 0 \\
& 0 \\
\lambda &
\end{pmatrix};
$$

for level 1, the setup policy affects the infinitesimal generator

$$
Q_{1,0} =
\begin{pmatrix}
0 & 2\mu_1 \\
& 0
\end{pmatrix},
\quad
Q_{1,1} =
\begin{pmatrix}
-b_1^{(1)} & \lambda & \\
2\mu_1 & -b_2^{(2)} & \lambda\\
& 2\mu_1 & -b_3^{(3)}
\end{pmatrix},
$$

$$
Q_{1,2} =
\begin{pmatrix}
a_{1,1}^{(1)} & & \\
& a_{1,2}^{(2)} & \\
& & a_{1,3}^{(3)}
\end{pmatrix}
\quad\text{and}\quad
Q_{1,3} =
\begin{pmatrix}
a_{2,2}^{(2)} & \\
& a_{2,3}^{(3)}
\end{pmatrix};
$$

for level 2, the sleep policy affects the infinitesimal generator

$$
Q_{2,0} =
\begin{pmatrix}
0 & \mu_2 \\
& 0
\end{pmatrix},
\quad
Q_{2,1} =
\begin{pmatrix}
a_{1,0,0}^{(1)} & & \\
\mu_2 & a_{1,1,0}^{(1)} & \\
& \mu_2 & a_{1,2,0}^{(1)}
\end{pmatrix}
\quad\text{and}\quad
Q_{2,2} =
\begin{pmatrix}
-b_{1,0}^{(0)} & \lambda & \\
2\mu_1 & -b_{1,1}^{(1)} & \lambda\\
& 2\mu_1 & -b_{1,2}^{(2)}
\end{pmatrix};
$$

for level 3, we have

$$
Q_{3,1} =
\begin{pmatrix}
0 & a_{2,0,0}^{(3)} & 0\\
0 & 0 & a_{2,1,0}^{(5)}
\end{pmatrix},
\quad
Q_{3,2} =
\begin{pmatrix}
2\mu_2 & a_{2,0,1}^{(3)} & 0\\
0 & 2\mu_2 & a_{2,1,1}^{(5)}
\end{pmatrix},
$$

$$
Q_{3,3} =
\begin{pmatrix}
-b_{2,0}^{(3)} & \lambda\\
2\mu_1 & -b_{2,1}^{(5)}
\end{pmatrix}
\quad\text{and}\quad
Q_{3,4} =
\begin{pmatrix}
0 & 0\\
\lambda & 0
\end{pmatrix};
$$

for level 4, we have

$$
Q_{4,3} =
\begin{pmatrix}
0 & 2\mu_1 + 2\mu_2\\
0 & 0
\end{pmatrix}
\quad\text{and}\quad
Q_{4,4} =
\begin{pmatrix}
-(\lambda + 2\mu_1 + 2\mu_2) & \lambda\\
2\mu_1 + 2\mu_2 & -(2\mu_1 + 2\mu_2)
\end{pmatrix}.
$$

Case A2. $m_1 = 2$, $m_2 = 3$ *and* $m_3 = 4$

To further understand the policy-based block-structured continuous-time Markov process $\left\{ \mathbf{X}^{(d)}(t) : t \geq 0 \right\}$, we take another example to illustrate that if the parameters m_2 and m_3 increase slightly, the complexity of the state transition relations will increase considerably.

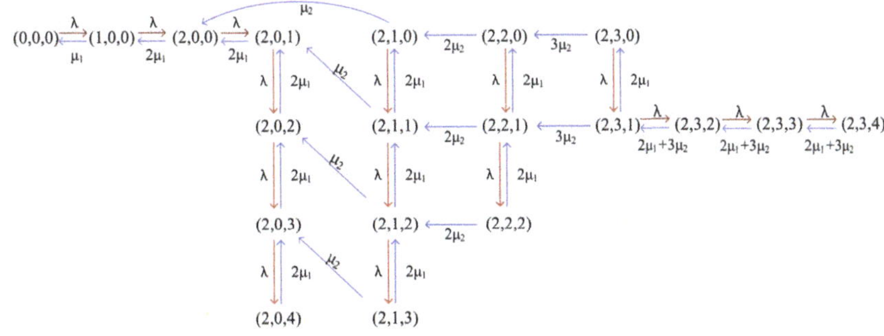

(a) The arrival and service processes

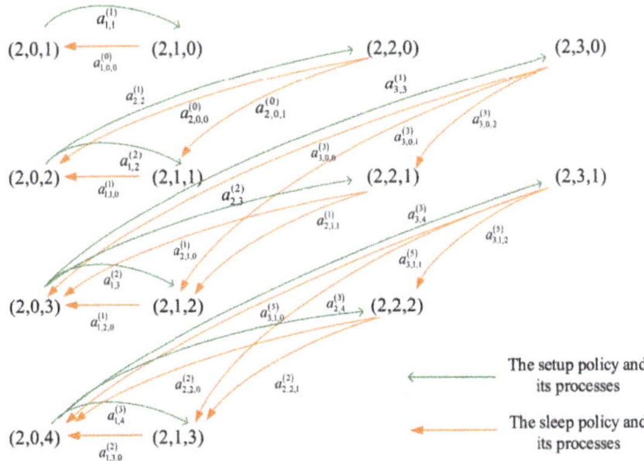

(b) The setup and sleep policies and their processes

Figure A2. State transition relations for the case of $m_1 = 2$, $m_2 = 3$ and $m_3 = 4$.

The number of servers in Group 2 and the buffer capacity both increase by one, which makes the number of state transitions affected by the setup and sleep policies increase from 5 to 9 and from 7 to 15, respectively. Compared with Figure A2, the state transition relations become more complicated. We divide the state transition rate of the Markov process into two parts, as shown in Figure A2: (a) the state transitions without any policy and (b) the state transitions by both setup and sleep policies. Similar to **Case 1**, for the state transitions in which the transfer rates and the policies are simultaneously involved in the Markov chain with sync jumps of multi-events, the transfer rate is equal to the total entering rate at a certain state.

Furthermore, its infinitesimal generator is given by

$$
\mathbf{Q}^{(d)} = \begin{pmatrix}
Q_{0,0} & Q_{0,1} & & & & \\
Q_{1,0} & Q_{1,1} & Q_{1,2} & Q_{1,3} & Q_{1,4} & \\
Q_{2,0} & Q_{2,1} & Q_{2,2} & & & \\
& Q_{3,1} & Q_{3,2} & Q_{3,3} & & \\
& Q_{4,1} & Q_{4,2} & Q_{4,3} & Q_{4,4} & Q_{4,5} \\
& & & & Q_{5,4} & Q_{5,5}
\end{pmatrix}.
$$

Similarly, we can obtain every element $Q_{i,j}$, $0 \leq i, j \leq 5$; here, we omit the computational details.

Appendix B. State Transition Relations

In this appendix, we provide a general expression for the state transition relations of the policy-based block-structured continuous-time Markov process $\left\{ \mathbf{X}^{(d)}(t) : t \geq 0 \right\}$. To express the state transition rates, in what follows, we need to introduce some notations.

Based on the state transition rates for both special cases that are related to either the arrival and service processes or the setup and sleep policies, Figure A3 provides the state transition relations of the Markov process $\left\{ \mathbf{X}^{(d)}(t) : t \geq 0 \right\}$ in general. Note that the figure is so complicated that we have to decompose it into three different parts: (a) the arrival and service processes, (b) the state transitions by the setup policy, and (c) the state transitions by the sleep policy. However, the three parts must be integrated as a whole.

(a) The arrival and service rates: The first type is ordinary for the arrival and service rates without any policy (see Figure A3a).

(b) The setup policy: For $k_2 = k_1$, we write

$$a_{k_1,k_1}^{(1)} = 1_{\left\{ d_{m_1,0,k_2}^{W} \geq k_1 \right\}} (\lambda + m_1\mu_1 + \mu_2),$$

for $k_2 = k_1 + 1, k_1 + 2, \ldots, m_3 - 1$,

$$a_{k_1,k_2}^{(2)} = 1_{\left\{ d_{m_1,0,k_2}^{W} = k_1 \right\}} (\lambda + m_1\mu_1 + \mu_2),$$

for $k_2 = m_3$,

$$a_{k_1,m_3}^{(3)} = 1_{\left\{ d_{m_1,0,k_2}^{W} = k_1 \right\}} \lambda.$$

Observing Figure A3b, what begins a setup policy at State $(m_1, 0, k)$ is the entering Poisson process to State $(m_1, 0, k)$, whose inter-entering times are i.i.d. and exponential with entering rates, namely either $\lambda + m_1\mu_1 + \mu_2$ for $0 \leq k \leq m_3 - 1$ or λ for $k = m_3$. Such an entering process is easy to see from Figure A3a.

Since the setup and sleep policies are asynchronous, $a_{k_1,k_2}^{(i)}$ will not contain any transition with the sleep policy because the sleep policy cannot be followed by the setup policy at the same time. To express the diagonal entries of $Q_{1,1}$ in Appendix C, we introduce

$$
\begin{aligned}
b_{k_2}^{(1)} &= \lambda + m_1\mu_1 + a_{k_1,k_1}^{(1)} + \sum_{k_1=1}^{k_2-1} a_{k_1,k_2}^{(2)}, \quad \text{if } k_2 = 1, 2, \ldots, m_2 - 1, \\
b_{k_2}^{(2)} &= \lambda + m_1\mu_1 + \sum_{k_1=1}^{m_2} a_{k_1,k_2}^{(2)}, \quad \text{if } k_2 = m_2, m_2 + 1, \ldots, m_3 - 1, \\
b_{k_2}^{(3)} &= m_1\mu_1 + \sum_{k_1=1}^{m_2} a_{k_1,m_3}^{(3)}, \quad \text{if } k_2 = m_3.
\end{aligned}
\tag{A1}
$$

(c) The sleep policy: For $k_3 = 1, 2, \ldots, m_2$, $k_4 = 0, 1, \ldots, m_3 - k_3$, and $k_5 = 0, 1, \ldots, m_2$, we write

$$a^{(0)}_{k_3,k_4,k_5} = 1_{\left\{ d^S_{m_1,k_3,k_4} = m_2 - k_5 \right\}} \left[m_1 \mu_1 + (k_3 + 1)\mu_2 \right],$$

$$a^{(1)}_{k_3,k_4,k_5} = 1_{\left\{ d^S_{m_1,k_3,k_4} = m_2 - k_5 \right\}} \left[\lambda + m_1 \mu_1 + (k_3 + 1)\mu_2 \right],$$

$$a^{(2)}_{k_3,k_4,k_5} = 1_{\left\{ d^S_{m_1,k_3,k_4} = m_2 - k_5 \right\}} \lambda,$$

$$a^{(3)}_{k_3,k_4,k_5} = 1_{\left\{ d^S_{m_1,k_3,k_4} = m_2 - k_5 \right\}} m_1 \mu_1, \qquad (A2)$$

$$a^{(4)}_{k_3,k_4,k_5} = 1_{\left\{ d^S_{m_1,k_3,k_4} = m_2 - k_5 \right\}} (\lambda + m_1 \mu_1),$$

$$a^{(5)}_{k_3,k_4,k_5} = 1_{\left\{ d^S_{m_1,k_3,k_4} = m_2 - k_5 \right\}} (\lambda + m_1 \mu_1 + m_2 \mu_2).$$

From Figure A3c, it is seen that there is a difference between the sleep and setup policies: State transitions with the sleep policy exist at many states (m_1, i, k) for $1 \leq i \leq m_2$. Clearly, the state transition with the sleep policy from State (m_1, i, k) is the entering Poisson processes, with the rate being the total entering rate to State (m_1, i, k). Note that the sleep policy cannot be followed by the setup policy at the same time. Thus, it is easy to check these state transition rates given in the above ones.

To express the diagonal entries of $Q_{i,i}$ for $2 \leq i \leq m_2 + 1$, we introduce

$$b^{(0)}_{k_3,k_4} = \lambda + k_3 \mu_2 + \sum_{k_5=0}^{k_3} a^{(0)}_{k_3,k_4,k_5},$$

$$b^{(1)}_{k_3,k_4} = \lambda + m_1 \mu_1 + k_3 \mu_2 + \sum_{k_5=0}^{k_3} a^{(1)}_{k_3,k_4,k_5},$$

$$b^{(2)}_{k_3,k_4} = m_1 \mu_1 + k_3 \mu_2 + \sum_{k_5=0}^{k_3} a^{(2)}_{k_3,k_4,k_5},$$

$$b^{(3)}_{k_3,k_4} = \lambda + m_2 \mu_2 + \sum_{k_5=0}^{k_3} a^{(3)}_{k_3,k_4,k_5}, \qquad (A3)$$

$$b^{(4)}_{k_3,k_4} = \lambda + m_1 \mu_1 + m_2 \mu_2 + \sum_{k_5=0}^{k_3} a^{(4)}_{k_3,k_4,k_5},$$

$$b^{(5)}_{k_3,k_4} = m_1 \mu_1 + m_2 \mu_2 + \sum_{k_5=0}^{k_3} a^{(5)}_{k_3,k_4,k_5}.$$

Remark A1. *The first key step in applications of the sensitivity-based optimization to the study of energy-efficient data centers is to draw the state transition relation figure (e.g., see Figure A3) and to write the infinitesimal generator of the policy-based block-structured Markov process. Although this paper has largely simplified the model assumptions, Figure A3 is still slightly complicated by its three separate parts: (a), (b), and (c). Obviously, if we consider some more general assumptions (for example, (i) the faster servers are not cheaper, (ii) Group 2 is not slower, (iii) there is no transfer rule, and so on), then the state transition relation figure will become more complicated, so that it is more difficult to write the infinitesimal generator of the policy-based block-structured Markov process and to solve the block-structured Poisson equation.*

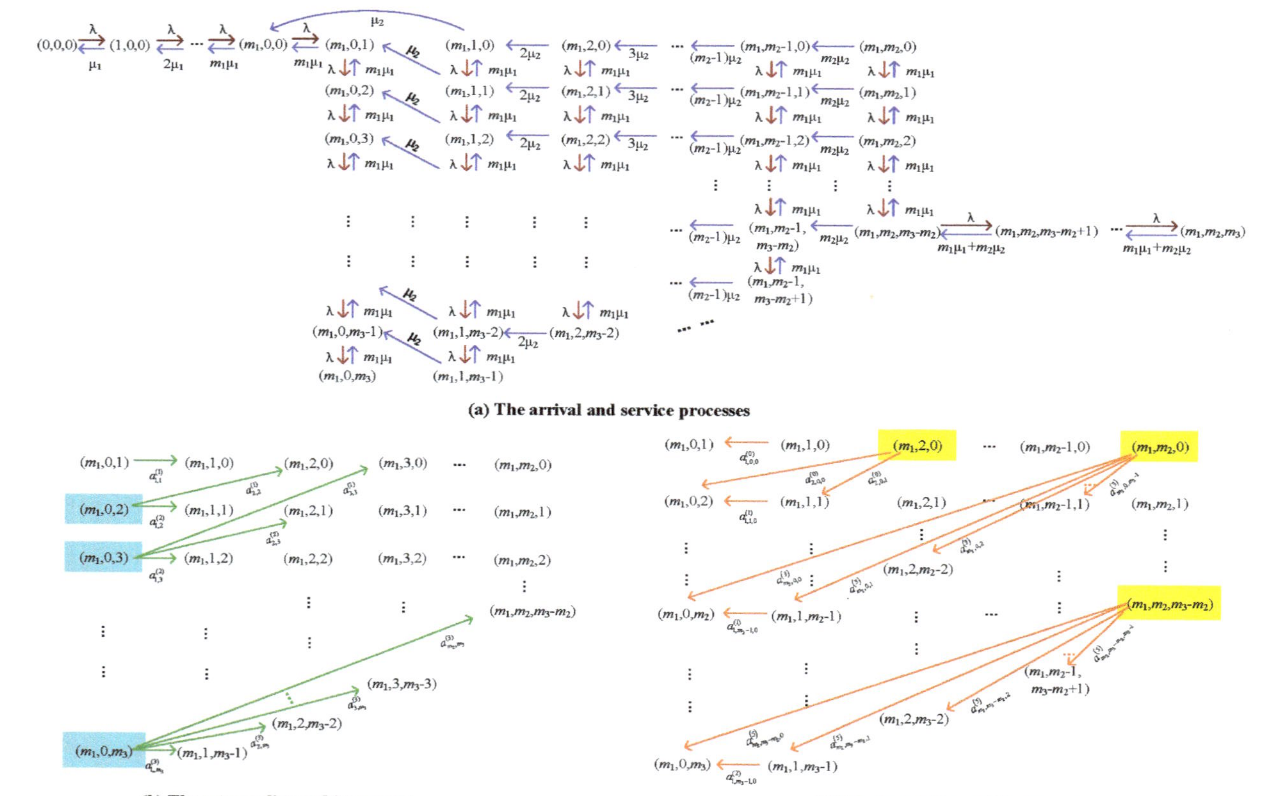

(a) The arrival and service processes

(b) The setup policy and its process

(c) The sleep policy and its process

Figure A3. State transition relations of the policy-based block-structured Markov process.

Remark A2. *Figure A3 shows that Part (a) expresses the arrival and service rates, while Parts (b) and (c) express the state transition rates caused by the setup and sleep policies, respectively. Note that the setup policy is started by only the arrival and service process at State $(m_1, 0, k)$ for $1 \leq k \leq m_3$ (see Part (b)), in which there is no setup rate because the setup time is so short that it is ignored. Similarly, it is easy to understand Part (c) for the sleep policy. It is worthwhile to note that an idle server may be at the work state, as seen in the idle servers with the work state in Group 1.*

Appendix C. Block Elements in $\mathbf{Q}^{(d)}$

In this appendix, we write each block element in the matrix $\mathbf{Q}^{(d)}$.

(a) For level 0, it is easy to see that

$$
Q_{0,0} = \begin{pmatrix}
-\lambda & \lambda & & & \\
\mu_1 & -(\lambda + \mu_1) & \lambda & & \\
& \ddots & \ddots & \ddots & \\
& (m_1 - 1)\mu_1 & -[\lambda + (m_1 - 1)\mu_1] & \lambda & \\
& & m_1 \mu_1 & -(\lambda + m_1 \mu_1)
\end{pmatrix}, \quad Q_{0,1} = \begin{pmatrix} & \\ & \\ & \\ & \lambda \end{pmatrix}.
$$

(b) For level 1, the setup policy affects the infinitesimal generator, and $Q_{1,0}$ is given by

$$
Q_{1,0} = \begin{pmatrix} & m_1 \mu_1 & \\ & & \end{pmatrix}, \quad Q_{1,k_1+1} = \Big(\underbrace{0, 0, \ldots, 0}_{(k_1-1)\ 0\text{s}}, A_{k_1} \Big)^T,
$$

where

$$
A_{k_1} = \begin{cases}
\operatorname{diag}\Big(a^{(1)}_{k_1,k_1}, a^{(2)}_{k_1,k_1+1}, \ldots, a^{(2)}_{k_1,m_3-1}, a^{(3)}_{k_1,m_3} \Big), & \text{if } 1 \leq k_1 \leq m_2 - 1, \\
\operatorname{diag}\Big(a^{(2)}_{k_1,k_1}, a^{(2)}_{k_1,k_1+1}, \ldots, a^{(2)}_{k_1,m_3-1}, a^{(3)}_{k_1,m_3} \Big), & \text{if } k_1 = m_2,
\end{cases}
$$

and $\mathbf{0}$ is a block of zeros with suitable size. From (A1), we have

$$
Q_{1,1} = \begin{pmatrix}
-b^{(1)}_1 & \lambda & & & & & & \\
m_1 \mu_1 & -b^{(1)}_2 & \lambda & & & & & \\
& \ddots & \ddots & \ddots & & & & \\
& & m_1 \mu_1 & -b^{(1)}_{m_2-1} & \lambda & & & \\
& & & m_1 \mu_1 & -b^{(2)}_{m_2} & \lambda & & \\
& & & & \ddots & \ddots & \ddots & \\
& & & & & m_1 \mu_1 & -b^{(2)}_{m_3-1} & \lambda \\
& & & & & & m_1 \mu_1 & -b^{(3)}_{m_3}
\end{pmatrix}. \tag{A4}
$$

(c) For level 2, i.e., $k_3 = 1$,

$$
Q_{2,0} = \begin{pmatrix} & \mu_2 & \\ & & \end{pmatrix}, \quad Q_{2,1} = \begin{pmatrix}
a^{(0)}_{1,0,0} & & & & \\
\mu_2 & a^{(1)}_{1,1,0} & & & \\
& \ddots & \ddots & & \\
& & \mu_2 & a^{(1)}_{1,m_3-2,0} & \\
& & & \mu_2 & a^{(2)}_{1,m_3-1,0}
\end{pmatrix},
$$

and

$$Q_{2,2} = \begin{pmatrix} -b_{1,0}^{(0)} & \lambda & & & \\ m_1\mu_1 & -b_{1,1}^{(1)} & \lambda & & \\ & \ddots & \ddots & \ddots & \\ & & m_1\mu_1 & -b_{1,m_3-2}^{(1)} & \lambda \\ & & & m_1\mu_1 & -b_{1,m_3-1}^{(2)} \end{pmatrix}.$$

(d) For level $k_3 + 1$, $k_3 = 2, 3, \ldots, m_2 - 2$, subdivided into three cases as follows: For $k_5 = 0, 1, \ldots, k_3 - 2$,

$$Q_{k_3+1,k_5+1} = \left(\underbrace{0, 0, \ldots, 0}_{(k_3-k_5)\ \mathbf{0s}}, A_{k_3,k_5} \right),$$

where

$$A_{k_3,k_5} = \mathrm{diag}\left(a_{k_3,0,k_5}^{(0)}, a_{k_3,1,k_5}^{(1)}, \ldots, a_{k_3,m_3-k_3,k_5}^{(1)}, a_{k_3,m_3-k_3,k_5}^{(2)} \right).$$

For $k_5 = k_3 - 1$,

$$Q_{k_3+1,k_3} = \begin{pmatrix} k_3\mu_2 & a_{k_3,0,k_3-1}^{(0)} & & & \\ & k_3\mu_2 & a_{k_3,1,k_3-1}^{(1)} & & \\ & & \ddots & \ddots & \\ & & & k_3\mu_2 & a_{k_3,m_3-k_3-1,k_3-1}^{(1)} \\ & & & & k_3\mu_2 & a_{k_3,m_3-k_3,k_3-1}^{(1)} \end{pmatrix}.$$

For $k_5 = k_3$,

$$Q_{k_3+1,k_3+1} = \begin{pmatrix} -b_{k_3,0}^{(0)} & \lambda & & & \\ m_1\mu_1 & -b_{k_3,1}^{(1)} & \lambda & & \\ & \ddots & \ddots & \ddots & \\ & & m_1\mu_1 & -b_{k_3,m_3-k_3-1}^{(1)} & \lambda \\ & & & m_1\mu_1 & -b_{k_3,m_3-k_3}^{(2)} \end{pmatrix}.$$

(e) For level $m_2 + 1$, i.e., $k_3 = m_2$,

$$Q_{m_2+1,k_5+1} = \left(\underbrace{0, 0, \ldots, 0}_{(m_2-k_5)\ \mathbf{0s}}, A_{m_2,k_5} \right), \quad k_5 = 0, 1, \ldots, m_2 - 2,$$

where

$$A_{m_2,k_5} = \mathrm{diag}\left(a_{m_2,0,k_5}^{(3)}, a_{m_2,1,k_5}^{(4)}, \ldots, a_{m_2,m_3-m_2-1,k_5}^{(4)}, a_{m_2,m_3-m_2,k_5}^{(5)} \right).$$

$$Q_{m_2+1,m_2} = \begin{pmatrix} m_2\mu_2 & a_{m_2,0,m_2-1}^{(3)} & & & \\ & m_2\mu_2 & a_{m_2,1,m_2-1}^{(4)} & & \\ & & \ddots & \ddots & \\ & & & m_2\mu_2 & a_{m_2,m_3-m_2-1,m_2-1}^{(4)} \\ & & & & m_2\mu_2 & a_{m_2,m_3-m_2,m_2-1}^{(5)} \end{pmatrix},$$

$$Q_{m_2+1,m_2+2} = \begin{pmatrix} & & \\ & & \\ \lambda & & \end{pmatrix},$$

and

$$Q_{m_2+1,m_2+1} = \begin{pmatrix} -b^{(3)}_{m_2,0} & \lambda & & & \\ m_1\mu_1 & -b^{(4)}_{m_2,1} & \lambda & & \\ & \ddots & \ddots & \ddots & \\ & & m_1\mu_1 & -b^{(4)}_{m_2,m_3-m_2-1} & \lambda \\ & & & m_1\mu_1 & -b^{(5)}_{m_2,m_3-m_2} \end{pmatrix}.$$

(f) For level $m_2 + 2$,

$$Q_{m_2+2,m_2+1} = \begin{pmatrix} & a & \\ & & \\ & & \end{pmatrix} \text{ and } Q_{m_2+2,m_2+2} = \begin{pmatrix} -(\lambda+a) & \lambda & & & \\ a & -(\lambda+a) & \lambda & & \\ & \ddots & \ddots & \ddots & \\ & & a & -(\lambda+a) & \lambda \\ & & & a & -a \end{pmatrix},$$

where $a = m_1\mu_1 + m_2\mu_2$.

References

1. Masanet, E.; Shehabi, A.; Lei, N.; Smith, S.; Koomey, J. Recalibrating global data center energy-use estimates. *Science* **2020**, *367*, 984–986. [CrossRef] [PubMed]
2. Zhang, Q.; Meng, Z.; Hong, X.; Zhan, Y.; Liu, J.; Dong, J.; Bai, T.; Niu, J.; Deen, M.J. A survey on data center cooling systems: Technology, power consumption modeling and control strategy optimization. *J. Syst. Archit.* **2021**, *119*, 102253. [CrossRef]
3. Nadjahi, C., Louahlia, H., Lemasson, S. A review of thermal management and innovative cooling strategies for data center. *Sustain. Comput. Inform. Syst.* **2018**, *19*, 14–28. [CrossRef]
4. Koot, M.; Wijnhoven, F. Usage impact on data center electricity needs: A system dynamic forecasting model. *Appl. Energy* **2021**, *291*, 116798. [CrossRef]
5. Shirmarz, A.; Ghaffari, A. Performance issues and solutions in SDN-based data center: A survey. *J. Supercomput.* **2020**, *76*, 7545–7593. [CrossRef]
6. Li, Q.L.; Ma, J.Y.; Xie, M.Z.; Xia, L. Group-server queues. In Proceedings of the International Conference on Queueing Theory and Network Applications, Qinhuangdao, China, 21–23 August 2017; pp. 49–72.
7. Harchol-Balter, M. Open problems in queueing theory inspired by data center computing. *Queueing Syst.* **2021**, *97*, 3–37. [CrossRef]
8. Barroso, L.A.; Hölzle, U. The case for energy-proportional computing. *Computer* **2007**, *40*, 33–37. [CrossRef]
9. Kuehn, P.J.; Mashaly, M.E. Automatic energy efficiency management of data center resources by load-dependent server activation and sleep modes. *Ad Hoc Netw.* **2015**, *25*, 497–504. [CrossRef]
10. Gandhi, A. Dynamic Server Provisioning for Data Center Power Management. Ph.D. Dissertation, School of Computer Science, Carnegie Mellon University, Pittsburgh, PA, USA, 2013.
11. Gandhi, A.; Doroudi, S.; Harchol-Balter, M.; Scheller-Wolf, A. Exact analysis of the M/M/k/setup class of Markov chains via recursive renewal reward. *Queueing Syst.* **2014**, *77*, 177–209. [CrossRef]
12. Gandhi, A.; Gupta, V.; Harchol-Balter, M.; Kozuch, M.A. Optimality analysis of energy-performance trade-off for server farm management. *Perform. Eval.* **2010**, *67*, 1155–1171. [CrossRef]
13. Gandhi, A.; Harchol-Balter, M.; Adan, I. Server farms with setup costs. *Perform. Eval.* **2010**, *67*, 1123–1138. [CrossRef]
14. Maccio, V.J.; Down, D.G. On optimal policies for energy-aware servers. *Perform. Eval.* **2015**, *90*, 36–52. [CrossRef]
15. Maccio, V.J.; Down, D.G. Exact analysis of energy-aware multiserver queueing systems with setup times. In Proceedings of the IEEE 24th International Symposium on Modeling, Analysis and Simulation of Computer and Telecommunication Systems, London, UK, 19–21 September 2016; pp. 11–20.
16. Phung-Duc, T. Exact solutions for M/M/c/setup queues. *Telecommun. Syst.* **2017**, *64*, 309–324. [CrossRef]
17. Phung-Duc, T.; Kawanishi, K.I. Energy-aware data centers with s-staggered setup and abandonment. In Proceedings of the International Conference on Analytical and Stochastic Modeling Techniques and Applications, Cardiff, UK, 24–26 August 2016; pp. 269–283.

18. Gebrehiwot, M.E.; Aalto, S.; Lassila, P. Optimal energy-aware control policies for FIFO servers. *Perform. Eval.* **2016**, *103*, 41–59. [CrossRef]

19. Gebrehiwot, M.E.; Aalto, S.; Lassila, P. Energy-performance trade-off for processor sharing queues with setup delay. *Oper. Res. Lett.* **2016**, *44*, 101–106. [CrossRef]

20. Gebrehiwot, M.E.; Aalto, S.; Lassila, P. Energy-aware SRPT server with batch arrivals: Analysis and optimization. *Perform. Eval.* **2017**, *15*, 92–107. [CrossRef]

21. Mitrani, I. Service center trade-offs between customer impatience and power consumption. *Perform. Eval.* **2011**, *68*, 1222–1231. [CrossRef]

22. Mitrani, I. Managing performance and power consumption in a server farm. *Ann. Oper. Res.* **2013**, *202*, 121–134. [CrossRef]

23. Kamitsos, I.; Ha, S.; Andrew, L.L.; Bawa, J.; Butnariu, D.; Kim, H.; Chiang, M. Optimal sleeping: Models and experiments for energy-delay tradeoff. *Int. J. Syst. Sci. Oper. Logist.* **2017**, *4*, 356–371. [CrossRef]

24. Hipp, S.K.; Holzbaur, U.D. Decision processes with monotone hysteretic policies. *Oper. Res.* **1988**, *36*, 585–588. [CrossRef]

25. Lu, F.V.; Serfozo, R.F. M/M/1 queueing decision processes with monotone hysteretic optimal policies. *Oper. Res.* **1984**, *32*, 1116–1132. [CrossRef]

26. Yang, J.; Zhang, S.; Wu, X.; Ran, Y.; Xi, H. Online learning-based server provisioning for electricity cost reduction in data center. *IEEE Trans. Control Syst. Technol.* **2017**, *25*, 1044–1051. [CrossRef]

27. Ding, D.; Fan, X.; Zhao, Y.; Kang, K.; Yin, Q.; Zeng, J. Q-learning based dynamic task scheduling for energy-efficient cloud computing. *Future Gener. Comput. Syst.* **2020**, *108*, 361–371. [CrossRef]

28. Liang, Y.; Lu, M.; Shen, Z.J.M.; Tang, R. Data center network design for internet-related services and cloud computing. *Prod. Oper. Manag.* **2021**, *30*, 2077–2101. [CrossRef]

29. Xia, L.; Chen, S. Dynamic pricing control for open queueing networks. *IEEE Trans. Autom. Control* **2018**, *63*, 3290–3300. [CrossRef]

30. Xia, L.; Miller, D.; Zhou, Z.; Bambos, N. Service rate control of tandem queues with power constraints. *IEEE Trans. Autom. Control* **2017**, *62*, 5111–5123. [CrossRef]

31. Ma, J.Y.; Xia, L.; Li, Q.L. Optimal energy-efficient policies for data centers through sensitivity-based optimization. *Discrete Event Dyn. Syst.* **2019**, *29*, 567–606. [CrossRef]

32. Chi, C.; Ji, K.; Marahatta, A.; Song, P.; Zhang, F.; Liu, Z. Jointly optimizing the IT and cooling systems for data center energy efficiency based on multi-agent deep reinforcement learning. In Proceedings of the 11th ACM International Conference on Future Energy Systems, Virtual Event, Australia, 22–26 June 2020; pp. 489–495.

33. Li, Q.L. *Constructive Computation in Stochastic Models with Applications: The RG-Factorizations*; Springer: Beijing, China, 2010.

34. Cao, X.R. *Stochastic Learning and Optimization—A Sensitivity-Based Approach*; Springer: New York, NY, USA, 2007.

35. Xia, L.; Zhang, Z.G.; Li, Q.L. A c/μ-rule for for job assignment in heterogeneous group-server queues. *Prod. Oper. Manag.* **2021**, 1–18. [CrossRef]

36. Li, Q.L.; Li, Y.M.; Ma, J.Y.; Liu, H.L. A complete algebraic transformational solution for the optimal dynamic policy in inventory rationing across two demand classes. *arXiv* **2019**, arxiv:1908.09295v1.

37. Ma, J.Y.; Li, Q.L. Sensitivity-based optimization for blockchain selfish mining. In Proceedings of the International Conference on Algorithmic Aspects of Information and Management, Dallas, TX, USA, 20–22 December 2021; pp. 329–343.

38. Xia, L. Risk-sensitive Markov decision processes with combined metrics of mean and variance. *Prod. Oper. Manag.* **2020**, *29*, 2808–2827. [CrossRef]

39. Puterman, M.L. *Markov Decision Processes: Discrete Stochastic Dynamic Programming*; John Wiley & Sons: New York, NY, USA, 2014.

40. Budhiraja, A.; Friedlander, E. Diffusion approximations for controlled weakly interacting large finite state systems with simultaneous jumps. *Ann. Appl. Probab.* **2018**, *28*, 204–249. [CrossRef]

Article

Simulation of Cooperation Scenarios of BRI-Related Countries Based on a GVC Network

Dawei Wang [1], Jun Guan [1], Chunxiu Liu [1], Chuke Jiang [2] and Lizhi Xing [1,3,*]

1   College of Economics and Management, Beijing University of Technology, Beijing 100124, China; wdw0116@163.com (D.W.); guanjun@bjut.edu.cn (J.G.); 17860508250@163.com (C.L.)
2   College of Economics & Management, Shenyang Ligong University, Shenyang 110170, China; jiangchuke@126.com
3   International Business School, Beijing Foreign Studies University, Beijing 100089, China
*   Correspondence: koken@bjut.edu.cn

Abstract: The inter-country input–output table is appropriate for presenting sophisticated inter-industry dependencies from a global perspective. Using the above table one can perceive the amount of production resources that sectors obtain from their upstream ones, as well as the number of productive capacities that sectors provide for their downstream ones. In other words, competition/collaboration occurs when sectors share the same providers/consumers because all sectors' products and services outputted to downstream ones are limited. Thus, inter-industry competition for inputs from upstream sectors, or collaboration on outputs to downstream sectors, may be quantified with input–output matrix transformation. In this paper, a novel analytical framework of inter-industry collaborative relations is established based on the bipartite graph theory and the resource allocation process. The Collaborative Opportunity Index and Collaborative Threat index are designed to quantitatively measure the industrial influence hidden in the topological structure of the global value chain (GVC) network. Scenario simulations are carried out to forecast the potential and trends of international capacity cooperation within Asian, European, and African nations related to the Belt and Road Initiative, respectively.

Keywords: global value chain; inter-country input-output table; the Belt and Road Initiative; Collaborative Opportunity Index; Collaborative Threat Index

Citation: Wang, D.; Guan, J.; Liu, C.; Jiang, C.; Xing, L. Simulation of Cooperation Scenarios of BRI-Related Countries Based on a GVC Network. *Systems* **2022**, *10*, 12. https:// doi.org/10.3390/systems10010012

Academic Editor: Stefano Armenia

Received: 20 December 2021
Accepted: 28 January 2022
Published: 3 February 2022

Publisher's Note: MDPI stays neutral with regard to jurisdictional claims in published maps and institutional affiliations.

1. Introduction

Belt and Road Initiative (BRI) is a transcontinental long-term policy and investment program which aims at infrastructure development and acceleration of the economic integration of countries along the route of the historic Silk Road. The initiative was unveiled in 2013 by China's President Xi Jinping.

At present, China has entered the **"New Normal"** development stage of the economy, and the BRI is being implemented in depth. In March 2015, China issued an action plan which described the main objectives of the BRI. The BRI-participating economies represent more than one-third of global GDP and over half of the world's population. In September 2016, General Secretary Jinping Xi pointed out in the keynote speech at the opening ceremony of the B20 Summit (a major support group for the G20 from industrial and commercial circles): "China's development benefits from the international community and is willing to provide more public goods to the international community. China has proposed the 'One Belt and One Road Initiative', which aims to share China's development opportunities and achieve common prosperity with countries along the routes." China is actively promoting the economic development of BRI-related nations, supporting and driving domestic superior and surplus production capacity to countries/regions that have fewer comparative advantages. Finally, to construct a fair and reasonable international order, China will offer both value ideas and institutional design ideas that reflect Chinese wisdom and China's plans.

BRI is based on the concept of mutually beneficial cooperation [1]. It plays an important role in foreign direct investment, infrastructure, international cooperation, and economic growth in countries along the route [2–6]. According to the report "Building the Belt and Road Initiative: Progress, Contribution and Prospects" released by the Ministry of Commerce in April 2019, the BRI has gained the support of 125 countries and 29 international organizations for its construction. According to the data published by the World Bank, the combined GDP of the countries along the BRI in the five years after its introduction (2013–2017) was 141 trillion dollars, which is 39.6% higher than that in the five years before the initiative (2008–2012). In addition, the World Bank Group's 2019 release "The Belt and Road Economy: Opportunities and Risks of Transport Corridors" quantifies the impact of the Belt and Road. According to the article, the Belt and Road can expand trade, increase foreign investment, and reduce poverty by reducing trade costs. Full implementation of the Belt and Road Initiative could increase world trade by 1.7% to 6.2% and global real income by 0.7% to 2.9%.

Rapidly promoting China's position and competitiveness on the **Global Value Chain (GVC)** and shaping new comparative advantages in the context of BRI is either the important guarantee for the continuous and in-depth development of **Global Cooperation on Production Capacity** strategy or the realistic requirement for China's industrial restructuring and factor allocation optimization at this stage. BRI aims to build an open, inclusive, and balanced regional economic cooperation architecture for the 21st century, strengthen the weak points of globalization, and transform partial globalization into an inclusive one. Under this context, there is an urgent need to rationally distribute international productivities to enhance the competitive advantages of both China and BRI-related nations. Therefore, how to realize the positive interaction between China's industrial structure and foreign trade upgrade in the process of international economic integration and BRI, and how to fully utilize the institutional dividend brought by such a cooperative strategy to build a new regional trade system, will definitely be the key research direction at the national strategic level for a long period in the future.

Based on a mature industrial system, the high-cost performance of equipment capacity, mighty construction abilities, enhancement on the international capacity cooperation with nations along the route, are feasible ways for China to achieve mutual benefits and the win-win goal. Nevertheless, the paradox is that international public opinion has many doubts and even dissatisfaction with China's BRI. Some countries believe that the BRI is China's economic plunder of countries along the route, and it is a Chinese version of the Marshall Plan. It has become an important driving factor for the threat inflation phenomenon of some Western countries' perception of China. In recent years, some countries have implemented a series of measures to weaken China's role and status on the GVC. The U.S. plans to start investing in 5 to 10 major infrastructure projects around the world in 2022 as part of the Group of Seven (G7) Build Back Better World (B3W) program to counter China's BRI. The EU and the UK have launched similar programs to invest in global infrastructure. The Global Gateway Initiative and the Clean Green Initiative have been proposed so that less developed countries will opt out of the streamlined and unconditional BRI. These approaches are equivalent to treating international trade as a zero-sum game, thus failing to achieve a win-win goal. Of course, under the perspective of systems science, it is impractical and will inevitably lead to negative impacts flowing along the GVC and ultimately leading to a decline in the global competitiveness of the industrial sectors within many countries. Accordingly, our econophysics framework will be adopted to simulate the international trade process under different policy backgrounds and development scenarios. We hope the network-based measurements proposed in this paper can be used as the evaluation criteria to deeply understand the policymaking of international trade and its long-term consequences.

As the result of globally economic integration, various countries and regions in the world have gradually formed a global economic system through increasingly deepened trade exchanges with each other. As the subsystems of this system, countries in the world have also evolved their internal **Industrial Value Chains (IVCs)** as an organic whole, and they play a specific role on a global scale. Therefore, economists began to systematically and comprehensively study global macroeconomic issues from the perspective of system theory. The GVC network reflects the topological structure of the global economic system, depicts the vertical division of labor and complex input–output relationships in global industrial sectors, and provides a theoretical framework and application basis for analyzing global economic issues. The popularization and promotion of the **Inter–Country Input–Output (ICIO)** database provides scientific and accurate data for studying the important role of various industries and economies on the GVC network. The IO table can be directly or slightly modified as an adjacency matrix to establish a network model [7], and measure the importance of industrial sectors through network characteristic indicators. **Input–Output Analysis (IOA)** can be performed on issues such as the impact of consumption shocks [8] on the economic system and the balance of supply and demand [9] in a free market economy. Sectoral production characteristics, such as the degree of industrial sector participation in vertical specialization trade [10], the number of stages required for production on the GVC and the stage between production and final consumption [11], are measured. In recent years, Wang, et al. proposed a series of indicators such as average production length and relative upstream degree [12] to measure the production structure of each economy. Xu et al. combined the input–output model with the social network analysis method to measure the economic structure in the global economic system [13]. Yang et al. established a community detection optimization algorithm based on biogeography to find the community structure in the network [14]. Piccardi et al. evaluated the importance of communities in the world input–output network based on the random walk Markov chain method [15]. Guan et al. established a global industrial resource competition network model based on the bibliographic coupling method to analyze the competitive relationship between industrial sectors [16]. Xing et al. used the Markov process to measure the degree of globalization [17] and industrial influence [18] of the industrial sectors and based on the Revised Floyd–Warshall Algorithm (RFWA), proposed the Strongest Relevance Path Length (SRPL) algorithm to measure the pivotability degree [19] of the input–output relationship between sectors and connectivity and tightness [20] of the value chain network. At the same time, the competition and cooperation relations [21] between industrial sectors are also described through the citation network, but the relationship generated in the citation network lacks precise similarity.

In the first section of the paper, the development status of the BRI and the research on the input–output relationship between various industrial sectors on the GVC are presented. In the second section, the **Global Production Capacity Collaboration Network (GPCCN)** model is formulated. The model is based on the bipartite graph theory, on the resource allocation process and on the selection of research data. In the third section are introduced two indicators: the **Collaborative Opportunity Index (*COI*)** and the **Collaborative Threat Index (*CTI*)** from the national and sectoral levels based on the GPCCN model. The indicators measure the state of cooperation between countries and sectors, and at the same time, measure the correlation between *COI* and **Competitive Strength Index (*CSI*)** at the national level. In the fourth section, scenario simulations are carried out in order to forecast the potential and trends of international capacity cooperation within Asian, European, and African nations related to the Belt and Road Initiative. In the fifth section, the laws and reasons for the differences in cooperation among BRI-related nations are explained. In the last section, the main contributions of the paper are summarized and some new directions for the research are indicated.

2. Methodology

2.1. Resource Allocation Process

In a bipartite graph G the node set V is divided in two nonempty sets P and O with no intersection in between. Let $G = (P, O, E)$ be a bipartite graph where E is the set of edges, $P = \{P_1, P_2, \cdots, P_n\}$ and $O = \{O_1, O_2, \cdots, O_m\}$ are the two sets of nodes. Of course, the intersection of P and O is empty. The nodes from the set P will be called participants and the nodes from the set O will be called objects. We take an example to describe its topological structure, where $n = 9$ and $m = 3$, as shown in Figure 1.

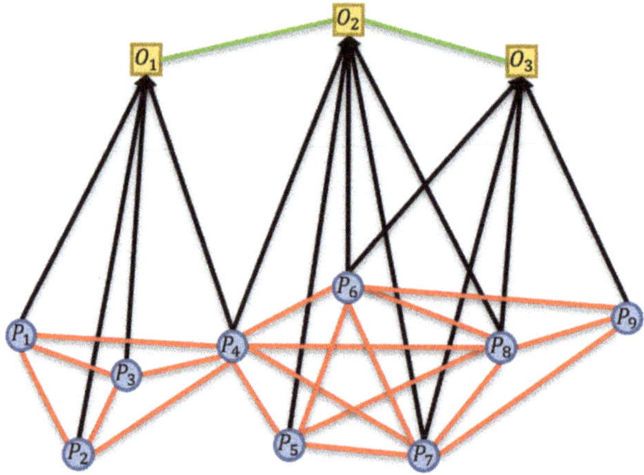

Figure 1. A two-mode network and its projection onto objects and participants.

In Figure 1, the squares in the upper part are the objects (denoted by O_1, O_2, ..., O_3), while the circles below are the participants (denoted by P_1, P_2, ..., P_9), and the edges in black belong to the two-mode network. It is more than common to project a two-mode network onto one kind of node, and the resulting edges have been granted the property to reflect a certain relationship. As we can see, the edges in red or green coming from the projection of two black edges constitute two one-mode networks, namely **Complete Object Subgraph** and **Complete Participant Subgraph**. Sometimes, there should be weights on the edges, which are gained through the definition of co-occurrences and used to measure the potential relationship of two participants in the same object, or that of two objects in the same participant. For instance, we can quantify the relationship between two scientists (participants) by the number of papers (objects) they wrote together; similarly, we can also quantify that of two papers (objects) by the number of the same scientists (participants) they have [22]. However, refined calculation on the weight of a projected edge is very difficult, and we must use a specific method to solve a specific problem.

The bipartite graph has a wide application in complex network analysis, including cooperation and competition networks (mainly dealt with through affiliation networks), for either cooperation or competition is the common existence in social networks consisting of units of people. Padrón believed that this modeling process could bring distinctive simulation on the potential cooperation or competition relation [23]. In the field of GVC-related studies, scholars and politicians all want to figure out the inter-country and inter-industry competition and collaboration for the purposes of academic research and policymaking. We have done a lot of work in the research of industrial competition [24–26], so we want to focus on the other side of it. If limited industrial resources lead to competition among downstream sectors, then limited market demand leads to cooperation among upstream sectors. Therefore, with the purpose of extracting the inter-industry collaborative relations, the **Resource Allocation Process (RAP)** is also adopted in this paper as the algorithm of projection [27]. The following is the specific derivation process.

Let $f : O \bigcup P \to \mathbb{R}_+$ be a function such that $f(O_h) = 1$ for all h in $\{1, 2, \ldots, m\}$. Firstly, the initial resources needed by the h-th object is $f(O_h) \geq 0$. We assume that the $O \to P$ primary distribution of initial demands is equal, as shown in Figure 2.

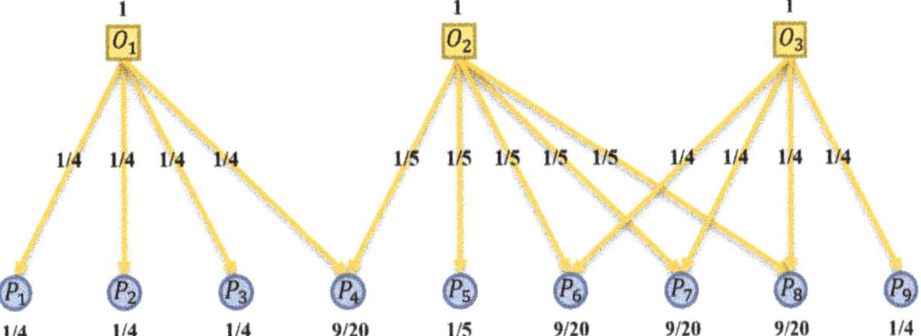

Figure 2. Primary distribution: initial demands from objects are equally sent to participants. Notes: For simplicity, we assume all objects here own an equal size of initial demands, i.e., $f(O_h) = 1$. As we can see, O_1 connects to P_1, P_2, P_3, and P_4, so $K(O_1) = 4$, $a_{11} = 1$, $a_{21} = 1$, $a_{31} = 1$, and $a_{41} = 1$. Within them, only P_4 is additionally connected to O_2, so $a_{42} = 1$ while $K(O_2) = 5$. Thus, $f(P_1) = \frac{1}{4}$, $f(P_2) = \frac{1}{4}$, $f(P_3) = \frac{1}{4}$, and $f(P_4) = \frac{1}{4} + \frac{1}{5} = \frac{9}{20}$.

The resources demand on the j-th node in P is:

$$f(P_j) = \sum_{h=1}^{m} \frac{a_{jh} f(O_h)}{K(O_h)} \tag{1}$$

where, $K(O_h)$ is the degree of O_h, $\left(a_{jh} \right)$ is a $n \times m$ matrix:

$$a_{jh} = \begin{cases} 1 & P_j O_h \in E \\ 0 & otherwise \end{cases} \tag{2}$$

With all the demand signals converging to set O, the required resources of objects are shown in Figure 3. Note that, the assumption of equal distribution still holds for the secondary distribution.

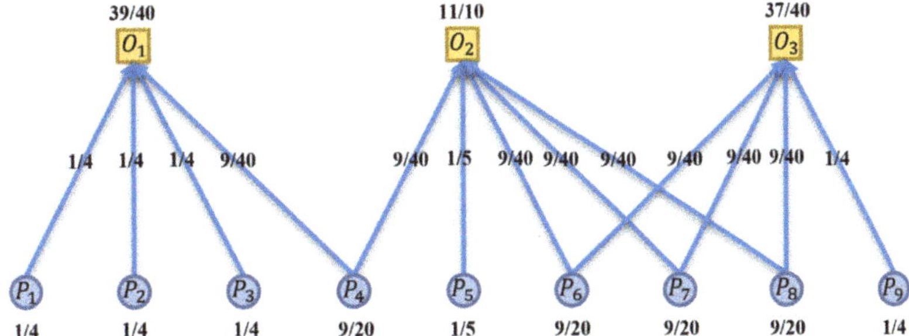

Figure 3. Secondary distribution: required resources from participants are equally allocated to objects. Notes: When a participant equally allocates its required resource to relevant objects, the secondary distribution depends on the number of relevant objects. Thus, the new amount of object's satisfied demand is equal to the sum of required resources back from all its participants, e.g., $f'(O_1) = \frac{f(P_1)}{K(P_1)} + \frac{f(P_2)}{K(P_2)} + \frac{f(P_3)}{K(P_3)} + \frac{f(P_4)}{K(P_4)} = \frac{1}{4} + \frac{1}{4} + \frac{1}{4} + \frac{9}{20} \times \frac{1}{2} = \frac{39}{40}$.

The satisfied demand on the resources of node O_h is:

$$f'(O_h) = \sum_{j=1}^{n} \frac{u_{jh}f(P_j)}{K(P_j)} = \sum_{j=1}^{n} \frac{a_{jh}}{K(P_j)} \sum_{k=1}^{m} \frac{a_{jk}f(O_k)}{K(O_k)} \tag{3}$$

Obviously, the satisfied demand for objects is not consistent with their initial one, i.e., $f'(O_h) \neq f(O_k)$, which means their status is different in the complete participant subgraph. This difference cannot be reflected just via the co-occurrence projection. Thus, the hidden collaborative relations among them can be expressed by:

$$f'(O_h) = \sum_{k=1}^{m} w_{hk}^{O} f(O_k) \tag{4}$$

where, w_{hk}^{O} is the measurement of difference produced in the process of two-times resources allocations, and describes the advantage of O_h in cooperating with O_k to allocate resources of their common participants.

The w_{hk}^{O} in Equation (4) could be written as:

$$w_{hk}^{O} = \frac{1}{K(O_k)} \sum_{j=1}^{n} \frac{a_{jh}a_{jk}}{K(P_j)} \tag{5}$$

Finally, we get the matrix $W^{O} = (w_{hk}^{O})_{m \times m}$ as the weight set of complete participant subgraph through RAP approach, as shown in Figure 4.

$$\begin{bmatrix} f'(O_1) \\ f'(O_2) \\ f'(O_3) \end{bmatrix} = \begin{bmatrix} w_{11}^{O} & w_{12}^{O} & w_{13}^{O} \\ w_{21}^{O} & w_{22}^{O} & w_{23}^{O} \\ w_{31}^{O} & w_{32}^{O} & w_{33}^{O} \end{bmatrix} \begin{bmatrix} f(O_1) \\ f(O_2) \\ f(O_3) \end{bmatrix}$$

$$= \begin{bmatrix} 7/8 & 1/10 & 0 \\ 1/8 & 3/5 & 3/8 \\ 0 & 3/10 & 5/8 \end{bmatrix} \begin{bmatrix} f(O_1) \\ f(O_2) \\ f(O_3) \end{bmatrix}$$

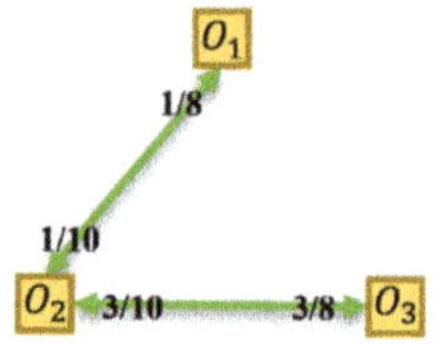

(*a*) Matrix-Form Linear Relation　　　　**(*b*) Complete Participant Subgraph**

Figure 4. Collaborative relations reflected by complete participant subgraph. Notes: In (**a**), the matrix W^{O} represents the linear relation between each object's satisfied demand and initial demand, whose different values reflect their different status in the resource allocation process. Therefore, the weighted and directed graph in (**b**) embodies the unsymmetrically and unequally collaborative relations among three objects, while the values on the diagonal of matrix W^{O} are useless. For example, according to Equation (5), the strength of cooperative relation from node O_2 to node O_3 is $w_{23}^{O} = \frac{1}{k(O_3)} \times \left(\frac{a_{62}a_{63}}{k(P_6)} + \frac{a_{72}a_{73}}{k(P_7)} + \frac{a_{82}a_{83}}{k(P_8)} \right) = \frac{1}{4} \times \left(\frac{1}{2} + \frac{1}{2} + \frac{1}{2} \right) = \frac{3}{8}$.

Furthermore, in the weighted two-mode network, Equation (5) is expanded to another form by replacing the adjacency matrix $A = \left(a_{jh} \right)$ with weight set $W = \left(w_{jh} \right)$:

$$w_{hk}^{O} = \frac{1}{S(O_k)} \sum_{k=1}^{n} \frac{w_{jh}w_{jk}}{S(P_j)} \tag{6}$$

where $S(O_k)$ is the strength of O_k, i.e., $S(O_k) = \sum_{j=1}^{n} w_{jk}$; $S(P_j)$ is the strength of P_j, $S(P_j) = \sum_{h=1}^{m} w_{jh}$; w_{jh} and w_{jk} are the weights on edges connecting O_h and O_k with P_j, respectively.

Therefore, the RAP approach reflects the scarcity of participant nodes resources and limits the resource allocation from participant nodes to object nodes. At the same time, the formation of a complete object subgraph through participant nodes mapping can clearly reflect cooperation relation between various industrial sectors on the GVC, thus providing a method for measuring the potential and trends of international capacity cooperation within Asian, European, and African nations related to the Belt and Road Initiative, respectively.

2.2. Database Selection

As we all know, BRI is a hotly debated topic in the field of the global economy, as well as GVC, which is a global development strategy proposed by the Chinese government involving infrastructure construction and investments in 152 countries and international organizations in Asia, Europe, Africa, the Middle East, and the Americas. "Belt" refers to the overland routes for road and rail transportation, called "the Silk Road Economic Belt"; "Road" refers to the sea routes or the 21st Century Maritime Silk Road. From the Chinese government's international viewpoints on politics and economy, BRI is supposed to be the developing blueprint that meets the demands of relevant countries and delivers mutual benefits. However, some observers see it as a push for Chinese dominance in global affairs with a China-centered trading network, and even consider BRI as a potential threat to countries involved [28].

For now (2021), there are 66 countries (including China) along the Road and Belt. In ICIO databases, Eora26 has the widest coverage of countries, including all the countries except Palestine, and that is why we use it to build **Eora 26-Based Global Industrial Value Chain Network (GIVCN–Eora26)** models and conduct an empirical analysis of capacity cooperation between BRI countries. In the analysis process, we further divide these countries into three categories according to their continents, namely Asian nations, European nations, and African nations, as shown in Tables 1–3.

Table 1. Thirty-six BRI-related Asian nations in Eora26.

Abbr.	Country	Abbr.	Country
AFG	Afghanistan	MNG	Mongolia
ARM	Armenia	MMR	Myanmar
AZE	Azerbaijan	NPL	Nepal
BHR	Bahrain	OMN	Oman
BGD	Bangladesh	PAK	Pakistan
BRN	Brunei	PHL	Philippines
KHM	Cambodia	QAT	Qatar
GEO	Georgia	KOR	Korea
IDN	Indonesia	SAU	Saudi Arabia
IRN	Iran	SGP	Singapore
IRQ	Iraq	LKA	Sri Lanka
KAZ	Kazakhstan	TJK	Tajikistan
KWT	Kuwait	THA	Thailand
KGZ	Kyrgyzstan	TUR	Turkey
LAO	Laos	ARE	UAE
LBN	Lebanon	UZB	Uzbekistan
MYS	Malaysia	VNM	Viet Nam
MDV	Maldives	YEM	Yemen

With the proposal and promotion of BRI, China is playing a leading role in the **Regional Value Chain (RVC)** networks constituted of Asian, European, and African nations. Accordingly, the global cooperation with China on production capacity will impact the economic development of BRI-related nations. Therefore, it is necessary to comparatively and empirically analyze how their status will change on the GVC and what kinds of influence the BRI will bring to them.

Before doing this, we need to extract three sub-networks out of the GVICN–Eora26 model, i.e., GIVCN–Eora26–AS, GIVCN–Eora26–EU, and GIVCN–Eora26–AF, reflecting the ICIO relations between China and other economies, respectively. Their brief topological structures in six different periods are as shown in Figures 5–7.

Table 2. Twenty-seven BRI-related European nations in Eora26.

Abbr.	Country	Abbr.	Country
ALB	Albania	LUX	Luxembourg
AUT	Austria	MLT	Malta
BLR	Belarus	MNE	Montenegro
BIH	Bosnia and Herzegovina	POL	Poland
BGR	Bulgaria	PRT	Portugal
HRV	Croatia	MDA	Moldova
CYP	Cyprus	ROU	Romania
CZE	Czech Republic	RUS	Russia
EST	Estonia	SRB	Serbia
GRC	Greece	SVK	Slovakia
HUN	Hungary	SVN	Slovenia
ITA	Italy	MKD	TFYR Macedonia
LVA	Latvia	UKR	Ukraine
LTU	Lithuania		

Table 3. Forty-four BRI-related African nations in Eora26.

Abbr.	Country	Abbr.	Country
DZA	Algeria	MDG	Madagascar
AGO	Angola	MLI	Mali
BEN	Benin	MRT	Mauritania
BWA	Botswana	MAR	Morocco
BDI	Burundi	MOZ	Mozambique
CMR	Cameroon	NAM	Namibia
CPV	Cape Verde	NER	Niger
TCD	Chad	NGA	Nigeria
COG	Congo	RWA	Rwanda
CIV	Cote d'Ivoire	SEN	Senegal
COD	DR Congo	SYC	Seychelles
DJI	Djibouti	SLE	Sierra Leone
EGY	Egypt	SOM	Somalia
ETH	Ethiopia	ZAF	South Africa
GAB	Gabon	SDS	South Sudan
GMB	Gambia	SUD	Sudan
GHA	Ghana	TGO	Togo
GIN	Guinea	TUN	Tunisia
KEN	Kenya	UGA	Uganda
LSO	Lesotho	TZA	Tanzania
LBR	Liberia	ZMB	Zambia
LBY	Libya	ZWE	Zimbabwe

To explore the core structure of the original network, this paper takes three sorts of GVICN–Eora26 models based on the ICIO databases of the latest version for example, which are GIVCN–Eora26–AS, GIVCN–Eora26–EU, and GIVCN–Eora26–AF, and statistics for the structural properties of three sub-networks after pruning [29] include **number of edges** $|E|$, **average distance** $\langle d \rangle$, **mean node degree** $\langle k \rangle$, **clustering coefficient** $\langle C \rangle$ and the **degree-degree correlation from out-degree source nodes to in-degree sink ones** $r(out, in)$ are shown in Tables 4–6.

We observe the changing trends of the network backbone over a 25-year period and find that: if the information of edge weight is not considered, all three sub-networks of GIVCN–Eora26 models own a high density of connections, which may give them access to the shorter average path and the greater clustering coefficient (see Tables 4–6). Therefore, both are strong evidence that these models belong to the small-world network, i.e., China has a great foundation to carry out BRI–related trade with these regions.

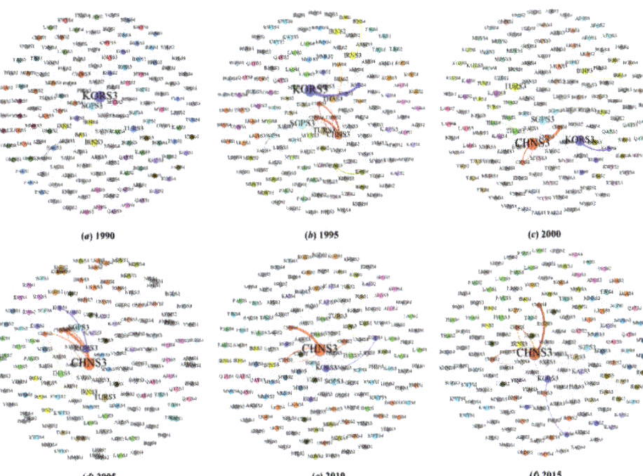

Figure 5. Topological structure of GIVCN–Eora26–AS models.

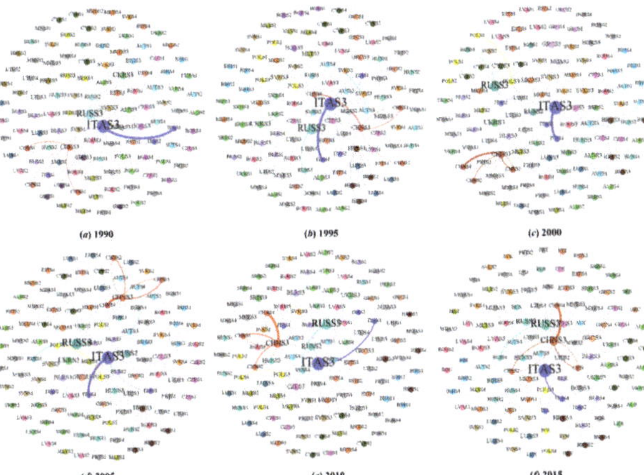

Figure 6. Topological structure GIVCN–Eora26–EU models.

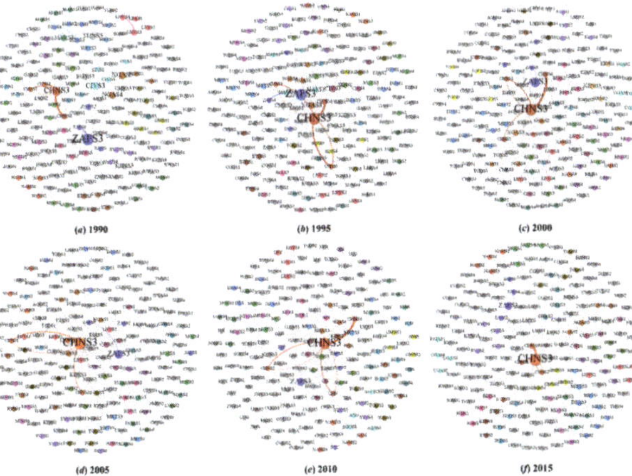

Figure 7. Topological structure GIVCN–Eora26–AF models.

Table 4. Illustration of properties of the GIVCN–Eora26–AS models.

GIVCN–Eora26–AS	1990	1995	2000	2005	2010	2015
$|E|$	367	366	373	363	362	362
$\langle d \rangle$	4.752	4.760	4.563	4.546	4.305	4.261
$\langle k \rangle$	2.480	2.473	2.520	2.453	2.446	2.446
$\langle C \rangle$	0.562	0.573	0.559	0.564	0.538	0.531
$r(out, in)$	0.517	0.536	0.542	0.546	0.502	0.489

Table 5. Illustration of properties of the GIVCN–Eora26–EU models.

GIVCN–Eora26–EU	1990	1995	2000	2005	2010	2015
$|E|$	272	270	273	270	277	276
$\langle d \rangle$	3.897	3.916	4.057	4.109	4.039	4.021
$\langle k \rangle$	2.429	2.411	2.438	2.411	2.473	2.464
$\langle C \rangle$	0.493	0.502	0.474	0.491	0.524	0.517
$r(out, in)$	0.463	0.453	0.464	0.466	0.452	0.450

Table 6. Illustration of properties of the GIVCN–Eora26–AF models.

GIVCN–Eora26–AF	1990	1995	2000	2005	2010	2015
$|E|$	499	492	468	482	491	480
$\langle d \rangle$	5.156	4.606	4.399	4.249	4.147	4.238
$\langle k \rangle$	2.772	2.733	2.600	2.678	2.728	2.667
$\langle C \rangle$	0.558	0.592	0.612	0.590	0.625	0.622
$r(out, in)$	0.540	0.495	0.457	0.421	0.428	0.441

2.3. Network Modeling

To reproduce the collaborative relations between industrial sectors on the GVC, we design a generation algorithm based on RAP approach:

$$
w_{ij}^{O} = \begin{cases} \frac{1}{\vec{w}_j} \sum_{k=1}^{N} \frac{w_{ik} w_{jk}}{\overleftarrow{w}_k} & , \quad i \neq j \\ 0 & , \quad i = j \end{cases}
\tag{7}
$$

where, w_{ik} (w_{jk}) is the i-th (j-th) row and k-th column element of the adjacency matrix of GIVCN model, representing the upstream sector i (j) and downstream sector k respectively; $\overleftarrow{w}_k$ is the gross inputs of downstream sector k, and it is numerically equal to the in-degree strength of node k in GIVCN model, say $\overleftarrow{w}_k = S^{IN}(k) = \sum_{i=1}^{N} w_{ik}$; $\vec{w}_j$ is the gross outputs of upstream sector j, i.e., $\vec{w}_j = S^{OUT}(j) = \sum_{k=1}^{N} w_{jk}$; w_{ij}^{O} measures the collaborative attraction from the sector i to j; upstream sectors i and j are connected by an edge denoted by e_{ij}^{O} in the complete participant subgraph.

Finally, the edge set $E^{O} = \left(e_{ij}^{O} \right)$ and weight set $W^{O} = (w_{ij}^{O})$ reflect all the collaborative relations among sectors in the global production system. We name the graph $G = (V, E^{O}, W^{O})$ as the **Global Production Capacity Collaboration Network (GPCCN)**. Accordingly, we separate three types of GPCCN–BRI models from the whole network, which are GPCCN–Eora26–AS, GPCCN–Eora26–EU, and GPCCN–Eora26–AF.

3. Measurement

According to our paper on the application of the complex network theory [18,30], network-based algorithms and indices have great potential to enhance the understanding of the industrial sector's position and function, given the network-form architecture of GVC. The inter-industry collaboration status has been embodied in the GPCCN model, and out-strength and in-strength as simple yet important tools are hence introduced to quantify industrial sectors' collaborative opportunity and threat on the GVC, based on which we further carry out econometric, static timing, and simulation analyses.

3.1. Sector-Level Indices

When discussing the cooperation relations in the GPCCN model, we define the sum of the collaborative attraction obtained from other sectors (out-strength) as one sector's **Collaborative Opportunity Index (COI)**, i.e., the greater the *COI*, the stronger the collaborative relations between this sector and the others. Additionally, the sum of the collaborative attraction is exerted to other sectors (in-strength) is defined as one sector's **Collaborative Threat Index (CTI)**. Once the collaboration degree declines, and the uncertainty of industrial development will go up, because a greater *CTI* indicates that the sector needs to rely on many collaborative relations to maintain its function and status on the GVC. Their statistical formula is as follows:

$$COI(i) = S^{OUT}(i) = \sum_{j=1}^{N} w_{ji}^{O} \tag{8}$$

$$CTI(i) = S^{IN}(i) = \sum_{j=1}^{N} w_{ij}^{O} \tag{9}$$

As the counterparts of *CSI* and *CWI*, we are interested in the distribution and correlation of *COI* and *CTI*. As shown in Figure 8, the heavy-tailed distribution of *COI* for all sectors in GPCCN–Eora26–2015 also follows the levy-stable distribution, but that of *CTI* is mainly concentrated in a narrow numerical range. The heterogeneity of *COI* and the homogeneity of *CTI* together result in no strong correlation between them (Pearson correlation coefficient is only −0.435).

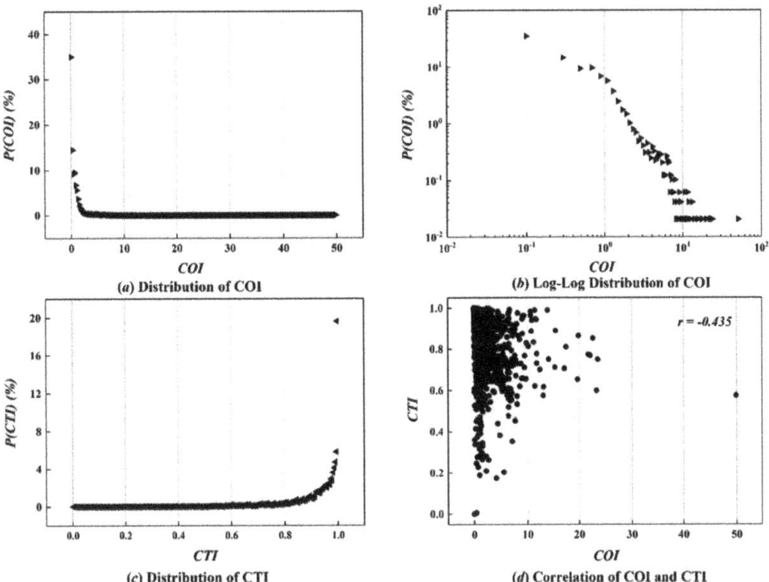

Figure 8. Distribution and correlation of *COI* and *CTI* in GPCCN–Eora26-2015.

In our opinion, the collaborative opportunities brought by global economic integration to various countries are very different, while the threats of collaboration are almost the same. The fundamental reason is that today's advanced information technology and convenient supply chains have greatly reduced the difficulty for the industrial sector to find partners.

3.2. Country-Level Indices

On the basis of *COI* and *CTI*, notions of **National Collaborative Opportunity Index** (*NCOI*) and **National Collaborative Threat Index** (*NCTI*) are here introduced:

$$NCOI(u) = \sum_{i \in \tau(u)} COI(i) \tag{10}$$

$$NCTI(u) = \sum_{i \in \tau(u)} CTI(i) \tag{11}$$

where $NCOI(u)$ and $NCTI(u)$ are used to measure the collaborative opportunity and threat of country u.

Since the difference in collaborative threats is not obvious, we focus on the changes in the *NCOIs* of BRI-related nations. Tables 7–9 list six countries with the highest *NCOI* in three sub-networks from 1990 to 2015. Obviously, China's *NCOI* rankings in the three sub-networks are constantly rising, which shows that the prospects for its cooperation with countries in multiple RVCs are as good as possible.

Table 7. Top five BRI-related Asian nations' *NCOIs* in GPCCN–Eora26 models.

Rank	1990		1995		2000		2005		2010		2015	
	Country	*NCOI*	Country	*NCOI*	Country	*NCOI*	Country	*NCOI*	Country	*NCOI*	Country	*NCOI*
1	KOR	37.992	KOR	42.409	CHN	49.699	CHN	58.448	CHN	74.766	CHN	77.275
2	THA	33.407	CHN	40.414	KOR	41.204	KOR	41.015	KOR	38.313	KOR	37.601
3	UZB	30.817	THA	38.097	THA	36.087	THA	37.827	THA	34.572	THA	34.614
4	CHN	30.440	IDN	31.849	IDN	29.806	IRN	31.301	IRN	31.483	IRN	31.685
5	IRN	28.771	SGP	30.356	IRN	29.651	SGP	30.130	SGP	29.588	SGP	28.839
6	IDN	28.642	IRN	29.870	SAU	29.614	SAU	28.181	IDN	28.238	IDN	27.845

Table 8. Top five BRI-related European nations' *NCOIs* in GPCCN–Eora26 models.

Rank	1990		1995		2000		2005		2010		2015	
	Country	*NCOI*	Country	*NCOI*	Country	*NCOI*	Country	*NCOI*	Country	*NCOI*	Country	*NCOI*
1	ITA	60.996	ITA	67.608	ITA	68.827	ITA	65.879	ITA	63.145	ITA	60.212
2	RUS	54.947	RUS	53.430	RUS	62.334	RUS	56.649	RUS	59.270	RUS	59.401
3	SRB	30.137	AUT	29.366	CHN	31.486	CHN	33.962	CHN	43.633	CHN	44.524
4	UKR	28.860	SRB	29.084	GRC	29.172	POL	28.177	SRB	33.570	SRB	37.034
5	AUT	26.898	CHN	28.763	AUT	28.746	AUT	27.745	POL	28.585	POL	28.103
6	GRC	24.992	GRC	27.841	POL	28.508	GRC	27.639	AUT	26.809	AUT	26.274

Table 9. Top five BRI-related African nations' *NCOIs* in GPCCN–Eora26 models.

Rank	1990		1995		2000		2005		2010		2015	
	Country	*NCOI*	Country	*NCOI*	Country	*NCOI*	Country	*NCOI*	Country	*NCOI*	Country	*NCOI*
1	ZAF	48.196	ZAF	59.683	ZAF	60.514	ZAF	59.005	CHN	70.547	CHN	72.044
2	LSO	39.341	CHN	40.776	CHN	46.740	CHN	56.222	ZAF	52.555	ZAF	50.861
3	CHN	33.572	KEN	27.445	KEN	29.039	KEN	27.919	AGO	25.332	AGO	25.585
4	CIV	28.893	NGA	26.717	NGA	26.535	DZA	25.889	KEN	25.289	KEN	25.577
5	MAR	26.592	LSO	26.425	DZA	26.354	AGO	25.860	DZA	24.708	DZA	24.523
6	KEN	25.619	DZA	25.788	CIV	25.528	SEN	25.529	EGY	24.047	MAR	24.292

3.3. Correlation Analysis between Competitive Strengths and Collaborative Opportunities

Xing et al. [26] defined **Competitive Strength Index** (*CSI*) as the competitive pressure that an industrial sector imposes on others. By observing the relation of *NCSI* and *NCOI* in different years, we find that there is a very significant positive correlation between them, as is shown in Figure 9. In our opinion, the latter is the leading index of the former.

As is well known, vertical specialization and international trade are the foundation and embodiment of global economic integration. In most cases, one country/nation's economic development is based on making full use of its own and the others' resource endowments, so reaching a consensus is more important than creating a conflict of interest. The world of the 21st century has long since gotten rid of the colonial and semi-colonial development model. Each nation uses its comparative advantages to engage in economic and political games on the world stage. Therefore, we believe that the reason why countries/nations' competitive strengths and collaborative opportunities are closely related is mainly because they first establish a connection with the world through cooperation, and then consolidate their industrial function and status on the GVC through competition.

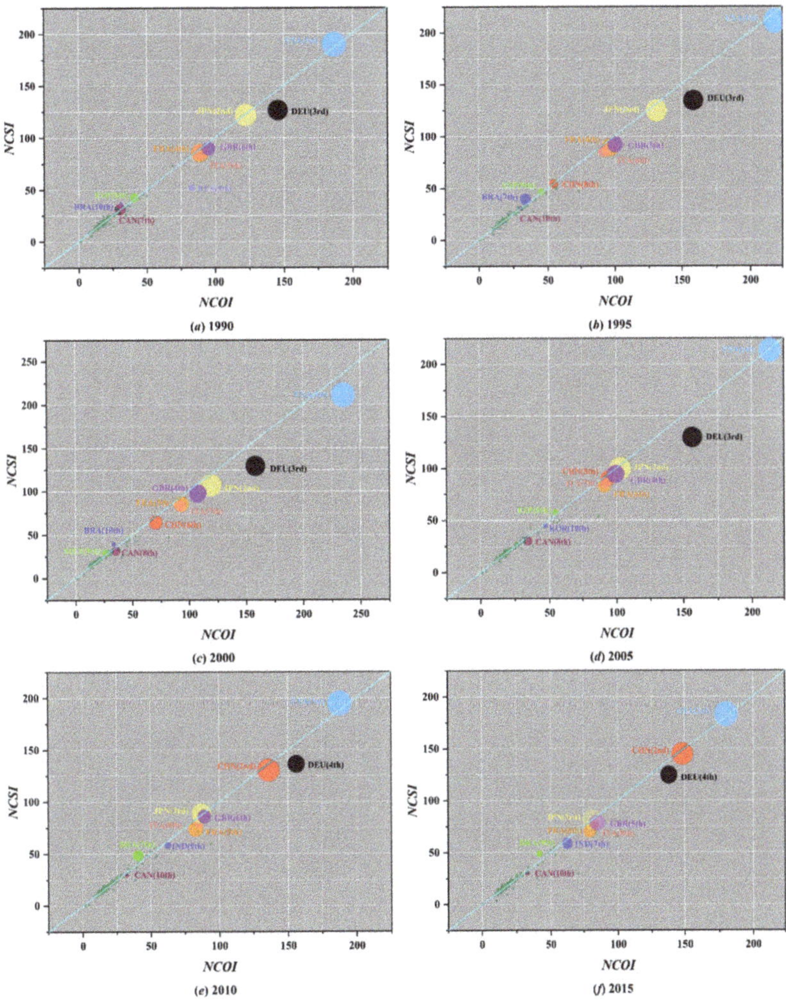

Figure 9. Correlation of *NCSI* and *NCOI* in GPCCN–Eora26 models. Notes: We use different colors to distinguish the top 10 nations in GDP in each year. In addition, the size of each point is proportional to the GDP of the corresponding nation.

4. Simulations

Three sets of scenario simulations have been designed to observe the effects on national collaborative opportunities of both China and main BRI-related nations under international trade fluctuation in GIVCN and GPCCN models. If the new trade policy was signed or the original trade one was withdrawn between two countries, there will be three possibilities for the volume of import and export trade between them as tariffs may change [30]:

Scenario I: X increases or decreases its export to Y while its counterpart remains stable.

Scenario II: Y increases or decreases its export to X while its counterpart remains stable.

Scenario III: Both parties increase or decrease its export to the counterpart simultaneously (there is no need to distinguish X and Y under this scenario).

Considering that the meaning of BRI is to promote interconnection and trade prosperity for all interested parties, we choose Scenario III as the only possibility. Then, simulations are carried out by increasing the volume of bilateral trade between two given nations from 0% (disruption of both import and export trades) of the initial value to 100% (gross value of trade in the ICIO table), and further up to 200% (both import and export trades doubled) in the specific GIVCN–BRI model, with every 10% as intervals. In the meanwhile, calculations on their *NCOIs* will be repeated in the corresponding GPCCN–BRI model, and simulation curves are acquired in this way for both parties of X and Y.

Despite trade volumes, we also need to consider the possible trade agreements, which have a more significant influence on international trade itself. We set three kinds of cases to observe how the collaborative status of China and BRI-related nations will change as shown in Figure 10.

(*a*) **Based on Pairs of China and Asian BRI-Related Nations**   (*b*) **Based on the Group of Asian BRI-Related Nations without China**   (*c*) **Based on the Group of Asian BRI-Related Nations with China**

Figure 10. Three cases of realization of collaboration at the scale of RVC. Notes: We take China and five main BRI-related nations in Asia as examples, and the basic settings are the same as those in the European group and African group.

Case A: China strengthens its trade collaboration with the others, respectively—As in the initial stage of BRI, China needs to establish mutually beneficial and collaborative relations with one country after another, in order to promote its transfer of excess production capacity.

Case B: Other nations bypass China to form an economic community—This is an undesirable situation for China since its economic development driven by foreign trade will be hindered, just like the TPP initiated by the United States.

Case C: All the nations strengthen the trade collaboration in between under a unified trade agreement—The newly formed RCV will be more beneficial to some nations than in other cases.

Based on these cases under Scenario III, this section lists the *NCOI* trends of the major economies (China and the top five countries in *NCOI*) in the GPCCN–BRI–2015 models with China as the core. As shown in Figures 11–19, the slope of the simulation result curve represents the elasticity of industrial collaborative ability to changes in trade volume. We try to find a better solution for both China and BRI-related nations in consideration of a win-win outcome.

4.1. Simulation on Asian Nations

According to the simulation results in Figures 11–13, we find that: (1) In Case A, China's *NCOI* sharply increases as the volume of international trade goes up, while other countries decrease to varying degrees; (2) In Case B, *NCOIs* of China, Iran (slightly), and Singapore decrease, while Indonesia (sharply), Korea (sharply) and Thailand (slightly after the trade rate is positive) increase; (3) In Case C, *NCOIs* of China (sharply), Indonesia (sharply), and Korea increase, while Iran, Singapore (sharply), and Thailand decrease.

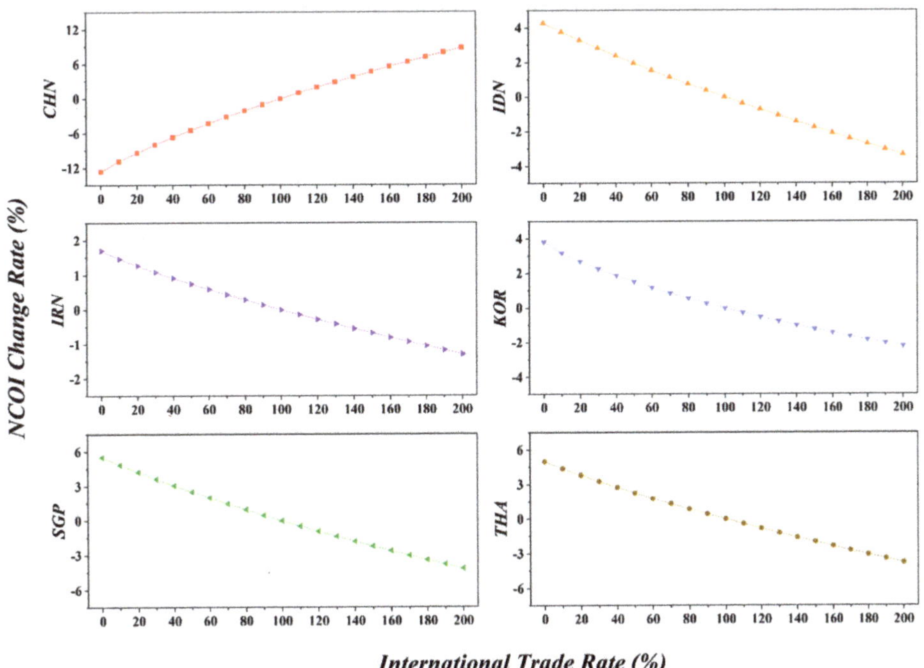

Figure 11. Influence on China and main Asian nations in Case A.

Next, we will specifically analyze the production capacity cooperation potential between China and major Asian countries.

Though hit by the financial crisis in 2008, Indonesia's economy managed to maintain a relatively fast growth rate, being the largest and fastest growing in Southeast Asia. Yet in recent years, its economic growth has slowed due to the shrunk volume of imports and exports significantly affected by global demand and prices. As a large agricultural country, Indonesia is the third largest producer of rice and the second largest producer and exporter of palm oil in the world. In the industrial sector, it is dominated by mining, oil and gas, textile, and light industry. China and Indonesia are highly complementary in various fields of industrialization and enjoy a wide scope for cooperation. Not only does Indonesia have a strong willingness to cooperate with China in production capacity, but Chinese companies are also quite attracted to the potential market and fastest growing economy in the southeast region. Chinese investment in Indonesia has mainly taken advantage of its infrastructure needs and labor force, focusing on infrastructure construction, energy development, palm oil plantation industry, and labor-intensive manufacturing industries such as textiles and cell phone assembly.

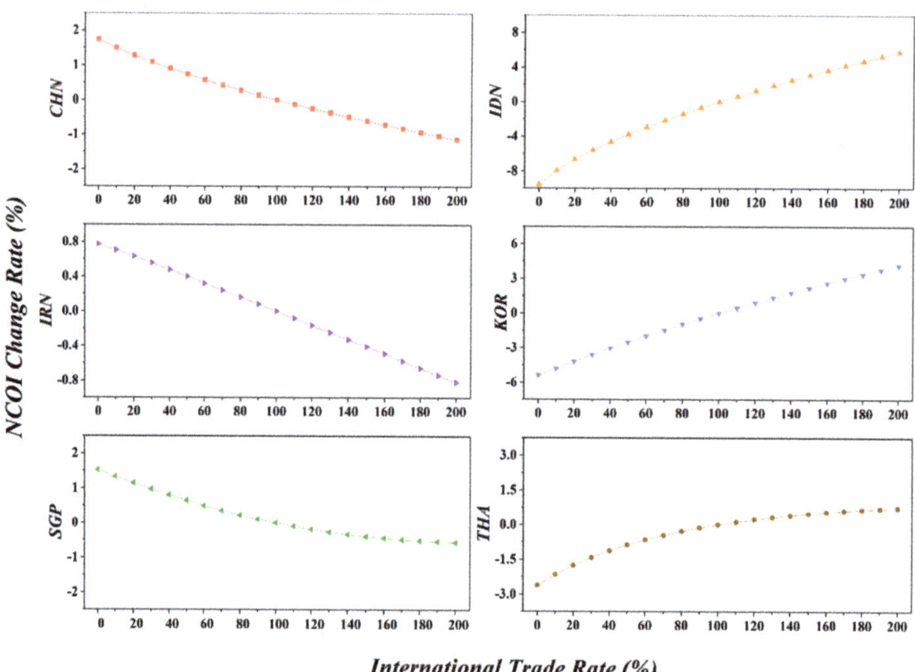

Figure 12. Influence on China and main Asian nations in Case B.

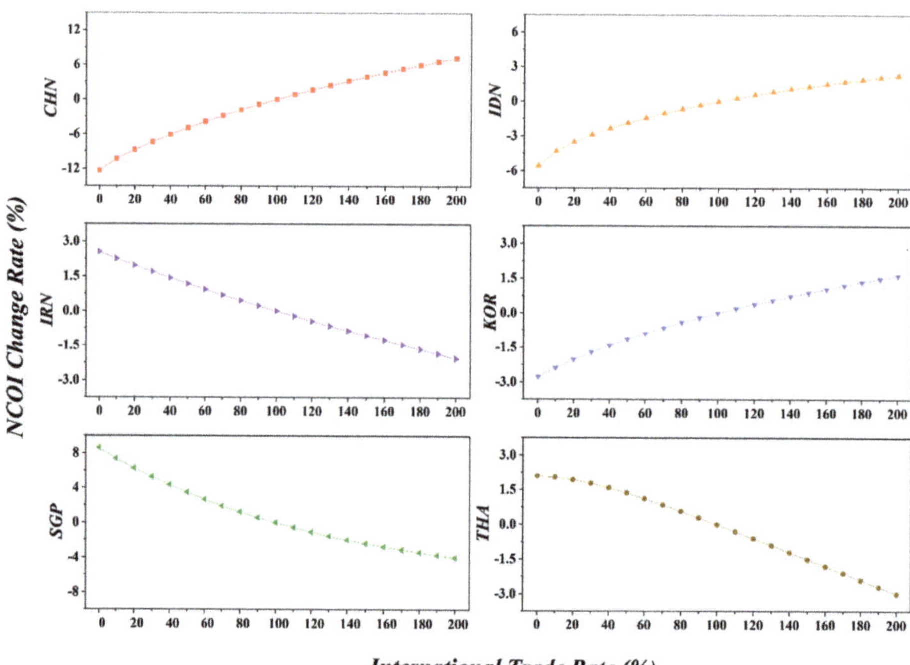

Figure 13. Influence on China and main Asian nations in Case C.

As one of the major economies in Asia, Iran's economic strength is second only to Saudi Arabia in the Middle East area, and its total population second only to Egypt, making it a pivotal regional power. The industrial structure of Iran is relatively simple, with the oil industry dominating the national economy. However, suffering from the long-lasting economic sanctions from Europe and the United States, its crude oil exports have been greatly restricted. The Iranian government has been increasing trade with other countries in recent years to revive the export trade volume, so as to free its economy from dependence on oil exports by increasing the income from non-oil products. China has been Iran's top trading partner for eleven consecutive years, and there is great room for economic cooperation between the two countries. In the energy sector, with its abundant oil and gas reserves, Iran has prioritized the attraction of foreign investment and technology in the oil and gas sector, and meanwhile, China boasts advanced technology and rich experience in energy exploration, exploitation, and equipment export. In the field of infrastructure construction, the current focus of economic and trade cooperation between China and Iran is closely based on interconnection and international production capacity cooperation, carrying out the construction of infrastructure, steel, electricity, railroads, and other projects. In the field of trade, Iran has been able to play the role of a trade hub in Eurasia thanks to its advantages as a transportation hub and a major re-exporting country, and the established free trade zones and special economic zones in Iran have provided convenient conditions and platforms for Chinese enterprises to make a direct investment. In the manufacturing sector, China is promoting the participation of China-invested enterprises with world-leading technologies in the construction of Iran's high-tech industries, such as high-speed rail, satellite, communications, and nuclear power, to meet the huge demand for manufacturing products in Iran's domestic market. Overall, the economies of China and Iran are highly complementary. As the BRI progresses, cooperation between the two countries in energy, infrastructure, transportation, communications, machinery manufacturing, and agriculture will be further deepened.

Korea witnessed an economic boom since the 1970s, and then was hit by the Asian financial crisis in 1997, dragging its economy into a stage of medium-rate growth. Due to the limited natural resources, its industrial structure is dominated by manufacturing and services industries. Its manufacturing industry is mainly technology- and knowledge-intensive, and has strong international competitiveness in shipbuilding, automobiles, electronics, and steel, yet the industrial materials of which are all dependent on imports. With China being Korea's top trading partner, the two countries enjoy broad prospects for cooperation in the manufacturing sector. As China's industrial structure gradually upgrades, China and Korea are mainly trading on high value-added electromechanical products, and the trading structures of the two countries are highly similar. The establishment of China–Korea FTA will further deepen the trade and investment between the two countries, form strong synergy in manufacturing industries, and become a new growth engine for multilateral cooperation in the Asian region, regional trade markets, and regional industrial development. China and Korea can cooperate more extensively in the future in emerging industries such as the Internet, and also in energy development, finance, and power grid construction in the Asian region.

With its well-developed services sector, Singapore is the fourth largest international financial center and the third largest foreign exchange trading center in the world. The three sectors account for less than 1%, 30%, and 70% of GDP, respectively. Singapore's unique geographical location has contributed to its status as a world powerhouse in the marine industry; high-quality logistics infrastructure wins it a reputation for reliability and speed of delivery; world-class port and airfreight facilities, excellent warehousing and delivery channels, and unparalleled regional and global connectivity gain it a firm foothold in global sourcing and integrated manufacturing. In terms of existing Sino–Singaporean economic and trade cooperation, Chinese investment in Singapore has seen a surge in recent years, mainly in contract labor, transportation, construction, energy, and other areas. Combined with Singapore's economic situation and the fruits of Sino–Singaporean economic and trade exchanges, Singapore has a limited role in absorbing and converting China's excess capacity in industries owing to its high stage of industrialization and services dominated industrial structure. However, the two countries have potentials for cooperation in capital-intensive manufacturing and services industries. Chinese enterprises can invest in high-end enterprises in Singapore's manufacturing industry chain to learn from its advanced management experience and technology; additionally, they can take advantage of Singapore's convenient transportation conditions and its status as an international financial center to develop trade and financial services, etc.

Located in the center of Southeast Asia, Thailand stands at the natural intersection of the ASEAN market and will become a booster rocket for the 21st Century Maritime Silk Road thanks to its relatively sound infrastructure. Thailand's economy is highly export-dependent, with exports accounting for about 2/3 of its GDP. Agriculture is the country's traditional economic sector, with agricultural products being one of the main sources of foreign exchange earnings. Thailand is the only net exporter of food in Asia, living up to its reputation as the "breadbasket of Southeast Asia". Among its top 10 export commodities, six are agricultural products, accounting for about 40% of the total export value. According to Chinese customs statistics, the total bilateral trade volume between China and Thailand accounts for 1/6 of the total bilateral trade between China and ten ASEAN countries, making Thailand China's fourth largest trading partner among ASEAN countries. Meanwhile, China is Thailand's largest trading partner, the largest source of imports, and the largest export destination. The trading structure between China and Thailand has been optimized in recent years, depicting a pattern featuring complementary advantages and mutual benefits. Among all the trading products, electrical and mechanical products take the largest share, and the proportion of plastics and their products is also increasing.

4.2. Simulation on European Nations

According to the simulation results in Figures 14–16, we find that: (1) In Case A, China's *NCOI* sharply increases as the volume of international trade goes up, while other countries decrease to varying degrees; (2) In Case B, *NCOIs* of China, Austria, and Poland (slightly) decrease, while Italy (sharply), Russia, and Serbia (sharply before the trade rate is positive and then slightly) increase; (3) In Case C, *NCOIs* of China (sharply), Italy, Russia (slightly), and Serbia (before the trade rate 60%) increase, while Austria (sharply), Poland, and Serbia (after the trade rate 60%) decrease.

Next, we will specifically analyze the production capacity cooperation potential between China and major European countries.

Lying at the south-end of Central Europe, Austria is an important transportation hub in Europe, with an economy growing faster than the EU average. Austria boasts an ample supply of mineral, forest, and hydraulic resources; in particular, its forest coverage accounts for nearly 50% of its total area. In recent years, as Austria's economy has been developing at a fast pace, the machinery industry is its largest industrial sector, its agriculture and tourism industries are well-developed, and the services industry occupies an important position. China is Austria's most important trading partner in Asia. China's rising living standards are attracting more and more Austrian companies to make investments, encouraging Sino–Austrian bilateral trade to continuously grow. With the unique advantages in the metal industry, mechanical engineering, food, chemical, automotive, and environmental protection industries, Austria exports high-tech products to China, and thus becomes an important technology importing source for China in the EU. Besides, Austria's position as a hub for China's interconnectivity with the CEE region is also noteworthy. In general, given the highly complementary bilateral trade, the cooperation between Austria and China in the fields of trade, finance, infrastructure construction, and culture will unlock significant potential.

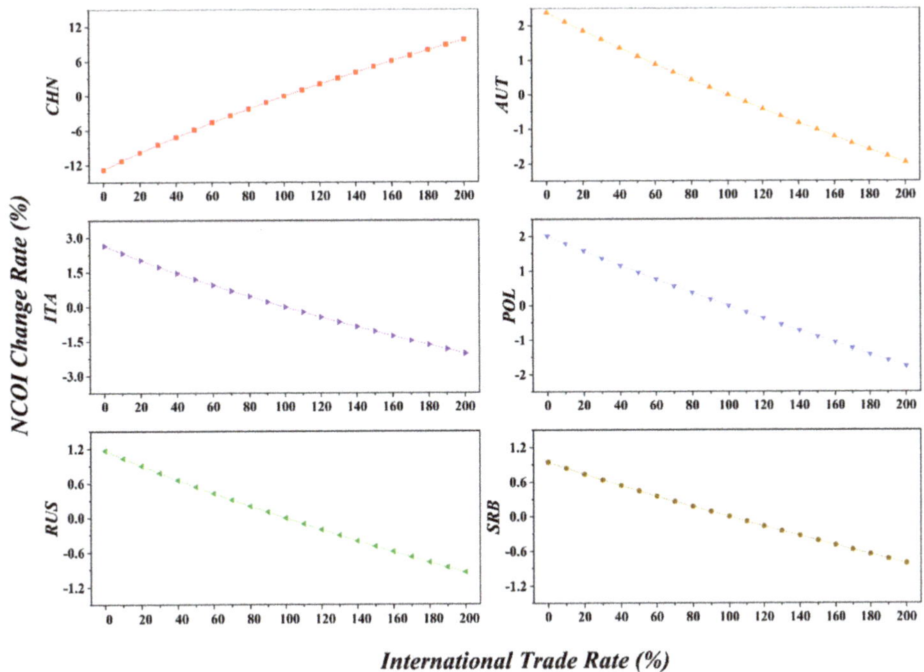

Figure 14. Influence on China and main European nations in Case A.

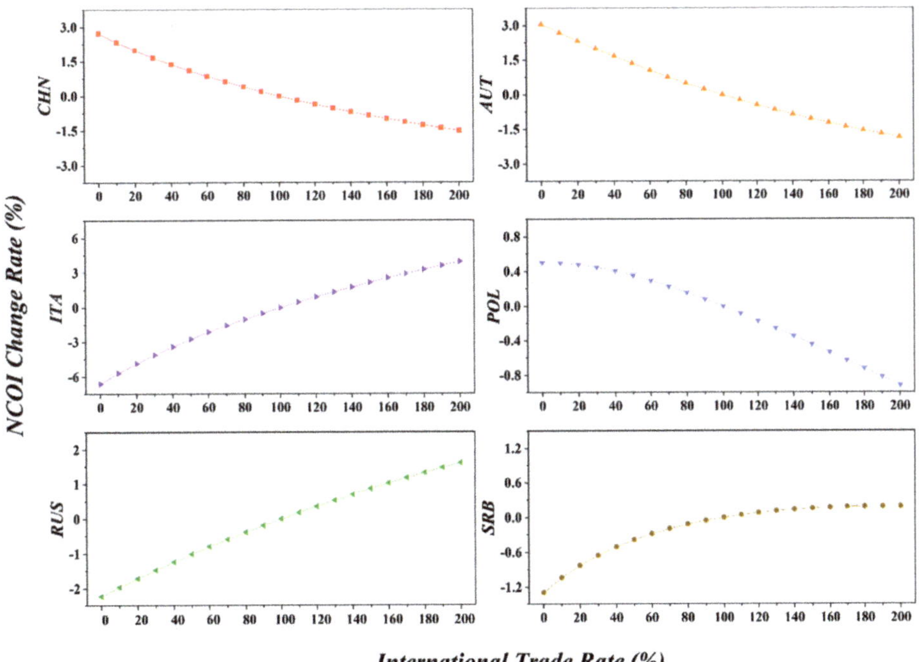

Figure 15. Influence on China and main European nations in Case B.

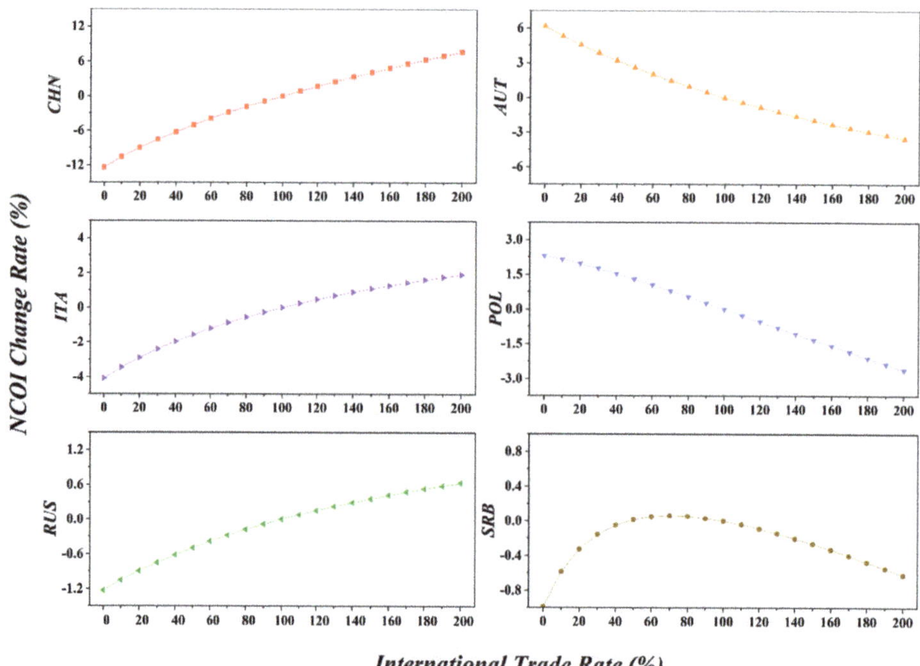

Figure 16. Influence on China and main European nations in Case C.

Italy is situated on the northern coast of the Mediterranean Sea in southern Europe. It is the second largest manufacturing country in the EU after Germany, the fourth largest economy in Europe and the eighth largest in the world. Known as the "Kingdom of SMEs", the number of Italy's small and medium-sized enterprises accounts for more than 98% of the total number of enterprises. However, in the short supply of natural resources, the country's oil and gas production can only meet a small portion of its domestic market demand. In addition, though being highly developed, its economy is facing unbalanced development, with a widening gap between the prosperous northern region and the relatively backward southern region, divided by the capital Rome. Italy was among the first batch of European countries to develop trade relations with China. The two countries signed a communiqué on the establishment of diplomatic relations as early as 1970. After the establishment of the China–Italy comprehensive strategic partnership in 2004, the economy and trade between the two countries has grown rapidly. As of 2018, Italy has become China's fourth largest trading partner, third largest export market, and source of imports in the European Union; likewise, China is Italy's top trading partner in Asia. Suffice it to say that the BRI between China and Italy can help bring into play the comparative advantages of both sides. To be specific, Italian companies have comparative advantages in high-end manufacturing and services industries, design, aerospace, biomedicine, etc., but lack capital liquidity, which can be complemented by Chinese companies which are seeking to transform and upgrade their value chains with relatively sufficient funds.

Located in Central Europe and south of the Baltic Sea, Poland is the largest and most populous country in Central and Eastern Europe. Poland's unique geographical advantage guarantees its important role in the Belt and Road. China and Poland have planned to use Poland as a hub for new logistics routes to build a logistics center in Central and Eastern Europe, thereby promoting the inflow and entry of Polish and Chinese products to the European region. As China's BRI and interconnectivity strategy progresses, a series of China–Europe freight trains by way of Poland have been launched to expand cooperation in trade and investment between the two countries. The economies of China and Poland are highly complementary and have the potential to develop together in the fields of infrastructure and high-tech industries, despite some obstacles such as limited trade volume, insufficient mutual investment, and a small number of large-scale cooperation projects, etc.

Russia, the largest country in the world, straddles the Eurasian continent and includes both the eastern half of Europe and the western part of Asia. Russia's industrial structure is homogeneous and its economic structure is overly dependent on energy exports. Its secondary industry is supported by heavy and chemical industries, while agriculture and services are relatively underdeveloped. China and Russia are each other's largest neighbors, and their unique geopolitical advantages facilitate economic and trade cooperation in the border areas of both countries. For a long time, China and Russia have been each other's important trade partners. China has been Russia's largest trading partner for eleven consecutive years, while Russia is the tenth largest trading partner of China. The trading structure of the two countries reflects complementarity. China imports Russia minerals, wood and wooden products, and other less processed primary products, while exporting to Russia electromechanical products, textiles, and raw materials; the various products in which the two countries have significant comparative advantages basically do not overlap.

Located in southeastern Europe, Serbia is a landlocked country in the middle of the Balkans that suffered severe damage to its industrial facilities in the 1990s when it was bombed by NATO during the Kosovo War. In the 21st century, with the introduction of privatization, Serbia's economy has gradually recovered, but the overall economic level is below the European average. The main economic obstacles are the high unemployment rate and large trade deficits. Serbia was the first country in Central and Eastern Europe to establish a strategic partnership with China, and since 2006 China has been Serbia's top trading partner in Asia and the fifth largest trading partner in the world. At present, though developing at a relatively fast pace, its overall economy is still underdeveloped, especially when it comes to the outdated infrastructure. In the future, China and Serbia have great potential for cooperation in infrastructure construction, energy, chemical industry, mineral products, and other fields.

4.3. Simulation on African Nations

According to the simulation results in Figures 17–19, we find that: (1) In Case A, China's *NCOI* sharply increases as the volume of international trade goes up, while Algeria, Angola (sharply when the trade rate is very low and then slightly), Kenya, Morocco, and South Africa decrease; (2) In Case B, *NCOIs* of China (slightly), Angola (slightly), and Kenya decrease, while Algeria (slightly), Morocco (slightly), and South Africa increase; (3) In Case C, *NCOIs* of China (sharply) and South Africa (slightly) increase, while Algeria, Angola (sharply when the trade rate is very low and then slightly), Kenya, and Morocco decrease.

Next, we will specifically analyze the production capacity cooperation potential between China and major African countries.

Located in northwest Africa, Algeria is the largest country in terms of area and the fourth largest economy in Africa. Rich in underground oil and gas resources, Algeria is the second largest gas exporter in the world, with the fifth largest reserves, so the oil and gas industry underpins its economic development. The industrial cooperation between China and Algeria can be mutually beneficial. Firstly, as Algeria's largest source of import, China mainly imports energy and mineral resources such as iron ore and LPG from Algeria, and invests in oil and gas, mining, aerospace, nuclear energy, and other fields. Secondly, the cooperation between China and Algeria in high-tech fields has strongly contributed to the economic growth and industrial development of both countries. Thirdly, China's overseas infrastructure capacity, which is high-level and cost-effective, can help build infrastructure such as roads, railroads, ports, and airports in Algeria.

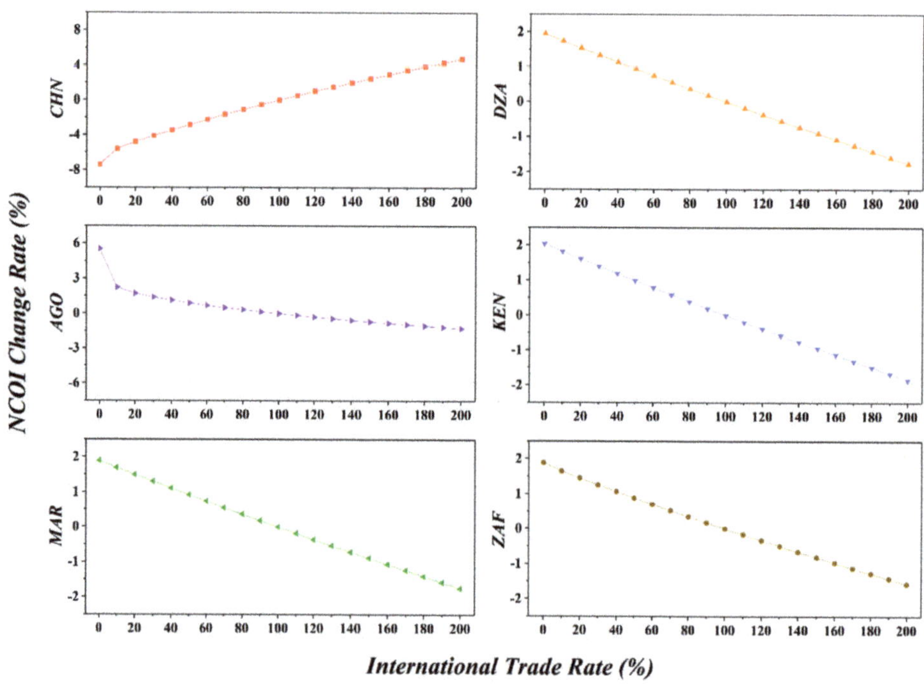

Figure 17. Influence on China and main African nations in Case A.

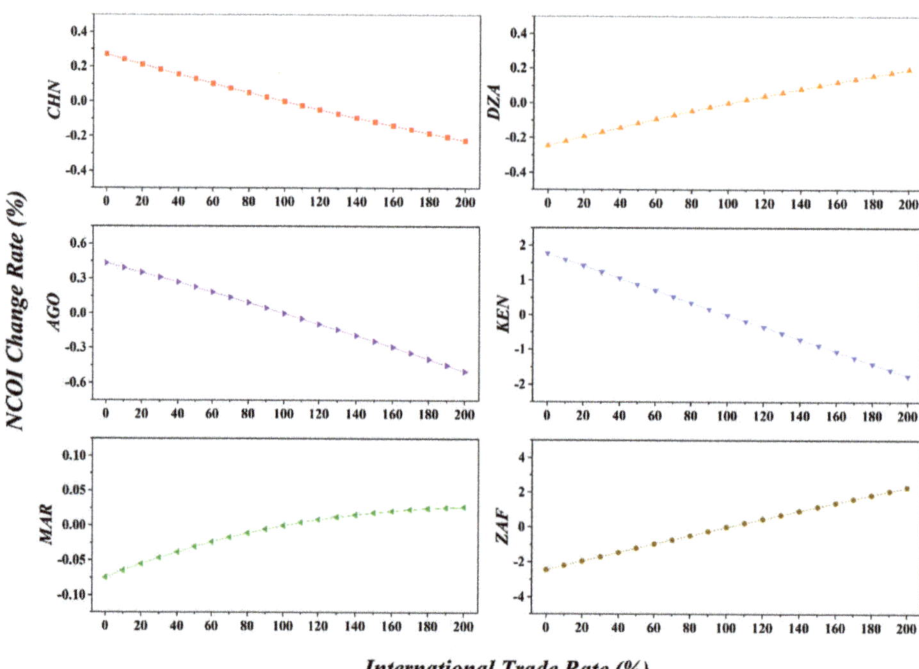

Figure 18. Influence on China and main African nations in Case B.

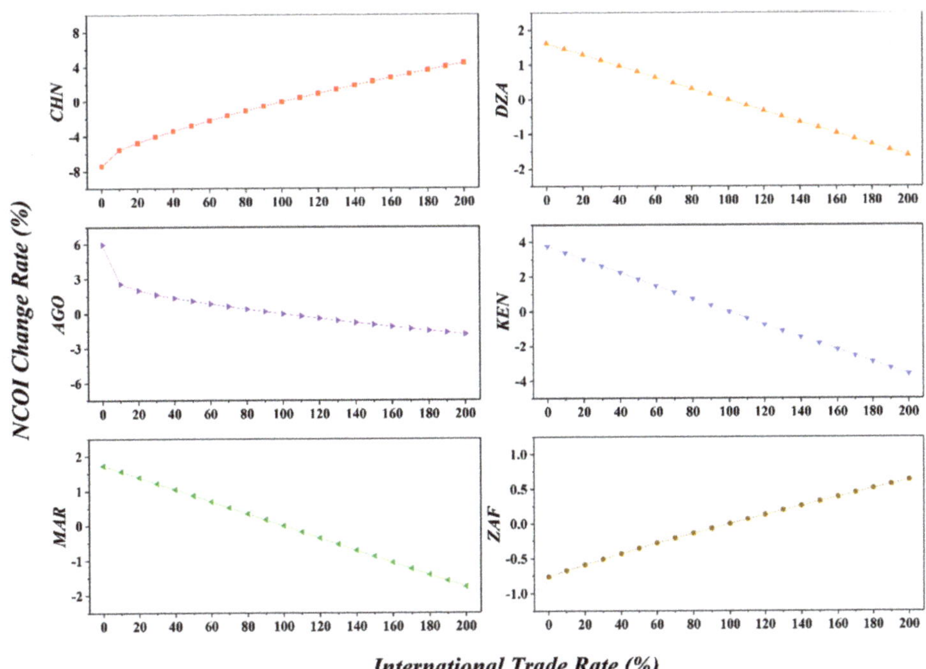

Figure 19. Influence on China and main African nations in Case C.

Situated in sub-Saharan Africa, Angola is the fourth largest economy and one of the largest capital attracting countries in Africa. It has ample oil, natural gas, and mineral resources, and also a large amount of hydroelectric power, as well as resources of agriculture, forestry, and fishery. Its hydropower generation accounts for 3/4 of the country's total power generation. Angola's economy is mainly based on agriculture and minerals, and oil, with oil being the mainstay industry. Although it has taken effective measures to promote economic diversification and reduce the dependence of the national economy on the oil industry, the country is still struggling with a low level of economic development and backward infrastructure. China's imports from Angola mainly include crude oil, natural gas, and other natural resources, and Angola's imports from China mainly include electromechanical, steel, automobile, and other products. China actively participates in investment in Angola and has obvious competitive advantages in infrastructure construction such as railroads, in addition to many other fields such as oil, construction, power grid, and telecommunication. In recent years, the two countries are making great efforts to promote capacity cooperation in areas such as electricity, ports, highways, agriculture, and manufacturing.

Kenya has the most developed and complete industrial sector in East Africa. Agriculture, services, and manufacturing are the three pillars of Kenya's national economy, and the oil, mineral extraction, agriculture, livestock and fisheries, and tourism industries are also developing well. Its natural port Mombasa connects East and Central African countries, with good water transport conditions. Kenya boasts well-operating infrastructure in communication, transportation, resources, and energy, rich natural resources, and huge market potential. However, Kenya's industrial sectors and regions varies greatly in terms of the level of development, so it needs to upgrade its industrialization development with reference to the success stories of other countries. China is Kenya's largest source of imports, and also has a number of maturely developed industries and redundant production capacity. China is now more than ready to make overseas investments and expand exports while sharing its best practices. China and Kenya enjoy high economic coupling—the capital, technology, and experience of the former can be fully utilized by the latter. The cooperation between the two countries will undoubtedly bring about mutual benefit and win-win.

Morocco is a coastal Arabian country in northwestern Africa and a hub connecting Europe, the Middle East, and Africa. Morocco's economy ranks fifth in Africa and third in North Africa. Phosphate exports, tourism, and remittances are the main pillars supporting Morocco's economy. It has a good foundation in agriculture but is not self-sufficient in food. Its rich fishery resources generate the highest production in Africa. But its industry is underdeveloped. The Moroccan government is committed to expanding domestic demand, strengthening infrastructure construction, supporting traditional industries such as textiles and tourism, developing new industries such as information and clean energy, actively attracting foreign investment, and promoting economic growth. As one of the first Arab countries to establish diplomatic relations with China, Morocco's superior geographical location, stable political environment, and perfect economic governance system provide conditions for further economic and trade cooperation between China and Morocco, and also serve as a bridge for Chinese enterprises to explore the African and European markets. In recent years, trade and investment between the two sides have continued to thrive, and production capacity cooperation in fisheries, infrastructure, telecommunications, automobiles, and other fields has been deepened.

As the second largest economy in Africa, South Africa is an important member of multilateral organizations such as BRICS, G20, and the United Nations. It maintains close relations with China in international organizations and multilateral mechanisms and is considered as China's important strategic partner. South Africa has abundant natural resources, low labor costs, and relatively complete infrastructure in transportation, electricity and information and communication. Mining and manufacturing are the most important pillar industries in its national economy. The cooperation between China and South Africa in the fields of manufacturing, investment, and trade is flourishing and of great significance. First, South Africa's manufacturing development has lagged in the past 20 years, and much of the manufacturing industry has been replaced by imports, mainly due to insufficient technology reserves, high factor costs, and insufficient economies of scale. The in-depth cooperation between South Africa and China in the manufacturing sector will enhance its own technological level and international competitiveness. Secondly, South Africa is China's largest trading partner and the most important investment destination in Africa, and China's investment in South Africa has promoted the development of its special economic zones. Thirdly, South Africa is now shifting from a mining and manufacturing-dependent economy to a technology- and services-oriented economy, whose domestic market can be further vitalized through its trade with China.

5. Results and Discussions

We try to explain the laws and reasons for the variation of collaboration among BRI-related nations from the following three perspectives.

Firstly, China can transfer its excess production capacity to other countries on the RCV through BRI, and then optimize its own industrial structure to move to the middle and high end of multiple IVCs, thereby enhancing its collaborative ability on the GVC, which is reflected by a substantial increase in the *NCOI* in Case A and Case B.

Secondly, by strengthening regional cooperation, some nations have made up for the shortcomings of their own industrial structure layout to some extent and enhanced the production transformation capacity within their NVCs. Among them: **Satisfying Effect** is observed when the collaborative potential of nations with a single industrial structure is satisfied, which is manifested as a decrease in *NCOI*; **Incentive Effect** is observed when the collaborative potential of nations with a diversified industrial structure is further stimulated, which is displayed as a rise in *NCOI*.

Thirdly, under the combined effect of satisfying effect and incentive effect, some nations (e.g., Thailand in Asia, Serbia in Europe, Angola in Africa) have more varying *NCOI* trends under different cooperation strategies—either rise or fall, which requires to be analyzed specifically on the causes at the sectoral level.

Regional collaboration can promote relevant economies to carry out production capacity cooperation, make full use of their comparative advantages to embed in the RVC network, and gradually achieve a rise in the GVC network. From a long-term perspective, the BRI initiated by China will help GVC restructure toward a win-win cooperation. In this chapter, our study provides a reference for how China can better implement the BRI. For example, in its cooperation with Asia, where most countries are rich in oil and gas and mineral resources, but have poor industrial systems, backward development technologies, and insufficient development capacity, China can cooperate with them in key areas such as oil and gas and mineral resources via helping them establish sound industrial, transportation, and infrastructure systems. In its cooperation with Europe, given the rapid development of the "construction" sector, China can take advantage of the rapid development cycle of European infrastructure, and use its experiences in rail–road industry to tap into the European rail transportation market. Meanwhile, China should also focus on the cooperation with European HMT sectors. In the cooperation with Africa, China should adhere to the humanitarian spirit, guide African industries to be more scientific and internationalized, and bring into play Africa's comparative advantages in the GVC network.

In the context of drastic changes in the international environment, the traditional countermeasures to the systemic crisis of the national economy have lost efficacy; and the priority is to optimize and upgrade industrial structure. With its complete industrial chain and supply chain and the vast domestic market, China should avoid the "industrial hollowing-out" similar to Japan, the United States, and other countries. From the perspective of economic security, while continuously encourage industrial sectors to "go global", China needs to respond to its dwindled competitiveness in the whole industrial chain and strengthen independent innovation to supplement the shortcomings. Against the backdrop of GVC reconstruction in the post-pandemic era, China shall explore a new development model, use the domestically economic circulation to drive the internationally economic circulation, take the BRI as the focus, and seize new foreign trade opportunities brought by RCEP. By doing so, it will embrace strengthened ties with other countries, and better integration into the GVC with a higher level of openness. This will be a favorable measure to promote global economic integration and counteract reverse globalization.

6. Conclusions

This paper measures the collaborative relations between industrial sectors and simulates that between countries in consideration of both the actual demand from downstream sectors and the potential industrial-capacity cooperation from upstream ones. We believe this chapter will be helpful to understand the trend of economic globalization and regional economic cooperation. Contributions of this paper are as follows:

(1)   Establish the GPCCN model to embody the collaborative relations among industrial sectors. In consideration of the scarcity of productive capacities, we use bipartite graphs to distinguish the roles of industrial sectors on the GVC as upstream and downstream ones. Then, we extract the collaborative relations hidden in the IO/ICIO table via RAP approach, transforming the GIVCN model into the GPCCN model. The latter depicts collaborations among countries and their industrial sectors.

(2)   Propose network-based measurement tools to reveal the collaboration status on the sectoral level and the national level. After getting the collaborative relations among industrial sectors, the summation of the collaborative attraction that one imposes on others is defined as the *COI*, and the summation of collaborative attraction that one receives from others is defined as the *CTI*, which are the out-strengths S^{OUT} and in-strengths S^{IN} of nodes, respectively, in the GPCCN model. As well, *NCOI* and *NCTI* standing for the country-level cooperation competence can be further calculated. Of

course, we pay more attention on the economies' collaborative opportunity measured by *NCOI* in the empirical analysis.

(3) Simulate collaborative opportunities of BRI-related nations. GVC is the most sophisticated economic system, whose relatedness, heterogeneity, and diversity deserve more attention from the relevant authorities when making international trade policies. Only by studying GVC can China and its trade partners benefit from the BRI. We believe the simulation framework in this paper possesses considerable reference value and will be a guide for analyzing globalization issues with physical statistics.

In this paper, we set three kinds of cases to observe how the collaborative status of China and BRI-related nations will change. The premise of global cooperation on production capacity is the complementarity and coupling of the two cooperating countries on the GVC, emphasizing the utilization of their respective advantages in technology, capital, and resources to achieve mutual benefits and win-win situations.

Empirical analysis has shown that China's BRI has indeed brought dividends to nations along the route. Especially for some less developed countries in Asia, Europe, and Africa, continued industrial-capacity cooperation with China in key areas has significantly improved their ability of globally synergic production. This further proves that BRI can provide good development opportunities for relevant countries through complementing advantages, resource sharing, and capacity cooperation, and can help achieve common prosperity.

In the next stage, more detailed analysis on the trade between China and BRI-related nations should be carried out from the perspective of their market sizes and industrial layouts.

Author Contributions: Conceptualization, L.X. and J.G.; methodology, L.X. and D.W.; software, L.X. and D.W.; validation, D.W. and C.J.; formal analysis, D.W., C.L. and C.J.; investigation, D.W. and C.L.; resources, D.W.; data curation, D.W.; writing—original draft preparation, L.X., D.W. and C.L.; writing—review and editing, L.X., D.W., C.L. and C.J.; visualization, D.W.; supervision, L.X. and J.G.; project administration, L.X.; funding acquisition, L.X. All authors have read and agreed to the published version of the manuscript.

Funding: This research was funded by National Natural Science Foundation of China (Grant No. 71971006, 71774008), Humanities and Social Science Foundation of Ministry of Education of the People's Republic of China (Grant No. 19YJCGJW014).

Institutional Review Board Statement: Not applicable.

Informed Consent Statement: Not applicable.

Data Availability Statement: The data presented in this study are openly available in The Eora Global Supply Chain Database, https://worldmrio.com/ access on 1 December 2021.

Conflicts of Interest: The authors declare no conflict of interest.

References

1. Jin, L. The "New Silk Road" Initiative: China's Marshall Plan? *Int. Stud.* **2015**, *165*, 88–99.
2. Zhang, Y.; Yu, J.; Li, D. Measuring and Analyzing the Trade Cost Elasticity between China and Countries of "One Belt, One Road": Based on Translog Gravity Model. *World Econ. Stud.* **2018**, *289*, 56–58.
3. Ramasamy, B.; Yeung, M.C. China's one belt one road initiative: The impact of trade facilitation versus physical infrastructure on exports. *World Econ.* **2019**, *42*, 1673–1694. [CrossRef]
4. Hsu, L. Asean and the Belt and Road Initiative: Trust-building in Trade and Investment. *China World* **2020**, *3*, 2050002. [CrossRef]
5. Xu, P.; Cheng, Q. Economic Growth Effects on B&R Sci-Tech Cooperation. *J. Financ. Econ.* **2020**, *5*, 140–154.
6. Wen, S.; Zhang, Y. Financial Development, FDI Spillover and Economic Growth Efficiency: An Empirical Study of the Countries Along the "One Belt and One Road". *World Econ. Stud.* **2020**, *11*, 41–50.
7. Leontief, W.; Dietzenbacher, E.; Lahr, M. *Wassily Leontief and Input-Output Economics*; Cambridge University Press: Cambridge, UK, 2004.
8. Raa, T.T. *The Economics of Input-Output Analysis*; Cambridge University Press: Cambridge, UK, 2005.
9. Raa, T.T. *Input-Output Economics: Theory and Applications*; World Scientific: Singapore, 2009.
10. Hummels, D.; Ishii, J.; Yi, K.M. The Nature and Growth of Vertical Specialization in World Trade. *J. Int. Econ.* **2001**, *54*, 75–96. [CrossRef]

11. Fally, T. *On the Fragmentation of Production in the US*; University of Colorado Working Paper; University of Colorado: Boulder, CO, USA, 2011.
12. Wang, Z.; Wei, S.J.; Yu, X.; Zhu, K. *Characterizing Global Value Chains: Production Length and Upstreamness*; NBER Working Papers; National Bureau of Economic Research: Cambridge, MA, USA, 2017.
13. Xu, M.; Ling, S. Input–output networks offer new insights of economic structure. *Phys. A Stat. Mech. Appl.* **2019**, *527*, 121178. [CrossRef]
14. Yang, B.; Cheng, W.; Hu, X.; Zhu, C.; Yu, X.; Li, X.; Huang, T. Seeking community structure in networks via biogeography-based optimization with consensus dynamics. *Phys. A Stat. Mech. Appl.* **2019**, *527*, 121188. [CrossRef]
15. Piccardi, C.; Riccaboni, M.; Tajoil, L.; Zhu, Z. Random walks on the world input–output network. *J. Complex Netw.* **2018**, *6*, 187–205. [CrossRef]
16. Guan, J.; Xu, X.; Xing, L. Analysis of inter-country input-output table based on bibliographic coupling network: How industrial sectors on the GVC compete for production resources. *Int. J. Mod. Phys. B Condens. Matter Phys.* **2018**, *32*, 1850063. [CrossRef]
17. Xing, L.; Guan, J.; Wu, S. Measuring the impact of final demand on global production system based on Markov process. *Phys. A Stat. Mech. Appl.* **2018**, *502*, 148–163. [CrossRef]
18. Xing, L.; Dong, X.; Guan, J. Global Industrial Impact Coefficient Based on Random Walk Process and Inter-Country Input-Output Table. *Phys. A Stat. Mech. Appl.* **2017**, *471*, 576–591. [CrossRef]
19. Xing, L.; Han, Y.; Wang, D. Measuring economies' pivotability on the global value chain under the perspective of inter-country input–output network. *Mod. Phys. Lett. B* **2021**, *35*, 2150289. [CrossRef]
20. Xing, L.; Dong, X.; Guan, J.; Qiao, X. Betweenness centrality for similarity-weight network and its application to measuring industrial sectors' pivotability on the global value chain. *Phys. A Stat. Mech. Appl.* **2019**, *516*, 19–36. [CrossRef]
21. Xing, L. Analysis of inter-country input-output table based on citation network: How to measure the competition and collaboration between industrial sectors on the global value chain. *PLoS ONE* **2017**, *12*, e0184055. [CrossRef] [PubMed]
22. Newman, M.E.J. The structure of scientific collaboration networks. *Proc. Natl. Acad. Sci. USA* **2001**, *98*, 404–409. [CrossRef]
23. Padrón, B.; Nogales, M.; Traveset, A. Alternative approaches of transforming bimodal in-to unimodal mutualistic networks. The usefulness of preserving weighted infor-mation. *Basic Appl. Ecol.* **2011**, *12*, 713–721. [CrossRef]
24. Xing, L.; Guan, J.; Dong, X.; Wu, S. Understanding the Competitive Advantage of TPP-Related Nations from an Econophysics Perspective: Influence Caused by China and the United States. *Phys. A Stat. Mech. Appl.* **2018**, *502*, 164–184. [CrossRef]
25. Guan, J.; Xu, X.; Wu, S.; Xing, L. Measurement and Simulation of the Relatively Competitive Advantages and Weaknesses between Economies Based on Bipartite Graph Theory. *PLoS ONE* **2018**, *13*, e0197575. [CrossRef]
26. Xing, L.; Wang, D.; Li, Y.; Guan, J.; Dong, X. Simulation Analysis of the Competitive Status between China and Portuguese-Speaking Countries under the Background of One Belt and One Road Initiative. *Phys. A Stat. Mech. Appl.* **2020**, *539*, 122895. [CrossRef]
27. Zhou, T.; Ren, J.; Medo, M.; Zhang, Y. Bipartite network projection and personal recommendation. *Phys. Rev. E* **2007**, *76*, 046115. [CrossRef] [PubMed]
28. Chohan, U. *What Is One Belt One Road? A Surplus Recycling Mechanism Approach*; Social Science Electronic Publishing: Rochester, NY, USA, 2018; pp. 205–219.
29. Xing, L.; Han, Y. Extracting the Backbone of Global Value Chain from High-Dimensional Inter-Country Input-Output Network. In Proceedings of the International Conference on Complex Networks and Their Applications, Madrid, Spain, 30 November–2 December 2021; Springer: Cham, Switzerland, 2021.
30. Xing, L.; Ye, Q.; Guan, J. Spreading effect in industrial complex network based on revised structural holes theory. *PLoS ONE* **2016**, *11*, e0156270. [CrossRef] [PubMed]

MDPI

St. Alban-Anlage 66

4052 Basel

Switzerland

www.mdpi.com

Systems Editorial Office

E-mail: systems@mdpi.com

www.mdpi.com/journal/systems